Lecture Notes in Chemistry

Volume 100

The Lecture Notes in Chemistry

The series Lecture Notes in Chemistry (LNC) reports new developments in chemistry and molecular science-quickly and informally, but with a high quality and the explicit aim to summarize and communicate current knowledge for teaching and training purposes. Books published in this series are conceived as bridging material between advanced graduate textbooks and the forefront of research. They will serve the following purposes:

- provide an accessible introduction to the field to postgraduate students and nonspecialist researchers from related areas,
- provide a source of advanced teaching material for specialized seminars, courses and schools, and
- be readily accessible in print and online.

The series covers all established fields of chemistry such as analytical chemistry, organic chemistry, inorganic chemistry, physical chemistry including electrochemistry, theoretical and computational chemistry, industrial chemistry, and catalysis. It is also a particularly suitable forum for volumes addressing the interfaces of chemistry with other disciplines, such as biology, medicine, physics, engineering, materials science including polymer and nanoscience, or earth and environmental science.

Both authored and edited volumes will be considered for publication. Edited volumes should however consist of a very limited number of contributions only. Proceedings will not be considered for LNC.

The year 2010 marks the relaunch of LNC.

More information about this series at http://www.springer.com/series/632

Jinlong Zhang • Baozhu Tian • Lingzhi Wang
Mingyang Xing • Juying Lei

Photocatalysis

Fundamentals, Materials and Applications

Springer

Jinlong Zhang
Key Laboratory for Advanced
Materials & Institute of Fine Chemicals
East China University of Science &
Technology
Shanghai, China

Baozhu Tian
Key Laboratory for Advanced
Materials & Institute of Fine Chemicals
East China University of Science &
Technology
Shanghai, China

Lingzhi Wang
Key Laboratory for Advanced
Materials & Institute of Fine Chemicals
East China University of Science &
Technology
Shanghai, China

Mingyang Xing
Key Laboratory for Advanced
Materials & Institute of Fine Chemicals
East China University of Science &
Technology
Shanghai, China

Juying Lei
Key Laboratory for Advanced
Materials & Institute of Fine Chemicals
East China University of Science &
Technology
Shanghai, China

ISSN 0342-4901 ISSN 2192-6603 (electronic)
Lecture Notes in Chemistry
ISBN 978-981-13-2112-2 ISBN 978-981-13-2113-9 (eBook)
https://doi.org/10.1007/978-981-13-2113-9

Library of Congress Control Number: 2018954494

This Springer imprint is published by the registered company Springer Nature Singapore Pte Ltd.
The registered company address is: 152 Beach Road, #21-01/04 Gateway East, Singapore 189721, Singapore

Foreword

In the twenty-first century, humanity is facing serious and urgent issues of energy depletion and environmental degradation. Due to rapid economic growth, we are experiencing the most extensive energy shortages and environmental pollution in history, also resulting in globally destructive climate change. While our industries are constantly providing a variety of innovative new products and materials, it is becoming imperative to focus on the recycling and treatment of waste materials as well. Moreover, we need to reduce the consumption of natural resources and raise awareness of the great impact of our consumerism so that the technologies, products, and materials we develop are environmentally harmonious and sustainable.

Many of the recent Nobel prizes in physics and chemistry are related to photochemistry and catalytic chemistry as well as materials or systems related to these areas. Photocatalysis which induces efficient and effective reactions at room temperature under sunlight irradiation has been the focus of much attention for its potential to establish ideal technologies which can convert clean, safe, and abundant solar energy into electrical and/or chemical energy.

In fact, in the last 45 years since the discovery of the sensitizing effect of a TiO_2 electrode for the electrolysis of water by Honda and Fujishima, much research has been carried out on the development of semiconductor photocatalysis using various metal oxide catalysts. New breakthrough technologies such as the separate production of H_2 and O_2 from water, CO_2 reduction with H_2O, the transformation or complete mineralization of organic molecules, reduction of heavy metal ions as well as nontoxic deodorization, antisepsis, decontamination, and sterilization processes will be essential in the reduction or elimination of harmful compounds from polluted air, water, and soil. This book brings to light much of the recent research in the development of such semiconductor photocatalytic systems.

In order to prepare highly efficient and functional photocatalysts, their meticulous design and construction are vital, and this book delves into molecular level investigations of their nanoscale structures as well as the mechanisms behind their reactivity. Modification methods to improve their performance including ion-doping, heterojunction structural development, noble metal deposition, and morphological control are also covered. In addition, dye sensitization processes to extend their light absorption ranges from UV to visible regions as well as to reduce electron-hole recombination are also explored. This book will thus be of interest to researchers in materials sciences, environmental science and technologies, energy storage and conversion, photocatalytic processes, and industrial applications.

The lead author, Prof. Jinlong Zhang, has studied fine chemical synthesis at the East China University of Science and Technology (Shanghai, China) where he received a PhD from the Department of Applied Chemistry in 1993. In 1996, he received the Japan Society for the Promotion of Science (JSPS) Award for innovative research. He then carried out research at the Department of Applied Chemistry of Osaka Prefecture University (Sakai City, Osaka, Japan) as a postdoctoral researcher for 4 years. After Dr. Zhang's return to Shanghai, he was promoted to full professor of East China University of Science and Technology in 2000. He has published over 380 original papers which have been cited more than 14,000 times. He serves as editor of the international journal, *Research on Chemical Intermediates* (Springer), and is also on the editorial boards of several international journals such as *Scientific Reports*, *Applied Catalysis B: Environmental Science and Technology*, *Dyes and Pigments*, and *Photographic Science and Photochemistry*.

We hope this book will contribute to the advancement of photocatalytic research for environmental purification and renewable energy technologies in order that future societies may thrive and prosper.

Prof. Masakazu Anpo

Emeritus Professor of Osaka Prefecture
University (Japan)
International Advisor and Special Honorary Professor
State Key Laboratory of Photocatalysis on Energy
and Environment
Fuzhou University (China)

Editor-in-Chief, *Res. Chem. Intermed.* (Springer)
Member, Academia Europae
Member, Science Council of Japan

Contents

Chapter 1
Mechanism of Photocatalysis

1.1 Introduction

Environmental pollution is becoming a big challenge for the new developing and developed world because of the industrial revolution, which on the one side is facilitating the life but on the other side destroying the environment. The chemical, textile, oil, and gas industries annually send billion of tons of hazardous effluents to the groundwater and upper environment which is affecting badly the marine life, human, and plant kingdom [1–3]. At present, there are a number of technologies that have been discovered to cater the fast-growing problem; the most commonly employed method of removing pollutants involves the use of various physical techniques, chemical and biological processes, and thermal techniques to remove commonly found organic and inorganic contaminants [4, 5]. Among many proposed processes being developed for the decontamination of the organic-based pollutants, biodegradation received the greatest attention. However, numerous organic chemicals, especially those which are toxic or refractory, were found very resistant to the biological degradation [6, 7]. Therefore, after the discovery of photocatalytic splitting of water by Fujishima and Honda [8] in 1972, the researchers turned their point of convergence toward semiconductor photocatalysis which proved very effective in the degradation of even those pollutants which are highly difficult to remove by the other means. In comparison with the traditional oxidation process, the photocatalysis has significant advantages, for example, they can be used to degrade and mineralize the dyes and chemicals completely to CO_2 and H_2O and are also helpful in the degradation of very stable compounds which cannot be easily degraded by the other processes, also they can work efficiently at ambient temperature and pressure conditions, and they do not need any special supply of oxygen. The other advantage of the photocatalysis is that it is a cheap process as compared to the other oxidation process having no waste disposal problems [3]. A large number of semiconductor photocatalysts have been investigated so far for the degradation of different pollutants, such as ZnO [9, 10], WO_3 [11, 12], Fe_2O_3 [13], CdSe [14], and

J. Zhang et al., *Photocatalysis*, Lecture Notes in Chemistry 100,
https://doi.org/10.1007/978-981-13-2113-9_1

$SrTiO_3$ [15]. In general, an ideal photocatalyst should have some basic properties, such as they must be active under UV, visible light, or solar light, they should have property of chemical and biological robustness, as well as they should be stable toward photocorrosion. The other most important characteristic they must have is that they should be nontoxic and must have low cost and easy availability [16]. These all are key factors for a good photocatalyst.

1.2 TiO_2 Photocatalysis

TiO_2 is the most common catalyst used in the field of photocatalysis. From the ancient times, TiO_2 has been used as a white pigment. It can be extracted from the ilmenite ($FeTiO_3$) [17], perovskite ($CaTiO_3$) [18], and titanite ($CaTiSiO_5$) [19] ores. TiO_2 is mechanically, thermally, and chemically very stable. It has a very high melting point of 1855 °C and is insoluble in water, HNO_3, HCl, and dilute H_2SO_4. It can be dissolved in hot concentrated H_2SO_4 and HF [20]. The reports on TiO_2 as photocatalysis were found back in 1921 where Renz [21] from the University of Lugano reported the partial reduction of titania during illumination under sunlight in the presence of glycerol and in 1929 where Keidel [22] found that the dyes which are quite stable to light and are not easily degradable are found to be decomposed when adsorbed on lithopone or white lead in contact with TiO_2 [23]. There they were called TiO_2 as photosensitizer. Titania is supposed to be an excellent photocatalyst due to its enormous properties such as high activity, low cost, photostability, nontoxicity, etc.

1.3 Mechanism of Photocatalytic Oxidation Reactions

In semiconductors the chemical reaction usually occurs due to the transfer of electrons from the valance band to the conduction band as shown in the Fig. 1.1 As the number of orbitals (N) in the HOMO (highest occupied molecular orbital (valance band)) and LUMO (lowest unoccupied molecular orbital (conduction band)) increases, the energy to shift the electrons from the valance band to the conduction band will decrease.

The detailed mechanism of the photocatalysis is quite a complex subject but the basic principle is the same. Upon irradiation of the light with the energy equal or more than the bandgap of the semiconductor photocatalyst, the electrons are excited from the valance band of the catalyst to the conduction band, leaving the holes behind in the valance band.

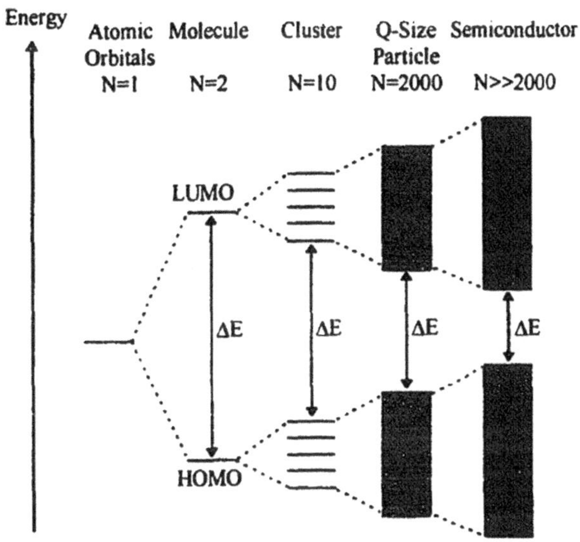

Fig. 1.1 Changes in the electronic structure of the compound as the number of atomic orbitals (L) increases. (Reprinted from Ref. [24], copyright 1997, with permission from Elsevier)

$$TiO_2 + h\nu \longrightarrow e^- + h^+ \tag{1.1}$$

These photo-generated electrons and holes can possibly recombine by releasing the heat between 10 nanoseconds and 100 nanoseconds as mentioned by Hofmann et al. by the measurement with the laser flash photolysis at room temperature [3]. This recombination of the electrons and holes causes the low quantum efficiency of the photocatalyst.

$$e^- + h^+ \longrightarrow \text{heat} \tag{1.2}$$

The reports on the mechanism of the photocatalysis indicate that the recombination of photo-generated electrons and holes mainly occurs in the bulk of the catalyst samples. This recombination process of the electrons and holes can be reduced significantly if these charge-carrier species are separated by the addition of suitable scavenger or incorporation of some of the trap sites on the surface as a result of producing defects, surface adsorbents, or other sites.

If we provide sufficient time to the holes and electrons before they recombine, then these produced charge carriers migrate to the surface of the catalyst and can undergo the charge transfer to initiate the redox reactions with the pollutants adsorbed on its surface which is shown in Fig. 1.2.

A valence band hole, h^+, contains a strong oxidation power by having a redox potential ranging from +1.0 to +3.5 V (measured vs normal hydrogen electrode (NHE) at room temperature), depending on the semiconductor and pH. Thus, the presence of the hole plays an important role in the photocatalytic degradation of the pollutant present on the surface of the catalyst. The oxidation can take place either due to the indirect oxidation via reaction with the surface-bound hydroxyl radical

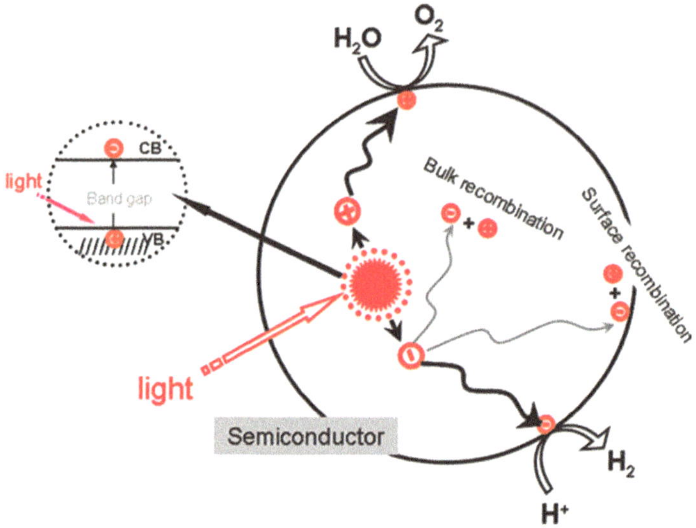

Fig. 1.2 Mechanism for the photocatalytic activity on the surface of the semiconductor under the irradiation of light. (Reprinted with permission from Ref. [25]. Copyright 2010 American Chemical Society)

(•OH) (i.e., surface-trapped h^+) or directly via h^+ attached on the metal surfaces, before it is trapped within the particle or at its surface.

$$h^+ + H_2O_{ads} \longrightarrow {}^{\bullet}OH_{ads} + h^+ \tag{1.3}$$

$$h^+ + HO^-{}_{ads} \longrightarrow {}^{\bullet}OH_{ads} \tag{1.4}$$

$$h^+ + M^{n+}{}_{ads} \longrightarrow M_{ads}{}^{(n+1)+} \tag{1.5}$$

These produced surface hydroxyl radicals contain enormous power to oxide the pollutants and hence mineralize them completely. Scheme 1.1 shows the formation of different oxide species by the reactions of electrons and holes through various pathways.

On the other hand, the electrons from the photoexcitation must also react to avoid accumulation of excess charges within the catalyst particles. Therefore, the efficient removal of the electrons can enhance the photocatalytic oxidation of the pollutants. Oxygen is commonly employed as electron scavenger if the overall reaction target is oxidation of dyes or organic compounds as it is very cost-effective and can be able to dissolve in aqueous and other solutions. After the attack of the electron on the oxygen, it can be reduced to the different oxygen activated species given by the following equations:

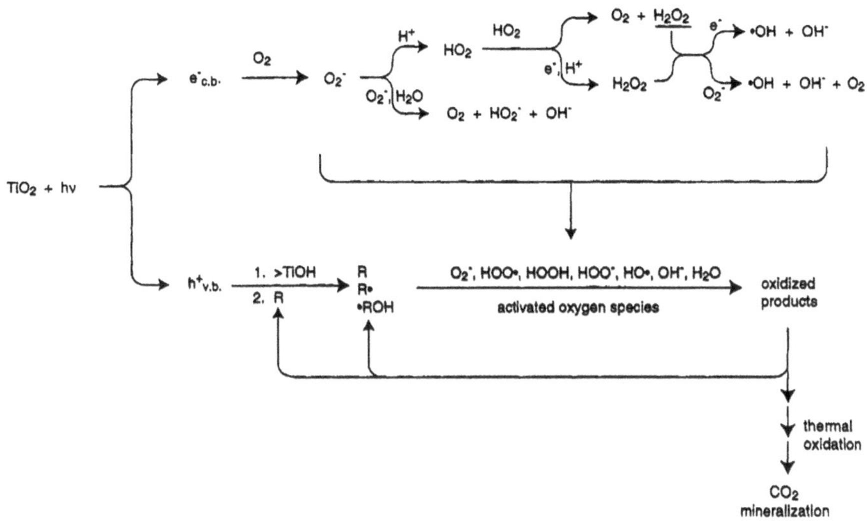

Scheme 1.1 Active oxygen species produced as a result of the irradiation of TiO_2. (Reprinted with permission from Ref. [3]. Copyright 1995 American Chemical Society)

$$O_2 + e^- \longrightarrow O_2^{\bullet -} \tag{1.6}$$

$$O_2^{\bullet -} + H^+ \longrightarrow HO_2^{\bullet} \tag{1.7}$$

$$2\,HO_2^{\bullet} \longrightarrow H_2O_2 \tag{1.8}$$

$$H_2O_2 + h\nu \longrightarrow {}^{\bullet}OH \tag{1.9}$$

$$H_2O_2 + O_2^- \longrightarrow {}^{\bullet}OH + {}^{-}OH + O_2 \tag{1.10}$$

$$H_2O_2 + e^- \longrightarrow {}^{\bullet}OH + {}^{-}OH \tag{1.11}$$

The generated hydroxyl radicals are the main oxidizing and reactive species with the very short lifetime and having the oxidation potential of 2.8 V. Hence, these produced hydroxyl radicals react with the substrate organic compounds and dye molecules very rapidly to mineralize them completely in a very nonselective manner as shown in Fig. 1.2.

1.4 Influence of Different Parameters on the Degradation of Pollutants

The parameters like pH of the reaction solution, surface area of the catalyst, concentration of the contaminant, catalytic loading, light wavelength, intensity, etc. play an important role in the degradation of the pollutants on the surface of

TiO_2. The changing in the different parameters can change the degradation results which will be discussed one by one.

1.4.1 Catalyst Loading

The catalyst concentration for photoactivity in the sample is one of the important parameters that control the degradation rate of the organic compounds. This is based upon the initial concentration of the pollutant and volume of the solution being examined. It has been generally noticed that the degradation rate of the dyes and pollutants on the surface of the catalyst increases with an increase in the loading of the catalyst [26]. The increased loading contributes to the high activity because of the presence of more active catalyst sites at higher concentrations, which helps in the formation of more hydroxyl species. Furthermore, when catalytic loading is increased above the optimum value, there observed no substantial increase in the degradation rate. This is due to the reason that the increased amounts of catalyst particles in the solution make it turbid and thus block and scatter UV radiation for the reaction to proceed, and therefore percentage degradation of the pollutant starts decreasing [27–29]. At high concentrations of photocatalysts, the chances for particle agglomeration increase, which results in the abatement of the available surface area for the light intake and hence reduces the generation of photoexcited electrons and holes which affects the photoactivity and reaction rate drops significantly. Table 1.1 shows the optimum catalyst concentration for the different dyes and organic compound degradations with different light sources.

1.4.2 pH Effect

The pH of the reaction solution is another very important and determining factor for the degradation of the pollutants in the presence of the catalyst. The degradation reactions are immensely sensitive to the pH of photocatalytic reaction mixture. The dependence of the reaction rate on pH is because of the reason that the pH controls the adsorption of the organic compound onto the photocatalyst surface, and hence, the degradation reactions rely on the ionization state and also on the surface charge of the photocatalyst and the pollutant compound. The surface of titania can be protonated or deprotonated under acidic or alkaline condition, respectively, according to the following reactions [26, 51]:

$$TiOH + H^+ \longrightarrow TiOH_2{}^+ \quad (1.12)$$

$$TiOH + {}^-OH \longrightarrow TiO^- + H_2O \quad (1.13)$$

Table 1.1 Effect of the catalytic loading on the degradation of pollutants [30–51]

Pollutant type	Light source	Photocatalyst	TiO_2 (g/L)	Optimum TiO_2 concentration, g/L
Reactive Yellow 14	UV	TiO_2	1.0–6.0	4
BPA	Solar	TiO_2	0–1.0	0.5
Chrysoidine R	UV	TiO_2	0.5–1.0	5.0
Cibacron Yellow LS-R	UV	TiO_2	0.1–4.0	4.0
Phenol	UV	TiO_2	0–4.0	2.0
Supra Blue BRL	Visible	K^+-TiO_2	0.25–2.0	1.5
Reactive Yellow 14	Solar	TiO_2	1.0–6.0	4.0
Acid Red 88	UV	ZnO	2–12	12
Remazol Brilliant Blue R	UV/solar	TiO_2	0.5–5.0	5.0
Disperse Blue 1	UV	TiO_2	0.5–4.0	3.0
Acridine orange	Visible	ZnO	0–0.35	0.25
Amaranth	UV	TiO_2	0.5–4.0	1.0
Bismarck	UV	TiO_2	0.5–4.0	1.0
Acid Orange 8	UV	TiO_2	0.5–5.0	2.0
Acid Blue 45	UV	TiO_2	0.5–3.0	2.0
Xylenol	UV	TiO_2	0.5–3.0	3.0
Acridine orange	UV	TiO_2	0.5–3.0	2.0
Bromothymol	UV	TiO_2	0.5–3.0	3.0
Fast green FCF	UV	TiO_2	0.5–4.0	4.0
Reactive Orange 4	UV	F-TiO_2	1.0–5.0	4.0
Acid Blue 80	Solar	TiO_2	0.3–0.4	2.0
Rhodamine B	Solar	ZnO	0.05–0.4	0.3
Chromotrope 2B	UV	TiO_2	0.5–5.0	5.0
Amido black 10B	UV	TiO_2	0.5–5.0	5.0

Thus, as these reaction equations show, the surface of the catalyst will remain positively charged in acidic medium and negatively charged in basic medium [26]. Table 1.2 shows some pollutants and their optimum pH at which they show the highest activity in terms of degradation/mineralization.

1.4.3 *Surface Area and Morphology*

The photocatalytic activity of semiconductor catalyst also decidedly confides on surface and structural properties, such as surface area, crystal composition, distribution, porosity, bandgap, particle size, and surface hydroxyl group density [52]. Average crystal size in the heterogeneous catalysis is of primary importance because it is directly related to its specific surface area which improves the efficiency of a catalyst. Lin et al. [60] prepared the high-quality brookite TiO_2 single-crystalline nanosheets

Table 1.2 Influence of pH on the photocatalytic degradation of pollutants [32, 36, 39–49, 53–59]

Pollutant type	Light source	Photocatalyst	pH range	Optimum pH
Phenol	UV	TiO_2	4.1–12.7	7.4
Chrysoidine Y	UV	TiO_2	3.0–9.0	9.0
m-Nitrophenol	UV	TiO_2	4.1–12.7	8.9
Reactive Blue 4	UV	Nd–ZnO	3.0–13.0	11
BPA	Solar	TiO_2	2.0–10.0	6
Methylene Blue	Visible	La^{3+}–TiO_2	2–10	10
Supra Blue BRL	Visible	K^{+}–TiO_2	4.5–11.8	7.2
Reactive Orange 4	UV	F–TiO_2	1.0–9.0	3.0
Acid Red 88	Visible	Ag–TiO_2	0.2–1.8	1.8
Remazol Brilliant Blue R	Solar	TiO_2	3.0–11.0	3.0
Disperse Blue 1	UV	TiO_2	3–11	3.0
Methyl orange	UV	Pt–TiO_2	2.5–11.0	2.5
Amaranth	UV	TiO_2	3.45–9.31	7.73
Bismarck	UV	TiO_2	3.25–7.85	3.25
Acid Orange 8	UV	TiO_2	3.0–11.0	9.0
Acid Blue 45	UV	TiO_2	2.05–10.05	5.8
Acridine orange	Visible	ZnO	2.9–7.1	7.1
Bromothymol	UV	TiO_2	2.2–9.0	4.35
Fast green FCF	UV	TiO_2	3.0–11.0	4.4
Methyl Red	UV	Ag–TiO_2	3.0–13.0	3–10.0
Acid Blue 80	Solar	TiO_2	2.0–10.0	10.0
Acid Red 29	UV	TiO_2	3.0–10.5	10.5

and tuned the catalyst activity from inert for the nanoflowers and nanospindles to the highly active nanosheets by the control of the morphology of the samples. Zuo et al. [61] prepared prism-shaped active faceted rutile Ti (III)-doped TiO_2 with the control of the morphology under acidic condition and hydrothermal treatment at 200 °C for 12 h. They proved that facet effect is an important factor for heterogeneous photocatalysts due to the surface atom arrangement and coordination which intrinsically determine the reactant adsorption on the surface of molecules, surface distribution between photoexcited electrons and reactant molecules, as well as the desorption of product molecules. Li et al. [62] prepared twist-like helix tungsten nitrogen co-doped TiO_2 for the enhanced degradation of phenol under visible light irradiation and attributed the enhancement of the photoactivity to the large surface area, special twist-like helix morphology, large pore size and better crystallinity of the anatase.

1.4.4 Effect of Temperature

Another factor that has a significant contribution in the photocatalytic degradation of organic dyes and compounds is temperature. The temperature effect is not profound if the reaction under observation has a little fluctuation; however, if the temperature will be excessively higher or lower, it might change the course of degradation reaction. Dissolved oxygen is one of the key elements that drive the photocatalysis, as it helps in scavenging CB electrons and leads to the formation of hydroxyl radicals. The higher or lower temperature causes the change in the percentage of the dissolved oxygen in the samples and hence changes the degradation rate of the reaction. Furthermore, the higher temperatures cause desorption of organic compounds from the photocatalyst surface. Because of the fact that the photocatalytic degradation of the compounds is a surface-based phenomenon, desorption of the pollutant before the reaction with the electrons or holes due to the high temperature results in a decrease of the reaction rate.

1.4.5 Effect of Contaminant Concentration

Another factor which can affect the degradation rate of the dyes and hazardous chemicals is concentration in the reaction mixture. Most of the pollutants for the degradation observe the pseudo-first-order kinetics which can be represented in terms of Langmuir–Hinshelwood equation modified for the reactions occurring at the liquid-solid interface:

$$\ln\,[C_o/C] = k_r K t = k^1 t \tag{1.14}$$

where k^1 is apparent first-order rate constant, t is the time required for the initial concentration of pollutant (C_o) to reduce to (C), K is the equilibrium constant for the adsorption of the pollutant on the surface of the catalyst, and K_r is the limiting rate of the reaction [63].

In the case of the colored compounds like dyes, commonly the degradation rate increases with the increase in the dye concentration, but after attaining a certain critical concentration level, it starts decreasing. This decrease of the rate of degradation with the increase of the concentration of pollutant can be attributed to the screening of UV–visible light radiations by the dye molecules before reaching to the surface of the catalyst. However, the concentration of the catalyst can be adjusted based on the organic compound concentration, so that organic compound can be considerably adsorbed on the photocatalyst surface and efficiently degraded. Most of the degradation studies have employed a concentration of the organic compounds or pollutant dyes in the range of 10–200 mg/L, which is similar with the concentration of the common pollutants found in real wastewater [26]. Table 1.3 illustrates the

Table 1.3 Effect of contaminant concentration on the photoactivity of the catalysts [30, 39, 40, 42, 44–49, 51, 53, 54, 64–76]

Pollutant type	Light source	Photocatalyst	Range of initial concentration, mM	Optimum concentration, mM
Phenol	UV	TiO_2	0.13–0.71	0.13
m-Nitrophenol	UV	TiO_2	0.13–0.71	0.13
Chrysoidine Y	UV	TiO_2	0.13–1.0	0.75
Acid Orange 7	UV	ZnO	0.003–0.009	0.003
Remazol Brilliant Blue R	Solar	TiO_2	0.12–0.5	0.12
Disperse Blue 1	UV	TiO_2	0.13–0.5	0.125
Amaranth	UV	TiO_2	0.3–0.6	0.5
Bismarck	UV	TiO_2	0.3–0.6	0.6
Reactive Orange 4	UV	F–TiO_2	0.015–0.035	0.3
Acid Blue 45	UV	TiO_2	0.3–0.6	0.3
Xylenol Orange	UV	TiO_2	0.3–0.6	0.5
Acridine orange	UV	TiO_2	0.1–0.5	0.25
Ethidium bromide	UV	TiO_2	0.1–0.4	0.1
Bromothymol	UV	TiO_2	0.15–0.5	0.35
Fast green FCF	UV	TiO_2	0.031–0.125	0.031
Phenol	UV	Pr–TiO_2	0.11–0.7	0.11
Acid Blue 80	Solar	TiO_2	0.03–0.2	0.03
Acid Red 18	UV	ZnO	0.2–1.0	0.2
Chromotrope 2B	UV	TiO_2	0.25–0.75	0.35
Amido black 10B	UV	TiO_2	0.25–0.75	0.25

optimum dye concentration at which they can have maximum degradation under solar and UV light.

1.4.6 *Influence of Calcination Temperature*

The calcination temperature is also an important factor which consummates the betterment of the efficiency of the prepared photocatalyst by affecting the physical and chemical properties of the catalyst. The calcination of the samples at high temperatures can increase the photoactivity of the samples because the increased temperature improves the crystallization of the sample and removes the unwanted loosely bound impurities. Furthermore, the uncalcined samples contain the excess number of water molecules on the surface which can reduce the surface active sites. However, the reports showed that the increase in the calcination decreases the surface area of the prepared catalyst because of the collapse of the pore at high temperature [67]. Ohtani et al. [68] suggested the two main reasons for the increase

Table 1.4 Effect of calcination temperature on the degradation of pollutants [52, 67, 70–86]

Pollutant	Light source	Photocatalyst	Range of temperature (°C)	Optimum temperature (°C)
Methyl orange	UV	Co^{2+}–TiO_2	150–700	200
Azo fuchsine	Visible	TiO_2	300–800	400
Methylene blue	UV	Ce^{4+}–TiO_2	300–500	400
Methyl orange	UV	N–TiO_2	0–400	300
Rhodamine B	UV	La^{3+}–TiO_2	600–900	700–800
Methylene blue	UV	Sm^{3+}–TiO_2	500–800	600
4-Chlorophenol	UV	N–TiO_2	300–700	500
o-Cresol	Visible	Pt–TiO_2	150–750	150
Methylene blue	UV–visible	WO_3–TiO_2	450–850	550
Acid orange 52	UV/solar	Zn^{2+}–TiO_2	400–550	450
BPA	UV	Zr–TiO_2	500–700	700
Methylene blue	Visible	N–TiO_2	400–700	500
Methyl orange	UV–visible	Co–TiO_2	500–850	650
Phenol	UV–visible	I–TiO_2	400–600	400
Rhodamine B	Visible	F–TiO_2	120–180	180
Phenol	UV	WO_X–TiO_2	400–800	700
Methyl orange	Solar	Fe^{3+}–TiO_2	200–600	300
2,4,6-Tribromophenol	UV	TiO_2	350–650	550

in the photoactivity of the samples: one is the high surface area of the prepared catalyst which gives the high room for the substrate dyes to adsorb on its surface and the other is crystallinity of the samples which allows less recombination of the photoexcited electrons and holes and increases the photocatalytic activity. Therefore, the increase in the calcination temperature increases the crystallinity but on the other side decreases the specific surface area of the samples. So, these two requirements are used to be partially fulfilled at a moderate calcination temperature [69]. Table 1.4 illustrates some catalysts with the optimum calcination temperature for the degradation of different types of pollutants.

References

1. Arakawa H, Aresta M, Armor JN et al (2001) Catalysis research of relevance to carbon management: progress, challenges, and opportunities. Chem Rev 101(4):953–996
2. Esswein AJ, Nocera DG (2007) Hydrogen production by molecular photocatalysis. Chem Rev 107(10):4022–4047
3. Hoffmann MR, Martin ST, Choi W et al (1995) Environmental applications of semiconductor photocatalysis. Chem Rev 95(1):69–96

4. Hashim MA, Mukhopadhyay S, Sahu JN et al (2011) Remediation technologies for heavy metal contaminated groundwater. J Environ Manag 92(10):2355–2388
5. Haji-Saeid SM, Sampa MH, Safrany A et al (2012) Radiation processing techniques in remediation of pollutants, and the role of the IAEA in supporting capacity building in developing countries. Radiat Phys Chem 81(8):1040–1044
6. Khan FI, Husain T, Hejazi R (2004) An overview and analysis of site remediation technologies. J Environ Manag 71(2):95–122
7. Yeung AT, Gu YY (2011) A review on techniques to enhance electrochemical remediation of contaminated soils. J Hazard Mater 195(0):11–29
8. Fujishima A, Honda K (1972) Electrochemical photolysis of water at a semiconductor electrode. Nature 238(0):37–38
9. Ba-Abbad MM, Kadhum AAH, Mohamad AB et al (2013) Visible light photocatalytic activity of Fe^{3+}-doped ZnO nanoparticle prepared via sol–gel technique. Chemosphere 91 (11):1604–1611
10. Burton P, Peterson E, Boyle T et al (2010) Synthesis of high surface area ZnO(0001) plates as novel oxide supports for heterogeneous catalysts. Catal Lett 139(1–2):26–32
11. Bignozzi CA, Caramori S, Cristino V et al (2013) Nanostructured photoelectrodes based on WO_3: applications to photooxidation of aqueous electrolytes. Chem Soc Rev 42(6):2228–2246
12. Tian L, Ye L, Liu J et al (2012) Solvothermal synthesis of CNTs–WO_3 hybrid nanostructures with high photocatalytic activity under visible light. Catal Commun 17(0):99–103
13. Franking R, Li L, Lukowski MA et al (2013) Facile post-growth doping of nanostructured hematite photoanodes for enhanced photoelectrochemical water oxidation. Energy Environ Sci 6(2):500–512
14. Bang JU, Lee SJ, Jang JS et al (2012) Geometric effect of single or double metal-tipped CdSe nanorods on photocatalytic H_2 generation. J Phys Chem Lett 3(24):3781–3785
15. Wang J, Yin S, Zhang Q et al (2003) Mechanochemical synthesis of fluorine-doped $SrTiO_3$ and its photo-oxidation properties. Chem Lett 32(6):540–541
16. Bhatkhande DS, Pangarkar VG, Beenackers AACM (2002) Photocatalytic degradation for environmental applications – a review. J Chem Technol Biotechnol 77(1):102–116
17. Ilmenite (2015) http://en.wikipedia.org/wiki/Ilmenite
18. Perovskite (2017) http://en.wikipedia.org/wiki/Perovskite
19. Titanite (2017) http://en.wikipedia.org/wiki/Titanite
20. Budavari S (1996) An encyclopedia of chemicals, drugs, and biologicals, merck index, 12th edn. Merck, Whitehouse Station, p 1617
21. Renz C (1921) Photoreactions of oxides of titanium, cerium and earth acids. Helv Chim Acta 4 (0):961–968
22. Keidel E (1929) Die Beeinflussung der Lichtechtheit von Teerfarblacken durch Titanweiss (The effect on the light fastness of coal tar dye paints by titanium white). Farben-Zeitung 34 (0):1242–1243
23. Serpone N, Emeline AV, Horikoshi S et al (2012) On the genesis of heterogeneous photocatalysis: a brief historical perspective in the period 1910 to the mid-1980s. Photochem Photobiol Sci 11(7):1121–1150
24. Mills A, Hunte SL (1997) An overview of semiconductor photocatalysis. J Photochem Photobiol A Chem 108(1):1–35
25. Chen X, Shen S, Guo L et al (2010) Semiconductor-based photocatalytic hydrogen generation. Chem Rev 110(11):6503–6570
26. Vinu R, Madras G (2011) Photocatalytic degradation of water pollutants using Nano-TiO2. In: Zang L (ed) Energy efficiency and renewable energy through nanotechnology. Springer, London, pp 625–677
27. Rauf M, Ashraf SS (2009) Fundamental principles and application of heterogeneous photocatalytic degradation of dyes in solution. Chem Eng J 151(1):10–18

28. Wang CC, Lee CK, Lyu MD et al (2008) Photocatalytic degradation of C.I. Basic Violet 10 using TiO_2 catalysts supported by Y zeolite: an investigation of the effects of operational parameters. Dyes Pigments 76(3):817–824
29. Macedo LC, Zaia DAM, Moore GJ et al (2007) Degradation of leather dye on TiO_2: a study of applied experimental parameters on photoelectrocatalysis. J Photochem Photobiol A Chem 185 (1):86–93
30. Ahmed S, Rasul MG, Martens W et al (2011) Advances in heterogeneous photocatalytic degradation of phenols and dyes in wastewater: a review. Water Air Soil Pollut 215(1–4):3–29
31. Muruganandham M, Swaminathan M (2006) Photocatalytic decolourisation and degradation of Reactive Orange 4 by TiO_2-UV process. Dyes Pigments 68(2):133–142
32. Kaneco S, Rahman MA, Suzuki T (2004) Optimization of solar photocatalytic degradation conditions of bisphenol A in water using titanium dioxide. J Photochem Photobiol A Chem 163 (3):419–424
33. Qamar M, Saquib M, Muneer M et al (2005) Titanium dioxide mediated photocatalytic degradation of two selected azo dye derivatives, chrysoidine R and acid red 29 (chromotrope 2R), in aqueous suspensions. Desalination 186(1–3):255–271
34. Kositzi M, Poulios I, Samara K et al (2007) Photocatalytic oxidation of Cibacron Yellow LS-R. J Hazard Mater 146(3):680–685
35. Chiou CH, Wu CY, Juang RS (2008) Influence of operating parameters on photocatalytic degradation of phenol in UV/TiO_2 process. Chem Eng J 139(2):322–329
36. Chen LC, Huang CM, Tsai FR (2007) Characterization and photocatalytic activity of K^+-doped TiO_2 photocatalysts. J Mol Catal A Chem 265(1–2):133–140
37. Muruganandham M, Shobana N, Swaminathan M (2006) Optimization of solar photocatalytic degradation conditions of Reactive Yellow 14 azo dye in aqueous TiO_2. J Mol Catal A Chem 246(1–2):154–161
38. Behnajady MA, Moghaddam SG, Modirshahla N (2009) Investigation of the effect of heat attachment method parameters at photocatalytic activity of immobilized ZnO nanoparticles on glass plate. Desalination 249(3):1371–1376
39. Saquib M, Muneer M (2002) Semiconductor mediated photocatalysed degradation of an anthraquinone dye, Remazol Brilliant Blue R under sunlight and artificial light source. Dyes Pigments 53(3):237–249
40. Saquib M, Abu Tariq M, Haque MM (2008) Photocatalytic degradation of disperse blue 1 using UV/TiO_2/H_2O_2 process. J Environ Manag 88(2):300–306
41. Pare B, Jonnalagadda SB, Tomar H (2008) ZnO assisted photocatalytic degradation of acridine orange in aqueous solution using visible irradiation. Desalination 232(1–3):80–90
42. Abu Tariq M, Faisal M, Muneer M (2005) Semiconductor-mediated photocatalysed degradation of two selected azo dye derivatives, amaranth and bismarck brown in aqueous suspension. J Hazard Mater 127(1–3):172–179
43. Saquib M, Muneer M (2003) Titanium dioxide mediated photocatalyzed degradation of a textile dye derivative, acid orange 8, in aqueous suspensions. Desalination 155(3):255–263
44. Abu Tariq M, Faisal M, Saquib M et al (2008) Heterogeneous photocatalytic degradation of an anthraquinone and a triphenylmethane dye derivative in aqueous suspensions of semiconductor. Dyes Pigments 76(2):358–365
45. Faisal M, Abu Tariq M, Muneer M (2007) Photocatalysed degradation of two selected dyes in UV-irradiated aqueous suspensions of titania. Dyes Pigments 72(2):233–239
46. Haque MM, Muneer M (2007) TiO_2-mediated photocatalytic degradation of a textile dye derivative, bromothymol blue, in aqueous suspensions. Dyes Pigments 75(2):443–448
47. Saquib M, Abu Tariq M, Faisal M (2008) Photocatalytic degradation of two selected dye derivatives in aqueous suspensions of titanium dioxide. Desalination 219(1–3):301–311
48. Vijayabalan A, Selvam K, Velmurugan R (2009) Photocatalytic activity of surface fluorinated TiO_2-P25 in the degradation of Reactive Orange 4. J Hazard Mater 172(2–3):914–921
49. Su Y, Deng L, Zhang N et al (2009) Photocatalytic degradation of C.I. Acid Blue 80 in aqueous suspensions of titanium dioxide under sunlight. React Kinet Catal Lett 98(2):227–240

50. Byrappa K, Subramani AK, Ananda S et al (2006) Photocatalytic degradation of rhodamine B dye using hydrothermally synthesized ZnO. Bull Mater Sci 29(5):433–438
51. Qamar M, Saquib M, Muneer M (2005) Photocatalytic degradation of two selected dye derivatives, chromotrope 2B and amido black 10B, in aqueous suspensions of titanium dioxide. Dyes Pigments 65(1):1–9
52. Ahmed S (2011) Impact of operating conditions and recent developments in heterogeneous photocatalytic water purification process. Crit Rev Environ Sci Technol 42(6):601–675
53. Chiou CH, Wu CY, Juang RS (2008) Photocatalytic degradation of phenol and m-nitrophenol using irradiated TiO_2 in aqueous solutions. Sep Purif Technol 62(3):559–564
54. Qamar M, Saquib M, Muneer M (2005) Semiconductor-mediated photocatalytic degradation of anazo dye, chrysoidine Y in aqueous suspensions. Desalination 171(2):185–193
55. Zhou Y, Lu SX, Xu WG (2009) Photocatalytic activity of Nd-doped ZnO for the degradation of C.I. Reactive Blue 4 in aqueous suspension. Environ Prog Sustain Energy 28(2):226–233
56. Parida KM, Sahu N (2008) Visible light induced photocatalytic activity of rare earth titania nanocomposites. J Mol Catal A Chem 287(1–2):151–158
57. Anandan S, Kumar PS, Pugazhenthiran N et al (2008) Effect of loaded silver nanoparticles on TiO_2 for photocatalytic degradation of Acid Red 88. Sol Energy Mater Sol Cells 92(8):929–937
58. Huang M, Xu C, Wu Z et al (2008) Photocatalytic discolorization of methyl orange solution by Pt modified TiO_2 loaded on natural zeolite. Dyes Pigments 77(2):327–334
59. Gupta AK, Pal A, Sahoo C (2006) Photocatalytic degradation of a mixture of Crystal Violet (Basic Violet 3) and Methyl Red dye in aqueous suspensions using Ag^+ doped TiO_2. Dyes Pigments 69(3):224–232
60. Lin H, Li L, Zhao M (2012) Synthesis of high-quality brookite TiO_2 single-crystalline nanosheets with specific facets exposed: tuning catalysts from inert to highly reactive. J Am Chem Soc 134(20):8328–8331
61. Zuo F, Bozhilov K, Dillon RJ (2012) Active facets on titanium(III)-doped TiO_2: an effective strategy to improve the visible-light photocatalytic activity. Angew Chem 124(25):6327–6330
62. Li J, Xu J, Dai WL et al (2008) One-pot synthesis of twist-like helix tungsten-nitrogen-codoped titania photocatalysts with highly improved visible light activity in the abatement of phenol. Appl Catal B Environ 82(3):233–243
63. Boarini P, Carassiti V, Maldotti A et al (1998) Photocatalytic oxygenation of cyclohexane on titanium dioxide suspensions: effect of the solvent and of oxygen. Langmuir 14(8):2080–2085
64. Daneshvar N, Rasoulifard MH, Khataee AR (2007) Removal of C.I. Acid Orange 7 from aqueous solution by UV irradiation in the presence of ZnO nanopowder. J Hazard Mater 143 (1–2):95–101
65. Sobana N, Swaminathan M (2007) The effect of operational parameters on the photocatalytic degradation of acid red 18 by ZnO. Sep Purif Technol 56(1):101–107
66. Chiou CH, Juang RS (2007) Photocatalytic degradation of phenol in aqueous solutions by Pr-doped TiO_2 nanoparticles. J Hazard Mater 149(1):1–7
67. An T, Liu J, Li G et al (2008) Structural and photocatalytic degradation characteristics of hydrothermally treated mesoporous TiO_2. Appl Catal A Gen 350(2):237–243
68. Ohtani B, Ogawa Y, Nishimoto SI (1997) Photocatalytic activity of amorphous–anatase mixture of titanium(IV) oxide particles suspended in aqueous solutions. J Phys Chem B 101 (19):3746–3752
69. Gupta S, Tripathi M (2011) A review of TiO_2 nanoparticles. Chin Sci Bull 56(16):1639–1657
70. Hua DS, Kunfeng Z, Daorong L et al (2009) Synthesis, characterization and photocatalytic activity of Co^{2+} -doped titania. J Phys Conf Ser 188(1):012–018
71. Wang J, Zhao G, Zhang Z et al (2007) Investigation on degradation of azo fuchsine using visible light in the presence of heat-treated anatase TiO_2 powder. Dyes Pigments 75(2):335–343
72. Du Y, Du M, Qiao Y et al (2007) Ce(IV) doped TiO_2 thin films: characterization and photocatalysis. Colloid J 69(6):695–699
73. Chen Y, Zhang S, Yu Y et al (2008) Synthesis, characterization, and photocatalytic activity of N-doped TiO_2 nanotubes. J Dispers Sci Technol 29(2):245–249

74. Quan X, Tan H, Zhao Q et al (2007) Preparation of lanthanum-doped TiO_2 photocatalysts by coprecipitation. J Mater Sci 42(15):6287–6296
75. Xiao Q, Si Z, Yu Z et al (2008) Characterization and photocatalytic activity of Sm^{3+}-doped TiO_2 nanocrystalline prepared by low temperature combustion method. J Alloys Compd 450 (1–2):426–431
76. Yu H, Zheng X, Yin Z et al (2007) Preparation of nitrogen-doped TiO_2 nanoparticle catalyst and its catalytic activity under visible light. Chin J Chem Eng 15(6):802–807
77. Chen HW, Ku Y, Kuo YL (2007) Effect of Pt/TiO_2 characteristics on temporal behavior of o-cresol decomposition by visible light-induced photocatalysis. Water Res 41(10):2069–2078
78. Saepurahman AMA, Chong FK (2010) Dual-effects of adsorption and photodegradation of methylene blue by tungsten-loaded titanium dioxide. Chem Eng J 158(3):418–425
79. Chen C, Wang Z, Ruan S (2008) Photocatalytic degradation of C.I. Acid Orange 52 in the presence of Zn-doped TiO_2 prepared by a stearic acid gel method. Dyes Pigments 77 (1):204–209
80. Gao B, Lim TM, Subagio DP et al (2010) Zr-doped TiO_2 for enhanced photocatalytic degradation of bisphenol A. Appl Catal A Gen 375(1):107–115
81. Sathish M, Viswanathan B, Viswanath RP (2007) Characterization and photocatalytic activity of N-doped TiO_2 prepared by thermal decomposition of Ti–melamine complex. Appl Catal B Environ 74(3–4):307–312
82. Sun CG, Tao L, Fan ML et al (2009) Replication route synthesis of mesoporous titanium–cobalt oxides and their photocatalytic activity in the degradation of methyl orange. Catal Lett 129 (1–2):26–38
83. Hong X, Wang Z, Cai W et al (2005) Visible-light-activated nanoparticle photocatalyst of iodine-doped titanium dioxide. Chem Mater 17(6):1548–1552
84. Ren G, Gao Y, Liu X et al (2010) Synthesis of high-activity F-doped TiO_2 photocatalyst via a simple one-step hydrothermal process. Reac Kinet Mech Catal 100(2):487–497
85. Piszcz M, Tryba B, Grzmil B (2009) Photocatalytic removal of phenol under UV irradiation on WO_x –TiO_2 prepared by sol–gel method. Catal Lett 128(1–2):190–196
86. Liu J, Qin W, Zuo S (2009) Solvothermal-induced phase transition and visible photocatalytic activity of nitrogen-doped titania. J Hazard Mater 163(1):273–278

Chapter 2
In Situ Characterization of Photocatalytic Activity

The photocatalytic reactions are initiated from the photoinduced electrons and holes on the surfaces of photocatalyst, which further react with water/oxygen to form reactive oxygen species (ROS) with high reactivity including superoxide anion radical ($^{\bullet}O_2^-$), hydrogen peroxide (H_2O_2), singlet oxygen ($^1O_2$), and hydroxyl radical ($^{\bullet}OH$). The determination of ROS formed from the semiconductor-based photocatalysis and the in situ tracking of the formation sites; trajectory and kinetics are essential for understanding the photocatalytic redox mechanism, thus helping to guide the structure design of the photocatalyst. This chapter focuses on the application of fluorescence, infrared, Raman, electron spin resonance (ESR), and surface photovoltage spectroscopies and atomic force microscopy (AFM) to reveal the above process on the surface of the photocatalyst regarding the surface heterogeneity, crystal facet, and molecule conformation. These studies have greatly promoted the design and application of photocatalyst.

2.1 Fluorescence

Fluorescence is one of the most important ways for the identification, quantification, and kinetics evaluation of ROS. Fluorophores including luminol, fluorescein, coumarin, and tetrafluoroborate are commonly used probes for ROS [1]. The dehydrogenation, esterification, and the intramolecular electron transferring may cause the on–off, off–on, or wavelength shift of the fluorescence. Such variation provides specific information about the ROS formation on the surface of photocatalysts.

J. Zhang et al., *Photocatalysis*, Lecture Notes in Chemistry 100,
https://doi.org/10.1007/978-981-13-2113-9_2

2.1.1 Types of Probing Agents for ROS

Hydrogen Peroxide Oxidized luminol, which is 5-amino-2, 3-diaza-1, 4-naphthoquinone, can react with H_2O_2 to generate the excited state of 3-aminophthalate ion, AP* [2]. It is also an active fluoregen to $^{\bullet}O_2^-$. For the exclusive detection of H_2O_2, the luminol solution was added to the reaction mixture, which was then kept in the dark for ca. 30 min to eliminate $^{\bullet}O_2^-$. Then hemoglobin (Hem) as an oxidant is added into the mixture to form fluorescence. A detection limit of about 1 nM can be achieved through this probe. Besides HEM, $K_3Fe(CN)_6$ could also be used as an oxidant to trigger the fluorescence [3]. The application of luminol is mainly restricted in basic system since it could be used only at higher pH ($pH > 10$).

Lucigenin (bis-*N*-methylacridinium nitrate) is another typical fluorescent probe for both H_2O_2 and $^{\bullet}O_2^-$ [4], which could be operated at moderate pH ($pH = 9$). A dioxetane can be produced after the oxidation, which decomposes to the excited state of *N*-methyl acridone. Using the detection of H_2O_2 as an example, lucigenin solution was injected into the suspension of TiO_2 after stopping UV irradiation on TiO_2 powder. On the injection, bluish-green fluorescence appeared and decayed in several seconds. The time integral of the fluorescence intensity is proportional to the amount of H_2O_2, which is used to estimate the H_2O_2 concentration with calibrations.

Superoxide Oxygen Anion As mentioned above, luminol is also widely used for probing $^{\bullet}O_2^-$. The detection limit of $^{\bullet}O_2^-$ with luminol chemiluminescence is estimated to be 0.1 nM. As a reference, MCLA (2-methyl-6-(4-methoxyphenyl)-3, 7-dihydroimidazo (1, 2-a) pyrazin-3-one, hydrochloride), a probe for both $^{\bullet}O_2^-$ and $^1O_2$, can be used to confirm the exclusive detection of $^{\bullet}O_2^-$ together with luminol [5]. The priority of MCLA is its applicability to neutral pH system.

Singlet Oxygen Sensor A singlet oxygen sensor green (SOSG) was employed to detect $^1O_2$, which produces an endoperoxide that fluoresces at 528 nm under the excitation at 488 nm [6]. SOSG shows no appreciable response to $^{\bullet}OH$ or $^{\bullet}O_2^-$. However, when it was used under light illumination, significant problems in photosensitized $^1O_2$ generation and photodecomposition were clearly demonstrated. $^1O_2$ generated in air can also be detected with the terrylene diimide (TDI) derivative, which produces fluorescent diepoxide [7].

Hydroxyl Radical Fluorescein derivatives are commonly used for the probing of $^{\bullet}OH$. For example, 3′-(*p*-hydroxyphenyl) fluorescein (HPF) selectively reacts with $^{\bullet}OH$ to form a strongly emissive fluorescein molecule but does not react with the other ROS [8]. Nonfluorescent Amplex Red is also a commonly used probe for $^{\bullet}OH$ and $^{\bullet}O_2^-$, forming the fluorescent product resorufin [9]. Other probes including terephthalic acid and coumarin derivatives were also active to generate the fluorescent product for detecting $^{\bullet}OH$ in solution, which have no absorption at the excitation wavelength for photocatalysts. Different from the detection methods for $^{\bullet}O_2^-$ and H_2O_2, the probe molecules were present during the photocatalytic reaction, and

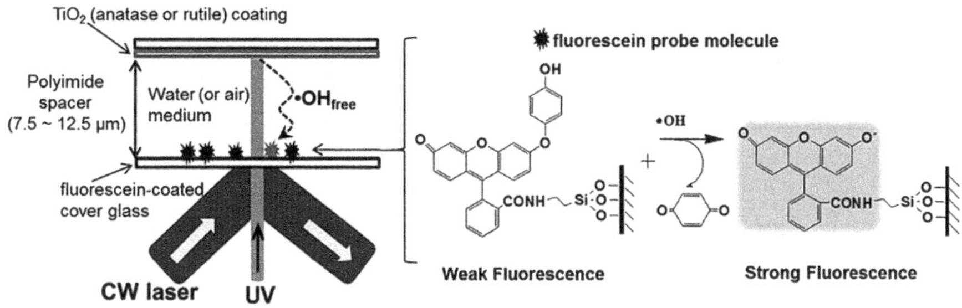

Fig. 2.1 Illustration of the experimental setup for the single-molecule detection of photogenerated $^{\bullet}$OH in water. Anatase or rutile was coated on the upper cover glass and the modified fluorescein (HPF) was anchored on the lower glass through a silanol group. The intervening gap was controlled using polyimide films. The gap was filled with air-saturated water. (Reproduced from Ref. [10] by permission of John Wiley & Sons Ltd)

after the reaction, the photocatalyst powders were removed from the suspension to improve the precision of the fluorescence spectrophotometry.

2.1.2 *Single-Molecule Spectroscopy*

Traditional fluorescent spectroscopy provides spectra regarding the intensity, lifetime, and polarization by averaging the emission from trillions of molecules. By contrast, single-molecule fluorescence spectroscopy rests upon the removal of ensemble average, thus allowing the detection of the heterogeneity of the local environment.

Single-molecule spectroscopy can be used to verify the diffusion of $^{\bullet}$OH from the surface of semiconductor to the solution (Fig. 2.1) [10]. HPF can selectively react with $^{\bullet}$OH generated on UV-irradiated TiO_2 instead of $O_2^{\bullet-}$, $^1O_2$, and H_2O_2, transforming into a strongly fluorescing product. HPF was anchored on a glass plate through the silanol group and separated from a TiO_2-coated glass plate with a spacer. The distance between the two glass plates was controlled using polyimide films, and the space was filled with air-saturated water. In this way, the generation and the subsequent diffusion of $^{\bullet}$OH from the illuminated TiO_2 surface to the solution bulk can be directly observed using the single-molecule fluorescence spectroscopy. Interestingly, bright fluorescent signals clearly emerged over the irradiated region of the anatase film after UV irradiation, whereas the signals generated over rutile were negligible. Therefore, the mobile $^{\bullet}$OH is generated on anatase but not on rutile. The photocatalytic oxidation on rutile is limited to adsorbed substrates whereas that on anatase is more facile and versatile owing to the presence of mobile $^{\bullet}$OH. This result partly explains the common observations that anatase has higher activity than rutile.

Facet-Selective Photocatalytic Reaction The differences in surface energy levels of the conduction and valence bands, surface structures, and adsorption energies of substrates on the exposed crystal faces may result in face-selective photocatalytic reaction. Fluorescence at the single-molecule and single-particle level has recently evolved as an important tool for studying selective photocatalytic reactions on different facets because of its high sensitivity, simplicity of data collection, and high spatial resolution in microscopic imaging techniques. Several organic dye probes have been successfully employed to detect the generated ROS and identify the active sites on individual TiO_2 nanoparticles by utilizing single-molecule fluorescence spectroscopy. A typical example is from boron dipyrromethene (BODIPY), which has a high extinction coefficient, high fluorescence quantum yield, and good chemical and photostability. It has evolved into versatile fluorescent sensors for biological and chemical detection. On the other hand, the reduction of aromatic nitro compounds to the corresponding hydroxylamines or amines has been widely used as a model system to investigate photocatalytic reduction reactions with semiconductor or metal nanoparticles. The major drawback for the development of a fluorogenic probe based on the reduction of a nitro-substituted benzene moiety for monitoring electron transfer (ET) process lies in the strong quenching effect of nitrobenzene and its reduction products (i.e., phenylhydroxylamine or aniline). Since mono-nitro-substituted BODIPY derivative fails to function as a fluorescent probe, 3, 4-dinitrophenyl modified BODIPY (DN-BODIPY) was designed as a redox-responsive fluorescent probe (Fig. 2.2), where the intramolecular ET process is suppressed when the produced electron-donating group encounters the second nitro group [11]. This probe was applied to both ensemble-averaged and single-molecule fluorescence-monitoring of photoinduced ET process on the TiO_2 surface.

Total internal reflection fluorescence microscopy (TIRFM) was used for monitoring the photocatalytic reduction of DN-BODIPY molecules over single TiO_2 particles. Figure 2.3 A shows typical fluorescence images from a single TiO_2 particle in Ar-saturated methanol containing DN-BODIPY (2 μM) under UV irradiation (middle and right images). A number of fluorescence bursts were generated from individual single particles. The locations of the fluorescence bursts, which were

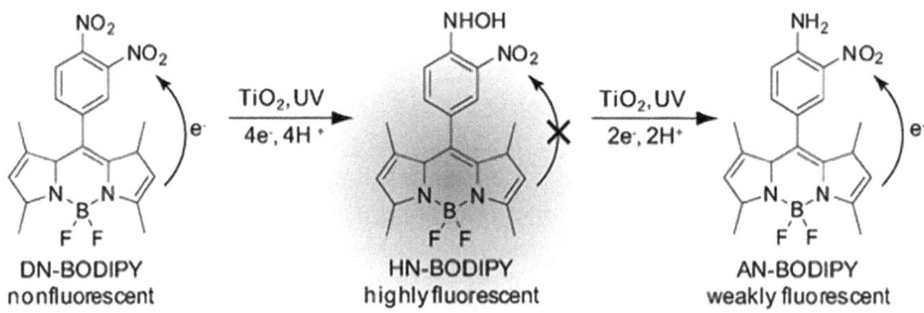

Fig. 2.2 Photocatalytic generation of fluorescent HN-BODIPY from nonfluorescent DN-BODIPY [11]. (Reproduced from Ref. [11] by permission of John Wiley & Sons Ltd)

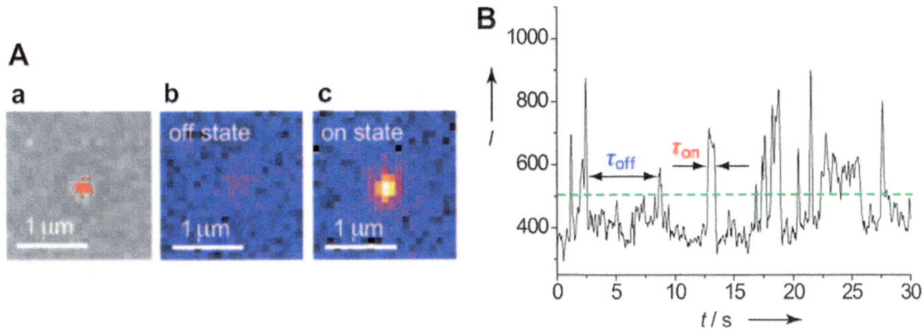

Fig. 2.3 (**a**) Photocatalytic generation of fluorescent HN-BODIPY from nonfluorescent DN-BODIPY. (**a**) Transmission (*a*) of a single TiO_2 particle on the cover glass and fluorescence images (*b* and *c*) of the same particle in Ar-saturated 2.0 mm DN-BODIPY solution under 488 nm laser and UV irradiation (0.5 Wcm^{-2} at the glass surface). The acquisition time of an image was 50 ms. The red dots in the transmission image indicate the location of fluorescence bursts. The accuracy of location was about 50 nm. (**b**) A typical fluorescence intensity trajectory observed for a single TiO_2 particle. The green dashed line indicates the threshold level separating the on and off states. (Reproduced from Ref. [11] by permission of John Wiley & Sons Ltd)

determined by fitting two-dimensional Gaussian functions to the intensity distribution of each fluorescence spot, are likely distributed over the particle according to the red dots in the transmission image. The fluorescence lifetimes of the in situ generated bursts over single TiO_2 particles were measured by combining confocal microscopy with a time-correlated single-photon counting (TCSPC) system. The fluorescence bursts exhibited a much longer lifetime than the background signal from DN-BODIPY in solution, thus suggesting that such a sudden intensity increase corresponds to the generation of fluorescent HN-BODIPY.

The precise mapping of photocatalytic activity in individual TiO_2 crystals at the nanometer scale were demonstrated by single-molecule fluorescence imaging (Figs. 2.3 and 2.4) [11, 12]. Interestingly, most fluorescence spots were found to be preferentially located on the (101) surface of the crystal (see red dots in image B). A similar tendency was observed for more than five individual examined crystals.

Metal–Semiconductor Metal–semiconductor heterostructures are promising visible light active photocatalysts for many chemical reactions. The photocatalytic behavior of metal–semiconductor heterostructures has been studied mainly at the ensemble level. The catalytic properties were considerably influenced by the individual particle sizes, structures, and so forth. To surmount the challenge arising from the intrinsic heterogeneity associated with ensemble-averaged measurements, it is highly desirable and necessary to employ photocatalytic measurements at the single-particle level.

The nature and photocatalytic properties of the surface reactive sites on single Au–CdS hybrid nano-catalysts were studied by high-resolution superlocalization fluorescence imaging (Fig. 2.5) [13]. The plasmon-induced hot electrons in Au are injected into the conduction band of the CdS semiconductor nanorod. The

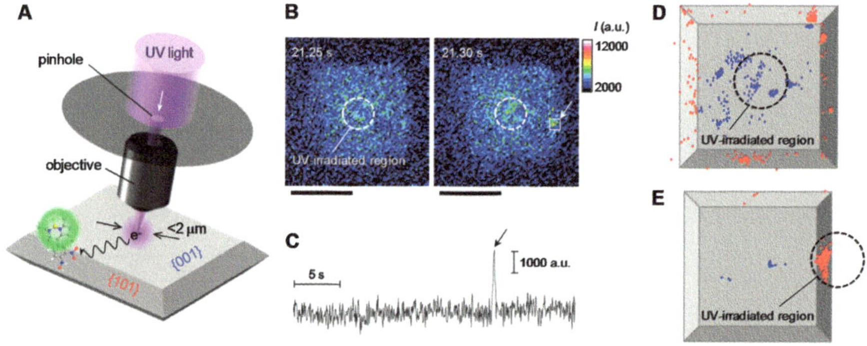

Fig. 2.4 (**a**) Illustration of the remote photocatalytic reaction on the {101} facets with DN-BODIPY during photoirradiation onto the {001} facets. The irradiated area was limited by a pinhole (the spot diameter is 2 μm on the crystal surface). (**b**) Fluorescence image of a TiO_2 crystal that is immobilized on the cover glass in Ar-saturated DN-BODIPY solution (1.0 μM, in methanol) under a 488 nm laser and UV irradiation. The scale bars are 4 μm. (**c**) Time trace of fluorescence intensity observed over the square region in panel B (see the arrow). The UV irradiation area is inside the white circle in the images. (**d** and **e**) Location of fluorescence bursts on the {001} (blue) and {101} (red) facets. The UV irradiation areas are inside the black circles (diameter 2 μm). (Reprinted with the permission from Ref. [12]. Copyright 2011 American Chemical Society)

specifically designed Au-tipped CdS heterostructures with a unique geometry (two Au nanoparticles at both ends of each CdS nanorod) provide more convincing high-resolution single-turnover mapping results and clearly prove the two charge-separation mechanisms. Engineering the direction of energy flow at the nanoscale can provide an efficient way to overcome important challenges in photocatalysis, such as controlling catalytic activity and selectivity. These results bear enormous potential impact on the development of better visible light photocatalysts for solar-to-chemical energy conversion.

Au-tipped CdS heterostructures were designed to clearly demonstrate the existence of two distinct photocatalysis mechanisms having the opposite direction of energy flow at the single-particle level, where two Au nanoparticles locate at both ends of each CdS nanorod. The relatively long distance between the two Au nanoparticles (186 nm on average) surmounts the obstacle caused by the single-molecule superlocalization accuracy (5–15 nm). Under the 532 nm laser irradiation, the h^+ reactive sites (circled blue cross) are positioned at the gold tips on both ends of the heterostructures, while the e^- reactive sites (circled red minus) are located along the inside length of the CdS nanorods within a distance of a few tens of nanometers from the Au tips. This result reveals the transfer of electron from Au to CdS. Opposite charge flowing direction was further verified when 405 nm laser was used since h^+ are distributed along the CdS nanorod, while the e^- reactive sites are located at both ends.

Imaging of Defect-Related Photocatalytic Activity Microporous titanosilicate ETS-10 is promising photocatalyst because of the inherent quantum nature of one-dimensional titania (–Ti–O–Ti–) wires in the framework and its high reaction

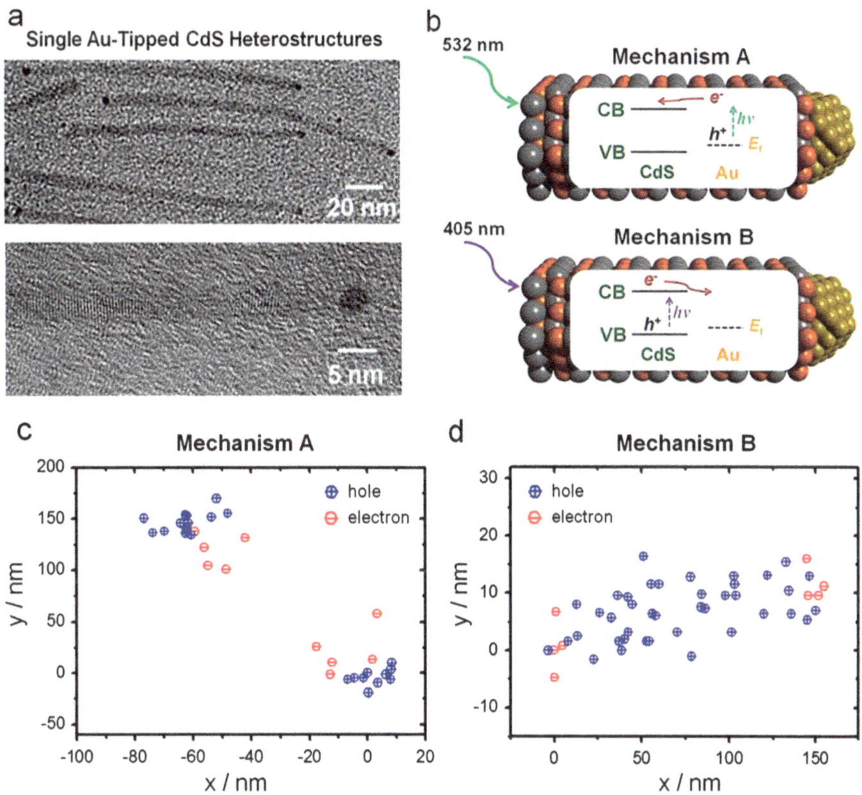

Fig. 2.5 (**a**) TEM image of typical single Au-tipped CdS nanorod. (**b**) Schematic illustrating two distinct photocatalysis mechanisms with the opposite direction of energy flow. In mechanism A at 532 nm, the photogenerated energetic electrons in Au are injected to the CB of the semiconductor. In mechanism B at 405 nm, the photogenerated electrons in the CB of the semiconductor are rapidly trapped by Au. (**c**) Super-resolution mapping of single reactive sites formed through mechanism A. (**d**) Super-resolution mapping of single reactive sites formed through mechanism B. (Reprinted with the permission from Ref. [13]. Copyright 2011 American Chemical Society)

selectivity owing to its nanoporous structure. The carriers formed from light irradiation migrate in the crystal through the wires. However, the wires may be randomly broken in the crystal because of inherent defects, which form active centers for molecular adsorption and redox reactions. To date, the influences of structural defects in ETS-10 on the adsorption and reaction dynamics of organic compounds have not been well characterized. The spatial heterogeneities of the surfaces and the inhomogeneous coupling interface between the adsorbed molecules and the nanoscale rough surfaces of the semiconductor make the dissecting of the complex interfacial ET processes highly difficult. The in situ fluorescence imaging of photocatalytic oxidation was carried out on single ETS-10 crystals using a redox-responsive fluorescent dye; 3′-(*p*-aminophenyl) fluorescein (APF), which was commonly used for the selective detection of $^{\bullet}$OH, was employed to identify the surface-

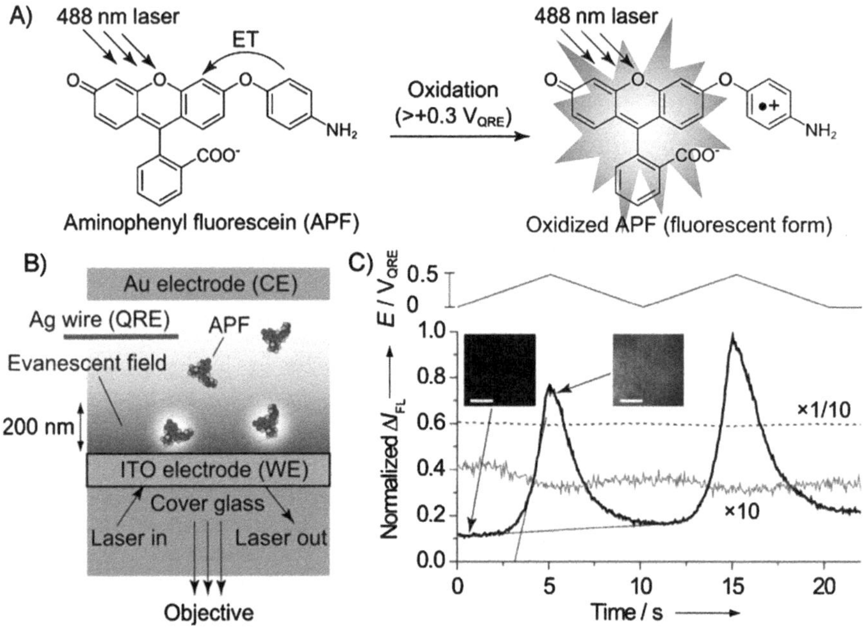

Fig. 2.6 (**a**) Reaction scheme for the one-electron oxidation of the p-aminophenyl moiety, which induces fluorescence in APF. ET = electron transfer. (**b**) Cell configuration for in situ spectroelectrochemical measurements under TIRFM. CE = counter electrode, QRE = quasi-reference electrode, WE = working electrode. (**c**) Applied potential dependence of normalized fluorescence intensity (IFL) obtained for the phosphate buffer solutions in the absence (gray line) and presence of APF (5 mm; solid line) or fluorescein (500 nm; dashed line). The images in (**c**) were acquired at a bin time of 50 ms. Scale bars are 5 mm. (Reproduced from Ref. [14] by permission of John Wiley & Sons Ltd)

active sites distributed over the ETS-10 (Fig. 2.6) [14]. It was revealed that surface treatment of ETS-10 with HF aqueous solution significantly increased not only the adsorption and reaction efficiencies but also the heterogeneity of photocatalytic activity among the crystals. The crystal defects serve as active sites during the photocatalytic reaction in aqueous solution. The fluorescence intensity of fluorescein increased gradually upon UV irradiation of phosphate buffer solutions containing APF and ETS-10 powder. In the absence of ETS-10, the fluorescence intensity did not increase at all, thus suggesting that fluorescein molecules might be generated by the reaction of photoexcited ETS-10 and APF molecules. Meanwhile, a significant increase in the fluorescence intensity was observed for HF-treated ETS-10 because of the exposure of surface-active sites such as titanols and larger micropores by partial dissolution of siliceous walls surrounding the titania wires (Fig. 2.7).

The Diffusion Distance of Radicals in a Single Particle Porous, cavitary, and tubular structured photocatalysts with a high specific surface area have been designed to improve the photocatalytic efficiency or achieve the selective transformation. These kinds of materials are supposed to have more active sites for the

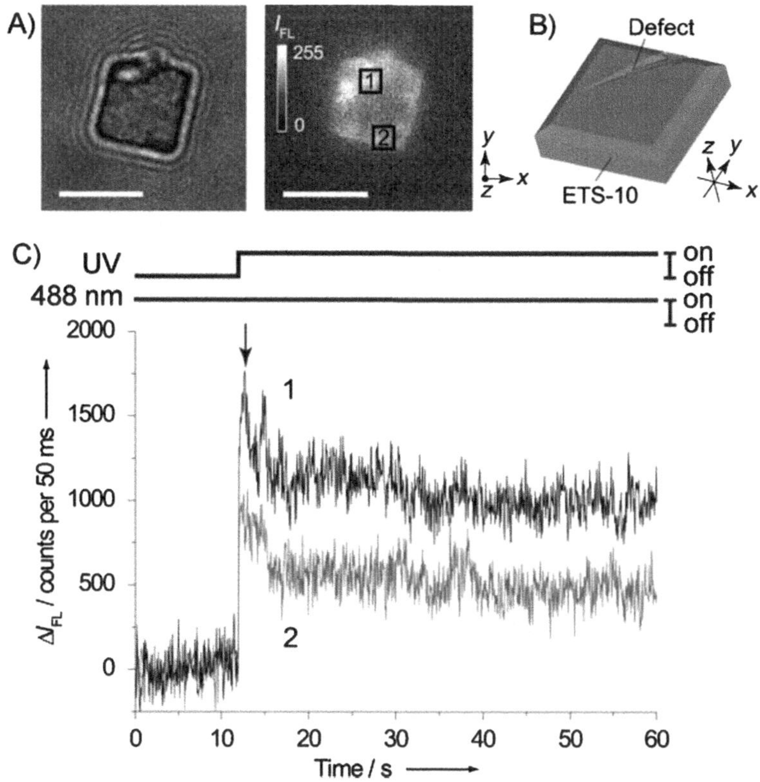

Fig. 2.7 (**a**) Transmission image (left) and fluorescence image under UV irradiation (right), where the imaging focal plane was located at the center of the crystal. The arrow in (**c**) denotes the time when the fluorescence image was acquired. Scale bars correspond to 5 mm. (**b**) Spatial configuration of the crystal (see the axes). (**c**) Time traces of the fluorescence intensity acquired at the defect (position 1, black line) and near the edge of the ETS-10 crystal (position 2, gray line). (Reproduced from Ref. [14] by permission of John Wiley & Sons Ltd)

adsorption and activation of reactants. However, the performance is highly dependent on the accessibility of reactant molecules to the active sites inside the bulk particles. The structural disorders or defects homogeneously distributed over the framework often hinder the efficient molecular transport. One of the examples for the single-molecule imaging of photocatalytic reactions in a porous TiO_2 nanotube was achieved using fluorescein and TIRFM [15]. Single TiO_2 nanotube was placed in a custom-made sample chamber, which is irradiated by UV light with a wavelength of 365 nm and laser light with a wavelength of 488 nm to excite the nanotube and fluorescein, respectively (Fig. 2.8). It is expected that the photocatalytic reaction will occur randomly on the entire nanotube. To examine the specific interaction between the wall of the TiO_2 nanotube and the fluorescein diffusing in the macropore, SiO_2 nanotube without mesopores was used as a control. Fluorescein was generated by auto-oxidation after intense UV irradiation. Both TiO_2 and SiO_2 nanotubes exhibit a

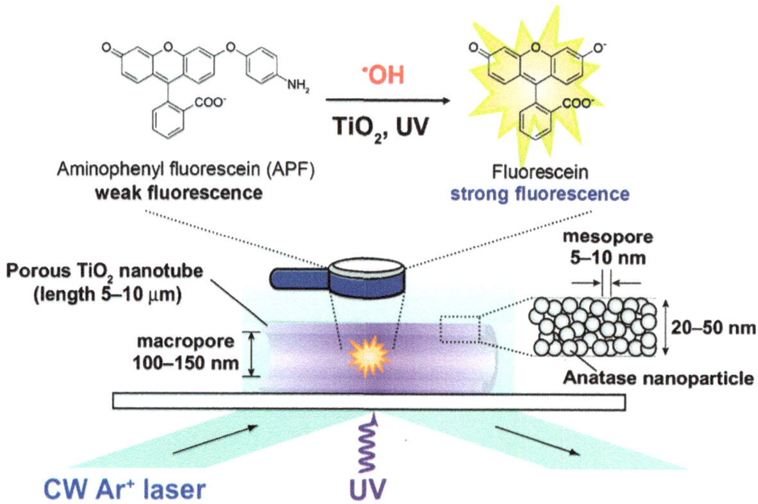

Fig. 2.8 Schematic representation of photocatalytic reaction occurring on a single nanotube. (Top) generation of emissive fluorescein induced by the photocatalytic reaction in the porous structure of the TiO_2 nanotube. (Bottom) the porous TiO_2 nanotube on the cover glass is simultaneously irradiated with UV light (wavelength: 365 nm) and evanescent light produced by a CW Ar^+ laser (wavelength: 488 nm) to excite the nanotube and fluorescein, respectively. (Reprinted with the permission from Ref. [15]. Copyright 2011 American Chemical Society)

similar distribution, confirming that the fluorescein in the macropore of the TiO_2 nanotube diffuses without interaction with the surface.

2.2 Infrared Spectroscopy

Infrared (IR) spectroscopy is a powerful method to study adsorbed species on solid surfaces, which has the ability to provide the vibrational spectrum of both reactants and photocatalyst. Abundant information can be real-time revealed regarding the molecular structure, orientation, and conformation of surface species through the in situ IR spectroscopy. Techniques including electron paramagnetic resonance (EPR), X-ray absorption near K-edge structure spectroscopy, absorption, and emission spectra only allow hypotheses to be formulated regarding the identity of surface intermediate species, but their chemical nature could not be directly specified. By contrast, the electric fields generated by photoexcited charge carriers and the further reaction of charge carrier with surface hydroxyl and adsorbed molecules may cause the intensity and wavelength shifts of stretching vibrations and the formation of new band in IR spectroscopy, thus allowing the inferring of the possible reaction path based on the intermediates found from the in situ FTIR study.

2.2.1 *Gas-Phase Photocatalysis*

Using the diffuse reflectance IR Fourier transform spectroscopy (DRIFTS), the gas-phase photocatalytic reaction can be well tracked. Szczepankiewicz et al. revealed a new band at 3716 cm^{-1} for UV-irradiated TiO_2 in the gas phase in the presence of a hole scavenger, ascribing to OH stretching for surface $Ti^{3+}-OH$ formed by electron capture at acidic $Ti^{4+}-OH$ centers [16]. Another band at 3683 cm^{-1} in an oxygen atmosphere in the absence of hole scavengers was attributed to a surface-bound OH radical formed by hole capture at $Ti^{4+}-OH$ centers. Intensity changes and corresponding wavelength shifts for ν(TiO−H) are proportional to the magnitude and polarity of the electric field. Wu et al. detected a large amount of surface peroxo species and OH groups on the TiO_2 and M (Cu, V, and Cr)/TiO_2 catalysts after a calcination treatment at 500 °C under airflow. The photocatalytic reaction of nitric oxide (NO) on TiO_2 and transition metal-loaded M/TiO_2 catalysts indicates nitric oxide is adsorbed on TiO_2 and M/TiO_2 in the form of bidentate nitrites and nitrates by reacting with OH groups, peroxo, or M = O species. In addition, NO can also be adsorbed on M^{n+} in the form of nitrosyls. Under UV irradiation, bidentate nitrite was oxidized to either monodentate or bidentate nitrate, which was induced by superoxo species oxidized from peroxo species by photogenerated holes.

2.2.2 *Aqueous-Phase Photocatalysis*

The application of IR spectroscopy in aqueous or vapor photocatalysis system is restricted by the strong IR absorption for water. To minimize the interference from water, the attenuated internal reflection (ATR) technique has been developed for dissecting the intermediate formed during photoirradiation, where evanescent IR waves penetrate intro a thin layer of an aqueous solution and reach the surface of semiconductor covered on the surface of the internal reflection element (IRE). The spectral cell for multiple internal reflection infrared (MIRIR) experiments uses a trapezoidal-shaped ZnSe as the internal reflection element (IRE), which was irradiated by an IR beam with 45° incident angle. Taking TiO_2 film as a demonstration, it is applied on one face of the IRE by a dip-coating method (Fig. 2.9), which was set in the spectral cell made by Diflon [17]. The inner volume of the cell was approximately 10 mL. The length and width of the exposed area of the IRE are 35.5 and 9.2 mm, respectively. The IR light was reflected about nine times in the IRE with TiO_2, as calculated from the geometry of the IRE. The spectral cell was placed in the sample chamber of an FTIR spectrometer with a deuterated triglycine sulfate (DTGS) detector.

The first direct in situ spectroscopic detection of primary intermediates for the photocatalytic O_2 reduction in aqueous solutions was achieved in TiO_2 system through IRE technology. Under the UV irradiation and in the presence of dissolved

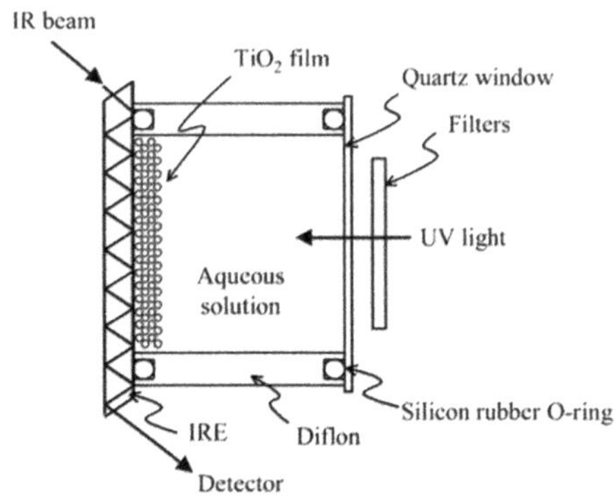

Fig. 2.9 Schematic illustration of a spectral cell used for in situ MIRIR measurements [17]. (Reprinted with the permission from Ref. [17]. Copyright 2011 American Chemical Society)

O_2, new peaks at 943, 838 and 1250–1120 cm^{-1} appeared together with intensity changes in other bands. Investigations of influences of the solution pH, the presence or absence of hole and electron scavengers, and isotopic exchange between H_2O and D_2O on the spectral changes have revealed that the primary step of photocatalytic O_2 reduction is the formation of the surface peroxo species, giving the 943 cm^{-1} band, probably with the surface superoxo species as a precursor, in neutral and acidic solutions. The surface peroxo species is then transformed to the surface hydroperoxo, TiOOH, giving the 838 and 1250–1120 cm^{-1} bands, by protonation in the dark.

To enhance the detection sensitivity, surface-enhanced IR absorption (SEIRA) has been utilized for molecules adsorbed on a thin noble metal (Au, Ag) film with abundant free electron. When SEIRA spectroscopy (SEIRAS) is carried out in ATR configurations, it has a great advantage that IR measurements can be carried out even in an aqueous solution, because an IR beam hardly passes through the solution. For example, SEIRAS was applied to in situ observation of surface species formed during the photocatalytic decomposition of acetic acid over bare TiO_2 [18], Au/TiO_2, and Au/Pt/TiO_2 films, on which an Au island film was deposited. For a gas-phase reaction, an evacuable IR cell was used, which was connected to a pumping and gas handling system. SEIRAS measurements were carried out in a backside reflection mode, in which an IR beam from FTIR was introduced from the CaF_2 side of the sample at the incident angle of ca. 50° and reflected at the interface between the TiO_2 and Au films. The reflected IR beam was collimated to an external MCT detector at liquid nitrogen temperature. SEIRA spectra were recorded with the resolution of 4 cm^{-1}. The adsorption states of acetic acid were determined in both the liquid and gas phase, and molecular adsorption was found to be dominant in the gas phase. Pt deposition on the TiO_2 film led to the formation of adsorbed methyl during the photodecomposition of gas-phase acetic acid, indicative of the formation of methyl radical as a reaction intermediate. The addition of gas-phase water to the system led to a significant increase in the coverage of adsorbed methyl as a result of enhancement in the reaction rate as reported before. Since the metal films used in SEIRAS are discontinuous and island-structure, SEIRAS can detect chemical species, which exist not only on the metal films but also in the vicinity of the metal islands. Water and acetic acid were deaerated by repeating a freeze-and-thaw cycle for a gas-phase reaction. The vapor of water or acetic acid was introduced into the cell from a gas handling system equipped with a Baratron pressure gauge. A 1:1 mixture of gas-phase water and acetic acid was prepared by mixing their vapor at the same pressure in a glass reservoir. The light source was a high-pressure Hg lamp (Ushio UIV-570) that was filtered through a band-pass filter (Toshiba UV-D33S, 240–400 nm) and a water filter (10 cm long) to remove heat.

Primary intermediates of oxygen photoevolution (water photooxidation) reaction in the interface of TiO_2 (rutile) and aqueous solution were investigated by in situ multiple internal reflection infrared (MIRIR) absorption and photoluminescence (PL) measurements [19]. UV irradiation of TiO_2 in the presence of 10 mM Fe^{3+} in the solution caused the appearance of a new peak at 838 cm^{-1} and a shoulder at 812 cm^{-1}. Detailed investigations of the effects of solution pH, the presence of methanol as a hole scavenger, and isotope exchange in water between $H_2{}^{16}O$ and $H_2{}^{18}O$ on the spectra have shown that the 838 and 812 cm^{-1} bands can be assigned to the O–O stretching mode of surface TiOOH and TiOOTi, respectively, produced as primary intermediates of the oxygen photoevolution reaction. The results give strong support to our previously proposed mechanism that the oxygen photoevolution is initiated by a nucleophilic attack of a H_2O molecule on a photogenerated hole at a surface lattice O site, not by oxidation of surface OH group by the hole. The conclusion is supported by PL measurements. A plausible reaction scheme is proposed for the oxygen photoevolution on TiO_2 (rutile) in aqueous solutions of pH less than about 12.

Semiconductor electrodes capable of using solar photons to drive water-splitting reactions have been the subject of tremendous interest over recent decades. The surface has been found to play a significant role in determining the efficiency of water oxidation. However, previous works have only allowed hypotheses to be formulated regarding the identity of relevant surface species. The first observation of a surface intermediate of oxygen evolution at an Ir oxide multi-electron catalyst was reported by Sivasankar et al. [20], where a surface hydroperoxide intermediate has been detected upon oxidation of water at an Ir oxide nanocluster catalyst system under pulsed excitation of a $[Ru(bpy)_3]^{2+}$ visible light sensitizer by recording of the O–O vibrational mode at 830 cm^{-1}. Rapid-scan FTIR spectroscopy of colloidal H_2O, D_2O, and $D_2{}^{18}O$ solutions in the attenuated total reflection mode allowed spectral assignment of IrOOH on the basis of an observed D shift of 30 cm^{-1} and ^{18}O shifts of 24 cm^{-1} ($^{16}O^{18}O$) and 46 cm^{-1} ($^{18}O^{18}O$) (Fig. 2.10). The laser pulse response of the infrared band is consistent with the kinetic relevancy of the intermediate.

For Co_3O_4, two surface intermediates of visible light-sensitized water oxidation were detected by rapid-scan FTIR spectroscopy. The agreement of the ^{18}O isotopic composition of a surface superoxide species and the final O_2 gas product provides evidence for the kinetic competency of the three-electron oxidation intermediate. At a fast catalytic site absorbing at 1013 cm^{-1}, the superoxide intermediate grows and O_2 evolves within a 300 ms photolysis pulse. By contrast, a slow site marked by a Co (IV) = O group does not advance beyond the one-electron intermediate absorbing at

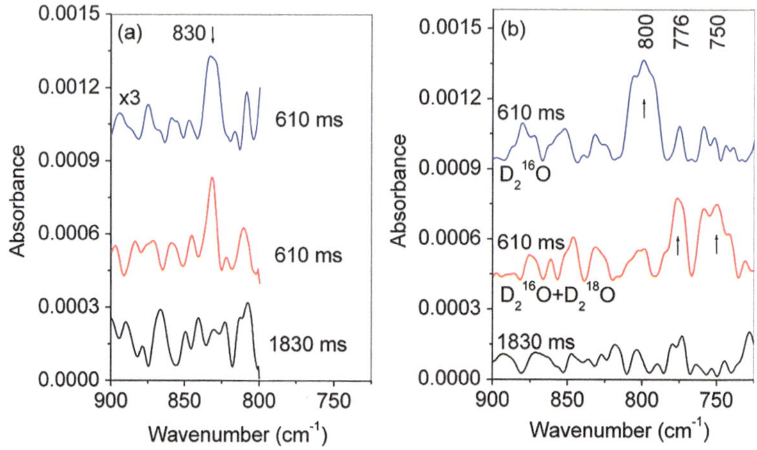

Fig. 2.10 Rapid-scan FTIR traces in the 900–700 cm^{-1} region. (**a**) 610 ms (light on) and 1830 ms spectra (light off) for the photooxidation of H_2O. The 610 ms slices from two separate experiments are shown to indicate the degree of uncertainty regarding band shape. The spectral region below 800 cm^{-1} exhibits high noise due to H_2O tumbling-mode absorption. (**b**) 610 ms slices of experiments in $D_2{}^{16}O$ (top, average of 140 runs) and $D_2{}^{16}O$ (33%) + $D_2{}^{18}O$ (66%) (middle, average of 68 runs), along with the 1830 ms slice of the $D_2{}^{16}O$ experiment (bottom). (Reprinted with the permission from ref. [20]. Copyright 2011 American Chemical Society)

840 cm^{-1} within the same 300 ms pulse. The widely different photocatalytic efficiency of the two types of sites is attributed to the presence/absence of adjacent Co(III)OH groups coupled via an oxygen bridge. Improvement of the sensitivity of the ATR method should allow time-resolved FTIR monitoring with shorter photolysis pulses, which, in turn, will enable detection and kinetic analysis of the early (one- and two-electron) surface intermediates of the fast site [21].

Despite of the successful determination of surface intermediates by IR technique, the role of the surface intermediate species in the photocatalytic water-oxidation reaction was rarely established definitively. α-Fe_2O_3 with excellent stability and elemental abundance has emerged as one of the most promising materials to carry out the light-driven water-oxidation half reaction for photoelectrochemical (PEC) water splitting. The combination of suitable optical and electrochemical properties enables the possibility of sustainable hydrogen production in a hematite-based PEC cell. However, so far the efficiencies measured with hematite have fallen well short of the theoretical maximum. For example, state-of-the-art systems produce only 30% of the maximum photocurrent that can be generated under water splitting with hematite. The cause of this poor performance has been the focus of extensive studies during the past decade. Bulk recombination and surface recombination have both been found to limit the quantum efficiency of photogenerated charge-carrier separation and the output power. Bulk recombination results in a near-zero minority carrier (hole) diffusion length, which limits the charge-collection length to the space-charge layer in the hematite–electrolyte interface. Surface recombination is in competition with the forward water-oxidation hole-transfer reactions and accounts

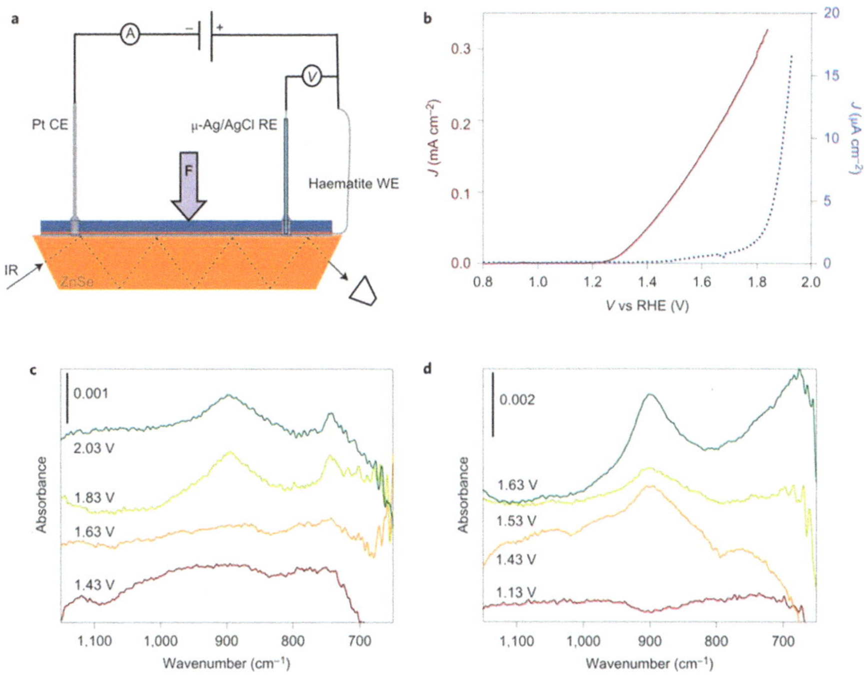

Fig. 2.11 (**a**) Experimental setup and results of infrared spectroscopy measurements during electrochemical and PEC water oxidation, where ZnSe is the ATR crystal. (**b**) J–V curves of a hematite electrode in the operando PEC infrared setup measured in contact with D_2O in the dark (blue) and under illumination (dark red). (**c**) Infrared spectra of hematite scanned at constant applied potentials, from 1.43 to 2.03 V versus RHE, in the dark. (**d**) Under illumination. (Reprinted from Ref. [22], with kind permission from Springer Science+Business Media)

for the loss of several hundred millivolts of photovoltage. A clear understanding of the surface chemistry of hematite during PEC water oxidation is thus crucial as it determines the extent of band bending and of Fermi-level pinning, which control the charge-separation and hole-collection (water-oxidation) efficiencies. The potential- and light-dependent water-oxidation reaction under photoelectrochemical (PEC) was monitored on α-Fe_2O_3 through MIRIR technology (Fig. 2.11). A prominent peak in the spectra was resolved reproducibly at 898 cm^{-1}, which grew in at applied potentials positive to the onset of the water-oxidation current (1.7 V and 1.25 V versus RHE in the dark and under illumination, respectively). The potential- and light-dependent peak evolution is consistent with electrochemically or photoelectrochemically generated species on the electrode surface. That the absorption peaks are only observed positive of the onset potential of the water-oxidation current indicates that this absorption peak is associated with species involved in the D_2O oxidation reaction. An additional peak at 743 cm^{-1} was observed with variable signal-to-noise ratios for the different experiments performed, which we attribute to

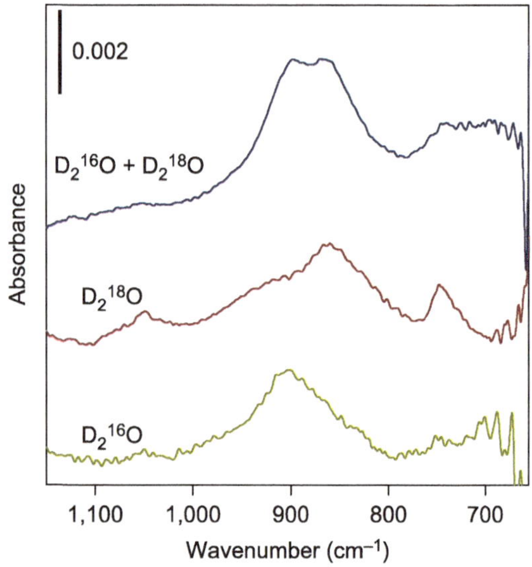

Fig. 2.12 Plot of the effect of oxygen isotopic variation on the infrared absorption spectra during PEC water oxidation with hematite electrodes. The spectra were measured at an applied potential of 1.63 V versus RHE under illumination in contact with $D_2{}^{16}O$, $D_2{}^{18}O$, or a 1:1 ratio of $D_2{}^{16}O/D_2{}^{18}O$. (Reprinted from Ref. [22], with kind permission from Springer Science +Business Media)

its proximity to the spectral cutoff region and is thus strongly affected by any traces of H_2O incorporated in the cell.

In the absence of temporal resolution, the kinetic relevancy or position of these species in the photocatalytic cycle could not be addressed. Yet, knowledge of individual events in the four-electron cycle of water oxidation on the catalyst surface through observation of structurally identified species and their temporal behavior is required to uncover kinetic bottlenecks and to provide insights regarding design changes to overcome them. Isotope-labeling experiments using ^{18}O-labeled and ^{16}O-labeled water were used for the peak assignments to distinguish Fe = O and Fe–O–OH groups with overlapped vibration frequency in the range of 750–900 cm^{-1} and 740–950 cm^{-1}, respectively (Fig. 2.12). The $^{16}O/^{18}O$ shift for Fe = O is expected to be ~40 cm^{-1}. The frequency shift associated with O–O isotope exchange is between 20 and 30 cm^{-1} for partially labeled (^{16}O–^{18}O) and between 44 and 61 cm^{-1} for fully labeled (^{18}O–^{18}O) groups. Operando spectra collected in contact with $D_2{}^{18}O$ exhibited absorption peaks at 743 and 857 cm^{-1}. The peak at 857 cm^{-1} in $D_2{}^{18}O$ is shifted by 42 cm^{-1} compared with the spectrum taken in $D_2{}^{16}O$. This shift is consistent with either Fe = ^{16}O to Fe = ^{18}O or Fe–^{16}O–^{16}O to Fe–^{18}O–^{18}O, both of which are possible intermediate species of water oxidation on hematite. Further study performed in a 1:1 ratio of $D_2{}^{16}O/D_2{}^{18}O$ produced an equally weighted doublet with peaks at 898 and 857 cm^{-1}, which are thus assigned as Fe = ^{16}O and Fe = ^{18}O, respectively, since the peroxide is supposed to produce three different vibrations that correspond to ^{16}O–^{16}O, ^{16}O–^{18}O, and ^{18}O–^{18}O stretching modes with a 1:2:1 ratio of intensities [22].

2.3 Raman

Raman spectroscopy utilizing the inelastic scattering of light was first developed in 1912 by the Indian physicist C. V. Raman [23]. Although Raman can provide the fingerprint spectrum interference-free from water, the tiny Raman cross-section area limits its broad application in chemical analysis for a long period since its invention. The integration of Fourier transform technology, charge-coupled device (CCD) detectors, compact spectrographs, effective laser rejection filters, near-infrared lasers, and small computers greatly push forward to the applicability of Raman spectroscopy in chemical analysis [24]. The enhancement of the inelastic scattering probability can be further achieved from resonance Raman (RR) scattering, where the incident laser is near an electronic transition of the molecule of interest, increasing the signal by an additional factor of 10^2–10^6. The surface-enhanced Raman spectroscopy (SERS) and tip-enhanced Raman spectroscopy (TERS) allow the possibility of the single-molecule detection [25], which use plasmonically enhanced Raman scattering to characterize the chemical information on single molecules.

Regarding the reaction tracking, the priority of Raman technology based on the molecular vibration and rotation lies in the fingerprint spectra interference-free from water. However, Raman spectroscopy usually gives rise to strong signals of inorganic catalyst instead of surface chemical probes, preventing it from the application in tracing the transformation path of the surficial species.

Currently, SERS has mainly used for in situ monitoring the kinetics of noble metal (Au, Ag)-catalyzed photocatalytic reaction on the basis of the surface plasmon resonance (SPR) effect [26]. In this way, SPR has the dual function of activating the chemical reaction and enhancing the Raman signal of surface species. O_2 have been found to be involved into the SPR-catalyzed oxidation reaction, but the activation mechanism/path of oxygen molecules is still obscure due to the extremely short lifetime of SPR-induced hole–electron pair. Using SPR-assisted selective oxidation of *p*-aminothiophenol (PATP) as a demonstration, SERS was used for the in situ tracking of the O_2 evolution (Fig. 2.13). Both experiments and DFT calculations reveal that O_2 were activated by accepting an electron from a metal nanoparticle under the excitation of SPR to form a strongly adsorbed oxygen molecule anion. The anion was then transformed to Au or Ag oxides or hydroxides on the surface to oxidize the surface species, which was also supported by the heating effect of the SPR (Fig. 2.14).

Using the PATP oxidation SPR-catalyzed by Ag nanoparticle as a model reaction, a radical-capturer-assisted SERS has been used as an in situ tracking technique to explore the primary active species determining the reaction path [27]. Hole is revealed to be directly responsible for the oxidation of PATP to DMAB and O_2 functions as an electron capturer to form isolated hole. The oxidation degree of PATP can be further enhanced through a joint utilization of electron capturers of $AgNO_3$ and atmospheric O_2, producing *p*-nitrothiophenol (PNTP) within 10 s due to the improved hole–electron separation efficiency (Fig. 2.15).

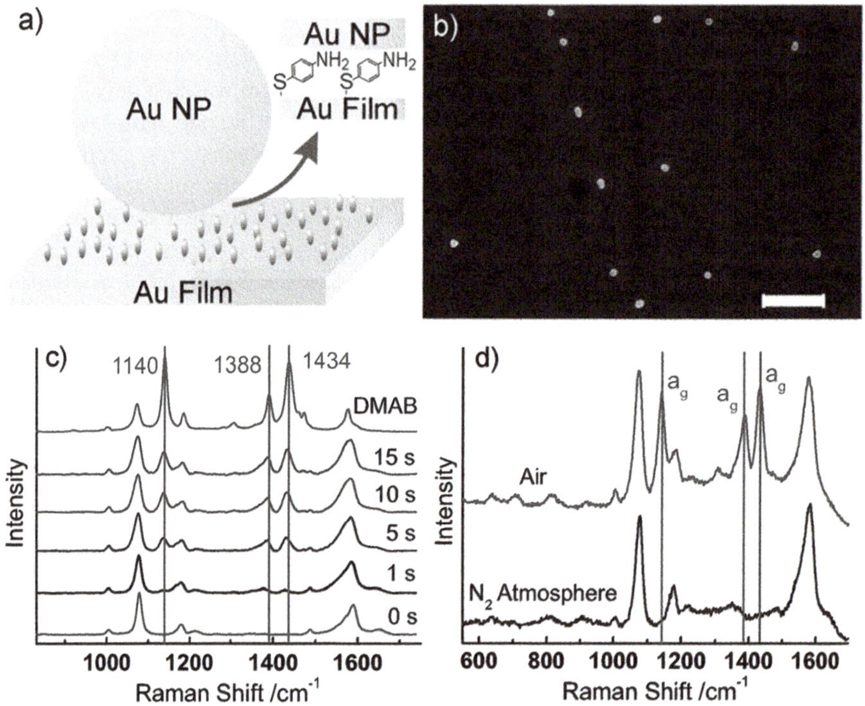

Fig. 2.13 (**a**) Au/PATP/Au nanoparticle, (**b**) SEM image (the scale bar is 500 nm), and (**c**) illumination-time-dependent Raman spectra and (**d**) comparative Raman spectra of Au/PATP/Au NPs recorded in air and N_2 atmosphere. (Reproduced from Ref. [26] by permission of John Wiley & Sons Ltd)

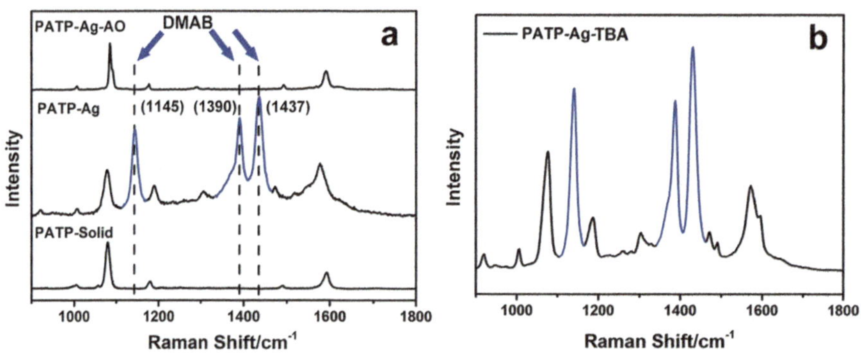

Fig. 2.14 (**a**) Raman signal of PATP solid (black line), SERS signals of PATP on Ag nanoparticle layer in the absence and presence of AO; (**b**) SERS signal of PATP on assembly Ag nanoparticle layer in the presence of TBA. (Reprinted from Ref. [27], with kind permission from Springer Science+Business Media)

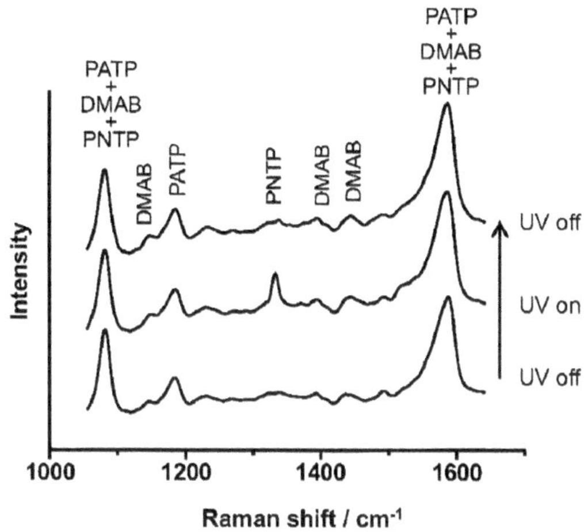

Fig. 2.15 SERS spectra recorded for TiO_2-Au NPs that had been functionalized with PATP: before UV illumination (bottom trace), under UV illumination (middle trace), and after the UV illumination was turned off (top trace). Before UV excitation, only peaks assigned to PATP were detected (DMAB peaks displayed very low intensities). Under UV exposure for 5 min, the formation of PNTP was detected. PNTP could be further reduced to DMAB as the UV illumination was removed (red trace). All spectra employed 1 mW and 1 $mWcm^{-2}$ as the laser and UV illumination power, respectively. (Reproduced from Ref. [28] by permission of John Wiley & Sons Ltd)

Although catalytic processes mediated by SPR excitation have emerged as a new frontier in catalysis, the selectivity of these processes remains poorly understood. The selectivity of the SPR-mediated oxidation of PATP employing Au NPs as catalysts was controlled by the choice of catalysts (Au or TiO_2-Au NPs) and by the modulation of the charge transfer from UV-excited TiO_2 to Au [28]. When Au NPs were employed as catalyst, the SPR-mediated oxidation of PATP yielded DMAB. When TiO_2-Au NPs were employed as catalysts under both UV illumination and SPR excitation, PNTP was formed from PATP in a single step. Interestingly, PNTP molecules were further reduced to DMAB after the UV illumination was removed. Therefore, the control over charge-transfer processes may play an important role to tune activity, product formation, and selectivity in SPR-mediated catalytic processes.

Semiconductors including TiO_2, Cu_2O, and MoO_{3-x} have also proven to be SERS active [29–32]. It is desirable to in situ and real-time (operando) reveal the interfacial information between semiconductor and surficial reactants through SERS, which is extremely significant for guiding the photocatalyst design. However, restricted by the low sensitivity, there is no study about SERS self-tracking of the photocatalytic process on semiconductor. A novel Ag-alumina hybrid SERS platform has been designed for the spectroscopic detection of surface reactions in the steady state [33]. Single crystalline and faceted silver (Ag) nanoparticles with strong light scattering were prepared in large quantity, which enables their reproducible self-assembly into large-scale monolayers of Raman sensor arrays by the Langmuir–

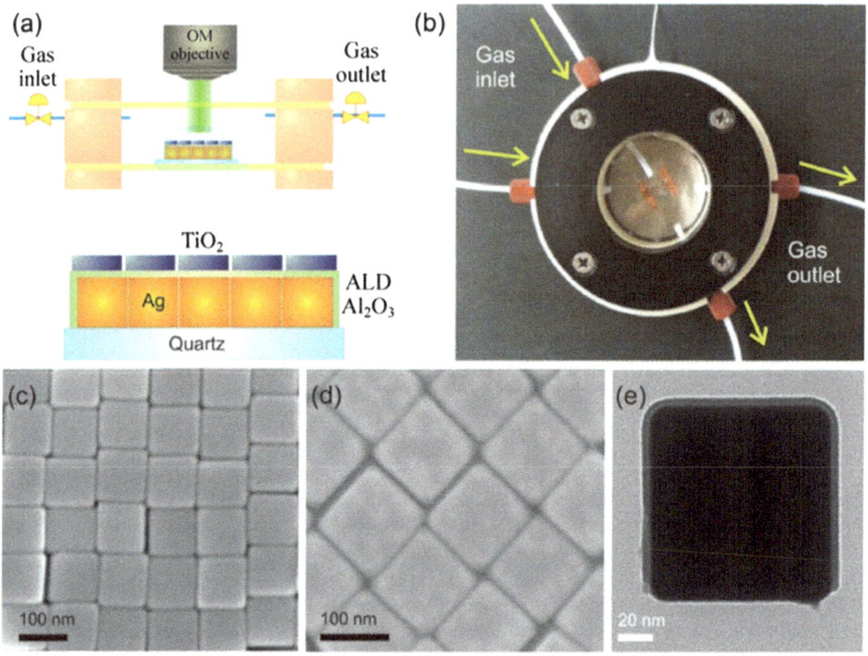

Fig. 2.16 (**a**) Scheme and (**b**) photograph of the experimental cell, where the Ag nanocube Langmuir–Blodgett film was first coated with a ~3 nm layer of Al_2O_3 by ALD deposition, follow by Langmuir–Blodgett assembly of square TiO_2 nanocrystals. The sample was placed on the bottom quartz slide, and deoxygenated solution was added. The chamber was closed and blown with Ar gas for at least 15 min prior to UV irradiation. SEM images of (**c**) the Ag nanocube film made by the Langmuir–Blodgett technique and (**d**) Ag nanocube film after Al_2O_3 coating. (**e**) TEM image of an Al_2O_3-coated Ag nanocube peeled off from the quartz slide. (Reprinted from Ref. [33], with kind permission from Springer Science+Business Media)

Blodgett technique. The closely packed sensor film contains high density of sub-nm gaps between sharp edges of Ag nanoparticles, which created large local electromagnetic fields that serve as "hot spots" for SERS enhancement. The SERS substrate was then coated with a thin layer of alumina by atomic layer deposition to prevent charge transfer between Ag and the reaction system (Fig. 2.16). The photocatalytic water-splitting reaction on a monolayer of anatase TiO_2 nanoplates decorated with Pt cocatalyst nanoparticles was employed as a model reaction system. Reaction intermediates of water photooxidation were observed at the TiO_2/solution interface under UV irradiation. The surface-enhanced Raman vibrations corresponding to peroxo, hydroperoxo, and hydroxo surface intermediate species were observed on the TiO_2 surface, suggesting that the photooxidation of water on these anatase TiO_2 nanosheets may be initiated by a nucleophilic attack mechanism.

Recently, TiO_2 with cavitary structure has been revealed to be highly SERS sensitive with an enhance factor over 10^4 due to the improved light scattering efficiency, which provides possibility for the operando self-monitoring of the

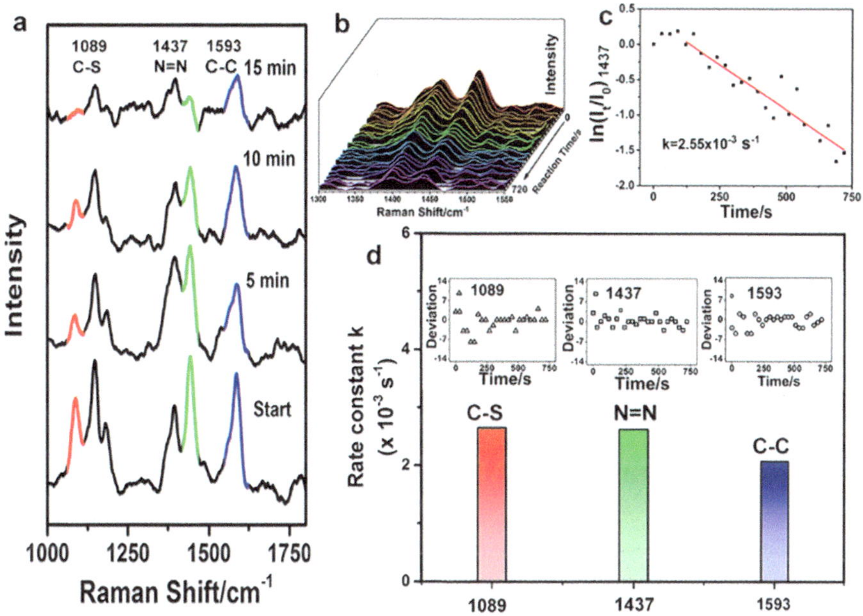

Fig. 2.17 (**a**) SERS spectra of PATP on TiO_2 IO under the irradiation of 532 nm laser for 15 min with a time interval of 5 min. (**b**) Evolution of $\nu_{N=N}$ peak recorded with a time interval of 30 s. (**c**) The logarithm of $(I_t/I_0)_{1437}$ processed using the normalized vibration peak intensity of TiO_2 at 146 cm^{-1} as an internal control. (**d**) Reaction rate constants for the cleavage of C–S, N=N, and C–C bonds under the long-time irradiation of 532 nm laser with a power of 5.0 mW and the corresponding wavenumber deviation collected at different time (inset). (Reprinted with the permission from Ref. [35]. Copyright 2011 American Chemical Society)

photocatalysis reaction [34]. The utilization of the fingerprint spectrum for the operando monitoring of a photocatalytic process is extremely desired to accurately understand the reaction mechanism but long remains challenging. Ordered macroporous TiO_2 that is concomitantly photocatalytically active and highly SERS sensitive was employed to self-track the photocatalytic reaction using the oxidation of PATP as the model [35]. The photocatalytic degradation under 532 nm laser irradiation initiated from the formation of the azo compound was explicitly revealed by finely resolved SERS spectra (Fig. 2.17). More importantly, the decomposition rates of different bonds including N=N, C–S, and C–C were, respectively, determined, following a first-order kinetics process with the rates in the range of 2.1 ~ 2.7 × 10^{-3} s^{-1}. Meanwhile, this self-monitoring strategy also provides an opportunity for gaining an insight into the effect of photothermal catalysis on selective formation of the azo compound.

Fundamental understanding of the energetic/electronic coupling properties of a molecule–semiconductor interface is of great importance. The changes in molecular conformations and vibrational modes can have significant impact on the interfacial charge-transfer reactions. Using single-hot-spot microscopic surface-enhanced

Raman spectroscopy (SMSERS), the change in the interface properties of alizarin–TiO_2 system can be probed as a result of the externally applied electric field [36]. The perturbation, caused by the external potential, has been observed as a shift and splitting of the 648 cm^{-1} peak, typical indicator of the strong coupling between alizarin and TiO_2. On the basis of the experimental results and DFT calculations, the presence of perturbed alizarin–TiO_2 coupling under interfacial electric potential may lead to changes in the interfacial electron transfer dynamics. Additionally, heterogeneously distributed dye molecules at the interface on nanometer length scale and different molecule–semiconductor binding interactions under charge accumulation associated interfacial electric field changes create intrinsically inhomogeneous interfacial ET dynamics associated with both static and dynamic disorders.

2.4 In Situ Atomic Force Microscopy and Fluorescence

Combining the single-molecule fluorescence spectroscopy approach with various other techniques such as computational studies, atomic force microscopy (AFM), electrochemistry, and Raman spectroscopy can facilitate inspection of multiple parameters with high chemical selectivity and wide temporal and spatial resolutions. Significant efforts made in this direction have already indicated the importance of various factors in determining injection dynamics and physical origins for interfacial ET rate fluctuations.

The fundamental information related to the energy flow between molecules and substrate surfaces as a function of surface site geometry and molecular structure is critical for understanding interfacial ET dynamics. The inhomogeneous nanoscale molecule–surface and molecule–molecule interactions are presumably the origins of the complexity in interfacial ET dynamics; thus, identifying the environment of molecules at nanoscale is crucial. It is challenging to correlate the observed heterogeneity of interfacial CT dynamics to the material and interfacial structures based on fluorescence measurement alone. The AFM correlated single-molecule fluorescence intensity/lifetime imaging microscopy (AFM-SMFLIM) is capable of identifying and characterizing individual molecules distributed across the heterogeneous surface at the nanometer length scale. Nanoscale morphology and interfacial ET dynamics at a single-molecule level can be observed. Moreover, the blinking behavior and lifetime of each molecule in combination with the topography of the environment at nanoscale provide the location of each molecule on the surface at nanoscale and the coupling strength of each molecule with TiO_2 nanoparticles.

Using single CdSe/CdS quantum dot (QD) functionalized atomic force microscopy (AFM) tips, the spatial dependence of photoinduced electron transfer dynamics from the single QD to TiO_2 nanoparticles can be controlled and probed with high spatial (sub diffraction-limited) and temporal (limited by fluorescence microscopy) resolutions (Fig. 2.18) [37]. CdSe/CdS QDs with a CdSe core (~1.2 nm in radius) and six monolayers of CdS (2.2 nm in thickness) were attached on the n-Si AFM tip

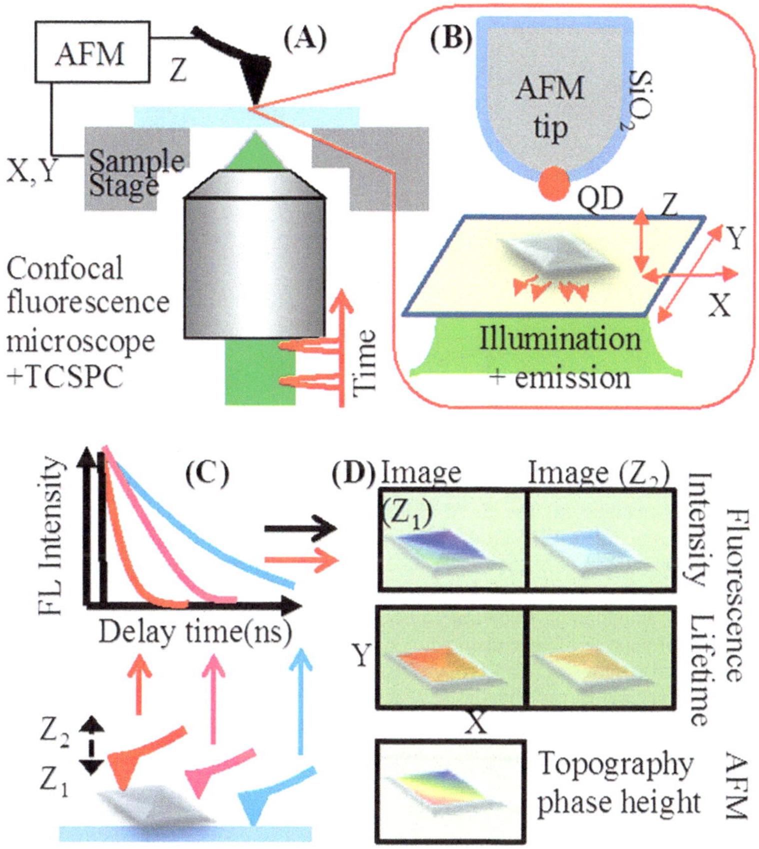

Fig. 2.18 Preparation of single QD-modified AFM tips. (**a**) Procedure for trapping a single QD by a MPTES functionalized AFM tip. (bi) Fluorescence images, (ci) AFM images, and (di) AFM line scan (along the line connecting QDs **a** and **b**) of a 3.5 μm × 3.5 μm area before ($i = 1$) and after ($i = 2$) the attachment of QD **b** to the AFM tip. Inset in c1 and c2 shows an expanded view of QD **b**. (Reprinted with the permission from Ref. [37]. Copyright 2011 American Chemical Society)

coated with a thin layer of SiO_2 (~10 nm) to reduce the quenching of QD emission by the AFM tips. The single QD-modified AFM tip is scanned over the TiO_2 to acquire AFM images, during which the XY positions are controlled by the sample stage and the Z by the AFM tip. The tip is illuminated by laser pulses at 400 nm and the resulting fluorescence is obtained. For every detected photon, the tip position and photon information are recorded using AFM and a time-correlated single-photon counting (TCSPC) board, respectively. The images of fluorescence lifetimes and intensities can be constructed, which can be correlated with the AFM topography/ phase images (Fig. 2.18d).

During the measurement, the tip was brought to contact with sample (glass surface or TiO_2) in contact mode. After withdrawing the tip and moving the sample stage laterally, the tip was brought to contact with the next position (Fig. 2.19 a),

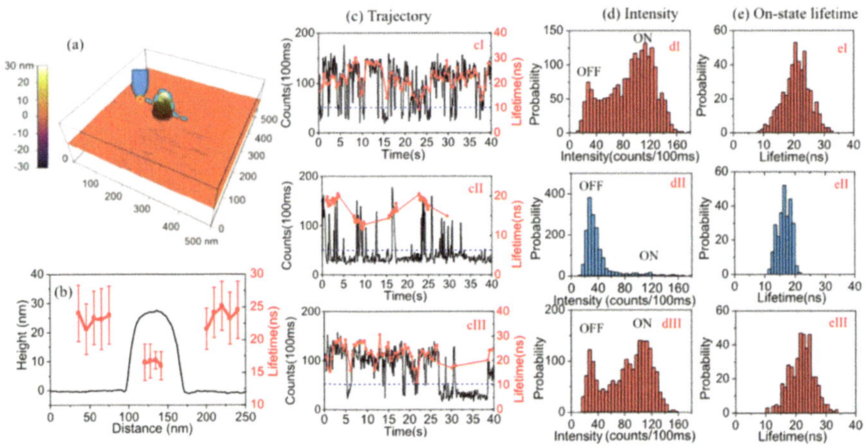

Fig. 2.19 Spatial dependence of electron transfer dynamics from a single QD (attached on a AFM tip) to a TiO_2 nanoparticle. (**a**) The 3D AFM image of a Degussa P25 TiO_2 nanoparticle acquired with a single QD-modified AFM tip operating in ac mode. (**b**) Cross-section line scan of AFM height (black solid line) and average QD on-state emission lifetimes at selected tip–sample contact positions (red circles) along the blue dashed line indicated in panel (**a**). (ci) Typical fluorescence intensity (black) and lifetime (red) trajectories, (di) histograms of intensity distribution, and (ei) histograms of on-state lifetime distribution of a QD-modified AFM tip at the left side (i = I), on top (i = II), and at right side (i = III) of the TiO_2 nanoparticle. The tip is in contact with the glass coverslip in region I and III and with the TiO_2 nanoparticle in region II. Blue dashed lines in cI-III indicate the fluorescence intensity threshold separating the on and off states. (Reprinted with the permission from Ref. [37]. Copyright 2011 American Chemical Society)

presenting the AFM height and the fluorescence information simultaneously (Fig. 2.19b). Representative fluorescence lifetime and intensity trajectories for regions at the left side (i = I), on top (i = II), and at right side (i = III) of the TiO_2 nanoparticle indicate the probability of dark states in region II is higher than regions I and III (Fig. 2.19cI – cIII), respectively. The corresponding intensity distribution histograms (Fig. 2.19dI – III) for points in regions I and III are similar, showing peaks of off and on states at 250 and 1100 counts per second. In comparison, in region II, the probability of on states is much smaller and the off state dominates. It is clear that the probability density of long on states decreases and the probability density of long off states increases in region II compared to regions I and III. Correspondingly, the on-state lifetimes are similar in regions I and III and are significantly shortened in region II (Fig. 2.19eI – III). This finding suggests the feasibility of using electron donor or acceptor modified AFM tips for simultaneous high-resolution imaging of morphology and photoinduced charge-transfer dynamics in nanomaterials.

2.5 In Situ NMR and ESR

The detailed structure–activity relationship of surface hydroxyl groups and adsorbed water on the surface of semiconductor is the key to clarifying the hole-transfer mechanism for photocatalytic water splitting. Owing to the lack of direct experimental evidence, the most controversial point concerning the nature of surface-active sites is whether the surface hydroxyl groups (Ti – OH) of TiO_2 can trap the photogenerated hole. Using the photocatalytic water splitting by Pt/TiO_2 as an example, one- (1D) and two-dimensional (2D) ^{1}H solid-state NMR together with in situ ESR techniques were employed to identify surface hydroxyl groups and adsorbed water molecules, as well as their spatial proximity/interaction on the surface of TiO_2 [38].

Two different types of Ti–OH including bridging hydroxyl (OH_B, 7.3 ppm) and terminal hydroxyl (OH_T, 1.8 ppm) groups were identified from 1D NMR spectroscopy (Fig. 2.20). 2D ^{1}H–^{1}H double-quantum (DQ) MAS NMR spectroscopy presents autocorrelation peaks along the diagonal (ω, 2ω) resulting from the dipolar interaction of protons with the same chemical shift spectroscopy, which can probe the spatial proximities of various Ti–OH groups and the adsorbed H_2O. The 2D ^{1}H–^{1}H DQ MAS NMR spectrum of Pt/TiO_2 dehydrated at 673 K shows only one diagonal peak at (1.8, 3.6) ppm due to the autocorrelation of OH_T groups, indicating that they are in close spatial proximity to each other. The presence of 0.9 μmol H_2O results in two diagonal peaks at (1.8, 3.6) and (7.3, 14.6) ppm, corresponding to the autocorrelations of OH_T and OH_B groups, respectively. The more introduction of H_2O resulted in new peaks. The new autocorrelation peak at (5.2, 10.4) ppm was ascribed to the spatial proximity of the hydrogen atom of adsorbed H_2O. Another

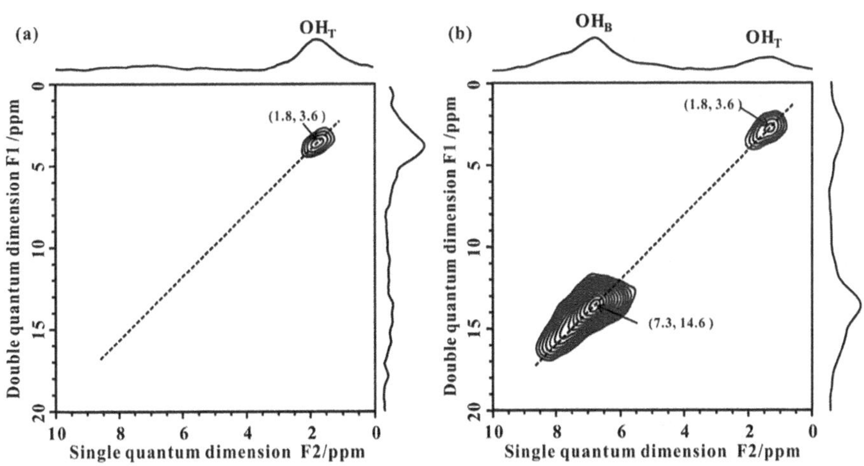

Fig. 2.20 2D ^{1}H–^{1}H DQ MAS NMR spectra of (**a**) bare PT-2 and (**b**) PT-2 loaded with H_2O (0.9 μmol). (Reprinted with the permission from Ref. [38]. Copyright 2011 American Chemical Society)

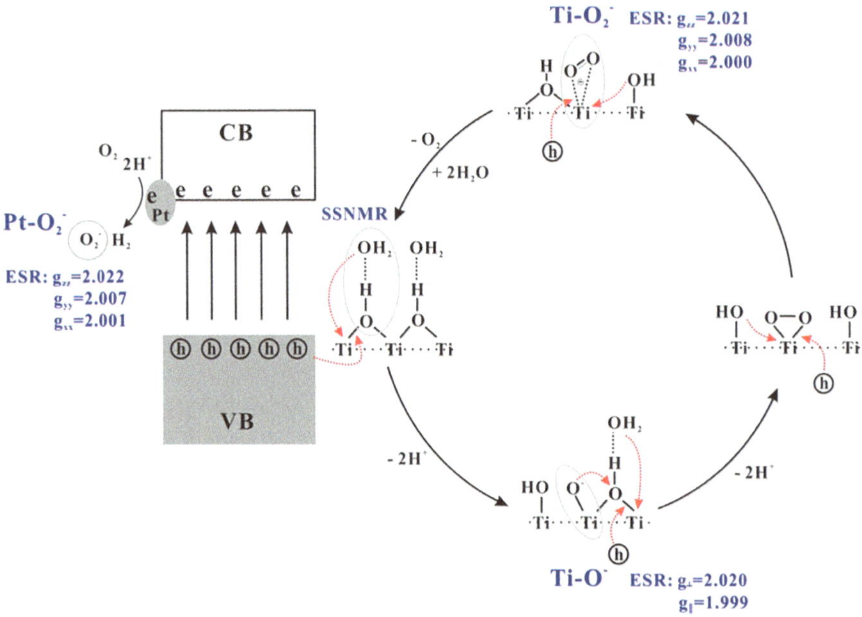

Scheme 2.1 Proposed hole-transfer mechanism for photocatalytic water splitting on the TiO_2 photocatalyst upon solar light irradiation. (Reprinted with the permission from Ref. [38]. Copyright 2011 American Chemical Society)

new autocorrelation peak appeared at (6.9, 13.8) ppm, and it grew with the consumption of the autocorrelation peak at (7.3, 14.6) ppm when the H_2O loading was gradually increased. The signal at 6.9 ppm was assigned to the ^1H signal of OH_B in hydrogen-bonding interaction with surface-adsorbed H_2O. Besides the autocorrelation peaks, the intense off-diagonal peak pair at (5.2, 12.1) and (6.9, 12.1) ppm correspond to the spatial correlation between H_2O and OH_B. All these findings demonstrated that H_2O only adsorbs on OH_B groups through hydrogen-bonding interaction, forming hydrated OH_B groups.

The in situ ^1H and ^{13}C NMR studies of the photocatalytic reaction on TiO_2 with different Ti–OH groups and different H_2O loadings illustrated that the enhanced activity was closely correlated to the amount of hydrated OH_B groups. The in situ ESR experiments performed with variable H_2O loading revealed that the hydrated OH_B offer a channel for the transfer of photogenerated holes in the photocatalytic reaction, and the adsorbed H_2O has a synergistic effect with the neighboring OH_B group to facilitate the formation and evolution of active paramagnetic intermediates. As shown in Scheme 2.1, upon the solar light irradiation, the hydrated OH_B groups trap a photoinduced hole to generate the $Ti–O^-$ intermediate. Meanwhile, a nucleophilic attack of adsorbed H_2O to the hole-trapped sites occurs, which hinders the recombination of photoinduced electrons and holes, and thereby stabilizes the formation of $Ti–O^-$ species. When another photoinduced hole is trapped by a neighboring hydrated OH_B group and the hole-trapped site is nucleophilically attacked by another adsorbed H_2O, two adjacent Ti – O^- centers can couple with each other to form surface peroxide species. The surface peroxide intermediate would be oxidized by photoinduced holes to form Ti–O_2^- species, which can be further oxidized by the photoinduced hole, forming an O_2 molecule followed by the recovery of OH_B group. The excessive H_2O molecules adsorb on the regenerated OH_B groups, which may hinder the interaction between the O_2 molecule and the TiO_2 surface and further prevent the O_2 molecule from trapping the photoinduced electron to form Ti–O_2^- species. Then the O_2 molecule would interact with the cocatalyst Pt and be reduced to form Pt–O_2^- species by the photoinduced electron. The reduction of O_2^- species by the photoinduced electron ($O_2^- + 2H^+ + e \rightarrow H_2O_2 + 2H^+ + 2e \rightarrow 2H_2O$) should be responsible for the low efficiency of the photocatalytic water splitting. Meanwhile, the reduction of H^+ from the above recycle by photoinduced electrons on the cocatalyst Pt produces the H_2 molecule.

References

1. Nosaka Y, Nosaka AY (2017) Generation and detection of reactive oxygen species in photocatalysis. Chem Rev 117:11302–11336
2. Hirakawa T, Nosaka Y (2008) Selective production of superoxide ions and hydrogen peroxide over nitrogen- and sulfur-doped TiO_2 photocatalysts with visible light in aqueous suspension systems. J Phys Chem C 112:15818–15823
3. Wang D, Zhao L, Guo LH et al (2014) Online detection of reactive oxygen species in ultraviolet (UV)-irradiated nano-TiO_2 suspensions by continuous flow chemiluminescence. Anal Chem 86:10535–10539
4. Kakuma Y, Nosaka AY, Nosaka Y (2015) Difference in TiO_2 photocatalytic mechanism between rutile and anatase studied by the detection of active oxygen and surface species in water. Phys Chem Chem Phys 17:18691–18698
5. Ishibashi K, Fujishima A, Watanabe T et al (2000) Hashimoto, K. Generation and deactivation processes of superoxide formed on TiO_2 film illuminated by very weak UV light in air or water. J Phys Chem B 104:4934–4938
6. Dimitrijevic NM, Rozhkova E, Rajh T (2009) Dynamics of localized charges in dopamine-modified TiO_2 and their effect on the formation of reactive oxygen species. J Am Chem Soc 131:2893–2899
7. Tachikawa T, Majima T et al (2010) Single-molecule, single-particle fluorescence imaging of TiO_2-based photocatalytic reactions. Chem Soc Rev 39:4802–4819
8. Naito K, Tachikawa T, Fujitsuka M et al (2008) Real-time single-molecule imaging of the spatial and temporal distribution of reactive oxygen species with fluorescent probes: applications to TiO_2 photocatalysts. J Phys Chem C 112:1048–1059
9. Xu W, Jain PK, Beberwyck BJ et al (2012) Probing redox photocatalysis of trapped electrons and holes on single Sb-doped titania nanorod surfaces. J Am Chem Soc 134:3946–3949
10. Kim W, Tachikawa T, Moon G et al (2014) Molecular-level understanding of the photocatalytic activity difference between anatase and rutile nanoparticles. Angew Chem Int Ed 53:14036–14041
11. Tachikawa T, Wang N, Yamashita S et al (2010) Design of a highly sensitive fluorescent probe for interfacial electron transfer on a TiO_2 surface. Angew Chem Int Ed 49:8593–8597
12. Tachikawa T, Yamashita S, Majima T (2011) Evidence for crystal-face-dependent TiO_2 photocatalysis from single-molecule imaging and kinetic analysis. J Am Chem Soc 133:7197–7204

13. Ha JW, Purnima T, Ruberu A et al (2014) Super-resolution mapping of photogenerated electron and hole separation in single metal–semiconductor manocatalysts. J Am Chem Soc 136:1398–1408
14. Tachikawa T, Yamashita S, Majima T (2010) Probing photocatalytic active sites on a single titanosilicate zeolite with a redox-responsive fluorescent dye. Angew Chem Int Ed 49:432–435
15. Naito KT, Tachikawa T, Fujitsuka M et al (2009) Single-molecule observation of photocatalytic reaction in TiO_2 nanotube: importance of molecular transport through porous structures. J Am Chem Soc 131:934–936
16. Szczepankiewicz SH, Colussi AJ, Hoffmann MR et al (2000) Infrared spectra of photoinduced species on hydroxylated titania surfaces. J Phys Chem B 104:9842–9850
17. Nakamura R, Imanishi A, Murakoshi K et al (2003) In situ FTIR studies of primary intermediates of photocatalytic reactions on nanocrystalline TiO_2 films in contact with aqueous solutions. J Am Chem Soc 125:7443–7450
18. Sato S, Ueda K, Kawasaki Y et al (2002) In situ IR observation of surface species during the photocatalytic decomposition of acetic acid over TiO_2 films. J Phys Chem B 106:9054–9058
19. Nakamura R, Nakato Y (2004) Primary intermediates of oxygen photoevolution reaction on TiO_2 (rutile) particles, revealed by in situ FTIR absorption and photoluminescence measurements. J Am Chem Soc 126:1290–1298
20. Sivasankar N, Weare WW, Frei H (2011) Direct observation of a hydroperoxide surface intermediate upon visible light-driven water oxidation at an Ir oxide nanocluster catalyst by rapid-scan FT-IR spectroscopy. J Am Chem Soc 133:12976–12979
21. Zhang M, Respinis MD, Frei H (2014) Time-resolved observations of water oxidation intermediates on a cobalt oxide nanoparticle catalyst. Nat Chem 6:362–367
22. Zandi O, Hamann TW (2016) Determination of photoelectrochemical water oxidation intermediates on haematite electrode surfaces using operando infrared spectroscopy. Nat Chem 8:778–783
23. Krishnan RS, Shankar RK (1981) Raman effect: history of the discovery. J Raman Spectrosc 10:1–8
24. McCreery RL (2000) Raman spectroscopy for chemical analysis. Wiley, New York
25. Nie SM, Emory SR (1997) Probing single molecules and single nanoparticles by surface-enhanced Raman scattering. Science 275:1102–1106
26. Huang Y, Zhang M, Zhao L et al (2014) Activation of oxygen on gold and silver nanoparticles assisted by surface plasmon resonances. Angew Chem Int Ed 53:2353–2357
27. Yan YF, Wang LZ, Tan XJ et al (2016) Surface-enhanced Raman spectroscopy assisted by radical capturer for tracking of plasmon-driven redox reaction. Sci Rep 6:30193
28. Wang J, Ando RA, Camargo PHC (2015) Controlling the selectivity of the surface plasmon resonance mediated oxidation of p-aminothiophenol on Au nanoparticles by charge transfer from UV-excited TiO_2. Angew Chem Int Ed 54:6909–6912
29. Qi D, Yan X, Wang L et al (2015) Plasmon-free SERS self-monitoring of catalysis reaction on Au nanoclusters/TiO_2 photonic microarray. Chem Commun 51:8813–8816
30. Tan X, Wang L, Cheng C et al (2016) Plasmonic MoO_{3-x}@MoO_3 nanosheets for highly sensitive SERS detection through nanoshell-isolated electromagnetic enhancement. Chem Commun 52:2893–2896
31. Alessandri I (2013) Enhancing Raman scattering without plasmons: unprecedented sensitivity achieved by TiO_2 Shell-based resonators. J Am Chem Soc 135:5541–5544
32. Qi D, Lu L, Wang L et al (2014) Improved SERS sensitivity on plasmon-free TiO_2 photonic microarray by enhancing light-matter coupling. J Am Chem Soc 136:9886–9889
33. Xing YL, Yan R, Lo S et al (2014) Alumina-coated Ag nanocrystal monolayers as surface-enhanced Raman spectroscopy platforms for the direct spectroscopic detection of water splitting reaction intermediates. Nano Res 7:132–143
34. Yan XF, Xu Y, Tian BZ et al (2017) Operando SERS self-monitoring photocatalytic oxidation of aminophenol on TiO_2 semiconductor. Appl Catal B: Environ. https://doi.org/10.1016/j.apcatb.2017.10.009

35. Sevinc PC, Wang X, Wang Y et al (2011) Simultaneous spectroscopic and topographic near-field imaging of TiO_2 single surface states and interfacial electronic coupling. Nano Lett 11:1490–1494
36. He Y, Rao G, Cao J et al (2016) Simultaneous spectroscopic and topographic imaging of single-molecule interfacial electron-transfer reactivity and local nanoscale environment. J Phys Chem Lett 7:2221–2227
37. Liu Z, Zhu H, Song N et al (2013) Probing spatially dependent photoinduced charge transfer dynamics to TiO_2 nanoparticles using single quantum dot modified atomic force microscopy tips. Nano Lett 13:5563–5569
38. Liu F, Feng N, Wang Q et al (2017) Transfer channel of photoinduced holes on a TiO_2 surface as revealed by solid-state nuclear magnetic resonance and electron spin resonance spectroscopy. J Am Chem Soc 139:10020–10028

Chapter 3
Titanium-Based Mesoporous Materials for Photocatalysis

3.1 The History of Mesoporous Materials

Up to date, it is well known that the rapid progress of society is closely associated with material science. Particularly in recent 30 or 40 years, there has been a boom of novel materials, including of mesoporous materials. Mesoporous molecular sieve belongs to a noticeable area owing to its excellent properties, such as high specific surface area and adjustable ordered mesoporous pore size. According to the rules of IUPAC, porous materials can be classified into three types: micropore materials (pore size <2 nm), mesopore materials (2 nm <pore size <50 nm), and macropore materials (pore size >50 nm), respectively [1]. In 1992, the Mobil's researchers successfully synthesized a novel ordered mesoporous silica materials called M41S with surfactants as templating agents, thus opening a new era of mesoporous materials [2]. On the basis of chemical composition, mesoporous molecular sieve can be divided into two categories: silicon-based materials and non-silicon-based materials. The most common one is M41S, which is a branch of silicon-based materials. Additionally, there are many other sorts of mesoporous molecular sieves, such as SBA-n series, MSU series, CMK series, HMS, KIT, and metallic or metal oxide series. Compared with microporous and macroporous molecular sieves, mesoporous materials possess many outstanding characteristics. But there are still some shortcomings exist, for instance, its low hydrothermal stability and weak acidity, which limit its application. Ryoo et al. [3] synthesized highly stable MCM-41 by adjusting the pH value of the solution and adding salt compounds. And Mokaya et al. [4] obtained ultrastable MCM-41 by post-processing. Robert Mokaya prepared restructured pure silica MCM-41 materials through seeded crystallization route. This method used MCM-41 as "silica source" for secondary synthesis with extending the reaction time of periods. Therefore, the pore wall thickness was increased systematically by extending the time. The remarkable stability is attributed to the combination of thicker pore walls and less strained silica frameworks.

J. Zhang et al., *Photocatalysis*, Lecture Notes in Chemistry 100,
https://doi.org/10.1007/978-981-13-2113-9_3

In general, developing new mesoporous structure, developing new synthetic methods, improving the crystallinity based on maintaining the structure, and seeking to optimize the functional structure and so on will continue to be hot spots for researchers in the future.

3.2 The Development of Mesoporous TiO_2 in Photocatalysis

3.2.1 The Preparation of Mesoporous TiO_2

Highly efficient mesoporous TiO_2 materials have been applied to the development of the sustainable, nonhazardous, and economical processes in different reaction systems and conditions (Fig. 3.1). It can be utilized in purification of polluted water and air, the decomposition of H_2O into H_2 and O_2, self-cleaning, and so on [5]. Currently, there are various preparation techniques of mesoporous titanium dioxide, such as sol–gel method [6–9], hydrothermal synthesis [10–14], atmospheric liquid phase [15], and radio frequency magnetron sputtering deposition method [16–19]. In addition to these methods, the removal of template for mesoporous titanium dioxide is also available now [20–24].

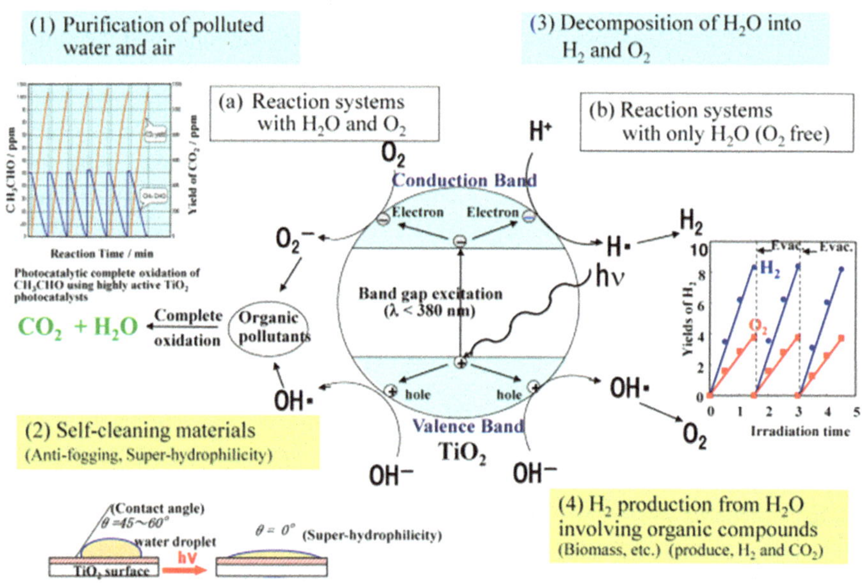

Fig. 3.1 (**a**) Applications of titanium oxide-based photocatalysts in the reaction system with H_2O and O_2. (**b**) Reaction systems with only H_2O (O_2 free). Inserted figure shows the production of electrons in the conduction band and holes in the valence band under UV light irradiation of TiO_2 semiconducting materials. (Reprinted with permission from Ref. [13].Copyright 2000, Royal Society of Chemistry)

Regarding the synthetic methods discussed above, we have listed some advantages and disadvantages among them. Sol–gel method exhibits good properties in terms of synthesized materials, such as low cost, high purity, good uniformity, and easy to be doped. In this part, organic and inorganic titanium precursors can be used to obtain photocatalysts, for example, TTIP and $TiCl_4$. Hydrothermal methods can prepare materials with high crystallinity; meanwhile, it can save time. In this method, there are various options of titanium precursors, such as $TiOSO_4$, TBOT, and TTIP. Compared with the previous two synthesis methods, the advantage of atmospheric liquid phase method is that it merely requires simple experimental equipment; thus it can achieve the possibility of industrialization. Adding the template agent during the reaction, and then stripped by heating or other methods to remove template, is a general method for the preparation of mesoporous materials. As a result, the heat treatment will significantly influence the products' structures, and inorganic skeleton is prone to collapse. Additionally, this process takes several hours, greatly extends the reaction time, and increases energy consumption. Hence, the adverse effect of template removal method is obvious. Moreover, the experimental conditions are not only harsh but also cumbersome. As to the radio frequency magnetron sputtering deposition method (RF-MS), the general way is using the TiO_2 plate, and then the depositions are carried out in a mixture of Ar and O_2. The thin films can be deposited on metal substrates such as Al, Fe, Ti, Zr, Pd, and Pt. Analyzing these methods, it is interesting to find the samples prepared by RF-MS methods have much smaller BET areas than those samples prepared by other three methods; therefore, each kind of catalysts has unique applications owing to their special properties.

3.2.2 Doping Modification on Mesoporous TiO_2

Since the discovery of mesoporous silica M41S, a variety of mesoporous materials have been synthesized [2]. These mesoporous materials exhibit widely potential applications in the industrial catalytic reactions, because they have high surface area, large pore size, and multidimensional framework and they are easy to be recycled. However, the widespread technological use of mesoporous TiO_2 is always to some extent constrained by its wide bandgap (3.2 eV), which requires ultraviolet irradiation for photocatalytic activation. And TiO_2 only absorbs 5% of the spectrum of the sunlight in the near ultraviolet region, which greatly limits its efficient application. To achieve efficient photocatalytic activity in the visible light range, one strategy is to reduce the bandgap of TiO_2. Doping of metal and nonmetal elements seems to be an effective method to enhance the photoactivity of mesoporous TiO_2.

In 2001, the metal ion implantation of TiO_2 with metal ions (V^+, Mn^+, Fe^+) at high energy acceleration was prepared by Yamashita et al. [25]. These catalysts exhibited photocatalytic reactivity for degradation of 2-propanol diluted in water under visible light irradiation ($\lambda > 450$ nm). After then, Zhang et al. [10] obtained the large mesoporous microspheres of titania and WO_3/TiO_2 composites by using $TiOSO_4$ as an inorganic precursor and P123 as structure directing agent via

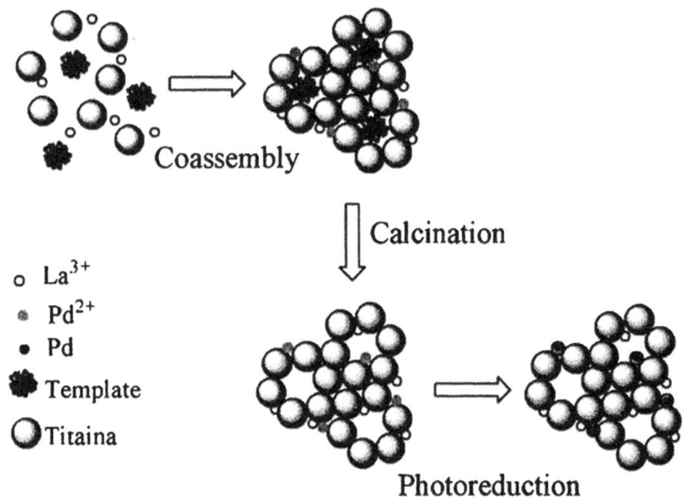

Fig. 3.2 Preparation of Pd nanoparticles in La-doped mesoporous titania by co-assembly and in situ photoreduction. (Reprinted with permission from Ref. [28]. Copyright 2006, Springer)

hydrothermal route. The workers also evaluated its photoactivity by the degradation of methyl orange, which is representative of stable azo dyes. Moreover, Tian et al. [26] synthesized gold loaded on TiO_2 (Au/TiO_2) catalysts using Au (I)–thiosulfate complex $(Au(S_2O_3)_2)^{3-}$ as the golden precursor for the first time, and its photoactivity was tested by the degradation of MO under visible light. Also, Yuan et al. [27] reported that La^{3+}-doped mesoporous TiO_2 with a highly crystallized framework and long-range order was prepared by using nano-anatase particles as nano-building units. Based on the previous work [12], Yuan and his coworkers also prepared highly dispersed Pd nanoparticles in La-doped mesoporous TiO_2 with crystalline framework via co-assembly and photoreduction method [28] in 2006. Figure 3.2 is an illustration of the process. By using this simple method, highly dispersed Pd nanoparticles were prepared in La-doped mesoporous titania with crystallized walls by in situ photoreducing of PdO at room temperature.

In 2010, Zhang et al. [29] prepared Fe^{3+}-doped mesoporous TiO_2 with ordered mesoporous structure via the solvent evaporation-induced self-assembly process through using P123 as soft template. They also successfully prepared the copper impregnated Ms-TiO_2 by using water immiscible room-temperature ionic liquid 1-butyl-3-methyl-imidazolium-tetrafluoroborate as a template and an effective additional solvent via the sol–gel method at low temperature [30]. As a result, the photoactivity of the sample Cu/Ms-TiO_2 is superior to P-25, Ms-TiO_2, and 2.0% Cu/P-25 in the case of 2,4-dichlorophenol and methyl orange under visible light ($\lambda > 420$ nm).

Not only the doping of metal can improve the photocatalytic activity, but also the doping of nonmental element can achieve the same goal. A thermal stable SiO_2-doped mesoporous TiO_2 with high crystallinity was prepared through a templating method proposed by Zhang et al. [31]. It was found that the 15% SiO_2-dopped mesoporous TiO_2 exhibited much higher photoactivity than P25, attributing to the high anatase crystallinity, large specific surface area, abundant preserved surface hydroxyl groups, and mesoporous channels. In 2010, N and F co-doped TiO_2 microspheres were prepared by ethanol solvothermal method, using tetrabutyl titanate as precursor, urea as a nitrogen source, and ammonium fluoride as a fluorine source [32]. Then Zhang and his group reported that the synergetic effect of nitrogen and fluorine doping is responsible for the enhancement of photodegradation activity of AO7 under the irradiation of visible light. Later, Zhang et al. [33] synthesized N, B, Si-tridoped mesoporous TiO_2 photocatalyst through a modified sol–gel method. In this process, dodecylamine not only acted as a pore template but also as a nitrogen dopant, and H_3BO_3 acted as a boron dopant. Moreover, it showed a strong absorption in the visible light region because the doping of N and B narrowed the bandgap.

Besides doping of either metal or nonmental elements, the co-doping of metal and nonmetal element is another effective way of modification. For instance, Zhang et al. [34] obtained the iron (III) and nitrogen co-doped mesoporous TiO2 for the first time by the modified sol–gel method. Ma et al. [35] synthesized well-ordered mesoporous TiO_2 co-doped with nitrogen and ytterbium by an evaporation-induced self-assembly process.

The merits of mesoporous TiO_2 have been listed aforementioned. And we have summarized various preparation methods reported in existing literature as well. By doping modification of TiO_2, the photoactivity of corresponding catalysts can be enhanced to some extent; however, the fast recombination of electron–hole pairs and scarce adsorptive sites undermined the further research. According to some reported literatures, the composition of Si or porous MOFs with the TiO_2 may be a feasible way in enhancing the separation of electrons and holes.

3.2.3 Mesoporous TiO_2–Graphene Materials

Graphene (GR) possesses large specific surface area, excellent conductive, mechanical, and hydrophobic properties, which allow it to be multifunctional materials with excellent capacity for carrying and conducting electrons and holes. Especially, three-dimensional (3D) graphene aerogels (GAs) compounded with mesoporous TiO_2 composites are ultralight massive catalysts, which display hydrophobic properties and facilitate photocatalytic recyclings. In 2014, Qiu et al. [36] studied TiO_2–graphene composites as solar light photocatalysts and electrode materials for lithium–ion batteries (LIBs). They used a one-step hydrothermal method to prepare 3D-structured TiO_2/GA composites. In this process, $Ti(SO_4)_2$ was firstly dissolved in aqueous solution to form crystal seeds before a known amount of glucose was adsorbed on the seeds, followed by fixation of the seeds on the surface of graphene oxides. The presence of glucose results in the exposure of (001) facets in the nanocrystals up to 50% (inset of Fig. 3.3b), achieving the sizes ranging from 15 to 20 nm and realizing mesoporous interface between TiO_2 and the GR (in Fig. 3.3c).

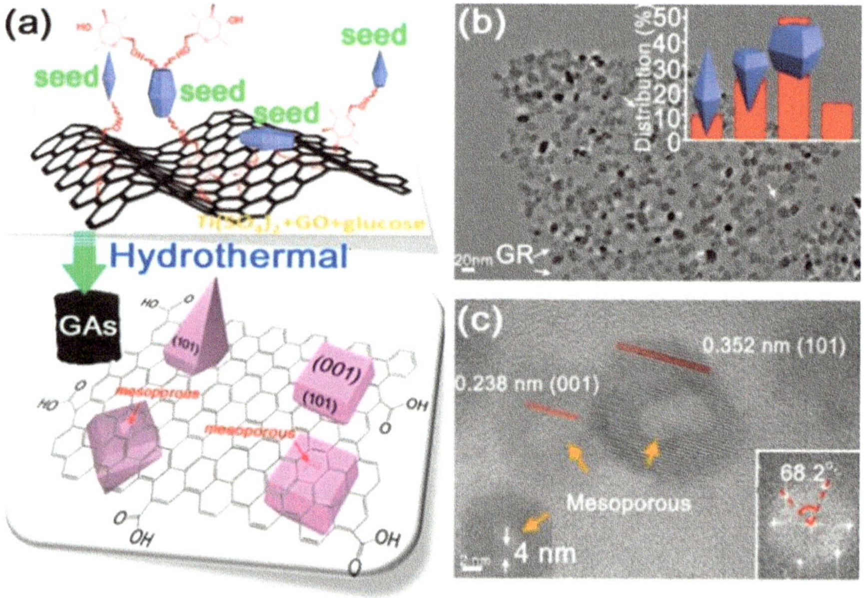

Fig. 3.3 (**a**) Glucose-linked transformation pathway for the in situ growth of TiO_2 nanocrystals with (001) facets on the GAs surface. (**b**) TEM image for TiO_2/GAs (67 wt % of TiO_2 in TiO_2/GAs). Inset (b) is the corresponding morphology distribution of the TiO_2 nanocrystals derived from 100 of TiO_2 crystals in image (**b**). (**c**) HRTEM image for TiO_2/GAs (67 wt %). Inset (**c**) is the corresponding fast Fourier transform (FFT) pattern. (Reprinted with permission from Ref. [36]. Copyright 2014, American Chemistry Society)

3.3 The Development of TiO_2–SiO_2 Mesoporous Materials

In this section, we briefly introduce the research of TiO_2–SiO_2 mesoporous materials. Although TiO_2 has some important properties, for instance, nontoxicity and excellent photostability, there are also some drawbacks constraining the performance of TiO_2 in photocatalytic process [37, 38]. To overcome the drawbacks, the researchers have prepared TiO_2 compounded materials that can provide large number of adsorptive sites by dispersion of TiO_2 species into a porous support with large surface area. Silica has been widely employed as the carrier, owing to its outstanding mechanical strength, high inner surface area, and uniform pore size. With the highly dispersion of TiO_2 into the porous SiO_2 support, the TiO_2–SiO_2 mixed oxide photocatalysts have shown significantly enhanced activities compared to pure TiO_2. On the one hand, TiO_2 and SiO_2 may be combined together to form a mixture of the two oxides, with interaction forces other than weak Vander Waals forces. On the other hand, they can integrate by means of the formation of Ti–O–Si bonds to form the composite oxides. When combined together through chemical bonding, the physical–chemical properties of TiO_2–SiO_2 differ from simple combination of each phase. In a word, homogeneity or dispersion largely depends on preparation methods and synthesis conditions. Those novel TiO_2–SiO_2 materials not only take advantages of both TiO_2 (an n-type semiconductor and an active catalytic support) and SiO_2 (high thermal stability and excellent mechanical strength) but also extend their applications through the generation of new catalytic active sites based on the interactions between TiO_2 and SiO_2 [39]. Hence, it is indeed a promising and new catalytic material in many research areas.

3.3.1 The Preparation of TiO_2–SiO_2 Mesoporous Materials

There are various methods of preparing TiO_2–SiO_2 mesoporous materials, such as sol–gel method [38, 40–42], hydrothermal method [43, 44], chemical vapor deposition [45], precipitation method [46], liquid phase deposition [47], microwave irradiation method [48], impregnation method [49, 50], and evaporation-induced self-assembly (EISA) method [51].

For instance, TiO_2–SiO_2 nanocomposite can be prepared by sol–gel method. Commonly, the first step is to formulate sol with all kinds of ingredients, and then the gelation process, and at last removal of the surfactant to obtain mesoporous materials. In 2003, titania–silica mixed oxides were prepared by the sol–gel method from tetraethylorthosilicate (TEOS) and titanium (IV) isopropoxide (TTIP) as precursors by Elizabeth et al. [42]. Li et al. [43] also obtained silica-modified titanium dioxides by a hydrothermal method. There was strong interaction between SiO_2 and TiO_2, and Ti–O–Si bonds formed during the hydrothermal process. In addition, He et al. [51] synthesized highly ordered bicontinuous cubic mesoporous titania–silica binary oxides via an evaporation-induced self-assembly (EISA) method. As illustrated in Scheme 3.1, TTIP and TEOS hydrolyzed simultaneously with the existence of HCl, then the condensation and polymerization of TTIP were slowed down, and later the hydrolysis of TEOS was accelerated owing to the large amount of HCl. In the aging section, titanium species and silica species co-assembly with F127, and ordered mesostructures were formed. In the calcination process, titanate oligomers and silicate oligomers can be cross-linked with each other through the Ti–O–Si bonds. Under this circumstance, silica acts as glue between TiO_2 nanocrystals; thus, the thermal stability of the mesostructures can be improved. Moreover, Li et al. [49] successfully introduced benzopyrylium salt S-2(2,4-diphenyl-5,6,7,8-tetrahedro-1-benzopyrylium perchlorate) into the channels of mesoporous molecular sieves Ti-HMS with different Ti content by impregnation method.

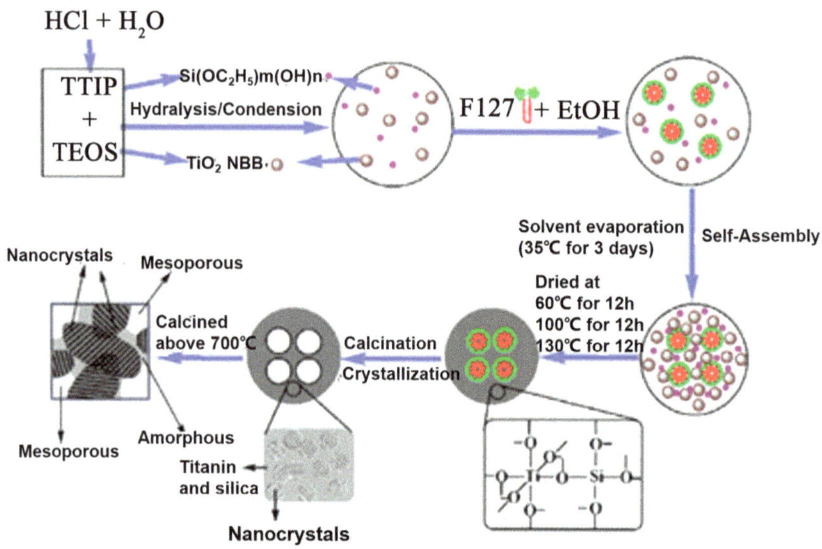

Scheme 3.1 Illustration for the self-assembly and structure evolution process of mesostructured titania–silica binary oxides. (Reprinted with permission from Ref. [50]. Copyright 2005, Elsevier)

3.3.2 The Application of TiO_2–SiO_2 Mesoporous Materials in Photocatalysis

3.3.2.1 Photodegradation of Organic Pollutants

We have briefly introduced the preparation methods of TiO_2–SiO_2 mesoporous materials. Next, we will explore its applications. As we all know, human's demand of energy will be much greater by the year of 2050. This increase poses an undue burden to our environment and the length of human's life. Besides that, with the prosperity of industrialization, the disposal of industrial waste poses a great threat to the environment, which is becoming the biggest concern for the sustainable development of human society. As previously mentioned, TiO_2 was considered as a catalyst for degradation lots of pollutants. However, its rapid recombination of photo-generated electron–hole pairs limits its application. According to previous reports, TiO_2–SiO_2 mesoporous materials are found to be effective in environmental remediation. Figure 3.4 briefly demonstrates the photocatalytic process of degradation of organic pollutants and dyes under visible light [52].

Aguado et al. [53] prepared titania-supported sample on different types of silica through a sol–gel method followed by hydrothermal processing. Afterward, the catalysts were tested by the degradation of iron (III) cyano complexes. In all cases, photoinduced CN^- released from the composite which happened by a homogeneous process. He et al. [59] applied cubic mesoporous titania–silica binary oxides to

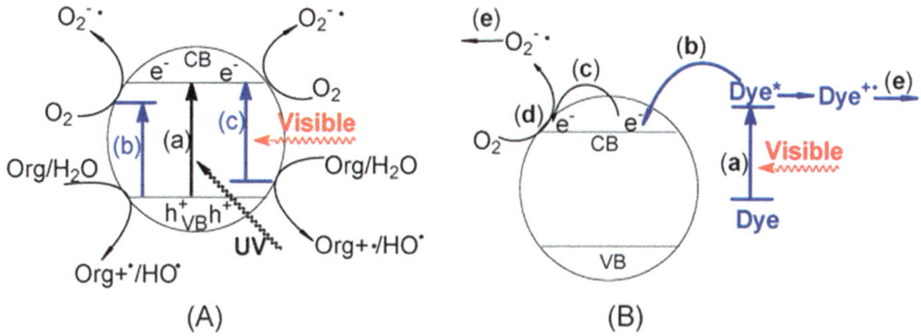

Fig. 3.4 Strategies to realize visible light-induced degradation of organic pollutants on a semiconductor with a wide bandgap. (**a**) The band–band excitation of the pure semiconductor under UV irradiation (*a*) and the bulk-doping to extend the photoresponse by forming electronic states below the conduction band (*b*) or above the valence band (*c*) of the semiconductor in the bandgap. (**b**) The semiconductor-mediated photodegradation initiated by the surface electron injection from the adsorbed dye molecular that harvest visible light. (Reprinted with permission from Ref. [51]. Copyright 2009, Elsevier)

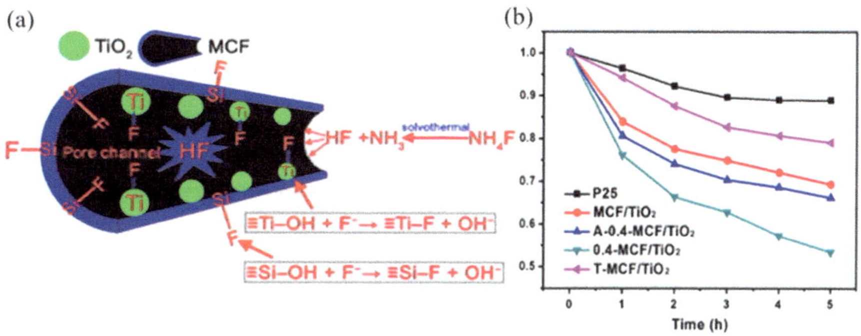

Fig. 3.5 (**a**) Illustration of the fluorination reaction occurred in the pore channels of MCF; (**b**) visible light photocatalytic activities of different samples. (Reprinted with permission from Ref. [53]. Copyright 2002, Elsevier)

degrade rhodamine B (RhB) under UV light irradiation. The result shows that the sample has a comparable photocatalytic activity with Degussa P25, and higher activity than pure TiO_2.

In terms of MCF materials, some papers have reported about it. Xing et al. [44] have obtained super-hydrophobic mesocellular foam (MCF), which is loaded with nano-sized TiO_2 photocatalysts in its pore channels, through a simple one-step solvothermal method followed by a low-temperature vacuum activation process to produce Ti^{3+}. And it can be well considered as an extractant for organics. In this method, NH_4F is used as hydrophobic modifier, and isopropanol is used as solvent to synthesize the super-hydrophobic mesoporous MCF loaded with highly dispersed and Ti^{3+} self-doped TiO_2 nanoparticles. Figures 3.5a illustrates the fluorination reaction occurred in the channels of MCF. In comparison with fluorine-containing silylation organic agent, NH_4F is easy to release HF during solvothermal process. And F ions will be adsorbed onto the MCF under acidic conditions owing to its mesoporous structure. TiO_2 particles are deposited into MCF's pore channels, indicating that the exchange between surface hydroxyl groups on TiO_2 and F ions to form the Ti–F bonds is effectively promoted. Simultaneously, the F ions adsorbed in channels also replace the surface hydroxyl groups on SiO_2 to generate the Si–F bonds. Ti^{3+} was generated in the vacuum drying process, which plays an important role in enhancing its visible light photocatalytic activity. During the degradation of RhB, the NH_4F-modified catalyst of 0.4-MCF/TiO_2 exhibited the optimal photocatalytic activity, indicating that the NH_4F modification and vacuum activation are beneficial to improving visible light photoactivity (Fig. 3.5b).

3.3.2.2 Water Splitting

Since the initial photocatalyst for water splitting into hydrogen and oxygen was developed in 1972 [54]. Various semiconductor-based catalyses either using UV or visible light have been investigated. As discussed in the above section, TiO_2–SiO_2 composite materials, which are combined with each other physically as well as chemically, can enhance photocatalytic activity; thus, these materials attracted much attention. In recent years, scientists deduced new two-step photoexcitation processes, the so-called Z-scheme [55], in order to realize overall water splitting (Fig. 3.6). This system consists of two visible light responsive semiconducting photocatalysts (A and B) and a redox mediator. Photocatalyst A is responsible for hydrogen evolution which is excited by visible light, and then, photo-formed electrons reduce H^+ into H_2 together with photo-formed holes oxidizing the redox

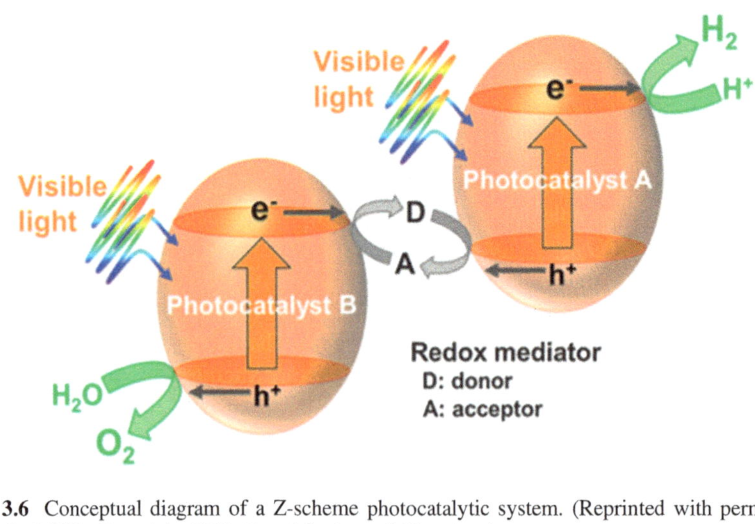

Fig. 3.6 Conceptual diagram of a Z-scheme photocatalytic system. (Reprinted with permission from Ref. [55]. Copyright 2013, Royal Society of Chemistry)

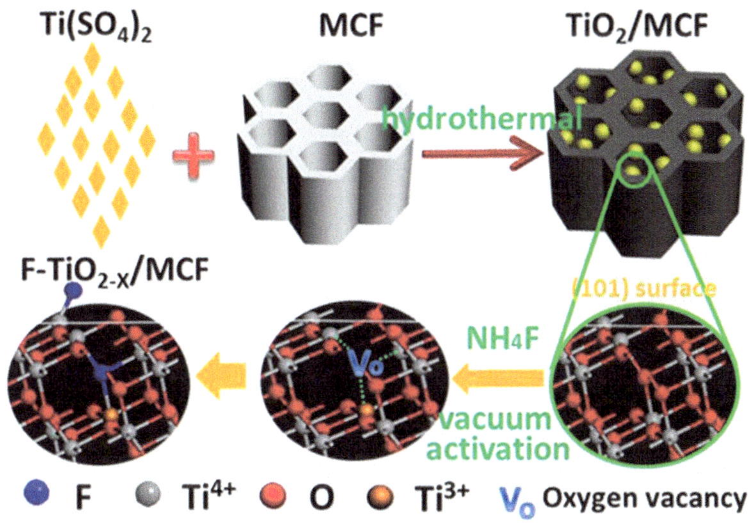

Scheme 3.2 Synthetic steps for the production of the fluorine-doped TiO_{2-x}/MCF composite and the displacement of lattice oxygen vacancies with F atoms during vacuum activation. (Reprinted with permission from Ref. [57]. Copyright 2014, Wiley)

mediator. At the same time, photocatalyst B is used for the water-oxidation reaction, during which photo-formed holes oxidize H_2O to produce O_2 together with photo-formed electrons reducing the redox mediator under visible light irradiation. Finally, water splitting into H_2 and O_2 is attained.

Niphadkar et al. [56] prepared TiO_2–SiO_2 mesoporous composite photocatalysts with different proportions of TiO_2 and SiO_2 by loading TiO_2 on as-synthesized Si–MCM-41 using simple sol–gel method. The photocatalytic evaluation of composite photocatalysts was carried out in production of hydrogen by water-splitting reaction under UV light. In 2014, Xing et al. [57] successfully prepared a brown mesoporous TiO_{2-x}/MCF composite with a high fluorine doping concentration (8.01 at%) by vacuum activation method. It displays an excellent solar absorption, a record-breaking quantum yield ($\Phi = 46\%$) and a high photon–hydrogen energy conversion efficiency ($\eta = 34\%$) in solar photocatalytic H_2 production process, which are all better than that of the black hydrogen-doped TiO_2 ($\Phi = 35\%, \eta = 24\%$). Scheme 3.2 illustrates the continuous steps for preparing F-TiO_{2-x}/MCF. Firstly, titanium source Ti $(SO_4)_2$ in situ transformed to TiO_2 nanocrystals in the pore walls of the MCF through hydrothermal method. Then NH_4F was added into the solution, mechanically mixed with the obtained TiO_2/MCF, followed by a vacuum activation treatment to produce oxygen vacancies in TiO_2 and the substitution of fluorine atoms for vacancies. The F-TiO_{2-x}/MCF exhibits much higher rate of H_2 generation than black H-TiO_{2-x}, P25 and other photocatalysts (Fig. 3.7a). In addition to H_2 evolution, the solar light, UV light, and visible light-driven photodegradation of dyes by F-TiO_{2-x}/MCF were also measured. It was shown that catalysts treated by vacuum activation exhibited better photocatalytic activity than the blank samples (Fig. 3.7b).

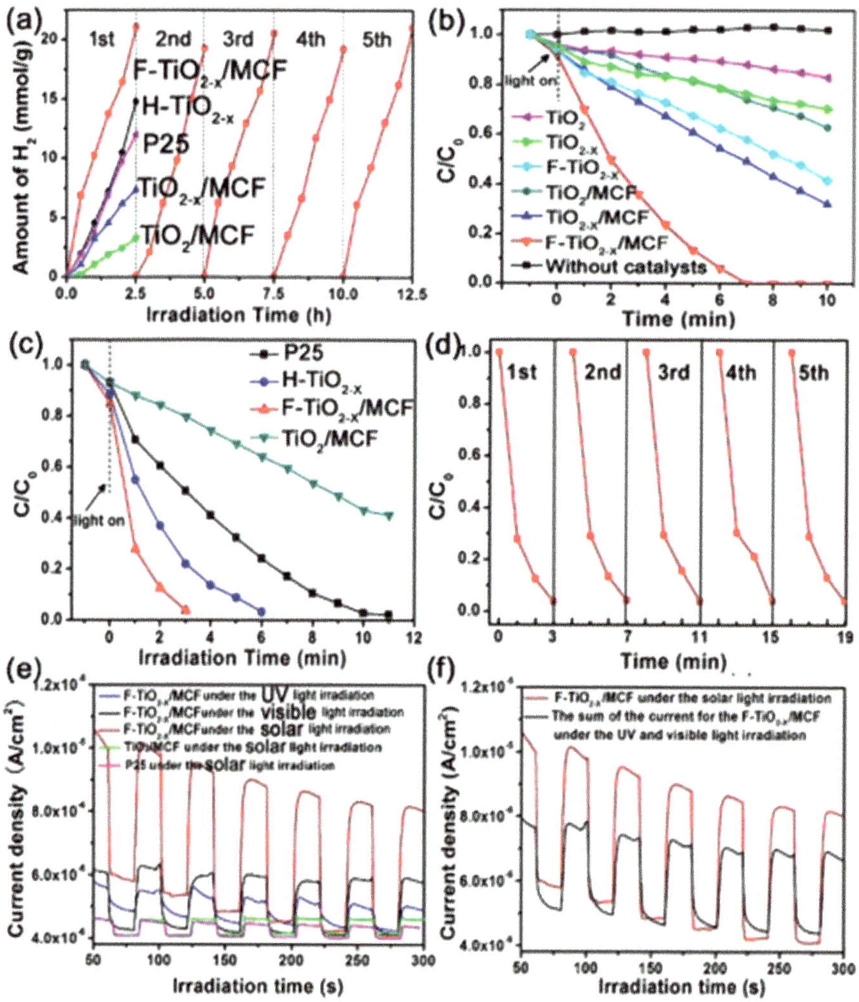

Fig. 3.7 (**a**) Solar light-driven (with an AM 1.5 air mass filter) photocatalytic water splitting for H_2 generation and the cycling measurements of F-TiO_{2-x}/MCF; (**b**) photocatalytic activities for degradation of MO induced by simulated solar light; (**c**) a comparison of photocatalytic decomposition of MB by F-TiO_{2-x}/MCF, blank H-TiO_{2-x}, and other catalysts under simulated solar light irradiation (with an AM 1.5 air mass filter); (**d**) cycling tests of solar-driven photocatalytic activity of F-TiO_{2-x}/MCF for the degradation of MB; (**e**) transient photocurrent responses of F-TiO_{2-x}/MCF in 0.5 M Na_2SO_4 aqueous solution under various irradiation conditions (UV light, < 380 nm filter; visible light, > 420 nm filter); (**f**) comparison between the solar light-driven photocurrent and the sum of the photocurrent of F-TiO_{2-x}/MCF under UV and visible light irradiation. (Reprinted with permission from Ref. [57]. Copyright 2014, Wiley)

To investigate the photoactivity of brown F-TiO_{2-x}/MCF, F-TiO_{2-x}/MCF, and hydrogen-doped TiO_2 (H-TiO_{2-x}), these samples were measured for the degradation of methylene blue (MB) under simulated solar light irradiation using an AM 1.5 air mass filter (Fig. 3.7c). What is more, cycling tests revealed that the brown F-TiO_{2-x}/MCF sample is especially stable after five photocatalytic cycles (Fig. 3.7d). The F-TiO_{2-x}/MCF was shown to produce electrons and exhibit a much higher photocurrent response than MCF/TiO_2 under solar light irradiation. Its solar light-driven current density is much higher than its UV and visible light-driven density as well (Fig. 3.7e). It is worth noting that the solar light-driven current density of F-TiO_{2-x}/MCF is much higher than the sum of the current densities of the catalyst under UV and visible light irradiation (Fig. 3.7f), which indicates that the lifetime of solar light-produced electrons exceeds those of UV- or visible light-produced electrons. It was concluded that the decrease of recombination sites induced by high concentration F doping and the synergistic effect between lattice Ti^{3+}-F and surface Ti^{3+}-F are responsible for the excellent absorption of solar light and photocatalytic production of H_2 of these catalysts.

In this section, we firstly introduce the application of mesoporous TiO_2–SiO_2 materials. It can be either applied to photodegradation of pollutants or used for hydrogen production. Then, we have briefly listed some notable literatures and analyzed them.

3.4 Visible Light Response Metal–Organic Frameworks (MOFs)

In recent years, visible light responsive porous metal–organic framework photocatalysts have been investigated deeply. Metal–organic frameworks (MOFs) are hybrid materials composed of organic linkers and metal–oxo clusters. These MOF materials are hot spots in research, and it can be utilized as adsorbents, separation materials, ion-conductive materials, and catalysts. Among various highly porous materials, metal–organic frameworks (MOFs) are unique in their degree of tunability, structural diversity, as well as their range of chemical and physical properties. Metal–organic frameworks (MOFs) are also known as coordination polymers, which are crystalline materials generated by the association of metal ions (nodes) and multitopic organic ligands (rods) [58–61]. Based on their structures, MOFs have been considered as a promising type of materials due to its unique attributes and open structures with periodic dual composition, which is amenable to bottom up assembly of secondary building blocks into a desired framework expanding or decorating a specific blueprint network topology [58, 62]. As a kind of porous material, metal–organic frameworks (MOFs) have shown semiconductor-like characteristics in photocatalysis [63–67]. In 2009, Kataoka et al. [67] have detected the first example of open porous metal–organic frameworks (MOFs) that functions as an activity site for the reduction of water into hydrogen molecules in the presence of $Ru(bpy)_3^{2+}$,MV^{2+}, and EDTA–2Na under visible light irradiation. Also, Zhou et al. [63] have synthesized and characterized a new metal–organic framework (MOF-253-Pt) material through immobilizing a platinum complex in 2,20-

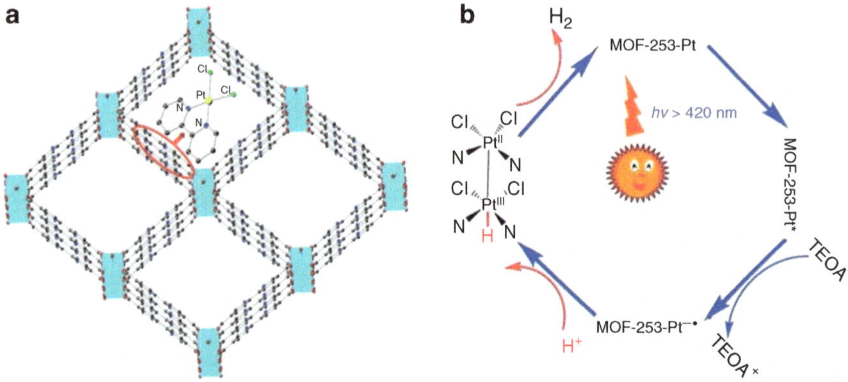

Fig. 3.8 (**a**) Model structure of MOF-253-Pt, through post-synthetic modification of MOF-253 with $PtCl_2$. Key: Cyan octahedron represents Al atoms, while yellow, green, red, blue, and blank circles represent Pt, Cl, O, N, and C atoms, respectively; H atoms are omitted for clarity. (**b**) Proposed reaction mechanism for the photocatalytic H_2 evolution over MOF-253-Pt under visible light irradiation. (Reprinted with permission from Ref. [63]. Copyright 2013, Royal Society of Chemistry)

bipyridine-based microporous MOF (MOF-253) using a post-synthesis modification strategy. The functionalized MOF-253-Pt serves both as a photosensitizer and a photocatalyst for hydrogen evolution under visible light irradiation. The structure and proposed mechanism are presented as shown in Fig. 3.8a, b, respectively. Upon light irradiation, the MOF with the presence of TEOA firstly generates a one-electron-reduced species MOF*(3MLCT) and holes. And then, the MOF*(^{3} MLCT) species is reductively quenched to form MOF^- with the electrons stored on the bpy^- ligands. After that, the reduced MOF^- further forms a Pt(III)-hydride intermediate via a proton-coupled electron transfer (PCET). The intermediates contribute to the formation of the hydride-diplatinum (II, III) intermediate by the synergistic effect of the nearing anchored $Pt(bpy)Cl_2$ complex on framework, leading to H_2 production by a heterolytic coupling pathway.

In 2016, it has been discovered that improving the efficiency of electron–hole separation and charge-carrier utilization plays a central role in photocatalysis. Jiang et al. [68] prepared a representative metal–organic framework (MOF) UiO-66-NH_2, in which Pt nanoparticles of ca. 3 nm are incorporated inside or supported. The resulting products are denoted as Pt@UiO-66-NH_2 and Pt/UiO-66-NH_2, respectively. Finally, these materials are especially applied in photocatalytic hydrogen production via water splitting. Scheme 3.3 simply clarified the pathway of synthesizing these two different materials. Pt@UiO-66-NH_2 greatly shortens the electron-transport distance and hence suppresses the electron–hole recombination, which is expected to have an enhanced catalytic activity compared to Pt/UiO-66-NH_2. In addition, the Pt NPs embedded in the MOF do not undergo aggregation or leaching during the reaction, which leads to better catalytic recyclability of Pt@UiO-66-NH_2 than that of Pt/UiO-66-NH_2.

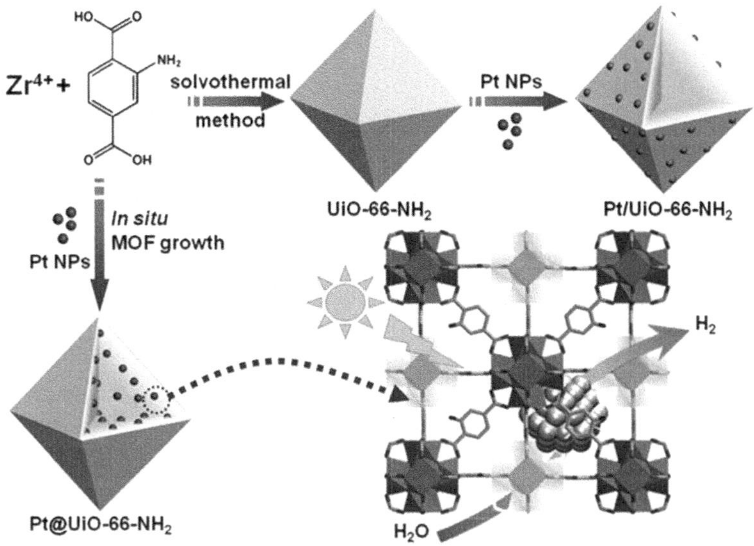

Scheme 3.3 Schematic illustration for the synthesis of Pt@UiO-66-NH_2 and Pt/UiO-66-NH_2, with the photocatalytic hydrogen production process over Pt@UiO-66-NH_2. (Reprinted with permission from Ref. [68]. Copyright 2016, Wiley)

In 2010, Garcia and his coworkers synthesized the Zr-containing metal–organic frameworks (MOFs) that exhibited photocatalytic activity for hydrogen generation upon irradiation at wavelength longer than 300 nm [69]. In 2012, Anpo et al. [64] employed 2-amino-benzenedicarboxylic acid as an organic linker to synthesize amino-functionalized Ti (IV) metal–organic framework (Ti–MOF–NH_2) by a facile solvothermal method, and it described the hydrogen production from an aqueous medium under visible light. The structure of the Ti–MOF–NH_2, its mechanism, and the yield of hydrogen are shown in Fig. 3.9. In summary, it is mentioned that Ti–MOF materials of semiconductor properties have potential in water splitting. All the reported literatures provide us with new ideas in the further development of water splitting.

With the exception of the application in water splitting, the Ti–MOFs can also be applied to CO_2 reduction. In the same year, Li et al. [70] successfully prepared a targeted photoactive catalyst $Ti_8O_8(OH)_4(bdc\text{-}NH_2)_6$ (NH_2-MIL-125 (Ti)) for the first time, which reduced CO_2 even under visible light irradiation. In Fig. 3.10a, MIL-125 (Ti) shows an absorption edge at 350 nm, whereas NH_2-MIL-125 (Ti) shows an extra absorption band in the visible light region with the absorption edge extending to around 550 nm, which is in agreement with the bright yellow color. An interesting photochromic phenomenon was observed over NH_2-MIL-125 (Ti) during the photocatalytic reaction. When the solution of NH_2-MIL-125 (Ti) and TEOA in MeCN was irradiated with visible light in the presence of N_2, the color of the solution changed from the original bright yellow to green. After CO_2 or O_2 was introduced into the reaction system, the green color of the solution changed gradually

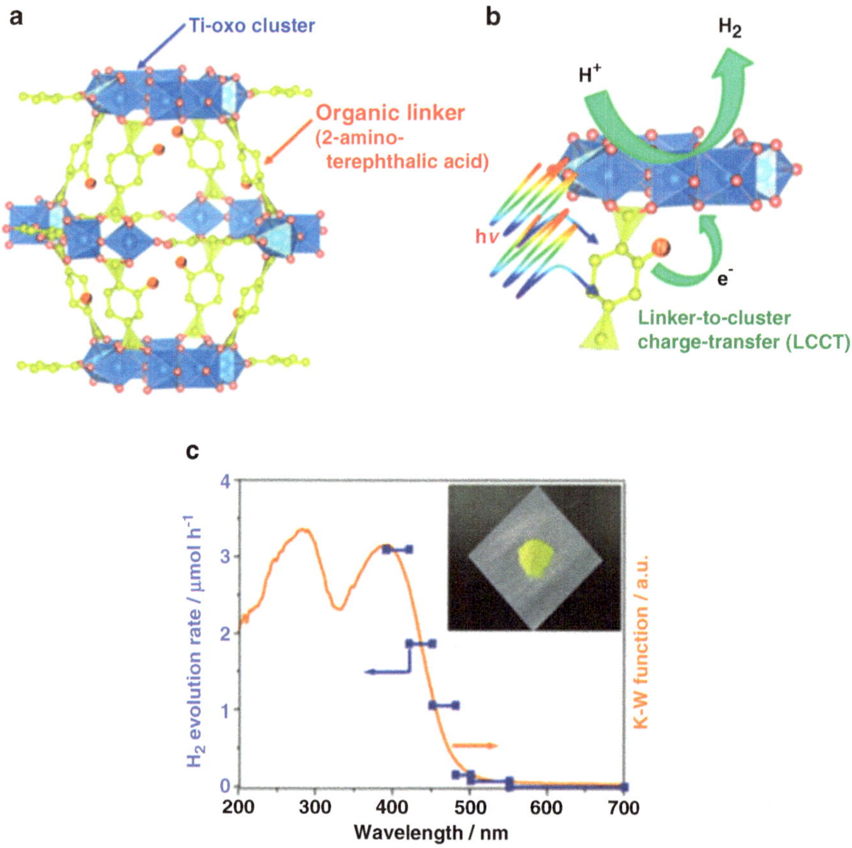

Fig. 3.9 Schematic illustrations of (**a**) the structure of Ti–MOF–NH_2 and (**b**) the reaction mechanism for hydrogen evolution over Ti–MOF–NH_2 induced by visible light irradiation. (**c**) Action spectrum for hydrogen evolution from water containing TEOA as a sacrificial electron donor over Ti–MOF–NH_2. Inset shows the photograph of Ti–MOF–NH_2. (Reprinted with permission from Ref. [64]. Copyright 2012, American Society Chemistry)

back to the original bright yellow (Fig. 3.10b). The photochromic phenomenon is ascribed to the presence of the inter-valence electron transfer from the optically induced hopping of electrons from Ti (III) to Ti (IV) sites in the titanium–oxo clusters.

Despite water splitting and CO_2 reduction, it was also reported by Garcia and Majima [71, 72] that the combination between the organic linkers and metal–oxo clusters would enable the development of dye-sensitized type MOF photocatalysts operating under visible light illumination. In 2012, Zhang et al. [73] adopted a doping strategy to tune the gas sorption and photocatalytic properties of a microporous material ZIF-67. It is worth mention that the Cu-doped phase integrated both structural

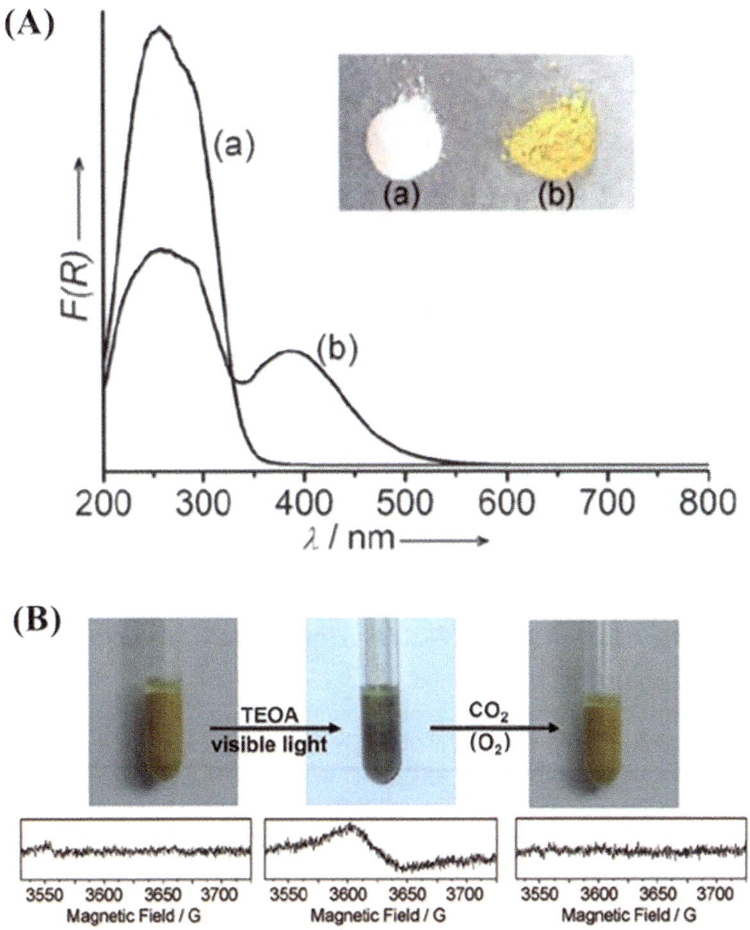

Fig. 3.10 (**a**) UV-vis spectra of (a) MIL-125(Ti) and (b) NH_2-MIL-125(Ti). The inset shows the samples. (**b**) Photos and corresponding ESR spectra of NH_2-MIL-125(Ti) under different conditions: (a) fresh NH_2-MIL-125(Ti), (b) TEOA, visible light, and N_2 and (c) after the introduction of CO_2 (or O_2). (Reprinted with permission from Ref. [70]. Copyright 2012, Wiley)

features and functions of ZIF-67 showed high gas uptake capacity and highly efficient visible light-driven photocatalytic property upon the degradation of methyl orange.

3.5 The Development of Mesoporous Ti–SiO$_2$ Materials in Photocatalysis

Up to now, we have grasped some points of TiO_2–SiO_2 mesoporous materials; however, the concept of Ti–SiO_2 materials is totally different from TiO_2–SiO_2 mesoporous materials because of its containing Ti-oxides species. In the pioneering work of Honda and Fujishima [54], they obtained the photo-assisted production of H_2 and O_2 from water with a photoelectrochemical cell consisting of a Pt and TiO_2 electrodes under a small electric bias. Moreover, TiO_2 photocatalysts with incorporated Ti-oxide species anchored onto supports such as SiO_2, glass, and zeolite exhibited selective photoactivity. The Ti-oxide species prepared within the supports, for example, SiO_2, glass, and zeolite, have revealed a unique local structure as well as high selectivity in the oxidation of organic substances with hydrogen peroxide. Ti–Si binary oxide powders with a low TiO_2 content prepared by sol–gel methods have been reported to include the fourfold coordinated Ti-oxide species highly dispersed within the SiO_2 matrices, showing a unique and characteristic photocatalytic performance for the hydrogenation of unsaturated hydrocarbons with H_2O; the decomposition of NO into N_2, O_2, and N_2O; as well as the reduction of CO_2 with H_2O to produce CH_3OH and CH_4 under UV light irradiation [74–79]. Moreover, many reports have mentioned that the specific photocatalytic reactivity of those catalysts was much higher than that for TiO_2 powder, which may be attributed to the tetrahedrally coordinated titanium oxide moieties. Thus, we will discuss it comprehensively in this section.

3.5.1 *The Preparation of Ti–SiO₂ Mesoporous Materials*

Many reports have been published on preparation of Ti–SiO_2 mesoporous materials, such as ionized cluster beam (ICB) method [76, 80], ion-exchange method [79, 81], anchored on substrate [74, 78, 82, 83], CVD method [84], hydrothermal synthesis method [85], solvent evaporation method [86], and metal ion implantation [87].

3.5.2 *The Application of Ti–SiO₂ Mesoporous Materials in Photocatalysis*

3.5.2.1 CO_2 Photoreduction

In 1992, Anpo et al. [82] prepared highly dispersed titanium oxide anchored onto Vycor glass through a facile reaction between surface OH groups of Vycor and $TiCl_4$. It was investigated that the photoreduction of CO_2 with H_2O should be linked to the high reactivity of the charge-transfer excited state, i.e., $(Ti^{3+}\text{-}O^-)^*$, owing to the presence of well-dispersed homogeneous titanium oxide species on the surface. Ti-MCM-41 and Ti-MCM-48 mesoporous zeolite catalysts synthesized by hydrothermal method exhibited high and unique photocatalytic reactivity for the reduction of CO_2 with H_2O to produce CH_4 and CH_3OH in the gas phase [74]. Keita Ikeue et al. [86] also prepared self-standing porous silica thin films with different pore structures by a solvent evaporation method. From 2002 to 2003, Anpo and his coworkers reported many works about photocatalytic reduction of CO_2 on

Ti-containing porous silica thin film [86, 88, 89]. Figure 3.11a shows the yields of products with the change of reaction time. The yields of CH_4 and CH_3OH in the photocatalytic reduction of CO_2 and H_2O on the Ti-oxides containing various porous materials are shown in Fig. 3.11b. They found that it is possible to determine a real quantum yield of the photocatalytic reduction of CO_2 with H_2O on tetrahedrally coordinated Ti-oxides. Ti-oxide was constructed within porous silica material, and its quantum yield could to be 0.3% at room temperature by the total number of photons absorbed by the catalyst.

3.5.2.2 NO/NO_2 Photoreduction

In 1985, Anpo et al. [77] carried out a research on photoluminescence studies of titanium oxide anchored onto porous Vycor glass. It is proposed that the photoluminescence quenching is closely associated with the electron transfer from the excited states of the catalyst to the added O_2 or N_2O molecules. In Fig. 3.12, it shows the comparison between the reduction of NO and the decomposition of CO_2. The structure of Ti-oxide single-site is presented on the left top of this picture. Similarly, the structure of TiO_2 particles is presented on the right bottom of this picture. The quantum yield of $CO_2 + H_2O \rightarrow CH_3OH + CH_4$ is much smaller than the decomposition of NO, due to its demanding of more configurations of co-adsorbed reactants involving six participating atoms [90]. In 1997, titanium oxide catalysts prepared within the Y-zeolite cavities via an ion-exchange method were reported by Anpo et al. [79], which exhibited high and unique photocatalytic reactivities for the decomposition of NO into N_2 and O_2. It was also found that the charge-transfer excited state of the titanium oxide species, $(Ti^{3+}\text{-}O^{-})^*$, plays a vital role in these unique photocatalytic reactions. Table 3.1 revealed the yields of the photo-formed N_2 and N_2O and its selectivity in the photocatalytic decomposition of NO. It is obvious that the efficiency and selectivity of the formation of N_2 strongly depend on the type of catalysts.

In 2000, Masato Takeuchi et al. [76] prepared transparent TiO_2 thin film photocatalysts on transparent porous Vycor glass (PVG) by an ionized cluster beam (ICB) method. These thin films worked with high efficiency as photocatalysts for the decomposition of NO into N_2, O_2, and N_2O under UV light irradiation at 275 K. When the film thickness increases, the photocatalytic reactivity decreases gradually. In 2004, Yamashita and Anpo [80] proposed a new concept of an ion beam technology using accelerated metal ions, a metal ion implantation, and an ionized cluster beam (ICB) (detailed schematic diagram is shown in Fig. 3.13a). The decomposition of NO into N_2, O_2, and N_2O can be occurred not only under UV light but also visible light, realizing the efficient use of solar beam energy. The samples were also characterized by XANES (as shown in Fig. 3.13b). It revealed the XAFS (XANES and FTEXAFS) spectra of the Cr ion–implanted TiO_2 powder catalyst. By analyzing on these spectra, we can tell that in the Cr ion–implanted TiO_2, the Cr ions are highly dispersed in the lattice of TiO_2 possessing octahedral coordination. These Cr ions are isolated and substitute the Ti^{4+} ions in the lattice positions of TiO_2.

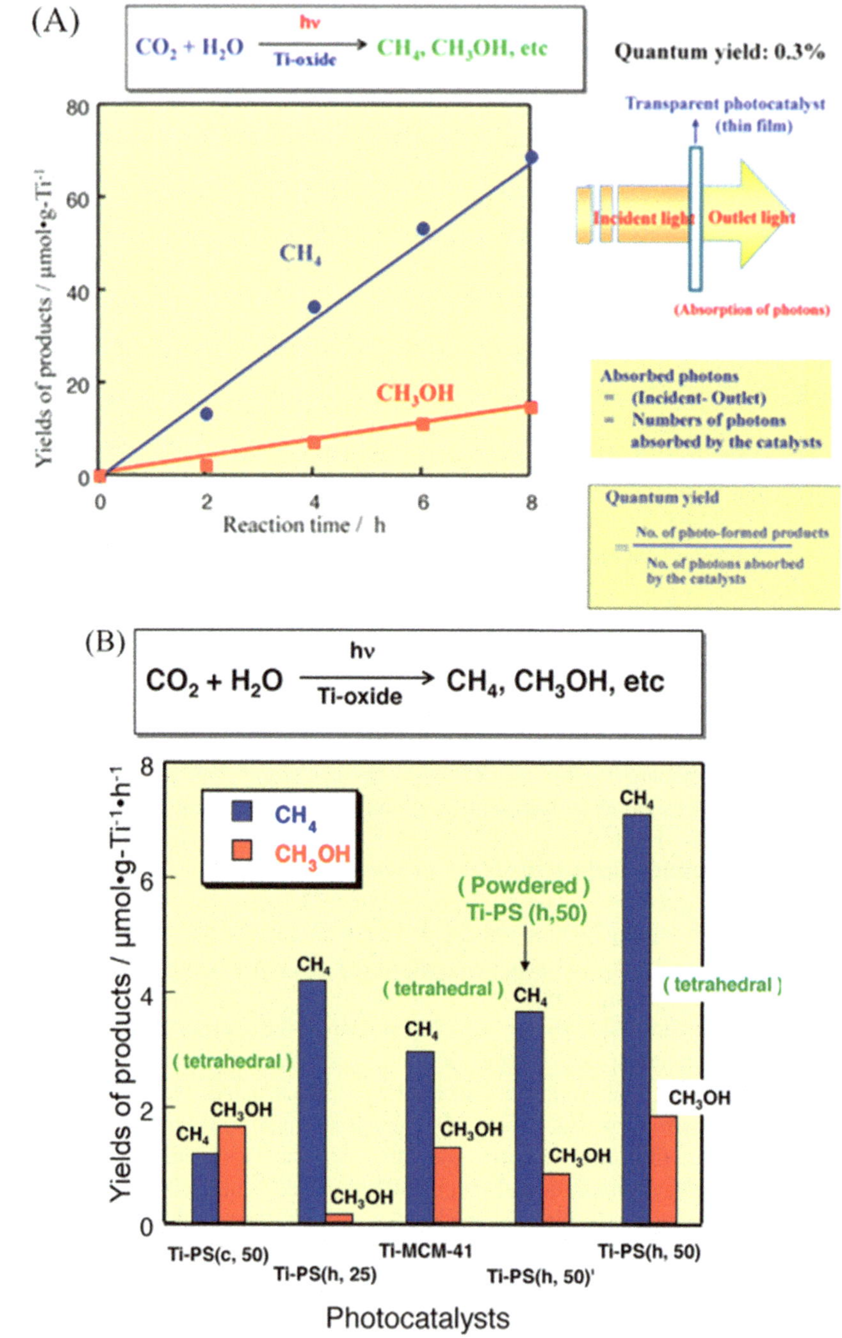

Fig. 3.11 (**a**) Reaction time profiles of the photocatalytic reduction of CO_2 with H_2O to produce CH_4 and CH_3OH on a Ti-oxide single-site containing mesoporous silica thin film photocatalyst at 298 K. Inserted figure shows how to measure the real quantum yields of the reaction. (**b**) The yields of CH_4 and CH_3OH in the photocatalytic reduction of CO_2 with H_2Oon Ti-PS (h, 25), Ti-PS(c, 50), Ti-MCM-41, powdered form of Ti-PS (h, 50) and Ti-PS (h, 50) photocatalysts at 295 K. (Reprinted with permission from [86, 88, 89]. Copyright 2002–2003, Elsevier)

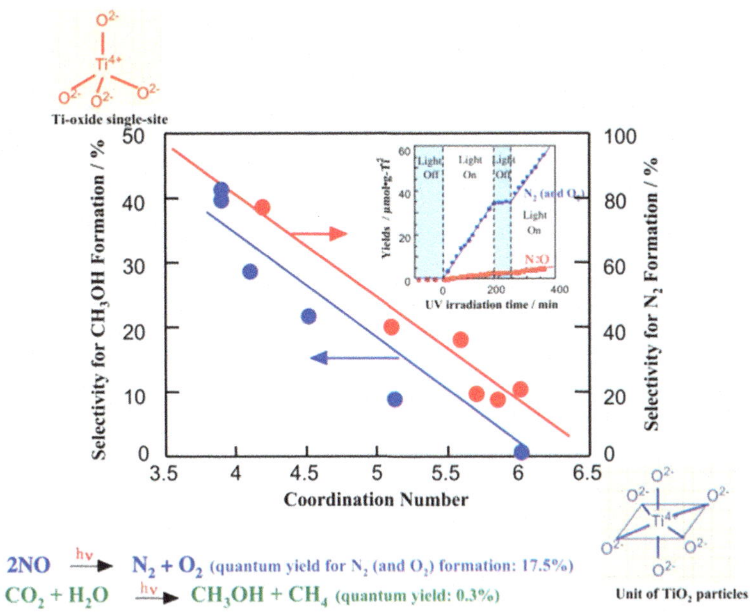

Fig. 3.12 Relationship between the coordination numbers and photocatalytic reactivity of titanium oxides. (Reprinted with permission from Ref. [78]. Copyright 1997, American Chemical Society)

Table 3.1 Comparisons of yields of N_2 and N_2O and their selectivities in the direct photocatalytic decomposition of NO at 275 K on various types of the Ti-oxide/zeolite catalysts and the powered bulk TiO_2 catalyst

Catalysts	Ti content (wt% of as TiO_2)	Yields (μmol/g-catal h)			Selectivity (%)	
		N_2	N_2O	total	N_2	N_2O
ex-Ti-oxide/Y-zeolite	1.1	14	1	15	91	9
imp-Ti-oxide/Y-zeolite	1.0	7	10	17	41	59
imp-Ti-oxide/Y-zeolite	10	5	22	27	19	81
Powered TiO_2	100	2	6	8	25	75

Besides that, the Cr-doped TiO_2 catalysts chemically prepared by impregnation or sol–gel method were found to have a mixture of the aggregated Cr-oxides in tetrahedral coordination similar to CrO_3 and octahedral coordination similar to Cr_2O_3.

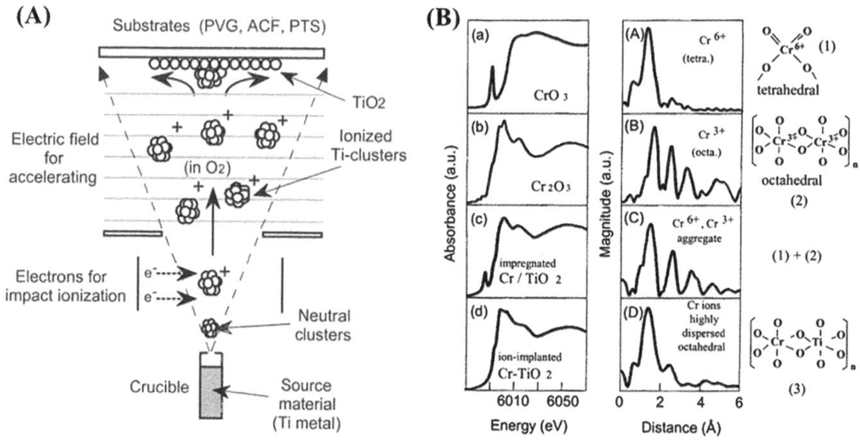

Fig. 3.13 (**a**) Schematic diagram of an ICB method. (**b**) XANES (a–d) and Fourier transforms of EXAFS (A–D) spectra of CrO_3 (A), Cr_2O_3 (B) and the Cr-impregnated TiO_2 (C) and Cr ion-implanted TiO_2 after calcination at 723 K (D). (Reprinted with permission from Ref. [81]. Copyright 2006, Springer)

3.6 Conclusion

Herein, to sum up, we have reviewed plenty of representative literatures about recent advances in photocatalysts over the highly dispersed TiO_2 in mesoporous materials. Although TiO_2 possesses specific properties, it is limited in photocatalysis because of its narrow bandgap. Thus, the doping modification on mesoporous TiO_2 materials proves to be an effective way to enhance its photocatalytic performance. Meanwhile, synthesizing mesoporous TiO_2–SiO_2, Ti–MOFs, and Ti–SiO_2 materials is another feasible way to achieve this goal. Based on the analysis of mechanism of these systems, it gives the researchers a promising future. The achieved progress in this field indicates that researchers can either be able to extend the photoresponse to the visible region or apply them to practical application, such as water splitting, degradation of pollutants and decomposition of greenhouse gas (CO_2, NO, NO_2).

References

1. Yuan S, Chen Y, Shi L, Fang J, Zhang J, Zhang J, Yamashita H (2007) Synthesis and characterization of Ce-doped mesoporous anatase with long-range ordered mesostructure. Mater Lett 61(21):4283–4286
2. Beck JS, Vartuli JC, Roth WJ, Leonowicz ME, Kresge CT, Schmitt KD, Chu CTW, Olson DH, Sheppard EW (1992) A new family of mesoporous molecular sieves prepared with liquid crystal templates. J Am Chem Soc 114(27):10834–10843
3. Ryoo R, Jun S (1997) Improvement of hydrothermal stability of MCM-41 using salt effects during the crystallization process. J Phys Chem B 101(3):317–320

4. Mokaya R (2001) Hydrothermally stable restructured mesoporous silica. Chem Commun 10(10):933–934
5. Anpo M (2013) Photocatalytic reduction of CO_2 with H_2O on highly dispersed Ti-oxide catalysts as a model of artificial photosynthesis. J CO2 Util 1:8–17
6. Antonelli DM, Ying J (1995) Synthesis of hexagonally packed mesoporous TiO_2 by a modified sol-gel method. Angew Chem Int Ed 34(18):2014–2017
7. Yang P, Zhao D, Margolese DI, Chmelka BF, Stucky GD (1998) Generalized syntheses of large-pore mesoporous metal oxides with semicrystalline frameworks. Nature 396 (6707):152–155
8. Zhang M, Bando Y, Wada K (2001) Sol-gel template preparation of TiO_2 nanotubes and nanorods. J Mater Sci Lett 20(2):167–170
9. Shibata H, Ogura T, Mukai T, Ohkubo T, Sakai H, Abe M (2005) Direct synthesis of mesoporous titania particles having a crystalline wall. J Am Chem Soc 127(47):16396–16397
10. Leghari SAK, Sajjad S, Zhang J (2013) Large mesoporous micro-spheres of WO_3/TiO_2 composite with enhanced visible light photo activity. RSC Adv 3(35):15354
11. Sheng Q, Yuan S, Zhang J, Chen F (2006) Synthesis of mesoporous titania with high photocatalytic activity by nanocrystalline particle assembly. Microporous Mesoporous Mater 87(3):177–184
12. Wang Y, Xu H, Wang X, Zhang X, Jia H, Zhang L, Qiu J (2006) A general approach to porous crystalline TiO_2, $SrTiO_3$, and $BaTiO_3$ spheres. J Phys Chem B 110(28):13835–13840
13. Yue Y, Gao Z (2000) Synthesis of mesoporous TiO_2 with a crystalline framework. Chem Commun 18(18):1755–1756
14. Zhao B, Chen F, Liu H, Zhang J (2011) Mesoporous TiO_2-B nanowires synthesized from tetrabutyl titanate. J Phys Chem Solids 72(3):201–206
15. Li Y, Lee N, Lee E, Song J, Kim SJ (2004) The characterization and photocatalytic properties of mesoporous rutile TiO_2 powder synthesized through self-assembly of nano crystals. Chem Phys Lett 389(1–3):124–128
16. Fukumoto S, Kitano M, Takeuchi M, Matsuoka M, Anpo M (2008) Photocatalytic hydrogen production from aqueous solutions of alcohol as model compounds of biomass using visible light-responsive TiO_2 thin films. Catal Lett 127(1–2):39–43
17. Kikuchi H, Kitano M, Takeuchi M, Matsuoka M, Anpo M, Kamat PV (2006) Extending the photoresponse of TiO_2 to the visible light region: photoelectrochemical behavior of TiO_2 thin films prepared by the radio frequency magnetron sputtering deposition method. J Phys Chem B 110(11):5537–5541
18. Kitano M, Takeuchi M, Matsuoka M, Thomas JM, Anpo M (2005) Preparation of visible light-responsive TiO_2 thin film photocatalysts by an RF magnetron sputtering deposition method and their photocatalytic reactivity. Chem Lett 34(4):616–617
19. Kitano M, Tsujimaru K, Anpo M (2006) Decomposition of water in the separate evolution of hydrogen and oxygen using visible light-responsive TiO_2 thin film photocatalysts: effect of the work function of the substrates on the yield of the reaction. Appl Catal A-Gen 314(2):179–183
20. Buchel G, Denoyel R, Llewellyn PL, Rouquerol J (2001) In situ surfactant removal from MCM-type mesostructures by ozone treatment. J Mater Chem 11(2):589–593
21. Hitz S, Prins R (1997) Influence of template extraction on structure, activity, and stability of MCM-41 catalysts. J Catal 168(2):194–206
22. Tian B, Liu X, Yu C, Gao F, Luo Q, Xie S, Tu B, Zhao D (2002) Microwave assisted template removal of siliceous porous materials. Chem Commun 11(11):1186–1187
23. Van Grieken R, Calleja G, Stucky GD, Melero JA, García RA, Iglesias J (2003) Supercritical fluid extraction of a nonionic surfactant template from SBA-15 materials and consequences on the porous structure. Langmuir 19(9):3966–3973
24. Yang C, Zibrowius B, Schmidt W, Schüth F (2004) Stepwise removal of the copolymer template from mesopores and micropores in SBA-15. Chem Mater 16(15):2918–2925

25. Yamashita H, Harada M, Misaka J, Takeuchi M, Ikeue K, Anpo M (2002) Degradation of propanol diluted in water under visible light irradiation using metal ion-implanted titanium dioxide photocatalysts. J Photoch Photobio A 148(1–3):257–261
26. Tian B, Zhang J, Tong T, Chen F (2008) Preparation of Au/TiO_2 catalysts from Au(I)-thiosulfate complex and study of their photocatalytic activity for the degradation of methyl orange. Appl Catal B-Environ 79(4):394–401
27. Yuan S, Sheng Q, Zhang J, Chen F, Anpo M, Zhang Q (2005) Synthesis of La^{3+} doped mesoporous titania with highly crystallized walls. Microporous Mesoporous Mater 79 (1–3):93–99
28. Yuan S, Sheng Q, Zhang J, Chen F, Anpo M, Dai W (2006) Synthesis of Pd nanoparticles in La-doped mesoporous titania with polycrystalline framework. Catal Lett 107(1–2):19–24
29. Yuan X, Zhang J, Anpo M, He D (2010) Synthesis of Fe^{3+} doped ordered mesoporous TiO_2 with enhanced visible light photocatalytic activity and highly crystallized anatase wall. Res Chem Intermed 36(1):83–93
30. Sajjad S, Leghari SAK, Zhang J (2013) Copper impregnated ionic liquid assisted mesoporous titania: visible light photocatalyst. RSC Adv 3(31):12678
31. He C, Tian B, Zhang J (2010) Thermally stable SiO_2-doped mesoporous anatase TiO_2 with large surface area and excellent photocatalytic activity. J Colloid Interface Sci 344(2):382–389
32. Wu Y, Xing M, Tian B, Zhang J, Chen F (2010) Preparation of nitrogen and fluorine co-doped mesoporous TiO_2 microsphere and photodegradation of acid orange 7 under visible light. Chem Eng J 162(2):710–717
33. He C, Tian B, Zhang J (2010) N, B, Si-tridoped mesoporous TiO_2 with high surface area and excellent visible-light photocatalytic activity. Res Chem Intermed 36(4):349–359
34. Hao H, Zhang J (2009) The study of Iron (III) and nitrogen co-doped mesoporous TiO_2 photocatalysts: synthesis, characterization and activity. Microporous Mesoporous Mater 121 (1–3):52–57
35. Ma Y, Xing M, Zhang J, Tian B, Chen F (2012) Synthesis of well ordered mesoporous Yb, N co-doped TiO_2 with superior visible photocatalytic activity. Microporous Mesoporous Mater 156:145–152
36. Qiu B, Xing M, Zhang J (2014) Mesoporous TiO_2 nanocrystals grown in situ on graphene aerogels for high photocatalysis and lithium-ion batteries. J Am Chem Soc 136(16):5852–5855
37. Dong C, Xing M, Zhang J (2016) Double-cocatalysts promote charge separation efficiency in CO_2 photoreduction: spatial location matters. Mater Horiz 3:608–612
38. Anderson C, Bard AJ (1995) An improved photocatalyst of TiO_2/SiO_2 prepared by a sol-gel synthesis. J Phys Chem 99(24):9882–9885
39. Gao X, Wachs IE (1999) Titania-silica as catalysts: molecular structural characteristics and physico-chemical properties. Catal Today 51(2):233–254
40. Dagan G, Sampath S, Lev O (1995) Preparation and utilization of organically modified silica-titania photocatalysts for decontamination of aquatic environments. Chem Mater 7(3):446–453
41. Klein S, Thorimbert S, Maier WF (1996) Amorphous microporous titania-silica mixed oxides: preparation, characterization, and catalytic redox properties. J Catal 163(2):476–488
42. Pabón E, Retuert J, Quijada R, Zarate A (2004) TiO_2-SiO_2 mixed oxides prepared by a combined sol-gel and polymer inclusion method. Microporous Mesoporous Mater 67 (2–3):195–203
43. Li Z, Hou B, Xu Y, Wu D, Sun Y (2005) Hydrothermal synthesis, characterization, and photocatalytic performance of silica-modified titanium dioxide nanoparticles. J Colloid Interface Sci 288(1):149–154
44. Xing M, Qi D, Zhang J, Chen F, Tian B, Bagwas S, Anpo M (2012) Super-hydrophobic fluorination mesoporous MCF/TiO_2 composite as a high-performance photocatalyst. J Catal 294:37–46
45. Gun'ko VM, Zarko VI, Turov VV, Leboda R, Chibowski E, Holysz L, Pakhlov EM, Voronin EF, Dudnik VV, Gornikov YI (1998) CVD-Titania on Fumed silica substrate. J Colloid Interface Sci 198(1):141–156

46. Fan G, Zou B, Cheng S, Zheng L (2010) Ligandless palladium supported on SiO_2-TiO_2 as effective catalyst for Suzuki reaction. J Ind Eng Chem 16(2):220–223
47. Mei F, Liu C, Zhang L, Ren F, Zhou L, Zhao W, Fang Y (2006) Microstructural study of binary TiO_2:SiO_2 nanocrystalline thin films. J Cryst Growth 292(1):87–91
48. Siddiquey IA, Furusawa T, Sato M, Honda K, Suzuki N (2008) Control of the photocatalytic activity of TiO_2 nanoparticles by silica coating with polydiethoxysiloxane. Dyes Pigments 76 (3):754–759
49. Li D (2004) Study on the fluorescence properties of benzopyrylium salt in Ti-HMS. Dyes Pigments 63(1):71–76
50. Zhao W, Li D, He B, Zhang J, Huang J, Zhang L (2005) The photoluminescence of coumarin derivative encapsulated in MCM-41 and Ti-MCM-41. Dyes Pigments 64(3):265–270
51. He C, Tian B, Zhang J (2009) Synthesis of thermally stable and highly ordered bicontinuous cubic mesoporous titania-silica binary oxides with crystalline framework. Microporous Mesoporous Mater 126(1–2):50–57
52. Chen C, Ma W, Zhao J (2010) Semiconductor-mediated photodegradation of pollutants under visible-light irradiation. Chem Soc Rev 39(11):4206–4219
53. Aguado J, van Grieken R, López-Muñoz MJ, Marugán J (2002) Removal of cyanides in wastewater by supported TiO_2-based photocatalysts. Catal Today 75(1–4):95–102
54. Fujishima A, Honda K (1972) Electrochemical photolysis of water at a semiconductor electrode. Nature 238(5358):37–38
55. Horiuchi Y, Toyao T, Takeuchi M, Matsuoka M, Anpo M (2013) Recent advances in visible-light-responsive photocatalysts for hydrogen production and solar energy conversion-from semiconducting TiO_2 to MOF/PCP photocatalysts. Phys Chem Chem Phys 15 (32):13243–13253
56. Niphadkar PS, Chitale SK, Sonar SK, Deshpande SS, Joshi PN, Awate SV (2014) Synthesis, characterization and photocatalytic behavior of TiO_2-SiO_2 mesoporous composites in hydrogen generation from water splitting. J Mater Sci 49(18):6383–6391
57. Xing M, Zhang J, Qiu B, Tian B, Anpo M, Che M (2014) A brown mesoporous TiO_{2-x}/MCF composite with an extremely high quantum yield of solar energy photocatalysis for H_2 evolution. Small 11(16):1920–1929
58. Hill RJ, Long DL, Champness NR, Hubberstey P, Schröder M (2005) New approaches to the analysis of high connectivity materials: design frameworks based upon 44- and 63-subnet tectons. Acc Chem Res 38(4):335–348
59. Yaghi OM, O'keeffe M, Ockwig NW, Chae HK, Eddaoudi M, Kim J (2003) Reticular synthesis and the design of new materials. Nature 423(6941):705–714
60. Kitagawa S, Kitaura R, Si N (2004) Functional porous coordination polymers. Angew Chem Int Ed 43(18):2334–2375
61. Moulton B, Zaworotko MJ (2001) From molecules to crystal engineering: supramolecular isomerism and polymorphism in network solids. Chem Rev 101(6):1629–1658
62. Seo JS, Whang D, Lee H, Im Jun S, Oh J, Jeon YJ, Kim K (2000) A homochiral metal-organic porous material for enantioselective separation and catalysis. Nature 404(6781):982–986
63. Zhou T, Du Y, Borgna A, Hong J, Wang Y, Han J, Zhang W, Xu R (2013) Post-synthesis modification of a metal-organic framework to construct a bifunctional photocatalyst for hydrogen production. Energy Environ Sci 6(11):3229–3234
64. Horiuchi Y, Toyao T, Saito M, Mochizuki K, Iwata M, Higashimura H, Anpo M, Matsuoka M (2012) Visible-light-promoted photocatalytic hydrogen production by using an amino-functionalized Ti (IV) metal-organic framework. J Phys Chem C 116(39):20848–20853
65. Wang C, DeKrafft KE, Lin W (2012) Pt nanoparticles@photoactive metal-organic frameworks: efficient hydrogen evolution via synergistic photoexcitation and electron injection. J Am Chem Soc 134(17):7211–7214
66. Fateeva A, Chater PA, Ireland CP, Tahir AA, Khimyak YZ, Wiper PV, Darwent JR, Rosseinsky MJ (2012) A water-stable porphyrin-based metal-organic framework active for visible-light photocatalysis. Angew Chem 124(30):7558–7562

67. Kataoka Y, Sato K, Miyazaki Y, Masuda K, Tanaka H, Naito S, Mori W (2009) Photocatalytic hydrogen production from water using porous material $[Ru_2(p\text{-}BDC)_2]_n$. Energy Environ Sci 2 (4):397–400
68. Xiao J, Shang Q, Xiong Y, Zhang Q, Luo Y, Yu S, Jiang H (2016) Boosting photocatalytic hydrogen production of a metal-organic framework decorated with platinum nanoparticles: the platinum location matters. Angew Chem 128(32):9535–9539
69. Gomes Silva C, Luz I, Llabrés i, Xamena FX, Corma A, García H (2010) Water stable Zr-benzenedicarboxylate metal-organic frameworks as photocatalysts for hydrogen generation. Chem-Eur J 16(36):11133–11138
70. Fu Y, Sun D, Chen Y, Huang R, Ding Z, Fu X, Li Z (2012) An amine-functionalized titanium metal-organic framework photocatalyst with visible-light-induced activity for CO_2 reduction. Angew Chem 124(14):3420–3423
71. Alvaro M, Carbonell E, Ferrer B, Llabrés i, Xamena FX, Garcia H (2007) Semiconductor behavior of a metal-organic framework (MOF). Chem-Eur J 13(18):5106–5112
72. Tachikawa T, Choi JR, Fujitsuka M, Majima T (2008) Photoinduced charge-transfer processes on MOF-5 nanoparticles: elucidating differences between metal-organic frameworks and semiconductor metal oxides. J Phys Chem C 112(36):14090–14101
73. Yang H, He X, Wang F, Kang Y, Zhang J (2012) Doping copper into ZIF-67 for enhancing gas uptake capacity and visible-light-driven photocatalytic degradation of organic dye. J Mater Chem 22(41):21849–21851
74. Anpo M, Yamashita H, Ikeue K, Fujii Y, Zhang S, Ichihashi Y, Park DR, Suzuki Y, Koyano K, Tatsumi T (1998) Photocatalytic reduction of CO_2 with H_2O on Ti-MCM-41 and Ti-MCM-48 mesoporous zeolite catalysts. Catal Today 44(1–4):327–332
75. Takeuchi M, Dohshi S, Eura T, Anpo M (2003) Preparation of titanium-silicon binary oxide thin film photocatalysts by an ionized cluster beam deposition method. Their photocatalytic activity and photoinduced super-hydrophilicity. J Phys Chem B 107(51):14278–14282
76. Takeuchi M, Yamashita H, Matsuoka M, Anpo M, Hirao T, Itoh N, Iwamoto N (2000) Photocatalytic decomposition of NO on titanium oxide thin film photocatalysts prepared by an ionized cluster beam technique. Catal Lett 66(3):185–187
77. Anpo M, Aikawa N, Kubokawa Y, Che M, Louis C, Giamello E (1985) Photoluminescence and photocatalytic activity of highly dispersed titanium oxide anchored onto porous Vycor glass. J Phys Chem 89(23):5017–5021
78. Anpo M, Yamashita H, Ichihashi Y, Fujii Y, Honda M (1997) Photocatalytic reduction of CO_2 with H_2O on titanium oxides anchored within micropores of zeolites: effects of the structure of the active sites and the addition of Pt. J Phys Chem B 101(14):2632–2636
79. Anpo M (1997) In situ characterization of highly dispersed catalysts included within zeolite cavities and their photocatalytic reactivities. Nouv Cim D 19(11):1641–1648
80. Yamashita H, Anpo M (2004) Application of an ion beam technique for the design of visible light-sensitive, highly efficient and highly selective photocatalysts: ion-implantation and ionized cluster beam methods. Catal Surv Jpn 8(1):35–45
81. Hu Y, Rakhmawaty D, Matsuoka M, Takeuchi M, Anpo M (2006) Synthesis, characterization and photocatalytic reactivity of Ti-containing micro- and mesoporous materials. J Porous Mater 13(3–4):335–340
82. Anpo M, Chiba K (1992) Photocatalytic reduction of CO_2 on anchored titanium oxide catalysts. J Mol Catal 74(1–3):207–212
83. Yamashita H, Shiga A, Kawasaki S-i, Ichihashi Y, Ehara S, Anpo M (1995) Photocatalytic synthesis of CH_4 and CH_3OH from CO_2 and H_2O on highly dispersed active titanium oxide catalysts. Energy Convers Manag 36(6–9):617–620
84. Ikeue K, Yamashita H, Anpo M (1999) Photocatalytic reduction of CO_2 with H_2O on titanium oxides prepared within the FSM-16 mesoporous zeolite. Chem Lett 28(11):1135–1136
85. Ikeue K, Yamashita H, Anpo M, Takewaki T (2001) Photocatalytic reduction of CO_2 with H_2O on Ti-β zeolite photocatalysts: effect of the hydrophobic and hydrophilic properties. J Phys Chem B 105(35):8350–8355

86. Ikeue K, Nozaki S, Ogawa M, Anpo M (2002) Characterization of self-standing Ti-containing porous silica thin films and their reactivity for the photocatalytic reduction of CO_2 with H_2O. Catal Today 74(3–4):241–248
87. Yamashita H, Honda M, Harada M, Ichihashi Y, Anpo M, Hirao T, Itoh N, Iwamoto N (1998) Preparation of titanium oxide photocatalysts anchored on porous silica glass by a metal ion-implantation method and their photocatalytic reactivities for the degradation of 2-propanol diluted in water. J Phys Chem B 102(52):10707–10711
88. Ikeue K, Nozaki S, Ogawa M, Anpo M (2002) Photocatalytic reduction of CO_2 with H_2O on Ti-containing porous silica thin film photocatalysts. Catal Lett 80(3–4):111–114
89. Shioya Y, Ikeue K, Ogawa M, Anpo M (2003) Synthesis of transparent Ti-containing mesoporous silica thin film materials and their unique photocatalytic activity for the reduction of CO_2 with H_2O. Appl Catal A-Gen 254(2):251–259
90. Anpo M, Thomas JM (2006) Single-site photocatalytic solids for the decomposition of undesirable molecules. Chem Commun 31(0):3273–3278

Chapter 4
Preparation of Reduced TiO_{2-x} for Photocatalysis

4.1 Introduction

Titanium dioxide (TiO_2) photocatalyst has received plenty of attention since it was applied for the photocatalytic hydrogen generation from water splitting [1]. Owing to its abundance, non-toxicity, and chemical stability, TiO_2 is widely used in areas such as pigments, cosmetics, paper, plastics, catalysis, solar cells, and antibacterial agents. As a catalyst, TiO_2 shows applications in photodegradation of organic pollutants, hydrogen evolution from water, CO_2 photoreduction, lithium-ion batteries (LIBs), and dye-sensitized solar cells (DSSCs).

However, the practical applications of TiO_2 are limited to UV region of solar light because of its wide bandgap (3.2 eV for anatase phase). It means that only a small part (~5%) of the solar energy can be well utilized by TiO_2 photocatalysts. In order to solve these problems, a lot of modification methods were employed to improve the solar light response of this catalyst and modify its bandgap engineering. Conventional approaches have been introduced for TiO_2 modification, such as nonmetal doping, noble metal grafting, dye-sensitization, special facet exposure, compositing with other materials, etc. However, new approaches are still required in performance improvement, to match the rising demands for energy consuming and environmental protection.

The concept of reduced TiO_2, or TiO_{2-x}, photocatalyst was early introduced decades ago [2–5], and it was usually used to describe TiO_2 catalysts with Ti^{3+} ions induced by doping, undercoordinated Ti atoms caused by oxygen vacancies, or hydrogen implantation [6]. Besides, plenty of works have been reported on reduced facets of TiO_2 materials [7], especially with the help of density functional theory (DFT) [8] and scanning tunneling microscope (STM) studies [3, 9, 10]. Recently, the investigation of TiO_{2-x} phtotocatalysts develops rapidly, owing to its superior photocatalytic and photoelectrochemical (PEC) performance. A lot of new preparations methods were employed for the synthesis of TiO_{2-x}, and the surface properties

J. Zhang et al., *Photocatalysis*, Lecture Notes in Chemistry 100,
https://doi.org/10.1007/978-981-13-2113-9_4

and bulk structure of TiO_{2-x} were carefully studied. Therefore, it is necessary to make a review of the current development of TiO_{2-x}.

In this chapter, we will first discuss the synthesis approaches of TiO_{2-x} photocatalysts, including annealing under H_2 atmosphere, vacuum activation, in situ reduction, metal reduction, electrochemical reduction, UV irradiation, plasma treatment, and partial oxidation starting from Ti, Ti(II), or Ti(III) precursors, etc. Then various characterization techniques are discussed to investigate the properties of TiO_{2-x}, such as the formation of structural distortions inherent, Ti^{3+} species, oxygen vacancies, or surface hydroxyl bonds. We also summarize the applications of reduced TiO_2 in the photocatalytic and photoelectric areas, point out further modification approaches for TiO_{2-x} in improving their performance, and demonstrate the sustainability of photocatalysis as a solution to environment pollution and energy crisis.

4.2 Synthesis of TiO_{2-x} Photocatalysts

Theoretically, reduced TiO_2 photocatalysts could be obtained either by the reduction of TiO_2 solid materials or Ti(IV) precursors or the oxidation of low-oxidation-state titanium compounds. Until now, a large number of methods have been developed for the synthesis of TiO_{2-x} materials, including thermal treatment under H_2 atmosphere or other reducing gases, in situ reduction, vacuum activation treatment, metal reduction, hydrogen or water plasma treatment, electrochemical reduction, UV light irradiation treatment, and partial oxidation starting from Ti, Ti(II), and Ti(III) precursors [11]. In general, all these reported methods for the synthesis of TiO_{2-x} could be classified by these two main aspects: under reducing or oxidizing atmosphere.

4.2.1 Reduction Method

4.2.1.1 Thermal Treatment Under Reducing Gases

Thermal treatment under H_2-gas atmosphere was early used for the synthesis of TiO_{2-x} materials, developed by Chen et al. [12]. The pristine TiO_2 nanocrystals were hydrogenated in a 20-bar H_2 atmosphere at 200 °C for 5 days. The obtained TiO_{2-x} nanoparticles became black colored (Fig. 4.1a) and disorder engineered (Fig. 4.1b).

Wang et al. investigated the hydrogenation of TiO_2 by H_2–gas heating at various temperatures [13]. The color of the obtained hydrogenated TiO_2 depends on the hydrogen annealing temperature, and it becomes darker with temperature increasing in the range of 200–500 °C [13]. The color implies the extended solar light absorption of TiO_2 as the result of hydrogenation. Besides, after thermal hydrogenation treatment, disorder surface shell is formed outside the crystalline TiO_2 core

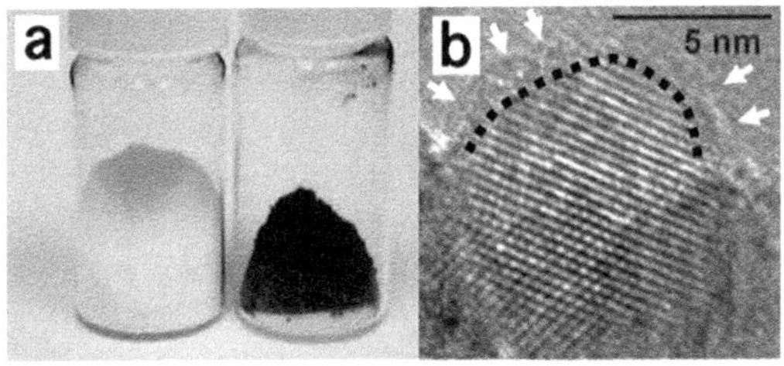

Fig. 4.1 (**a**) Photos of the pristine unmodified TiO_2 samples and the black TiO_2 nanocrystals treated by gas hydrogenation. (**b**) HR-TEM of the gas-hydrogenated TiO_{2-x} nanocrystals. Disordered layers were found outside the crystalized TiO_2 core. (Reprinted from Ref. [12], with kind permission from AAAS)

[14]. Hu et al. demonstrated that no Ti^{3+} species was observed on the surface of the hydrogenated TiO_2 catalysts but with the formation of Ti-OH species on the TiO_2 surface after hydrogenation by H_2–gas thermal treatment [15]. The formation of hydroxyl group on TiO_2 surface after hydrogen treatment was also confirmed by other literatures [13]. The generated Ti–H or O–H species lead to the formation of surface disorder structure, which introduce midgap states above the valance band maximum (VBM) and cause the bandgap narrowing of TiO_2 photocatalyst [15]. It seems that high-temperature hydrogenation is likely to introduce disordered layers outside the crystalline TiO_2 core and Ti^{3+} species in the bulk of TiO_2 catalysts [15]. However, it is possible that TiO_2 is hydrogenated at room temperature. Lu et al. treated commercial Evonik P25 under 35 bar hydrogen gas at room temperature for 20 days, and a similar crystalline-disordered core-shell structure was also obtained [16].

Thermal treatment under hydrogen atmosphere is an efficient way to introduce disordered structure and bandgap narrowing of TiO_2. However, there are also concerns about the potential safety impacts of the using of hydrogen gas in practical applications. In order to avoid the direct usage of hydrogen gas, Xing et al. developed a new approach for the synthesis of TiO_2 using $NaBH_4$ as the reductant, which would release hydrogen during the preparation process [17]. During the hydrothermal process, $NaBH_4$ is likely to hydrolyze and release the hydrogen gas and active atomic hydrogen in the solution. Ti^{3+} was introduced into the lattice of TiO_2 without changing the unit cell [18]. However, by-products especially boron oxide species were generated, while Ti^{4+} was reduced by $NaBH_4$. In order to eliminate the negative effect of carbon and boron impurities which were generated during the synthesis, Fang et al. developed the $NaBH_4$ reduction method with calcination treatment and post-washing with HCl aqueous solution [19]. Except for in situ reaction starting from titanium precursors and $NaBH_4$, solid-state reaction of $NaBH_4$ with TiO_2 nanoparticles also promotes the surface defect implantation of TiO_2 [20]. This solid-state reaction leads to the generation of oxygen vacancies and the subsequent formation of disordered TiO_{2-x} shell on the surface of TiO_2 nanocrystals [21]. The introduction of Ti^{3+} using $NaBH_4$ as the reductant was also applied in the fabrication of Ni^{2+} doped porous TiO_{2-x} [22], 3D mesoporous black $TiO_2/MoS_2/TiO_2$ nanosheets [23], and gold grafted TiO_2 sphere [24].

The fabrication of TiO_{2-x} photocatalysts by thermal treatment under other reducing atmospheres was also employed. Zuo et al. reduced the pristine TiO_2 by CO and NO gases which were released from the oxidative decomposition of 2-ethylimidazole. Besides, hydrazine hydrate was proved to be efficient to promote the formation of Ti^{3+} species in TiO_2 catalysts [25, 26]. In addition, the reduction of TiO_2 can also be realized by the reducing species released from the decomposition of C_3N_4 at high temperature [27].

4.2.1.2 Vacuum Activation Treatment

Xing et al. developed a simple approach for the generation of Ti^{3+} species into commercial TiO_2 nanomaterials (Evonik P25), and the obtained TiO_{2-x} showed enhanced solar light absorption, improved photodegradation of methyl orange, and promoted photocatalytic hydrogen evolution from water splitting [28]. Electron paramagnetic resonance (EPR) confirmed the generation of Ti^{3+} and oxygen vacancies in the bulk of TiO_2, and the obtained Ti^{3+} and oxygen defects were demonstrated to be responsible for the high photocatalytic performance of the reduced TiO_2 catalyst under visible light irradiation. The oxygen vacancies created by heating in ultrahigh vacuum (UHV) were early confirmed by STM (Fig. 4.2), reported by Diebold et al. [10]. Lu et al. demonstrated that the surface oxygen vacancies were probably created by removal of bridging oxygen atoms under vacuum, and each oxygen vacancy resulted into two Ti^{3+} sites [29].

Fang et al. employed the vacuum activation method on the co-doping of TiO_2 with nitrogen [30]. The thermal treatment under vacuum was proved to be beneficial for the migration of nitrogen atoms from the surface to the bulk of TiO_2 and also the generation of oxygen vacancies in the subsurface of TiO_2 [30], thus contributing to the substitution of nitrogen for oxygen vacancies [31].

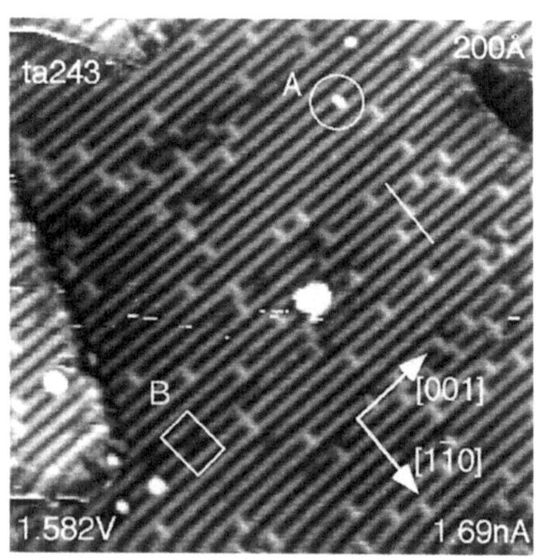

Fig. 4.2 STM image showing defects of rutile $TiO_2(110)$ surface. Oxygen vacancies are created by heating in UHV, shown as bright features centered on *dark rows* in STM. (Reprinted from Ref. [10], copyright 1998, with permission from Elsevier)

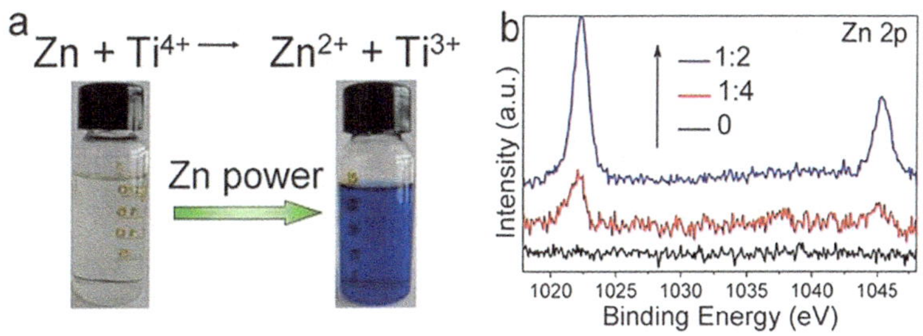

Fig. 4.3 (**a**) Photos of the $TiCl_4$ solution before and after the adding of Zn powder. (**b**) XPS Zn 2p spectra of the samples prepared with different Zn/Ti molar ratio. It shows the presence of Zn element in the obtained Zn-assisted TiO_{2-x} samples. (Reprinted from Ref. [36] by permission of The Royal Society of Chemistry)

4.2.1.3 Metal Reduction

Wang et al. reported an aluminum reduction method through high-temperature thermal treatment for the synthesis of TiO_{2-x} photocatalyst [32]. The process was operated in a two-zone furnace and post-annealing at Ar atmosphere. As a result, a unique core-shell structured TiO_2/TiO_{2-x} was formed, with wild solar light absorption and efficient H_2 generation from water splitting. Yang et al. treated pristine TiO_2 samples with the same process however with post-annealing at H_2S atmosphere, in order to introduce Ti^{3+} and surface S dopants simultaneously [33]. Cui et al. synthesized TiO_2 nanotube arrays by electrochemical anodization process and subsequently reduced the as-prepared TiO_2 nanomaterials by this two-zone thermal treatment using aluminum as the reductant [34]. Zheng et al. directly mixed TiO_2 nanosheets and metallic Al nanoparticles together and then treated the mixture at 500 °C for 3 h under nitrogen protection [35]. The Al nanoparticles not only act as the reductant of Ti(IV) species but also lead to the stabilization of the obtained Ti^{3+} species and oxygen vacancies on the surface of TiO_2 through Al–O bonds.

Metallic Zn powder is an efficient reductant for the formation of Ti^{3+} species in TiO_2. The color of Ti^{4+}-containing solution turns from light yellow to blue after the adding of Zn powder (Fig. 4.3a). Zheng et al. demonstrated that the surface Ti^{3+} species can be stabilized by Zn doping (Fig. 4.3b) [36]. The formation of ZnO clusters on the surface of TiO_{2-x} was reported in other literatures [37, 38]. However, Zhao et al. washed the Zn-assisted TiO_{2-x} photocatalyst with HCl aqueous solution to remove the residual Zn powder, and the final TiO_{2-x} sample still kept deep blue colored [39]. No Zn signal was observed on the obtained TiO_{2-x} materials in Zhao's work [39].

In addition, Chen et al. prepared TiO_{2-x} nanorods by solvothermal method using magnesium powder as the reductant and $TiCl_3$ as the titanium source [40]. Mg-doping of TiO_{2-x} could be eliminated due to the absence of Mg in any form in the final products. Sinhamahapatra et al. treated the mixture of TiO_2 nanoparticles and Mg powder with heating at high temperature in the flow of 5% H_2/Ar and

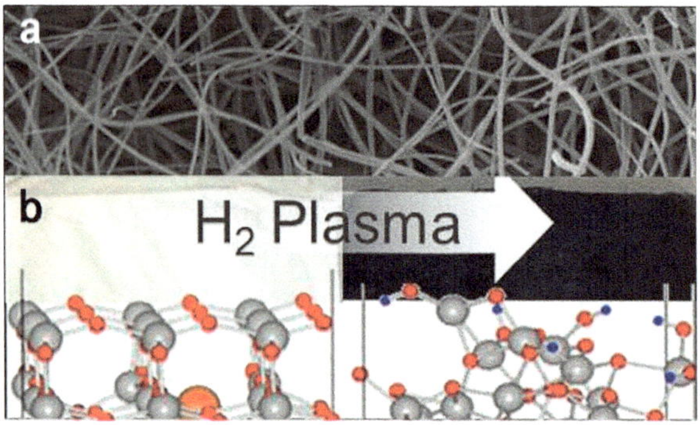

Fig. 4.4 (**a**) SEM images of hydrogen plasma-treated TiO_2 nanofibers. (**b**) TiO_2 slab with oxygen vacancies (left) and unique TiO_2 disordered structure with hydrogen interstitials after hydrogen plasma treatment (right). Ti, O, oxygen vacancies, and H are shown in gray, red, orange, and blue, respectively. (Reprinted with permission from Ref. [45], copyright 2015 American Chemical Society)

washed the residual Mg away by acid treatment [41]. The obtained TiO_{2-x} photocatalyst showed an excellent H_2 production rate from water splitting.

4.2.1.4 Plasma Reduction

The in-flight reduction of TiO_2 by H_2 plasma was early reported two decades ago [42]. Compared to the conventional thermal processing, plasma provides improved thermodynamics and kinetics [43]. And the obtained highly active species such as atomic hydrogen in the process are well-suited to the reduction of metal oxides [44]. Therefore, under H_2 plasma treatment, metallic state or the suboxides are obtained from TiO_2 vaporization [42, 43]. Lepcha et al. reported that H_2 plasma treatment induced the formation of oxygen vacancies and hydrogen interstitials on the synthesized TiO_2 nanofibers (Fig. 4.4a, b), which led to the reduction of neighboring Ti^{4+} to Ti^{3+} [45]. These oxygen vacancies are mainly formed within a small region near the surface of the TiO_2 material, due to the limited plasma penetration in to TiO_2. The presence of oxygen vacancies after hydrogen plasma treatment was evidenced by EPR characterization [46]. Wang et al. reported that unique core-shell structure was also formed after hydrogen plasma process, and the obtained TiO_{2-x} is superior to the sample prepared by thermal treatment under high-pressure H_2 [47]. In a word, the hydrogenation leads to the formation of Ti^{3+}, oxygen vacancies, disordered layers, decrease resistance, and enhanced solar light absorption, which account for the high PEC performance of the catalyst [48], and improved efficiency of DSSCs [49, 50]. In addition, water plasma was also proved to be efficient for the synthesis of TiO_{2-x} spheres through Ti electrodes at low temperature and atmospheric pressure [51].

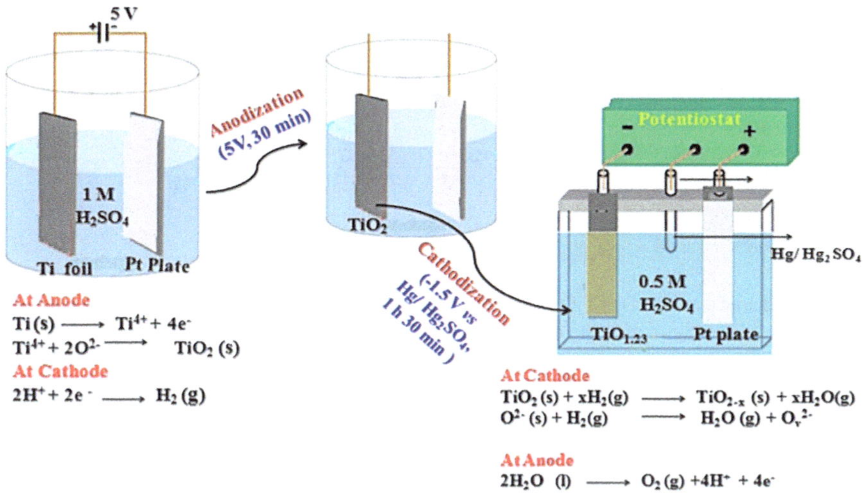

Fig. 4.5 Schematic illustration of TiO_{2-x} photocatalyst through cathodic reduction of the anodized TiO_2 samples. (Reprinted with permission from Ref. [55], copyright 2016 American Chemical Society)

4.2.1.5 Electrochemical Reduction

Electrochemical reduction was reported to be an easy and simple method to introduce Ti^{3+} species into TiO_2 photocatalysts. Compared with the high requirement of traditional approaches, electrochemical reduction method is easy to achieve, and it avoids expensive facilities, harsh experimental conditions, or polluting by-products [52]. Zhang et al. reported the TiO_{2-x} nanotube arrays induced by electrochemical reduction process [53]. Unlike the direct synthesis of TiO_{2-x} via anodization, Swaminathan et al. reported the cathodic reduction of the anodized TiO_2 sample. Firstly, TiO_2 was obtained by the anodization of Ti, then neutral oxygen atoms were removed in the cathodization process, and oxygen vacancies were formed (Fig. 4.5). The formation of oxygen vacancies in TiO_{2-x} nanomaterial was influenced by the duration of anodization and the post-annealing treatment [54].

Mo et al. demonstrated that during electrochemical reduction process, atomic H was generated on the Ag electrode and then diffused into TiO_2 as interstitial hydrogen (H_i), while in the gas hydrogenation process, the hydrogen was like to occupy the position of oxygen vacancies and became negatively charged (H_o), based on the results of DFT calculations (Fig. 4.6) [56]. However, the weak bonding between H_i and the closet oxygen atoms facilitates to break, making the obtained TiO_{2-x} samples sensitive to aging at room temperature. Therefore, a combination of electrochemical reduction and post-heating treatment was employed to further stabilize the Ti^{3+} in TiO_{2-x} [54]. In addition, Liu et al. demonstrated that cathodic reduction led to the activation of Ti^{3+} species in TiO_2, and a combination of cathodic reduction and H_2-gas heating treatment resulted into a synergistic effect in

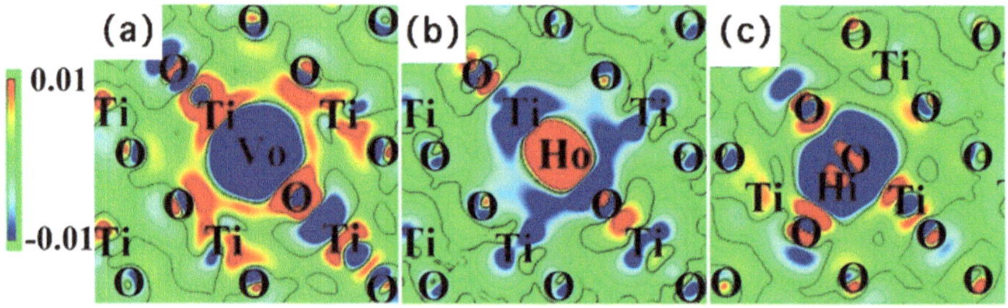

Fig. 4.6 Charge density of (**a**) TiO_2 with oxygen vacancies, (**b**) TiO_2 with H_o, and (**c**) TiO_2 with H_i, respectively. The color represents the electron density, as shown on the left. (Ref. [56])

photocatalytic applications [57]. Except for reduced TiO_2, reduced Nb_2O_5 and WO_3 polycrystals were also electrochemically hydrogenated, reported by Chen et al. [58].

4.2.1.6 UV Light Irradiation Reduction

UV light irradiation was early reported to introduce surface oxygen vacancies at bridging sites, which was favorable for the conversion of Ti^{4+} to Ti^{3+} in TiO_2 catalysts [59]. The photogeneration of Ti^{3+} sites on TiO_2 surface promoted the development of highly amphiphilic TiO_2 materials [60]. Based on the use of STM [61], oxygen defects along <001> direction of TiO_2 were detected, and the driving force of oxygen vacancy formation was considered to be the UV-generated excitons: oxygen anions trap hole components of the excitons and become neutralized, finally resulting into the transformation of oxygen anions to molecular oxygen gas and subsequently its release into vacuum.

However, the Ti^{3+} generated by UV irradiation is usually located on the surface of TiO_2, and the oxygen defects are easy to be healed while exposing under O_2 gas [62]. EPR also confirmed the Ti^{3+} species on the surface or few surface monolayers of TiO_2 and their extinction in oxygen [63]. Furthermore, vacuum condition should be maintained while employing UV treatment on the TiO_2 samples. In order to stabilize the UV-induced Ti^{3+} species of TiO_{2-x} nanoparticles, Li et al. combined UV irradiation with post-annealing process at Ar atmosphere [64]. Besides, Wu et al. synthesized yellow TiO_{2-x} nanoparticles via UV irradiation-assisted sol-gel method, and the obtained TiO_{2-x} was stable upon exposure in air for several months [65].

4.2.2 Oxidation Method

Except for reduction treatment of TiO_2 nanoparticles or Ti(IV) precursors, the preparation of TiO_{2-x} photocatalysts was realized by partial oxidation of low-oxidation-state titanium compounds, such as metallic Ti powder and Ti(II) and Ti

(III) precursors. The oxidation process should be well controlled, and complete oxidation of these low-oxidation-state titanium compounds should be avoided.

4.2.2.1 Starting from Metallic Ti Powder

According to the best of our knowledge, Zuo et al. firstly reported rutile TiO_{2-x} photocatalysts with active {111} and {110} facets via a simple hydrothermal method starting from the mixture of titanium powder and hydrochloric acid [66]. The Ti^{3+} species in the final TiO_{2-x} samples resulted from the reduction of Ti^{4+} by H_2 gas, which was induced by the reaction between Ti and HCl at the high vapor pressure and high-temperature atmosphere. The desired dual {111} and {110} were reported to collect holes and electrons, respectively, which resulted into the separation of photogenerated electron and hole pairs. Liu et al. reported similar experiment starting with Ti and HCl, however, with additional injection of HF solution [67]. Because of the adding of HF solution, the obtained Ti^{3+} species were further stabilized by F-termination of the TiO_{2-x} samples. In addition, the adding of HF facilitated to the formation of {001} facets of TiO_{2-x} catalysts [67, 68], owing to the low energy of (001) surface decreased by high Ti-F bonding energy [69].

4.2.2.2 Starting Form Ti(II) Precursors

Using TiH_2 as the starting material for the synthesis of TiO_{2-x} photocatalysts was investigated by Huang and co-workers [70–73]. Hydrothermal treatment was employed on the mixture of TiH_2 and H_2O_2 aqueous solution [70]. The using of TiH_2 can guarantee a high doping concentration of Ti^{3+} in the final TiO_{2-x} samples. The choosing of H_2O_2 as the oxidation agent is the key to the innovation because the oxidation process could be controlled and new chemical contaminations could be avoided [74]. During the hydrothermal process, TiH_2 was oxidized by H_2O_2, and a "solid interface diffusion–redox" reaction was proposed by Liu et al. for the generation of Ti^{3+} species [70]. Apart from hydrothermal treatment, heating at high temperature under argon atmosphere was also employed on the gel dried from the mixture of TiH_2 and H_2O_2 solution [71, 74], as shown in Fig. 4.7. Ar–gas protection during the thermal treatment process is necessary for the maintaining of Ti^{3+} species. Otherwise, the obtained Ti^{3+} species will be oxidized to Ti^{4+} by oxygen while annealing in air [74]. Xin et al. combined the hydrothermal and post-annealing processes together for the synthesis of TiO_{2-x} catalyst, and they demonstrated that the distribution of surface/subsurface defects could be controlled by the further annealing of the as-prepared TiO_{2-x} photocatalysts [75]. Zhu et al. prepared TiO_{2-x} via thermal treatment on the mixture of commercial Evonik P25 and TiH_2 under vacuum [76]. In this case, active hydrogen atoms were proposed to release from the decomposition of TiH_2 while heating over 400 °C during this process, resulting into the corporation of H on the final obtained TiO_{2-x} samples, rather than the generation of Ti^{3+} species [76].

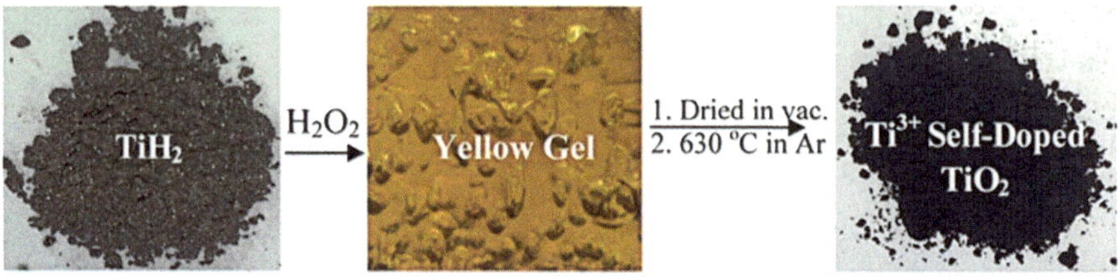

Fig. 4.7 Process of the reaction between TiH_2 and H_2O_2 to synthesize TiO_{2-x} nanomaterials and the corresponding photos. (Reprinted with permission from Ref. [74], copyright 2013 American Chemical Society)

4.2.2.3 Starting from Ti(III) Precursors

In order to introduce Ti^{3+} to TiO_2 catalyst, Liu et al. mixed Ti_2O_3 and TiO_2 powders together and then heated them at high temperature (900 °C) [77]. The ratio between Ti_2O_3 and TiO_2, which is related to the degree of oxidation states of the TiO_{2-x} samples, is the key for the successful doping of Ti^{3+}. Liu et al. and Wang et al. also chose Ti_2O_3 as the raw material to provide high concentration of Ti^{3+} into the bulk of TiO_{2-x}. Ti_2O_3 was firstly reacted with strong alkaline solution via hydrothermal process, and then the obtained titanate was annealed under argon atmosphere to avoid Ti^{3+} oxidation.

Fang et al. obtained stable deep-blue-colored TiO_{2-x} nanocrystals (Fig. 4.8a, b) using $TiCl_3$ as the starting material [78]. In order to avoid the oxidation of the obtained Ti^{3+} species, hydrofluoric acid was injected during the preparation process, which also contributed to the formation of disordered layers outside the TiO_{2-x} materials. Having Ti^{3+} species located in the bulk of TiO_{2-x} materials avoided the oxidation of Ti^{3+} and contributed to the stabilization of these catalysts. As a result, the obtained TiO_{2-x} photocatalysts showed enhanced photocatalytic CO_2 reduction for selective CH_4 evolution under solar light irradiation, owning to the improved solar light absorption and fast electron-hole separation induced by special dual {001}-{101} facet exposure (Fig. 4.8c).

Zhu et al. fabricated TiO_{2-x} nanoparticles via solvothermal method on the mixture of $TiCl_3$ and TiF_4 solutions [79]. The introduction of Ti^{3+} species originated from the injection of $TiCl_3$, and the inhibition of Ti^{3+} oxidation was achieved by Le Chatelier's principle [79]. Similarly, $TiCl_3$ and $(NH_4)_2TiF_6$ were used as the source of Ti^{3+} and Ti^{4+}, to synthesize TiO_{2-x} nanomaterials [80].

4.3 Properties of TiO_{2-x} Photocatalysts

The structure of TiO_{2-x} photocatalysts can be regarded as the incorporation of H atoms or the removal of oxygen atoms on the surface and/or in the bulk of TiO_2 photocatalysts. According to the result of DFT calculations, Mo et al. reported that in

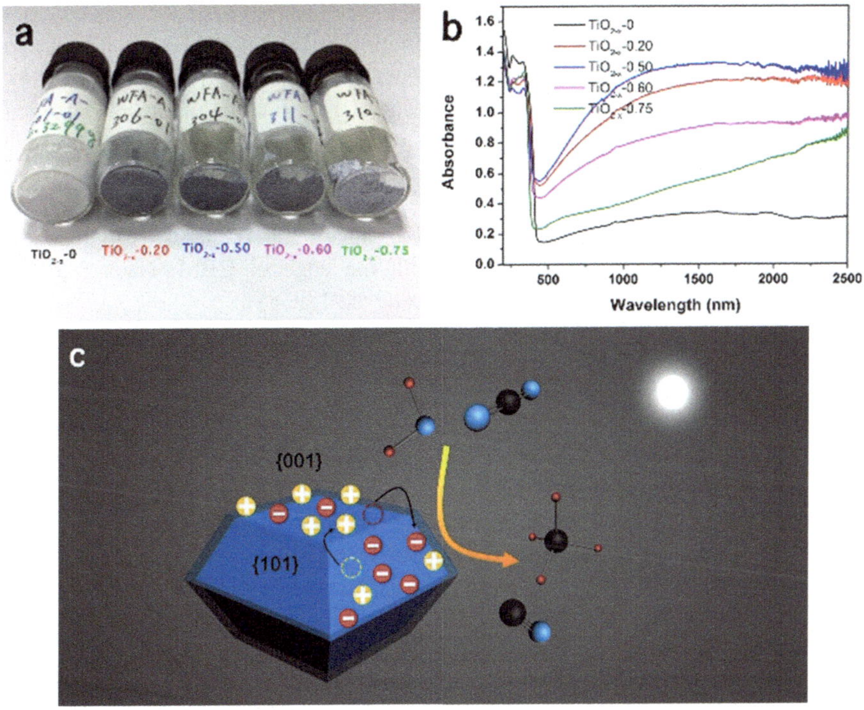

Fig. 4.8 (**a**) Photos of the TiO_{2-x} samples treated with different amount of HF solution and (**b**) their corresponding light absorption spectra. The catalysts presented strong solar light absorption among the visible and near-IR regions. (**c**) Schematic illustration of charge transfer on fluoride-treated TiO_{2-x} with {001} and {101} facet exposure and its photocatalytic CO_2 reduction for CH_4 and CO evolution process under solar light irradiation. The black, blue, and red balls represent carbon, oxygen, and hydrogen atoms, respectively. (Reprinted from Ref. [78] by permission of The Royal Society of Chemistry)

TiO_{2-x} thermally treated by H_2 gas, the H atoms would like to occupy the vacancies which were left by the oxygen, and then the H atoms became negatively charged [56]. However, Chen et al. demonstrated that the gas hydrogenation contributed to the formation of disordered layers outside the crystallized TiO_2 core, rather than Ti^{3+} generation [12]. The obtained disordered layers after hydrogenation lead to its enhanced hydrogen mobility and electronic structural changes in TiO_{2-x} [81].

It was reported that hydrogenation through thermal treatment under H_2 gas leads to the formation of midgap states above the VBM [15, 81, 82]. Besides, the formation of impurity states below the conduction band minimum (CBM) is also observed in TiO_{2-x} catalysts [32, 51]. In general, the tail below CBM is attributed to the formation of Ti^{3+} species and oxygen vacancies, while the uplift of VBM contributed from the TiO_{2-x} disordered layers [21, 83].

However, Liu et al. considered that the introduction of Ti^{3+} resulted into the generation of isolated states between the forbidden gap, and the bandgap did not change although the solar light absorption enhanced after Ti^{3+} doping [77]. This phenomenon could be caused by the low doping concentration of Ti^{3+} species. Zhu et al. and Liu et al. regarded that the concentration of Ti^{3+} should be high enough to introduce a continuous impurity state below the conduction band of TiO_{2-x} [79, 84], because low concentration of Ti^{3+} doping leads to the formation of localized oxygen defect states, which is harmful for the mobility of electron and photocatalytic activities of the catalysts [79]. That is, Ti^{3+} species and oxygen vacancies in a proper concentration would promote the charge separation of the photogenerated carriers [21]. With continuous formation of oxygen vacancies and Ti^{3+} doping, the isolated impurity states will enhance and mix with the edge of CBM, leading to the narrowing of the bandgap of TiO_{2-x} photocatalysts [79, 85].

It is well known that Ti^{3+} species on the surface or top few layers of TiO_2 are usually not stable, and they are easy to be oxidized by oxygen in air or in water [30, 62]. Surface oxygen vacancies could be healed by treating reduced TiO_2 with oxygen exposure [86]. However, XPS characterization can only reach few layers of TiO_2 surface, which limits the depth of its detection to ~10 nm. In order to test the information of titanium chemical states in the bulk of TiO_{2-x}, an argon sputtering treatment was introduced to remove the top few layer of TiO_2 before the XPS measurement was carried out [53]. TiO_{2-x} shows a blue shift of the locations of Ti $2p_{3/2}$ and Ti $2p_{1/2}$ in XPS spectra, because of the introduction of Ti^{3+} species [20, 40, 87]. The generation of oxygen vacancies could also be evidenced by O 1 s XPS tests [51, 64, 80].

EPR is efficient for the detection of Ti oxidation states of TiO_2, especially in the bulk of TiO_2. The g-factors, which are independent of the microwave frequency in EPR characterization, are usually used for the identification of a compound. Generally speaking, the EPR absorption of a solid with an isotropic magnetic moment shows a sharp peak, where the peak location in the absorption derivative spectra is associated with the g tensor ($g_x = g_y = g_z$) [87], as shown in Fig. 4.9, left side. The EPR absorption of a solid with an axial magnetic moment will show doublet peaks, which can be fit into two peaks in the absorption derivative spectra corresponding to $g_\perp$ ($g_x = g_y$) and $g_\parallel$ (g_z) [87], Fig. 4.9, right side. EPR signals with $g_\perp = 1.992$ and $g_\parallel = 1.962$ were reported as the typical Ti^{3+} centers in the lattice of anatase TiO_2 photocatalysts [88–91]. EPR signals with $g_\perp = 1.976$ and $g_\parallel = 1.945$ have been reported by massive works, which should be attributed to Ti^{3+} ions in the bulk or subsurface of TiO_{2-x} [15, 26, 70, 74, 87].

In addition, other characterization techniques are introduced for the identification of structural and chemical changes in TiO_{2-x} catalysts. With the help of STM, oxygen vacancies can be visually displayed [61, 92, 93]. And the oxidation state of Ti and concentration of Ti^{3+} species are studied by X-ray absorption near-edge structure (XANES) [81, 94], extended X-ray absorption fine structures (EXAFS) [94], and superconducting quantum interference device (SQUID) measurement [45], respectively. The structural information for TiO_{2-x} can be further investigated by Raman spectroscopy [16, 51, 95].

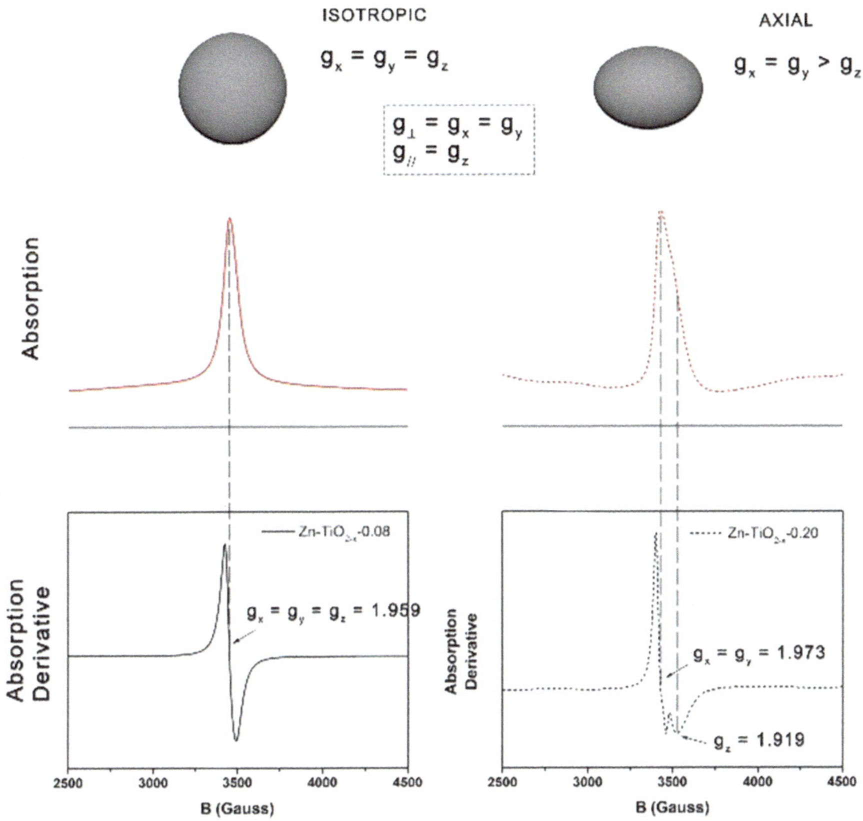

Fig. 4.9 Schematic illustration of the g tensor and the corresponding EPR spectra. Two common TiO_2 solid body models associated with isotropic and axial magnetic moments are shown at the top. (Reprinted with the permission from Ref. [87], Copyright 2017 American Chemical Society)

4.4 Applications of TiO_{2-x} Photocatalysts

Based on its extended solar light absorption, narrowed bandgap, and improved photogenerated electron and hole pairs, TiO_{2-x} shows enhanced photocatalytic performance compared to pure TiO_2 materials. It is reported that TiO_{2-x} is well-suited for the photodegradation of organic pollutants, such as phenol [28, 96], bisphenol A [97], formic acid [87], rhodamine B [19, 98], methyl orange [23, 47], methylene blue [99], reactive black 5 [46], 4-chlorophenol [100], etc. The performance of photocatalytic hydrogen production from water splitting improves after the doping of Ti^{3+} or hydrogenation of TiO_2 (Fig. 4.10a) [101–103]. TiO_{2-x} materials show photocatalytic H_2 evolution activities under visible light, while pure TiO_2 catalysts do not show any visible light-driven activities (Fig. 4.10b) [15]. Besides, TiO_{2-x} shows high activities for photocatalytic reduction of CO_2 to CH_4 [78, 104, 105], because oxygen vacancies promote the trapping of CO_2 molecules [106] and

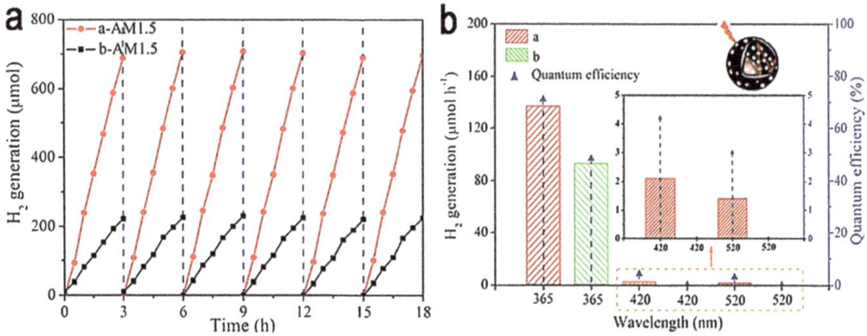

Fig. 4.10 (**a**) Photocatalytic hydrogen evolution from water splitting of mesoporous TiO_{2-x} hollow spheres (sample a) and mesoporous TiO_2 hollow spheres (sample b). (**b**) The photocatalytic H_2 generation rates under single-wavelength light irradiation and the corresponding quantum efficiency (QE) of the catalysts. (Reproduced from Ref. [15] by permission of The Royal Society of Chemistry)

the formation of CO_2^- species [104]. It means that the surface oxygen vacancies facilitate the adsorption and chemical activation of CO_2 molecules [107]. Furthermore, TiO_{2-x} also shows high PEC performance under solar light irradiation, such as linear sweep voltammograms [45, 108], photocurrent responses [30, 109–111], electrochemical impedance spectroscopy (EIS) Nyquist plots [83, 112, 113], Mott–Schottky plots [13, 25], incident photon-to-current conversion efficiency (IPCE) [34, 114], cyclic voltammograms (CV) [95, 114], galvanostatic charge/discharge voltage profiles [115, 116], and specific discharge capacity, which makes it a good candidate in applications of PEC water splitting [13, 48, 117], DSSCs [49, 118, 119], lithium-ion capacitor (LIC) [120], memory capacitor [121], LIBs [40, 115, 116], sodium-ion batteries (SIBs) [122, 123], and oxygen reduction reaction (ORR) [37]. TiO_{2-x} catalysts have been proven to be efficient energy conversion and storage materials.

4.5 Modification on TiO_{2-x} Photocatalysts

Although the hydrogenation or the self-doping of TiO_2 has gained a lot of improvement of its photocatalytic performance compared to pure TiO_2, requirement for the further enhancement of its activities is raised in industrial applications. Conventional methods which are usually used for the modification of TiO_2 materials are also introduced to further improve the activities of TiO_{2-x} catalysts, such as doping with nonmetal elements, grafting with metals, compositing with other materials, designing of ordered morphology, special facet exposure, etc.

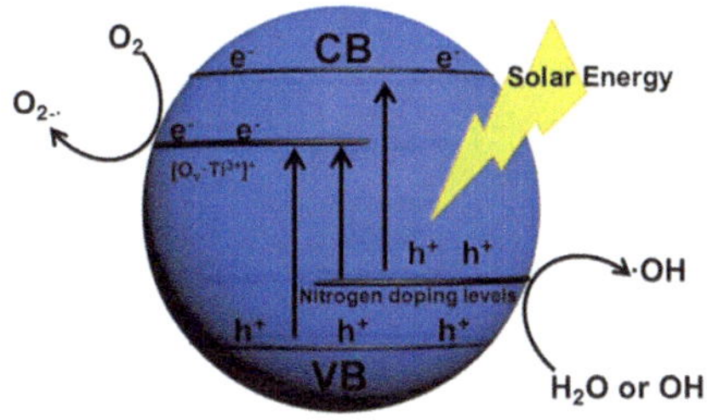

Fig. 4.11 Proposed mechanism of electron excitation in nitrogen-doped TiO_{2-x} photocatalyst under solar light irradiation. Ti^{3+} impurity states and N doping levels are proposed to generate within the bandgap of TiO_2, leading to the bandgap narrowing of the catalysts. (Reprinted from Ref. [30], Copyright 2016, with permission from Elsevier)

4.5.1 TiO_{2-x} Doped with Nonmetal Elements

Nitrogen is one of the most investigated elements for the doping of TiO_2. As a nonmetal element, nitrogen is beneficial for the bandgap narrowing and enhanced photocatalytic activities under visible light [124]. The combination of Ti^{3+} introduction and N doping will lead to the formation of Ti^{3+} impurity states below the CBM [85] and upshift of VBM [125, 126], respectively, acting as an efficient approach for further modification of reduced TiO_2 nanomaterials (Fig. 4.11).

Hoang et al. fabricated TiO_{2-x} with N doping by heating treatment at H_2 and NH_3 atmosphere [127], and it was demonstrated that the formation of bulk N species was more favorable in the presence of Ti^{3+}. As proposed by Fang et al. [30], the generation of oxygen vacancies in the bulk of TiO_2 will accelerate the migration of N species from the surface to the bulk of TiO_2. The replacement of oxygen defects with nitrogen results in the decrease of surface Ti–O–N bonds and increase of Ti–N–O bonds, evidenced by XPS spectra [30, 128]. This phenomenon was also observed by Hoang et al. [127]; the broadening and blue-shifted XPS N 1 s feature indicated the electronic interaction between Ti^{3+} and substitutional N in TiO_2. The interaction between substitutional N and Ti^{3+} species in TiO_2 has been addressed, and it is responsible for the enhanced organic pollutant removal [30, 129], photocatalytic hydrogen evolution [130], and high water photooxidation performance [127].

NH_3 gas is usually used as the source providing nitrogen and hydrogen active species during the preparation of N-doped TiO_{2-x} materials. This process could be achieved by heating at relatively high temperature [129] or nonthermal dielectric barrier discharge (DBD) plasma treatment [131]. Besides, as a traditional reducing agent, hydrazine hydrate consists of nitrogen, hydrogen, and oxygen elements. Thus, treating with hydrazine hydrate could induce simultaneously the formation of Ti^{3+} species and doping of N in the bulk of TiO_2 [26].

In addition, other nitrogen sources such as tripolycyanamide [130], diethylenetriamine [132], and ethanolamine [126] were also reported in the synthesis of N-doped TiO_{2-x} materials. Using diethylenetriamine as the nitrogen source, Li et al. synthesized N-doped 3D hierarchical structured TiO_{2-x} catalysts composed of

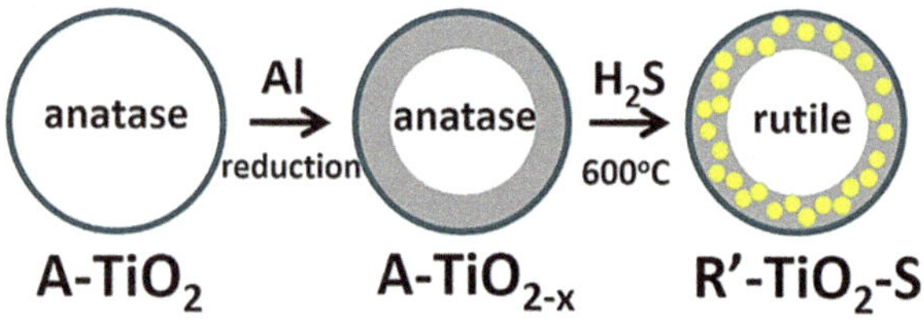

Fig. 4.12 Schematic illustration of the synthesis of S-doped TiO_{2-x} synthesized by Al reduction process. (Reprinted with the permission from Ref. [33]. Copyright 2013 American Chemical Society)

ultrathin nanosheets, which not only supply high surface area for the adsorption of the pollutants but also promote multi-reflection of the light [132]. Except for nitrogen source, tripolycyanamide also acted as structure-directing agent, leading to the formation of flower-like hierarchical architecture of the obtained N-doped TiO_{2-x} nanoplates [130].

Yang et al. reduced TiO_2 with molten aluminum in a two-zone furnace and subsequently heated the obtained TiO_{2-x} in H_2S atmosphere [33]. After sulfidation, both Ti^{3+} and S were doped on the surface disordered layer of the catalyst (Fig. 4.12), contributing to the extended solar light absorption from UV to near-IR regions and improved photodegradation activities and photoelectrochemical properties. By adjusting the nonmetal sources, this method could be applied to other nonmetal incorporation into core-shell structured TiO_{2-x} such as H, N, and I [133].

Feng et al. reported boron-doped TiO_{2-x} with high and stable Ti^{3+} species located at the surface disordered layer [134]. And it was demonstrated that B doping could promote the formation and stabilization of Ti^{3+} species due to the coupling of B atoms to the neighboring lattice Ti and O atoms. The doping of boron was stressed to improve the visible light absorption, narrow the bandgap, and promote electron-hole separation of TiO_2 photocatalyst. However, Sayed et al. reported that TiO_{2-x} co-doped with N and B showed lower photoactivity than the TiO_{2-x} catalyst doped with single N element [126]. Besides, Xing et al. reported that F substitution for oxygen vacancies efficiently narrowed the bandgap of TiO_2 by introducing impurity states among the bandgap of TiO_2, and the recombination of electrons and holes was inhibited due to prolonged lifetime of charges captured by Ti^{3+}–F levels [135].

4.5.2 *TiO_{2-x} Grafted with Metals*

Grafting with noble or non-noble metal on TiO_2 photocatalysts is a conventional approach to improve their photocatalytic performance. For example, noble metals such as Pt nanoparticles are usually used as cocatalysts with TiO_2 in the photocatalytic H_2 evolution test from water splitting, due to the effect of surface plasmon resonance (SPR), and promoted separation of photogenerated electrons and

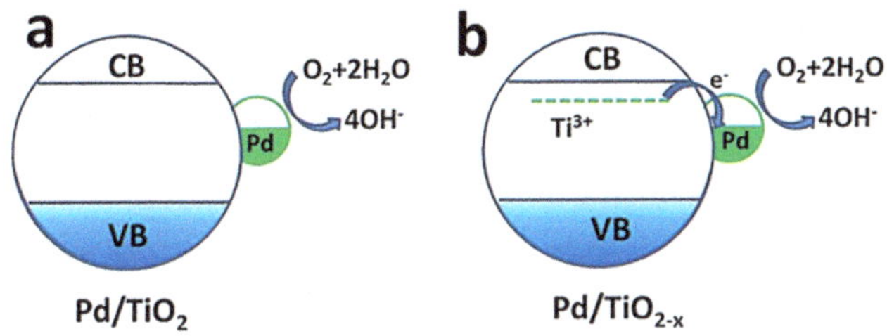

Fig. 4.13 Electronic charge transfer in ORR catalytic process of Pd/TiO_2 (**a**) and Pd/TiO_{2-x} (**b**). The electron transfer between Pd and Ti^{3+} species leads to an electron-rich Pd surface. (Reprinted with the permission from Ref. [137]. Copyright 2016 American Chemical Society)

holes. Most semiconductor photocatalysts do not exhibit a good photocatalytic performance without the loading of suitable noble metal cocatalyst.

It is reported that the loaded noble metals act as active sites for trapping electrons and protons [136]. The electron transfer between noble metal and Ti^{3+} species results in effect of strong metal-support interaction (SMSI) and leads to an electron-rich noble metal surface (Fig. 4.13), which finally enhances the catalytic performance of these materials [137]. The electron-rich effect of Pt on TiO_{2-x} was evidenced by low-temperature EPR [138]. And the charge transfer from defect sites on reduced TiO_2(110) to Au cluster was also evidenced by XPS [7]. Lian et al. reported that the Pt species in the framework of TiO_2 facilitated the photogenerated electron transfer from the bulk to the surface of TiO_2 through Pt^{n+}–O bonds [139]. Besides, DFT calculations show that on reduced TiO_2(110) surface, the presence of oxygen vacancies promotes the adsorption and diffusion of Au particles, and small Au particles are preferable on reduced TiO_2(110) [140].

The formation of oxygen vacancies on TiO_2 is favored with the loading of noble metal dopants, supported by DFT calculations [141]. It was reported that the Au loading induced the formation of oxygen vacancies in the Au–TiO_2 interface [24]. Bennett et al. reported that with the deposition of gold nanoclusters, a high concentration of Ti^{3+} species was generated on TiO_2 under X-ray and UV irradiation [142]. Similar like noble metal loading, the grafting of metallic Cu particles was also reported to promote the formation of oxygen vacancies in TiO_2 [143].

Noble metal nanomaterials are likely to accept the photogenerated electrons from TiO_2 catalysts. Therefore, while designing the Au-grafted TiO_{2-x} nanoparticles, the loading of Au could be achieved by the reduction of gold ions by electrons located at the oxygen vacancies. The electrons are captured by gold ions and then reduce the ions into gold nanoparticles, leading to a fast and spontaneous formation of Au nanoparticles on TiO_{2-x} [144]. Besides, Ti^{3+} species and oxygen vacancies were proved to be efficient for the in situ reduction of Pt^{4+} to form metallic Pt in the bulk of TiO_{2-x} [139].

Compared to other noble metals such as Au, Pt, or Pd, Ag nanoparticles are cheaper and easier to synthesize. Metallic Ag nanoparticles usually exist in the forms of Ag^0 and Ag_2O nanocrystalline on TiO_2 photocatalyst. The formation of Ag could be achieved by UV reduction [145], sonochemical deposition [146], or KBH_4 reduction [147]. The SPR effect of Ag, enhanced solar light absorption and promoted charge separation of Ag/TiO_{2-x} are responsible for the high visible-light-driven photocatalytic degradation of organic pollutants [147, 148].

In order to reduce the high cost of noble metals, researchers are seeking new approaches for the replacement with non-noble metals. Among all the non-noble metal elements, Cu is regarded as a competitive candidate to substitute noble metals and works as cocatalyst with TiO_2 photocatalysts, because of its low cost and considerable catalytic performance. In general, nanosized Cu nanoparticles are not stable, and they are easy to be oxidized to Cu_2O or CuO while being exposed in air. Zhao et al. synthesized TiO_{2-x}@Cu particles, in order to prevent the oxidation of metallic Cu in the photocatalytic reaction process [143]. However, it is also reported that the Cu^+ species are more active than Cu or Cu^{2+}, and the high photocatalytic performance of the Cu-grafted TiO_2 catalysts is related to the presence of stabilized Cu^+ species [149]. In addition, the generation of oxygen vacancies in the preparation process could promote the stabilization of Cu_2O species on the surface of TiO_2 [149].

Liu et al. ascribed the high photocatalytic CO_2 reduction performance of Cu/TiO_2 catalysts to the synergistic effect of Ti^{3+} species and Cu^+/Cu^0 couples, which were formed during the thermal treatment in H_2 atmosphere [150]. The formation of Ti^{3+} species contributes to the enhanced adsorption of CO_2 and improved separation of electrons and holes, while the Cu^+ generation also promotes the electron transfer during the photocatalytic reaction process. Besides, the Cu^+ species were also reported to be the main active phase for CO oxidation, where the Cu^+ species resulted from the reduction of Cu^{2+} by Ti^{3+} species on TiO_{2-x} [94].

4.5.3 TiO_{2-x} Composited with Carbon

Graphene is a kind of single-layered carbon material [151] and basic building block for creating other graphitic materials: wrapped into 0D fullerenes, rolled into 1D carbon nanotubes, and stacked into 3D graphite [152]. Recently, it has attracted plenty of interest for compositing with other semiconductors because of its superior conductivity property and unique structure.

Xing et al. synthesized TiO_{2-x}/graphene composites via a simple hydrothermal process using $NaBH_4$ as the reducing agent and also boron dopant source [99] (Fig. 4.14a). Electron transfer was achieved from TiO_{2-x} nanorods to graphene through Ti–O–C bonds. As the result, electrons were collected on the graphene, while holes were lying on the surface of TiO_{2-x}, promoting the efficient separation of photogenerated electrons and holes. Qiu et al. employed vacuum activation process to achieve self-doping of TiO_2, reduction of graphene, and the compositing of TiO_{2-x} on the surface of graphene simultaneously [90]. It was demonstrated that the self-doping of Ti^{3+} and the bonding between TiO_{2-x} and graphene were responsible for

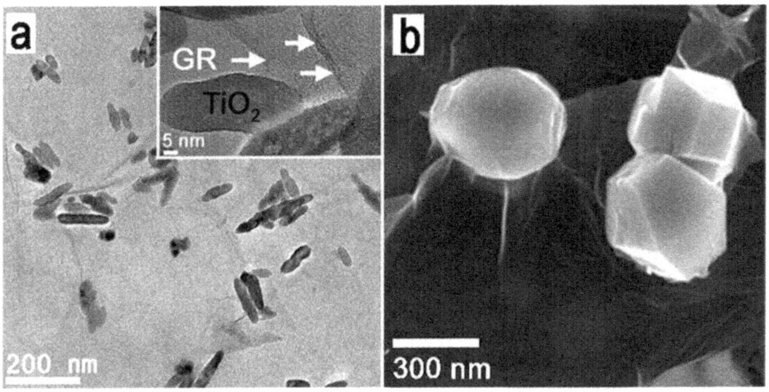

Fig. 4.14 (**a**) TEM images of TiO_{2-x} nanorods decorated on graphene sheets. Reprinted from Ref. [99], with kind permission from Nature Publishing Group. (**b**) SEM image of TiO_{2-x} nanocrystals wrapped by graphene. (Reprinted from Ref. [153], with kind permission from WILEY-VCH)

its high photodegradation activity of organic compounds and enhanced hydrogen evolution performance from water splitting under visible light irradiation.

Wang et al. wrapped graphene sheets on the surface of TiO_{2-x} nanocrystals with exposed {001} and {101} facets [153] (Fig. 4.14b). The wrapping was realized by the electrostatic attraction interaction between the positively charged TiO_2 (treated with 3-aminopropyltriethoxysilane to introduce amine groups) and negatively charged graphene oxide. Besides, it was reported that the formation of Ti–C bonds between TiO_2 and graphene could be promoted by laser ablation in liquid (LAL) method [154]. Considering that LAL also contributes to the self-reduction of TiO_2 nanoparticles, the synthesis of TiO_{2-x}/graphene composites could be realized by the one-step LAL method without any chemical reducing agent. In addition, a UV pre-catalytic microwave approach was also reported for the synthesis of TiO_{2-x}/graphene hybrids [155].

Due to the excellent charge transporting property of carbon nanotubes (CNTs), TiO_{2-x}/CNTs composite materials were fabricated [95]. The obtained composite was successfully applied as counter electrode in DSSCs. Fu et al. coated carbon on the surface of TiO_{2-x} to stabilize the formed Ti^{3+} species [156]. Besides, the co-doping of Ti^{3+} species and graphite-like carbon [157] or coke carbon [158] in TiO_2 photocatalysts was also reported.

4.5.4 TiO_{2-x} Composited with Other Compounds

Graphitic carbon nitride (g-C_3N_4) is a novel metal-free and environment-friendly material with excellent response to visible light. It was reported that the heterojunctions of g-C_3N_4 and TiO_2 can promote the separation of electrons and holes and accelerate the charge transfer between these two compounds [159]. Liao et al. confirmed that the efficient charge separation was caused by the Z-scheme process of g-C_3N_4/TiO_{2-x} composites [96]. Due to its excellent visible light response promoted by Ti^{3+} and g-C_3N_4, and efficient electron-hole separation and transportation, g-C_3N_4/TiO_{2-x} photocatalyst shows great performance of photodegradation of organic pollution and Cr(VI) reduction [160]. Other composite materials such as N-doped TiO_{2-x}/MoS_2 [161], 3D mesoporous TiO_{2-x}/MoS_2/TiO_{2-x} nanosheets [23], and Pd-MgNi_x nanospheres/TiO_{2-x} composite films [103] were also reported.

4.5.5 TiO_{2-x} with Ordered Morphology

TiO_{2-x} photocatalysts with ordered morphology were synthesized, such as hollow porous structure [116], mesoporous [162], core shell [163], yolk shell [164], nanocage [165], inverse opals [166, 167], etc. Hu et al. fabricated mesoporous TiO_{2-x} hollow spheres to enhance the solar light utilization by extended absorption edge by Ti^{3+} doping and multi-refraction of solar light within the hollow structure [15]. The unique hollow structure also allows better charge separation and transfer properties of TiO_{2-x}. Wang et al. demonstrated that the increased conductivity of TiO_{2-x} and its hollow porous structure were beneficial for lithium-ion electron diffusion while using TiO_{2-x} hollow porous sphere as the anode materials in LIBs [116]. Besides, yolk-shell structure of TiO_{2-x} contributes to the multiple reflection of solar light in the chamber and provides high surface area of the catalysts [164] and thus makes the TiO_{2-x} catalyst superior photocatalytic activities under visible light irradiation.

Qi et al. reported inverse-opal-structured TiO_{2-x} photocatalyst using polystyrene sphere (PS) as the hard template [166]. The introduction of Ti^{3+} was realized by vacuum activation process (Fig. 4.15). The intrinsic optical response was improved because of the slow light effect provided by the inverse opal structure. As the result, the improvement of its photocatalytic performance was achieved by the combination of physical route (slow light effect) and chemical route (Ti^{3+} self-doping). Xin et al. prepared TiO_{2-x} inverse opals by in situ H_2 reduction of the pristine TiO_2 inverse opals [167]. The combination of hydrogenation and ordered structure provides this material-enhanced solar light absorption and improved photocatalytic activities. In addition, well-shaped TiO_{2-x} film was reported to show superior solar desalination

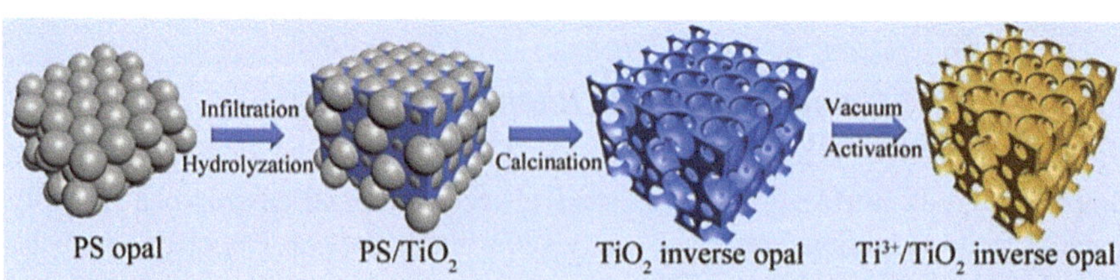

Fig. 4.15 Preparation process of inverse-opal-structured TiO_{2-x} photocatalysts. (Reprinted from Ref. [166], Copyright 2014, with permission from Elsevier)

performance, because the enhancement of solar light harvesting could increase the temperature localized at the water–air interface rather than heating the whole bulk of water [165].

4.5.6 TiO_{2-x} with Special Facets Exposed

The formation of high percentage Ti^{3+} in TiO_2 is a great challenge due to its instability in air, and the obtained Ti^{3+} species may also act as electron and hole recombination sites, which are harmful for the photocatalytic performance of the catalysts [68]. In order to solve these problems, Fang et al. successfully fabricated deep-blue TiO_{2-x} photocatalysts with dual {001} and {101} facets exposed, starting from the $TiCl_3$ precursor [78]. The spin intensity in EPR spectra reaches as high as 24.6×10^{22} spins per g. Besides, the partial oxidation of Ti^{3+} was controlled and stabilized by the adding of aqueous HF solution. In a dual {101}-{001} facets system, the holes are like to gather at the {001} facets, while the electrons prefer to migrate to the {101} facets, which contributes to the separation of photogenerated electrons and holes [78, 168, 169].

During the preparation of {001} facet-dominated TiO_2 nanomaterials, HF is usually injected as the structure-directing agent [78, 87, 131, 170]. The introduction of HF may promote the formation of oxygen vacancies. However, this effect is not strong enough to introduce high concentrated Ti^{3+} in the bulk of TiO_2 [171]. The subsequent hydrogenation of F-modified TiO_2 with exposed {001} facets by thermal treatment under high-pressure H_2 atmosphere could produce a large number of Ti^{3+} species and oxygen vacancies, resulting in enhanced solar light absorption and improved photocatalytic activities [172]. Chen et al. compared the effect of hydrogenation treatments on TiO_2 with exposed {001}, {110}, and {101} facets, respectively, and demonstrated that the electric field formed between the reduced subsurface and stoichiometric surface is the key for the high concentrated Ti^{3+} species and F^{1+} color centers of hydrogenated TiO_{2-x} with exposed {001} facets [173].

Li et al. synthesized TiO_2 nanosheets with exposed {101}-{001} facets first and then employed low-temperature plasma treatment to introduced Ti^{3+} and unique core-shell structure on the obtained TiO_2 [131]. By adjusting the working gases, N co-doping could be realized by choosing NH_3 as the plasma source [131]. In addition, other approaches for {001} facet-dominated TiO_{2-x} were also reported: metallic Zn [170] or Ti [68] powders were added in the HF-containing solution, acting as the reductants for the synthesis of TiO_{2-x}.

4.6 Summary and Outlook

In this chapter, the synthesis of TiO_{2-x} photocatalysts was carefully discussed and scientifically classified into two main categories, where the reactions take place under reducing or oxidizing atmosphere. These methods include reducing gas heat treatment, in situ reduction, metal reduction, vacuum activation treatment, hydrogen plasma reduction, electrochemical reduction, UV light irradiation treatment, and oxidation methods starting from Ti, Ti(II), or Ti(III) precursors. The obtained TiO_{2-x} photocatalysts, resulting from hydrogen incorporation or oxygen removal, are proved to achieve efficient photocatalytic performance such as photodegradation of organic compound, hydrogen generation from water splitting, photocatalytic CO_2 reduction, dye-sensitized solar cells, lithium-ion batteries, etc. The development of TiO_{2-x} materials has provided a promising way to solve the increasing environment and energy problems in the twenty-first century. In order to further improve the activities of TiO_{2-x} and better prepare this catalyst in industrial applications, some approaches are introduced for the modification of TiO_{2-x} materials, such as doping with nonmetal elements (N, S, I, B, and F), grafting with metals (Pt, Au, Pd, Ag, and Cu), compositing with other materials (graphene, CNTs, g-C_3N_4, and MoS_2), designing of ordered morphology (hollow porous, mesoporous, core shell, yolk shell, nanocage, and inverse opal structures), and special facet exposure ({101}, {001}, and {110} facets).

However, there are still new challenges we are facing on the investigation of TiO_{2-x} material and its industrial applications. Although we have already studied the incorporation of hydrogen or the removal of oxygen at atomic levels with the help of highly advanced instruments, the process of the reaction is now still difficult to record and visualize. The low solar utilization and the high cost of solar energy systems also greatly limit the applications of this material. Cleaner, higher efficient, and more affordable systems are still greatly required in the future.

References

1. Fujishima A, Honda K (1972) Electrochemical photolysis of water at a semiconductor electrode. Nature 238(5358):37–38
2. Komaguchi K, Nakano H, Araki A, Harima Y (2006) Photoinduced electron transfer from anatase to rutile in partially reduced TiO_2 (P-25) nanoparticles: an ESR study. Chem Phys Lett 428(4–6):338–342
3. Diebold U, Anderson JF, Ng KO, Vanderbilt D (1996) Evidence for the tunneling site on transition-metal oxides: TiO_2(110). Phys Rev Lett 77(7):1322–1325
4. Cronemeyer DC (1959) Infrared absorption of reduced rutile TiO_2 single crystals. Phys Rev 113(5):1222–1226
5. Epling WS, Peden CHF, Henderson MA, Diebold U (1998) Evidence for oxygen adatoms on TiO_2 (110) resulting from O_2 dissociation at vacancy sites. Surf Sci 412-413(0):333–343
6. Di Valentin C, Pacchioni G, Selloni A (2009) Reduced and n-type doped TiO_2: nature of Ti^{3+} species. J Phys Chem C 113(48):20543–20552

7. Jiang Z, Zhang W, Jin L, Yang X, Xu F, Zhu J, Huang W (2007) Direct XPS evidence for charge transfer from a reduced rutile $TiO_2(110)$ surface to Au clusters. J Phys Chem C 111 (33):12434–12439
8. Deskins NA, Rousseau R, Dupuis M (2011) Distribution of Ti^{3+} surface sites in reduced TiO_2. J Phys Chem C 115(15):7562–7572
9. Petrik NG, Zhang Z, Du Y, Dohnálek Z, Lyubinetsky I, Kimmel GA (2009) Chemical reactivity of reduced $TiO_2(110)$: the dominant role of surface defects in oxygen chemisorption. J Phys Chem C 113(28):12407–12411
10. Diebold U, Lehman J, Mahmoud T, Kuhn M, Leonardelli G, Hebenstreit W, Schmid M, Varga P (1998) Intrinsic defects on a $TiO_2(110)(1\times1)$ surface and their reaction with oxygen: a scanning tunneling microscopy study. Surf Sci 411(1–2):137–153
11. Fang W, Xing M, Zhang J (2017) Modifications on reduced titanium dioxide photocatalysts: a review. J Photochem Photobiol C 32:21–39
12. Chen X, Liu L, Yu P, Mao S (2011) Increasing solar absorption for photocatalysis with black hydrogenated titanium dioxide nanocrystals. Science 331(6018):746–750
13. Wang G, Wang H, Ling Y, Tang Y, Yang X, Fitzmorris RC, Wang C, Zhang J, Li Y (2011) Hydrogen-treated TiO_2 nanowire arrays for photoelectrochemical water splitting. Nano Lett 11 (7):3026–3033
14. Cai J, Wang Y, Zhu Y, Wu M, Zhang H, Li X, Jiang Z, Meng M (2015) In situ formation of disorder-engineered $TiO_2(B)$-anatase heterophase junction for enhanced photocatalytic hydrogen evolution. ACS Appl Mater Inter 7(45):24987–24992
15. Hu W, Zhou W, Zhang K, Zhang X, Wang L, Jiang B, Tian G, Zhao D, Fu H (2016) Facile strategy for controllable synthesis of stable mesoporous black TiO_2 hollow spheres with efficient solar-driven photocatalytic hydrogen evolution. J Mater Chem A 4(19):7495–7502
16. Lu H, Zhao B, Pan R, Yao J, Qiu J, Luo L, Liu Y (2014) Safe and facile hydrogenation of commercial Degussa P25 at room temperature with enhanced photocatalytic activity. RSC Adv 4(3):1128–1132
17. Xing M, Fang W, Nasir M, Ma Y, Zhang J, Anpo M (2013) Self-doped Ti^{3+}-enhanced TiO_2 nanoparticles with a high-performance photocatalysis. J Catal 297(0):236–243
18. Ren R, Wen Z, Cui S, Hou Y, Guo X, Chen J (2015) Controllable synthesis and tunable photocatalytic properties of Ti^{3+}-doped TiO_2. Sci Rep 5:10714
19. Fang W, Xing M, Zhang J (2014) A new approach to prepare Ti3+ self-doped TiO2 via NaBH4 reduction and hydrochloric acid treatment. Appl Catal B 160(0):240–246
20. Zhao Z, Zhang X, Zhang G, Liu Z, Qu D, Miao X, Feng P, Sun Z (2015) Effect of defects on photocatalytic activity of rutile TiO_2 nanorods. Nano Res 8(12):4061–4071
21. Tan H, Zhao Z, Niu M, Mao C, Cao D, Cheng D, Feng P, Sun Z (2014) A facile and versatile method for preparation of colored TiO_2 with enhanced solar-driven photocatalytic activity. Nanoscale 6(17):10216–10223
22. Zhang H, Xing Z, Zhang Y, Li Z, Wu X, Liu C, Zhu Q, Zhou W (2015) Ni^{2+} and Ti^{3+} co-doped porous black anatase TiO_2 with unprecedented-high visible-light-driven photocatalytic degradation performance. RSC Adv 5(129):107150–107157
23. Liu X, Xing Z, Zhang H, Wang W, Zhang Y, Li Z, Wu X, Yu X, Zhou W (2016) Fabrication of 3D mesoporous black $TiO_2/MoS_2/TiO_2$ nanosheets for visible-light-driven photocatalysis. ChemSusChem 9(10):1118–1124
24. Ma C, Pang G, He G, Li Y, He C, Hao Z (2016) Layered sphere-shaped TiO_2 capped with gold nanoparticles on structural defects and their catalysis of formaldehyde oxidation. J Environ Sci 39:77–85
25. Mao C, Zuo F, Hou Y, Bu X, Feng P (2014) In situ preparation of a Ti^{3+} self-doped TiO_2 film with enhanced activity as photoanode by N_2H_4 reduction. Angew Chem 126 (39):10653–10657
26. Cheng X, Cheng Q, Li B, Deng X, Li J, Wang P, Zhang B, Liu H, Wang X (2015) One-step construction of N/Ti^{3+} codoped TiO_2 nanotubes photoelectrode with high photoelectrochemical and photoelectrocatalytic performance. Electrochim Acta 186:442–448

27. Su J, Zou X, Zou Y, Li G, Wang P, Chen J (2013) Porous titania with heavily self-doped Ti^{3+} for specific sensing of CO at room temperature. Inorg Chem 52(10):5924–5930
28. Xing M, Zhang J, Chen F, Tian B (2011) An economic method to prepare vacuum activated photocatalysts with high photo-activities and photosensitivities. Chem Commun 47 (17):4947–4949
29. Lu G, Linsebigler A, Yates JT (1994) Ti^{3+} defect sites on TiO_2(110): production and chemical detection of active sites. J Phys Chem 98(45):11733–11738
30. Fang W, Zhou Y, Dong C, Xing M, Zhang J (2016) Enhanced photocatalytic activities of vacuum activated TiO_2 catalysts with Ti^{3+} and N co-doped. Catal Today 266:188–196
31. Zhou Y, Liu Y, Liu P, Zhang W, Xing M, Zhang J (2015) A facile approach to further improve the substitution of nitrogen into reduced TiO_{2-x} with an enhanced photocatalytic activity. Appl Catal B 170(0):66–73
32. Wang Z, Yang C, Lin T, Yin H, Chen P, Wan D, Xu F, Huang F, Lin J, Xie X, Jiang M (2013) Visible-light photocatalytic, solar thermal and photoelectrochemical properties of aluminium-reduced black titania. Energy Environ Sci 6(10):3007–3014
33. Yang C, Wang Z, Lin T, Yin H, Lü X, Wan D, Xu T, Zheng C, Lin J, Huang F, Xie X, Jiang M (2013) Core-shell nanostructured "Black" rutile titania as excellent catalyst for hydrogen production enhanced by sulfur doping. J Am Chem Soc 135(47):17831–17838
34. Cui H, Zhao W, Yang C, Yin H, Lin T, Shan Y, Xie Y, Gu H, Huang F (2014) Black TiO_2 nanotube arrays for high-efficiency photoelectrochemical water-splitting. J Mater Chem A 2 (23):8612–8616
35. Zheng J, Ji G, Zhang P, Cao X, Wang B, Yu L, Xu Z (2015) Facile aluminum reduction synthesis of blue TiO_2 with oxygen deficiency for lithium-ion batteries. Chemistry 21 (50):18309–18315
36. Zheng Z, Huang B, Meng X, Wang J, Wang S, Lou Z, Wang Z, Qin X, Zhang X, Dai Y (2013) Metallic zinc- assisted synthesis of Ti^{3+} self-doped TiO_2 with tunable phase composition and visible-light photocatalytic activity. Chem Commun 49(9):868–870
37. Pei D, Gong L, Zhang A, Zhang X, Chen J, Mu Y, Yu H (2015) Defective titanium dioxide single crystals exposed by high-energy {001} facets for efficient oxygen reduction. Nat Commun 6:8696
38. Fu R, Gao S, Xu H, Wang Q, Wang Z, Huang B, Dai Y (2014) Fabrication of Ti^{3+} self-doped TiO_2(A) nanoparticle/TiO_2(R) nanorod heterojunctions with enhanced visible-light-driven photocatalytic properties. RSC Adv 4(70):37061–37069
39. Zhao Z, Tan H, Zhao H, Lv Y, Zhou LJ, Song Y, Sun Z (2014) Reduced TiO_2 rutile nanorods with well-defined facets and their visible-light photocatalytic activity. Chem Commun 50 (21):2755–2757
40. Chen J, Song W, Hou H, Zhang Y, Jing M, Jia X, Ji X (2015) Ti^{3+} self-doped dark rutile TiO_2 ultrafine nanorods with durable high-rate capability for lithium-ion batteries. Adv Funct Mater 25(43):6793–6801
41. Sinhamahapatra A, Jeon JP, Yu JS (2015) A new approach to prepare highly active and stable black titania for visible light-assisted hydrogen production. Energy Environ Sci 8 (12):3539–3544
42. Kitamura T, Shibata K, Takeda K (1993) In-flight reduction of Fe_2O_3, Cr_2O_3, TiO_2 and Al_2O_3 by Ar-H_2 and Ar-CH_4 plasma. ISIJ Int 33(11):1150–1158
43. Bullard D, Lynch D (1997) Reduction of titanium dioxide in a nonequilibrium hydrogen plasma. Metall Mater Trans B Process Metall Mater Process Sci 28(6):1069–1080
44. Palmer RA, Doan TM, Lloyd PG, Jarvis BL, Ahmed NU (2002) Reduction of TiO_2 with hydrogen plasma. Plasma Chem Plasma Process 22(3):335–350
45. Lepcha A, Maccato C, Mettenbörger A, Andreu T, Mayrhofer L, Walter M, Olthof S, Ruoko TP, Klein A, Moseler M, Meerholz K, Morante JR, Barreca D, Mathur S (2015) Electrospun black titania nanofibers: influence of hydrogen plasma-induced disorder on the electronic structure and photoelectrochemical performance. J Phys Chem C 119(33):18835–18842

46. An HR, Park SY, Kim H, Lee CY, Choi S, Lee SC, Seo S, Park EC, Oh YK, Song CG, Won J, Kim YJ, Lee J, Lee HU, Lee YC (2016) Advanced nanoporous TiO_2 photocatalysts by hydrogen plasma for efficient solar-light photocatalytic application. Sci Rep 6:29683
47. Wang Z, Yang C, Lin T, Yin H, Chen P, Wan D, Xu F, Huang F, Lin J, Xie X, Jiang M (2013) H-doped black titania with very high solar absorption and excellent photocatalysis enhanced by localized surface plasmon resonance. Adv Funct Mater 23(43):5444–5450
48. Tian Z, Cui H, Zhu G, Zhao W, Xu J, Shao F, He J, Huang F (2016) Hydrogen plasma reduced black TiO_2-B nanowires for enhanced photoelectrochemical water-splitting. J Power Sources 325:697–705
49. Kim HJ, Kim J, Hong B (2013) Effect of hydrogen plasma treatment on nano-structured TiO_2 films for the enhanced performance of dye-sensitized solar cell. Appl Surf Sci 274:171–175
50. Siuzdak K, Szkoda M, Lisowska-Oleksiak A, Karczewski J, Ryl J (2016) Highly stable organic–inorganic junction composed of hydrogenated titania nanotubes infiltrated by a conducting polymer. RSC Adv 6(39):33101–33110
51. Panomsuwan G, Watthanaphanit A, Ishizaki T, Saito N (2015) Water-plasma-assisted synthesis of black titania spheres with efficient visible-light photocatalytic activity. Phys Chem Chem Phys 17(21):13794–13799
52. Zhu W, Wang C, Chen J, Li Y, Wang J (2014) Enhanced field emission from Ti^{3+} self-doped TiO_2 nanotube arrays synthesized by a facile cathodic reduction process. Appl Surf Sci 301:525–529
53. Zhang Z, Hedhili MN, Zhu H, Wang P (2013) Electrochemical reduction induced self-doping of Ti^{3+} for efficient water splitting performance on TiO_2 based photoelectrodes. Phys Chem Chem Phys 15(37):15637–15644
54. Zhang Z, Tan X, Yu T, Jia L, Huang X (2016) Time-dependent formation of oxygen vacancies in black TiO_2 nanotube arrays and the effect on photoelectrocatalytic and photoelectrochemical properties. Int J Hydrog Energy 41(27):11634–11643
55. Swaminathan J, Subbiah R, Singaram V (2016) Defect-rich metallic titania ($TiO_{1.23}$)–an efficient hydrogen evolution catalyst for electrochemical water splitting. ACS Catal 6 (4):2222–2229
56. Mo LB, Wang Y, Bai Y, Xiang Q, Li Q, Yao W, Wang J, Ibrahim K, Wang H, Wan C, Cao J (2015) Hydrogen impurity defects in rutile TiO_2. Sci Rep 5:17634
57. Liu N, Schneider C, Freitag D, Zolnhofer EM, Meyer K, Schmuki P (2016) Noble-metal-free photocatalytic H_2 generation: active and inactive 'black' TiO_2 nanotubes and synergistic effects. Chemistry 22(39):13810–13814
58. Chen W, He KF, Wang Y, Chan HLW, Yan Z (2013) Highly mobile and reactive state of hydrogen in metal oxide semiconductors at room temperature. Sci Rep 3:3149
59. Wang R, Hashimoto K, Fujishima A, Chikuni M, Kojima E, Kitamura A, Shimohigoshi M, Watanabe T (1997) Light-induced amphiphilic surfaces. Nature 388(6641):431–432
60. Wang R, Hashimoto K, Fujishima A, Chikuni M, Kojima E, Kitamura A, Shimohigoshi M, Watanabe T (1998) Photogeneration of highly amphiphilic TiO_2 surfaces. Adv Mater 10 (2):135–138
61. Mezhenny S, Maksymovych P, Thompson TL, Diwald O, Stahl D, Walck SD, Yates JT Jr (2003) STM studies of defect production on the $TiO_2(110)$-(1×1) and $TiO_2(110)$-(1×2) surfaces induced by UV irradiation. Chem Phys Lett 369(1–2):152–158
62. Shultz AN, Jang W, Hetherington WM, Baer DR, Wang L, Engelhard MH (1995) Comparative second harmonic generation and X-ray photoelectron spectroscopy studies of the UV creation and O_2 healing of Ti^{3+} defects on (110) rutile TiO_2 surfaces. Surf Sci 339 (1–2):114–124
63. Coronado JM, Maira AJ, Conesa JC, Yeung KL, Augugliaro V, Soria J (2001) EPR study of the surface characteristics of nanostructured TiO_2 under UV irradiation. Langmuir 17 (17):5368–5374
64. Li L, Chen Y, Jiao S, Fang Z, Liu X, Xu Y, Pang G, Feng S (2016) Synthesis, microstructure, and properties of black anatase and B phase TiO_2 nanoparticles. Mater Design 100:235–240

65. Wu Q, Huang F, Zhao M, Xu J, Zhou J, Wang Y (2016) Ultra-small yellow defective TiO_2 nanoparticles for co-catalyst free photocatalytic hydrogen production. Nano Energy 24:63–71
66. Zuo F, Bozhilov K, Dillon RJ, Wang L, Smith P, Zhao X, Bardeen C, Feng P (2012) Active facets on titanium(III)-doped TiO_2: an effective strategy to improve the visible-light photocatalytic activity. Angew Chem Int Ed 51(25):6223–6226
67. Liu Y, Quan B, Ji G, Zhang H (2016) One-step synthesis of Ti^{3+} doped TiO_2 single anatase crystals with enhanced photocatalytic activity towards degradation of methylene blue. Mater Lett 162:138–141
68. Cai J, Za H, Lv K, Sun J, Deng K (2014) Ti powder-assisted synthesis of Ti^{3+} self-doped TiO_2 nanosheets with enhanced visible-light photoactivity. RSC Adv 4(38):19588–19593
69. Yang H, Sun C, Qiao S, Zou J, Liu G, Smith SC, Cheng H, Lu G (2008) Anatase TiO_2 single crystals with a large percentage of reactive facets. Nature 453(7195):638–641
70. Liu X, Gao S, Xu H, Lou Z, Wang W, Huang B, Dai Y (2013) Green synthetic approach for Ti^{3+} self-doped TiO_{2-x} nanoparticles with efficient visible light photocatalytic activity. Nanoscale 5(5):1870–1875
71. Liu X, Xu H, Grabstanowicz LR, Gao S, Lou Z, Wang W, Huang B, Dai Y, Xu T (2014) Ti^{3+} self-doped TiO_{2-x} anatase nanoparticles via oxidation of TiH_2 in H_2O_2. Catal Today 225 (0):80–89
72. Wang X, Li Y, Liu X, Gao S, Huang B, Dai Y (2015) Preparation of Ti^{3+} self-doped TiO_2 nanoparticles and their visible light photocatalytic activity. Chin J Catal 36(3):389–399
73. Wu C, Gao Z, Gao S, Wang Q, Xu H, Wang Z, Huang B, Dai Y (2016) Ti^{3+} self-doped TiO_2 photoelectrodes for photoelectrochemical water splitting and photoelectrocatalytic pollutant degradation. J Energ Chem 25(4):726–733
74. Grabstanowicz LR, Gao S, Li T, Rickard RM, Rajh T, Liu D, Xu T (2013) Facile oxidative conversion of TiH_2 to high-concentration Ti^{3+}-self-doped rutile TiO_2 with visible-light photoactivity. Inorg Chem 52(7):3884–3890
75. Xin X, Xu T, Yin J, Wang L, Wang C (2015) Management on the location and concentration of Ti^{3+} in anatase TiO_2 for defects-induced visible-light photocatalysis. Appl Catal B 176-177:354–362
76. Zhu G, Shan Y, Lin T, Zhao W, Xu J, Tian Z, Zhang H, Zheng C, Huang F (2016) Hydrogenated blue titania with high solar absorption and greatly improved photocatalysis. Nanoscale 8(8):4705–4712
77. Liu M, Qiu X, Miyauchi M, Hashimoto K (2011) Cu(II) oxide amorphous nanoclusters grafted Ti^{3+} self-doped TiO_2: an efficient visible light photocatalyst. Chem Mater 23(23):5282–5286
78. Fang W, Khrouz L, Zhou Y, Shen B, Dong C, Xing M, Mishra S, Daniele S, Zhang J (2017) Reduced {001}-TiO_{2-x} photocatalysts: noble-metal-free CO_2 photoreduction for selective CH_4 evolution. Phys Chem Chem Phys 19(21):13875–13881
79. Zhu Q, Peng Y, Lin L, Fan C, Gao G, Wang R, Xu A (2014) Stable blue TiO_{2-x} nanoparticles for efficient visible light photocatalysts. J Mater Chem A 2(12):4429–4437
80. Qiu M, Tian Y, Chen Z, Yang Z, Li W, Wang K, Wang L, Wang K, Zhang W (2016) Synthesis of Ti^{3+} self-doped TiO_2 nanocrystals based on Le Chatelier's principle and their application in solar light photocatalysis. RSC Adv 6(78):74376–74383
81. Chen X, Liu L, Liu Z, Marcus MA, Wang W, Oyler NA, Grass ME, Mao B, Glans PA, Yu P, Guo J, Mao S (2013) Properties of disorder-engineered black titanium dioxide nanoparticles through hydrogenation. Sci Rep 3:1510
82. Naldoni A, Allieta M, Santangelo S, Marelli M, Fabbri F, Cappelli S, Bianchi CL, Psaro R, Dal Santo V (2012) Effect of nature and location of defects on bandgap narrowing in black TiO_2 nanoparticles. J Am Chem Soc 134(18):7600–7603
83. Khan MM, Ansari SA, Pradhan D, Ansari MO, Han DH, Lee J, Cho MH (2014) Band gap engineered TiO_2 nanoparticles for visible light induced photoelectrochemical and photocatalytic studies. J Mater Chem A 2(3):637–644

84. Liu Y, Wang J, Yang P, Matras-Postolek K (2015) Self-modification of TiO_2 one-dimensional nano-materials by Ti^{3+} and oxygen vacancy using Ti_2O_3 as precursor. RSC Adv 5 (76):61657–61663
85. Zuo F, Wang L, Wu T, Zhang Z, Borchardt D, Feng P (2010) Self-doped Ti^{3+} enhanced photocatalyst for hydrogen production under visible light. J Am Chem Soc 132 (34):11856–11857
86. Diebold U (2003) The surface science of titanium dioxide. Surf Sci Rep 48(5–8):53–229
87. Fang W, Dappozze F, Guillard C, Zhou Y, Xing M, Mishra S, Daniele S, Zhang J (2017) Zn-assisted TiO_{2-x} photocatalyst with efficient charge separation for enhanced photocatalytic activities. J Phys Chem C 121:17068
88. Livraghi S, Chiesa M, Paganini MC, Giamello E (2011) On the nature of reduced states in titanium dioxide as monitored by electron paramagnetic resonance. I: the anatase case. J Phys Chem C 115(51):25413–25421
89. Sekiya T, Kurita S (2008) Defects in anatase titanium dioxide. In: Ohno K, Tanaka M, Takeda J, Kawazoe Y (eds) Nano- and micromaterials, vol 9. Springer, Berlin/Heidelberg, pp 121–141
90. Qiu B, Zhou Y, Ma Y, Yang X, Sheng W, Xing M, Zhang J (2015) Facile synthesis of the Ti^{3+} self-doped TiO_2-graphene nanosheet composites with enhanced photocatalysis. Sci Rep 5:8591
91. Giannakas AE, Antonopoulou M, Deligiannakis Y, Konstantinou I (2013) Preparation, characterization of N-I co-doped TiO_2 and catalytic performance toward simultaneous Cr (VI) reduction and benzoic acid oxidation. Appl Catal B 140:636–645
92. Wendt S, Schaub R, Matthiesen J, Vestergaard EK, Wahlström E, Rasmussen MD, Thostrup P, Molina LM, Lægsgaard E, Stensgaard I, Hammer B, Besenbacher F (2005) Oxygen vacancies on TiO_2(110) and their interaction with H_2O and O_2: a combined high-resolution STM and DFT study. Surf Sci 598(1–3):226–245
93. Fischer S, Schierbaum K-D, Göpel W (1997) Surface defects and platinum overlayers on TiO_2(110) surfaces: STM and photoemission studies. Vacuum 48(7–9):601–605
94. Chen C, Chen T, Chen C, Lai Y, You J, Chou T, Chen C, Lee J (2012) Effect of Ti^{3+} on TiO_2-supported Cu catalysts used for CO oxidation. Langmuir 28(26):9996–10006
95. Zhang C, Xie Y, Ma J, Hu J, Zhang C (2015) A composite catalyst of reduced black TiO_{2-x}/CNT: a highly efficient counter electrode for ZnO-based dye-sensitized solar cells. Chem Commun 51(98):17459–17462
96. Liao W, Murugananthan M, Zhang Y (2015) Synthesis of Z-scheme g-C_3N_4-Ti^{3+}/TiO_2 material: an efficient visible light photoelectrocatalyst for degradation of phenol. Phys Chem Chem Phys 17(14):8877–8884
97. Ioannidou E, Ioannidi A, Frontistis Z, Antonopoulou M, Tselios C, Tsikritzis D, Konstantinou I, Kennou S, Kondarides DI, Mantzavinos D (2016) Correlating the properties of hydrogenated titania to reaction kinetics and mechanism for the photocatalytic degradation of bisphenol A under solar irradiation. Appl Catal B 188:65–76
98. Qin X, He F, Chen L, Meng Y, Liu J, Zhao N, Huang Y (2016) Oxygen-vacancy modified TiO_2 nanoparticles as enhanced visible-light driven photocatalysts by wrapping and chemically bonding with graphite-like carbon. RSC Adv 6(13):10887–10894
99. Xing M, Li X, Zhang J (2014) Synergistic effect on the visible light activity of Ti^{3+} doped TiO_2 nanorods/boron doped graphene composite. Sci Rep 4:5493
100. Wang J, Yang P, Huang B (2015) Self-doped TiO_{2-x} nanowires with enhanced photocatalytic activity: facile synthesis and effects of the Ti^{3+}. Appl Surf Sci 356:391–398
101. Liu N, Schneider C, Freitag D, Hartmann M, Venkatesan U, Müller J, Spiecker E, Schmuki P (2014) Black TiO_2 nanotubes: cocatalyst-free open-circuit hydrogen generation. Nano Lett 14 (6):3309–3313
102. Yan X, Xing Z, Cao Y, Hu M, Li Z, Wu X, Zhu Q, Yang S, Zhou W (2017) In-situ C-N-S-tridoped single crystal black TiO_2 nanosheets with exposed {001} facets as efficient visible-light-driven photocatalysts. Appl Catal B 219:572–579

103. Zheng J, Bao S, Zhang X, Wu H, Chen R, Jin P (2016) Pd–$MgNi_x$ nanospheres/black-TiO_2 porous films with highly efficient hydrogen production by near-complete suppression of surface recombination. Appl Catal B 183:69–74
104. Liu L, Zhao C, Li Y (2012) Spontaneous dissociation of CO_2 to CO on defective surface of Cu (I)/TiO_{2-x} nanoparticles at room temperature. J Phys Chem C 116(14):7904–7912
105. Sasan K, Zuo F, Wang Y, Feng P (2015) Self-doped Ti^{3+}-TiO_2 as a photocatalyst for the reduction of CO_2 into a hydrocarbon fuel under visible light irradiation. Nanoscale 7 (32):13369–13372
106. Zhang L, Wang W, Jiang D, Gao E, Sun S (2015) Photoreduction of CO_2 on BiOCl nanoplates with the assistance of photoinduced oxygen vacancies. Nano Res 8(3):821–831
107. Yin G, Bi Q, Zhao W, Xu J, Lin T, Huang F (2017) Efficient conversion of CO_2 to methane Photocatalyzed by conductive black titania. ChemCatChem:n/a-n/a
108. Fu R, Wang Q, Gao S, Wang Z, Huang B, Dai Y, Lu J (2015) Effect of different processes and Ti/Zn molar ratios on the structure, morphology, and enhanced photoelectrochemical and photocatalytic performance of Ti^{3+} self-doped titanium-zinc hybrid oxides. J Power Sources 285:449–459
109. Li K, Huang Z, Zeng X, Huang B, Gao S, Lu J (2017) Synergetic effect of Ti^{3+} and oxygen doping on enhancing photoelectrochemical and photocatalytic properties of TiO_2/g-C_3N_4 heterojunctions. ACS Appl Mater Interfaces 9(13):11577–11586
110. Zhu Y, Shah MW, Wang C (2017) Insight into the role of Ti^{3+} in photocatalytic performance of shuriken-shaped $BiVO_4$/TiO_{2-x} heterojunction. Appl Catal B 203:526–532
111. Zhang X, Zuo G, Lu X, Tang C, Cao S, Yu M (2017) Anatase TiO_2 sheet-assisted synthesis of Ti^{3+} self-doped mixed phase TiO_2 sheet with superior visible-light photocatalytic performance: roles of anatase TiO_2 sheet. J Colloid Interface Sci 490:774–782
112. Duan Y, Zhang M, Wang L, Wang F, Yang L, Li X, Wang C (2017) Plasmonic Ag-TiO_{2-x} nanocomposites for the photocatalytic removal of NO under visible light with high selectivity: the role of oxygen vacancies. Appl Catal B 204:67–77
113. Singh AP, Kodan N, Mehta BR, Dey A, Krishnamurthy S (2016) In-situ plasma hydrogenated TiO_2 thin films for enhanced photoelectrochemical properties. Mater Res Bull 76:284–291
114. Zhang Q, Wang L, Feng J, Xu H, Yan W (2014) Enhanced photoelectrochemical performance by synthesizing CdS decorated reduced TiO_2 nanotube arrays. Phys Chem Chem Phys 16 (42):23431–23439
115. Zheng J, Liu Y, Ji G, Zhang P, Cao X, Wang B, Zhang C, Zhou X, Zhu Y, Shi D (2015) Hydrogenated oxygen-deficient blue anatase as anode for high-performance lithium batteries. ACS Appl Mater Inter 7(42):23431–23438
116. Wang C, Wang F, Zhao Y, Li Y, Yue Q, Liu Y, Liu Y, Elzatahry AA, Al-Enizi A, Wu Y, Deng Y, Zhao D (2016) Hollow TiO_{2-x} porous microspheres composed of well-crystalline nanocrystals for high-performance lithium-ion batteries. Nano Res 9(1):165–173
117. Shang M, Hu H, Lu G, Bi Y (2016) Synergistic effects of $SrTiO_3$ nanocubes and Ti^{3+} dual-doping for highly improved photoelectrochemical performance of TiO_2 nanotube arrays under visible light. J Mater Chem A 4(16):5849–5853
118. Su T, Yang Y, Na Y, Fan R, Li L, Wei L, Yang B, Cao W (2015) An insight into the role of oxygen vacancy in hydrogenated TiO_2 nanocrystals in the performance of dye-sensitized solar cells. ACS Appl Mater Inter 7(6):3754–3763
119. Pan S, Liu X, Guo M, Sf Y, Huang H, Fan H, Li G (2015) Engineering the intermediate band states in amorphous Ti^{3+}-doped TiO_2 for hybrid dye-sensitized solar cell applications. J Mater Chem A 3(21):11437–11443
120. Byeon A, Boota M, Beidaghi M, Aken KV, Lee JW, Gogotsi Y (2015) Effect of hydrogenation on performance of TiO_2(B) nanowire for lithium ion capacitors. Electrochem Commun 60:199–203
121. Sang-Joon P, Jeong-Pyo L, Jong Shik J, Hyun R, Hyunung Y, Byung Youn Y, Chang Soo K, Kyung Joong K, Yong Jai C, Sunggi B, Woo L (2013) In situ control of oxygen vacancies in

TiO_2 by atomic layer deposition for resistive switching devices. Nanotechnology 24 (29):295202
122. Yabuuchi N, Kubota K, Dahbi M, Komaba S (2014) Research development on sodium-ion batteries. Chem Rev 114(23):11636–11682
123. Chen J, Ding Z, Wang C, Hou H, Zhang Y, Wang C, Zou G, Ji X (2016) Black anatase titania with ultrafast sodium-storage performances stimulated by oxygen vacancies. ACS Appl Mater Inter 8(14):9142–9151
124. Asahi R, Morikawa T, Ohwaki T, Aoki K, Taga Y (2001) Visible-light photocatalysis in nitrogen-doped titanium oxides. Science 293(5528):269–271
125. Di Valentin C, Pacchioni G, Selloni A, Livraghi S, Giamello E (2005) Characterization of paramagnetic species in N-doped TiO_2 powders by EPR spectroscopy and DFT calculations. J Phys Chem B 109(23):11414–11419
126. Sayed FN, Jayakumar OD, Sasikala R, Kadam RM, Bharadwaj SR, Kienle L, Schürmann U, Kaps S, Adelung R, Mittal JP, Tyagi AK (2012) Photochemical hydrogen generation using nitrogen-doped TiO_2-Pd nanoparticles: facile synthesis and effect of Ti^{3+} incorporation. J Phys Chem C 116(23):12462–12467
127. Hoang S, Berglund SP, Hahn NT, Bard AJ, Mullins CB (2012) Enhancing visible light photo-oxidation of water with TiO_2 nanowire arrays via cotreatment with H_2 and NH_3: synergistic effects between Ti^{3+} and N. J Am Chem Soc 134(8):3659–3662
128. Zhou Y, Yi Q, Xing M, Shang L, Zhang T, Zhang J (2016) Graphene modified mesoporous titania single crystals with controlled and selective photoredox surfaces. Chem Commun 52 (8):1689–1692
129. Chen Y, Cao X, Lin B, Gao B (2013) Origin of the visible-light photoactivity of NH_3-treated TiO_2: effect of nitrogen doping and oxygen vacancies. Appl Surf Sci 264:845–852
130. Zhang K, Zhou W, Chi L, Zhang X, Hu W, Jiang B, Pan K, Tian G, Jiang Z (2016) Black N/H-TiO_2 nanoplates with a flower-like hierarchical architecture for photocatalytic hydrogen evolution. ChemSusChem 9(19):2841–2848
131. Li B, Zhao Z, Zhou Q, Meng B, Meng X, Qiu J (2014) Highly efficient low-temperature plasma-assisted modification of TiO_2 nanosheets with exposed {001} facets for enhanced visible-light photocatalytic activity. Chem Eur J 20(45):14763–14770
132. Li G, Li J, Li G, Jiang G (2015) N and Ti^{3+} co-doped 3D anatase TiO_2 superstructures composed of ultrathin nanosheets with enhanced visible light photocatalytic activity. J Mater Chem A 3(44):22073–22080
133. Lin T, Yang C, Wang Z, Yin H, Lu X, Huang F, Lin J, Xie X, Jiang M (2014) Effective nonmetal incorporation in black titania with enhanced solar energy utilization. Energy Environ Sci 7(3):967–972
134. Feng N, Liu F, Huang M, Zheng A, Wang Q, Chen T, Cao G, Xu J, Fan J, Deng F (2016) Unravelling the efficient photocatalytic activity of boron-induced Ti^{3+} species in the surface layer of TiO_2. Sci Rep 6:34765
135. Xing M, Zhang J, Qiu B, Tian B, Anpo M, Che M (2015) A brown mesoporous TiO_{2-x}/MCF composite with an extremely high quantum yield of solar energy photocatalysis for H_2 evolution. Small 11(16):1920–1929
136. Xing J, Chen J, Li Y, Yuan W, Zhou Y, Zheng L, Wang H, Hu P, Wang Y, Zhao H, Wang Y, Yang H (2014) Stable isolated metal atoms as active sites for photocatalytic hydrogen evolution. Chem Eur J 20(8):2138–2144
137. Yuan X, Wang X, Liu X, Ge H, Yin G, Dong C, Huang F (2016) Ti^{3+}-promoted high oxygen-reduction activity of Pd nanodots supported by black titania nanobelts. ACS Appl Mater Inter 8(41):27654–27660
138. Bonneviot L, Haller GL (1988) EPR characterization of Ti^{3+} ions at the metal-support interface in $PtTiO_2$ catalysts. J Catal 113(1):96–105
139. Lian Z, Wang W, Li G, Tian F, Schanze KS, Li H (2016) Pt-enhanced mesoporous Ti^{3+}/TiO_2 with rapid bulk to surface electron transfer for photocatalytic hydrogen evolution. ACS Appl Mater Inter

140. Pillay D, Hwang GS (2005) Growth and structure of small gold particles on rutile TiO_2(110). Phys Rev B 72(20):205422
141. Albuquerque AR, Bruix A, dos Santos IMG, Sambrano JR, Illas F (2014) DFT study on Ce-doped anatase TiO_2: nature of Ce^{3+} and Ti^{3+} centers triggered by oxygen vacancy formation. J Phys Chem C 118(18):9677–9689
142. Bennett T, Adnan RH, Alvino JF, Kler R, Golovko VB, Metha GF, Andersson GG (2015) Effect of gold nanoclusters on the production of Ti^{3+} defect sites in titanium dioxide nanoparticles under ultraviolet and soft X-ray radiation. J Phys Chem C 119(20):11171–11177
143. Zhao J, Li Y, Zhu Y, Wang Y, Wang C (2016) Enhanced CO_2 photoreduction activity of black TiO_2-coated Cu nanoparticles under visible light irradiation: role of metallic Cu. Appl Catal A-Gen 510:34–41
144. Pan X, Xu YJ (2013) Fast and spontaneous reduction of gold ions over oxygen-vacancy-rich TiO_2: a novel strategy to design defect-based composite photocatalyst. Appl Catal A-Gen 459:34–40
145. Chen P (2016) A novel synthesis of Ti^{3+} self-doped Ag_2O/TiO_2(p–n) nanoheterojunctions for enhanced visible photocatalytic activity. Mater Lett 163:130–133
146. Li M, Liu H, Liu T, Qin Y (2017) Design of a novel dual Z-scheme photocatalytic system composited of Ag_2O modified Ti^{3+} self doped TiO_2 nanocrystals with individual exposed (001) and (101) facets. Mater Charact 124:136–144
147. Cui Y, Ma Q, Deng X, Meng Q, Cheng X, Xie M, Li X, Cheng Q, Liu H (2017) Fabrication of Ag-Ag_2O/reduced TiO_2 nanophotocatalyst and its enhanced visible light driven photocatalytic performance for degradation of diclofenac solution. Appl Catal B 206:136–145
148. Yin H, Wang X, Wang L, Nie Q, Zhang Y, Yuan Q, Wu W (2016) Ag/AgCl modified self-doped TiO_2 hollow sphere with enhanced visible light photocatalytic activity. J Alloys Compd 657:44–52
149. Colón G, Maicu M, Hidalgo MC, Navío JA (2006) Cu-doped TiO_2 systems with improved photocatalytic activity. Appl Catal B 67(1–2):41–51
150. Liu L, Gao F, Zhao L, Li Y (2013) Tailoring Cu valence and oxygen vacancy in Cu/TiO_2 catalysts for enhanced CO_2 photoreduction efficiency. Appl Catal B 134(0):349–358
151. Novoselov KS, Geim AK, Morozov SV, Jiang D, Zhang Y, Dubonos SV, Grigorieva IV, Firsov AA (2004) Electric field effect in atomically thin carbon films. Science 306 (5696):666–669
152. Geim AK, Novoselov KS (2007) The rise of graphene. Nat Mater 6(3):183–191
153. Wang C, Meng D, Sun J, Memon J, Huang Y, Geng J (2014) Graphene wrapped TiO_2 based catalysts with enhanced photocatalytic activity. Adv Mater Interfaces 1(4):1300150
154. Li L, Yu L, Lin Z, Yang G (2016) Reduced TiO_2-graphene oxide heterostructure as broad spectrum-driven efficient water-splitting photocatalysts. ACS Appl Mater Inter 8 (13):8536–8545
155. Cao S, Liu T, Tsang Y, Chen C (2016) Role of hydroxylation modification on the structure and property of reduced graphene oxide/TiO_2 hybrids. Appl Surf Sci 382:225–238
156. Fu G, Zhou P, Zhao M, Zhu W, Yan S, Yu T, Zou Z (2015) Carbon coating stabilized Ti^{3+}-doped TiO_2 for photocatalytic hydrogen generation under visible light irradiation. Dalton Trans 44(28):12812–12817
157. Liu Y, Xing M, Zhang J (2014) Ti^{3+} and carbon co-doped TiO_2 with improved visible light photocatalytic activity. Chin J Catal 35(9):1511–1519
158. Yi Q, Zhou Y, Xing M, Zhang J (2016) Vacuum activation-induced Ti^{3+} and carbon co-doped TiO_2 with enhanced solar light photo-catalytic activity. Res Chem Intermed 42(5):4181–4189
159. Li K, Gao S, Wang Q, Xu H, Wang Z, Huang B, Dai Y, Lu J (2015) In-situ-reduced synthesis of Ti^{3+} self-doped TiO_2/g-C_3N_4 heterojunctions with high photocatalytic performance under LED light irradiation. ACS Appl Mater Inter 7(17):9023–9030
160. Lu D, Zhang G, Wan Z (2015) Visible-light-driven g-C_3N_4/Ti^{3+}-TiO_2 photocatalyst co-exposed {001} and {101} facets and its enhanced photocatalytic activities for organic pollutant degradation and Cr(VI) reduction. Appl Surf Sci 358:223–230

161. Liu X, Xing Z, Zhang Y, Li Z, Wu X, Tan S, Yu X, Zhu Q, Zhou W (2017) Fabrication of 3D flower-like black N-TiO_{2-x}@MoS_2 for unprecedented-high visible-light-driven photocatalytic performance. Appl Catal B 201:119–127
162. Wen M, Zhang S, Dai W, Li G, Zhang D (2015) In situ synthesis of Ti^{3+} self-doped mesoporous TiO_2 as a durable photocatalyst for environmental remediation. Chin J Catal 36 (12):2095–2102
163. Wei S, Wu R, Xu X, Jian J, Wang H, Sun Y (2016) One-step synthetic approach for core-shelled black anatase titania with high visible light photocatalytic performance. Chem Eng J 299:120–125
164. Wang S, Yang X, Wang Y, Liu L, Guo Y, Guo H (2014) Morphology-controlled synthesis of Ti^{3+} self-doped yolk-shell structure titanium oxide with superior photocatalytic activity under visible light. J Solid State Chem 213:98–103
165. Zhu G, Xu J, Zhao W, Huang F (2016) Constructing black titania with unique nanocage structure for solar desalination. ACS Appl Mater Inter 8(46):31716–31721
166. Qi D, Lu L, Xi Z, Wang L, Zhang J (2014) Enhanced photocatalytic performance of TiO_2 based on synergistic effect of Ti^{3+} self-doping and slow light effect. Appl Catal B 160:621–628
167. Xin L, Liu X (2015) Black TiO_2 inverse opals for visible-light photocatalysis. RSC Adv 5 (88):71547–71550
168. Yu J, Low J, Xiao W, Zhou P, Jaroniec M (2014) Enhanced photocatalytic CO_2-reduction activity of anatase TiO_2 by coexposed {001} and {101} facets. J Am Chem Soc 136 (25):8839–8842
169. Xing M, Yang B, Yu H, Tian B, Bagwasi S, Zhang J, Gong X (2013) Enhanced photocatalysis by Au nanoparticle loading on TiO_2 single-crystal (001) and (110) facets. J Phys Chem Lett 4 (22):3910–3917
170. Si L, Huang Z, Lv K, Tang D, Yang C (2014) Facile preparation of Ti^{3+} self-doped TiO_2 nanosheets with dominant {001} facets using zinc powder as reductant. J Alloys Compd 601:88–93
171. Wang W, Lu C, Ni Y, Song J, Su M, Xu Z (2012) Enhanced visible-light photoactivity of {001} facets dominated TiO_2 nanosheets with even distributed bulk oxygen vacancy and Ti^{3+}. Catal Commun 22(0):19–23
172. Wang W, Ni Y, Lu C, Xu Z (2012) Hydrogenation of TiO_2 nanosheets with exposed {001} facets for enhanced photocatalytic activity. RSC Adv 2(22):8286–8288
173. Chen S, Li D, Liu Y, Huang W (2016) Morphology-dependent defect structures and photocatalytic performance of hydrogenated anatase TiO_2 nanocrystals. J Catal 341:126–135

Chapter 5
Graphene-Modified TiO_2 with Enhanced Visible Light Photocatalytic Activities

5.1 Introduction

With the development of nanotechnology, nanomaterials have attracted plenty of attentions in the field of photocatalysis. Especially, semiconductor gradually becomes a hot spot for researchers due to its low cost and high activity. In 1972, Fujishima and Hoda reported that the H_2 could be produced by the TiO_2 electrodes under the light irradiation [1]. Since then, TiO_2 has become one of the most promising oxide semiconductors and been used in solar energy cells and air and wastewater purifiers [2–5]. Due to its high photostability, nontoxic, low cost, high activity, and so on [6–8], its research in the field of photocatalysis has made a great progress in recent years. However, for the heterogeneous photocatalysis, there are still some limitations in the applications of TiO_2, which could not be ignored. For instance, the photon flux of TiO_2 in the process of photoreaction is very easy to reach saturation under the weak light irradiation, which will significantly reduce the energy efficiency of the whole process during wastewater purification [9–12]. What is more, the largest limitation of TiO_2 application is that it can only absorb the UV light less than 387 nm [13]. In other words, the TiO_2 can only effectively utilize less than 6% of the energy derived from the sunlight, which suggests its low potential of sustainable development in the photocatalysis.

Due to the existence of these mentioned drawbacks, many modifications have been done on TiO_2, including nonmetal doping [9, 14–17], metal doping [11, 18–24], semiconductor compound modification [25–29], and organic photosensitization [17, 30]. However, these modifications also have some limitations. For instance, the nonmetal doping and metal doping on the surface of TiO_2 will lead to the decrease of thermal stability of crystals and also introduce the traps to capture the photo-generated carriers on the surface or in the bulk of catalysts. As a result, it can produce a large number of electron–hole recombination centers [19], which will reduce its photocatalytic activities. Until the introduction of graphene, a visible light-

J. Zhang et al., *Photocatalysis*, Lecture Notes in Chemistry 100,
https://doi.org/10.1007/978-981-13-2113-9_5

responsive carbon material, in the TiO2, photocatalytic activity has been greatly improved.

When the excellent performance of graphene was found, such as high surface area, excellent mechanical property, outstanding electronic conductivity, special photoelectrochemical property, and controllable bandgap [31–34], graphene as a superstar in the material science has attracted much more attention in the past several years. These properties determine its large potential for applications in electrochemistry and photocatalysis. Moreover, graphene can also potentially act as a support material, which allows semiconductor particles such as TiO_2 to anchor on its surface. Recently, the application of graphene in the electrochemical modification on TiO_2 has become a hot spot in the field of photocatalysis [35–40]. The compounding between graphene and TiO_2 makes the successful connection between nanometer modules with excellent properties, which greatly improves the transmission efficiency of optical carriers. Undoubtedly, the design and preparation of a series of new types of efficient TiO_2/graphene photocatalysts with excellent performance are the new focus in the photocatalysis.

5.2 TiO_2/Graphene Composite

5.2.1 Two-Dimensional TiO_2/Graphene Composites

In 2009, the TiO_2/graphene was firstly used in the photocatalysis for the photodegradation of methylene blue (MB) reported by Li and his coworkers [41]. They successfully synthesized a chemically bonded TiO_2–graphene composite with 2D graphene and nanosized P25. The as-prepared TiO_2/graphene photocatalyst exhibited extended light absorption range, excellent adsorption of MB, and efficient photocatalytic activity. From then on, many researchers have reported many kinds of TiO_2/graphene composites by using different methods and used them in the photocatalysis for the degradation of organic pollutants, water splitting, and CO_2 photoreduction. Here, we will discuss the preparation methods, the characterization methods, and the applications of TiO_2/graphene in photocatalysis.

5.2.1.1 Preparations

Since the TiO_2/graphene composite was discovered by the researchers, the 2D-structured TiO_2/graphene with various morphologies could be synthesized by different methods, such as hydrothermal, mechanical mixing, pyrolysis, sol–gel, CVD, UV light irradiation, microwave, vacuum activation, and so on. Hence, according to the reported articles, we will introduce several important methods for the synthesis of the TiO_2/graphene composites.

The hydrothermal method involves reactions under controlled temperature and/or pressure, which is usually performed in stainless steel autoclaves. Liang et al. [42]

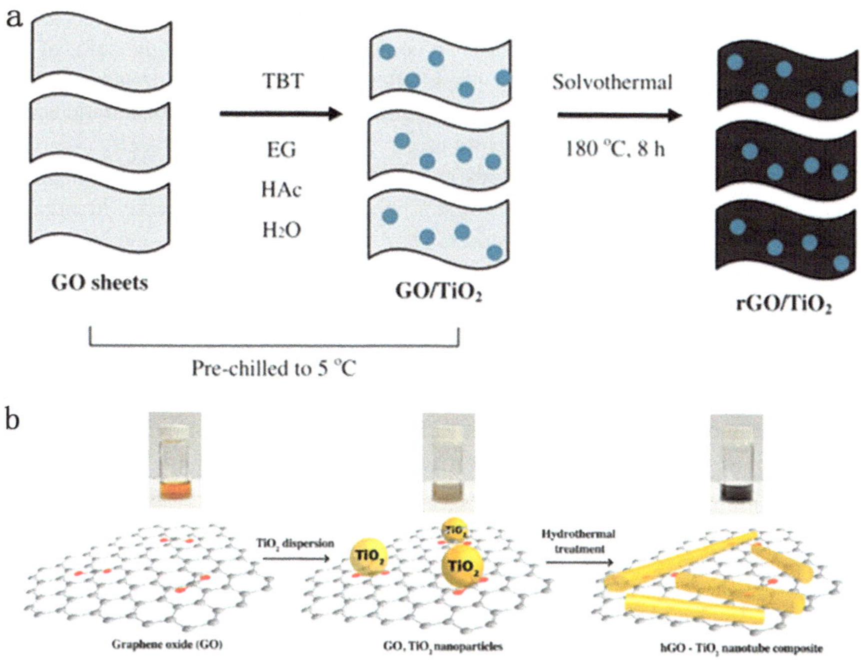

Fig. 5.1 (**a**) Procedure for the synthesis of rGO–TiO_2 nanocomposites [44]. (**b**) Graphical illustration of the synthesis of TiO_2 nanotubes on hGO sheets. (Reprinted with permission from Ref. [45]. Copyright 2012, American Chemical Society)

used the GO as the substrate for the TiO_2 growth to obtain a TiO_2/GO hybrid. The functional groups of hydroxyl and carboxyl on GO provided reactive and anchoring sites for nucleation and growth of nanomaterials. In 2012, the TiO_2 spheres were compounded with the graphene by a hydrothermal method to prepare core–shell-like TiO_2/graphene composites by Park and his coworkers [43]. They successfully synthesized the graphene–anatase TiO_2 hybrid NPs by wrapping amorphous TiO_2 NPs with GO, during which APTMS-modified TiO_2 spheres were chosen as the precursor followed by a one-step GO reduction and TiO_2 crystallization via a hydrothermal treatment. Tan et al. [44] fabricated the reduced graphene oxide (rGO)–TiO_2 hybrid nanocrystals through a hydrothermal method. As shown in Fig. 5.1a, in the preparation, the TBT can be readily grafted onto the surface of GO through chemical adsorption at the molecular level, owing to the rich assortment of oxygen-containing groups, such as epoxide, hydroxyl, carbonyl, and carboxylic groups on the graphene surface. EG and HAc were introduced into the mixture to co-control the hydrolysis rate of TBT. Furthermore, the mixtures were prechilled in an ice bath to further reduce the hydrolysis rate, which would hinder the agglomeration of the TiO_2 nanoparticles on the graphene. During the hydrothermal process, the reduction of GO and the loading of TiO_2 nanoparticles on the rGO surface were occurring simultaneously. Sanjaya et al. [45] reported a facile route for the growth of

TiO_2 nanotubes on reduced graphene oxide sheets via hydrothermal synthesis under basic conditions. As shown in Fig. 5.1b, it could be clearly seen that the color of the TiO_2/GO dispersion became dark black after the hydrothermal treatment. The prepared TiO_2 nanotubes/GO composite exhibits much higher photocatalytic activity toward the degradation of malachite green than TiO_2 nanotubes itself.

Among all the preparation methods, the simplest route is mixing and sonication, owing to that the GO would be exfoliated when added to an aqueous or organic solution. Although the mixing method is very simple, in most cases the interaction between the two phases is very weak, since chemical bonding is not expected [46]. In our previous work, we designed ternary P25/GO/Pt hybrid photocatalysts, which were prepared in different orders by employing different methods and using Degussa's TiO_2 (P25), GO, and chloroplatinic acid as precursors [47]. For instance, we prepared the (Pt/P25) + mGO and (Pt/mGO) + P25 by using the mechanical mixing method. A certain amount of Pt/P25 particles were dispersed into an aqueous suspension of GO, stirred for 24 h at room temperature, centrifuged, washed, and dried, which was denoted as (Pt/P25) + mGO, where "m" represents the quality ratio of GO/P25. 0.05 g chloroplatinic acid was dissolved into a solution containing GO, ethanol, and deionized water, stirred for 30 min at room temperature, and illuminated for 3 h. Then 0.5 g P25 was dispersed into the above mixture and stirred for 24 h at room temperature, centrifuged, washed, and dried. The obtained catalyst was denoted as (Pt/mGO) + P25. Though there is absence of a strong chemical bonding between TiO_2 and GO by using a physical mixing technology, the photo-generated electrons also can transfer from TiO_2 to the GO owing to the intermolecular forces. In our investigation, the (Pt/mGO) + P25 gave a better photocatalytic activity than other catalysts prepared by different methods. After that, we continued employing the simple mixing method to load the P25 nanoparticles on the surface of boron-doped graphene nanosheets [48]. Firstly, the vacuum activation method was employed to prepare the boron-doped graphene nanoribbons, and then the TiO_2 was successfully embedded into the boron-doped graphene nanosheets by a mechanical stirring and ultrasonic treatment. The loading of TiO_2 on graphene surface gave a distinct pressure stress on the surface of graphene, which caused the rupture of C–C bonds that is close to Ti–O–C in B–GR nanoribbons by a shear force from the ultraphonic treatment. The ultraphonic energy could cut some large-sized graphene nanoribbons into smaller nanosheets.

Guo et al. [49] reported a simple method by mechanical mixing TiO_2 nanosheets and reduced graphene oxide (rGO) powder to prepare a series of TiO_2@rGO composite electrodes for the first time, as shown in Fig. 5.2a. Yamashita and his coworkers reported that the TiO_2 nanoparticles supported on a mesoporous silica surface (TiO_2/MCM-41) were selectively coated with graphene through the formation of surface complexes between TiO_2 nanoparticles and 2,3-dihydroxynaphthalene [50]. And then the complex was carbonization under N_2 flow at a high temperature of 1073 K, as shown in Fig. 5.2b.

Sol–gel is an advanced technology, which usually employs the compounds with high activity as the precursors. Chen et al. [36] prepared the GO/TiO_2 composites by using $TiCl_3$ and GO as reactants through a sol–gel method. The authors found that

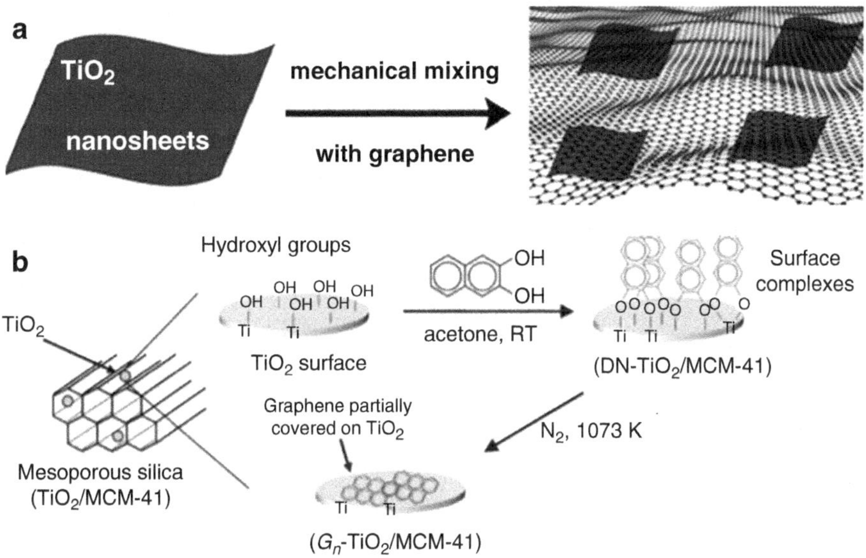

Fig. 5.2 (**a**) Illustration of the formation of TiO_2 nanosheet mechanical mixing with graphene. Reprinted with permission from Ref. [49]. Copyright 2016, John Wiley and Sons. (**b**) Schematic diagram of the procedures for the preparation of Gn–TiO_2/MCM-41. (Reprinted with permission from Ref. [50]. Copyright 2010, American Chemical Society)

the concentration of graphene oxide in starting solution played an important role in photoelectronic and photocatalytic activity of composites. Li et al. [51] reported a sol–gel method to synthesis the ultradispersed TiO_2 nanoparticles which were grown on graphene sheets. The preparation process was schematically illustrated in Fig. 5.3. The amorphous TiO_2 nanoparticles were deposited on GO sheets via the hydrolysis and condensation of TBOT in pure ethanol with a low content of concentrated ammonia, which would lead to the selective formation of amorphous TiO_2/GO sheets. After a thermal treatment process, amorphous TiO_2 nanoparticles are crystallized into uniform anatase nanoparticles accompanied by the reduction of GO sheets, leading to the formation of TiO_2 nanocrystals/RGO sheets.

In addition to the methods mentioned above, there are also some other methods which could be used in the preparation of 2D TiO_2/graphene composites, such as UV light irradiation, microwave-assisted method, chemical vapor deposition [52], and vacuum activation methods [53]. Mohamed et al. [54] prepared the TiO_2-reduced RGO composite by UV-assisted photocatalytic reduction of graphite oxide mixed with TiO_2 nanoparticles in the ethanol. The mixture was illuminated under a 250 W high-voltage Hg lamp with the main wave crest at 254 nm for 24 h under ambient conditions and magnetic stirring. After UV irradiation, it was obvious that the color of suspension changed into grayish black, which indicated the successful reduction of graphene oxide sheets. Pu et al. [55] synthesized the GO–TiO_2 hybrids (GTHs) by a one-pot microwave-assisted combustion method. The precursor solution was heated on a hot plate at 100 °C to remove the solvent until a gel-like precursor

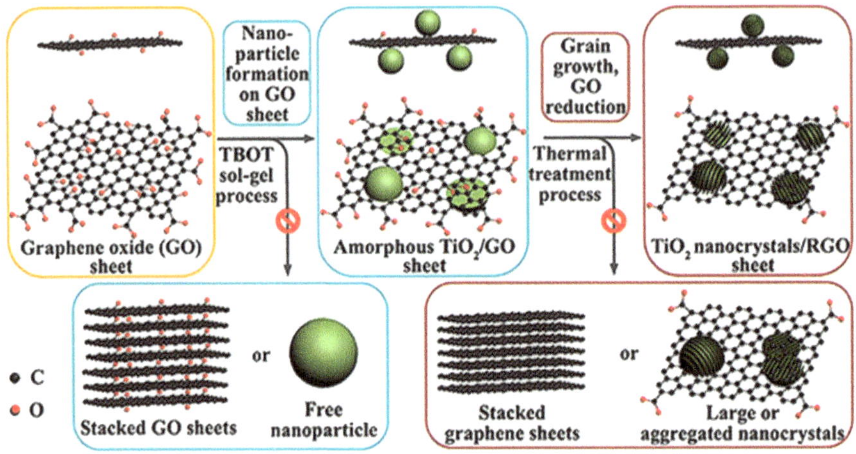

Fig. 5.3 Schematic representation of the sol–gel design strategy toward ultradispersed TiO_2 nanocrystals on graphene. (Reprinted with permission from Ref. [51]. Copyright 2013, American Chemical Society)

was obtained. And then the precursor was introduced into a domestic microwave oven (700 W) and irradiated for different length of time, during which the combustion took place and black foamlike GTHs were obtained.

Recently, we also employ the vacuum activation in the preparation of graphene-based composites. Ti^{3+} self-doped TiO_2–graphene photocatalyst was successfully prepared by a one-step vacuum activation which involved a relative lower temperature and was facile to reproduce on large scale [53]. Compared with other traditional preparation methods, the vacuum activation exhibits advantages of low temperature and low costing, which can achieve the reduction of GO, the self-doping of Ti^{3+} in TiO_2, and the loading of TiO_2 nanoparticles on GR surface at the same time. In conclusion, considering the applications of simple vacuum activation method to prepare different kinds of photocatalysts with high photocatalytic activities, we anticipate our research to be a starting point for the synthesis of new materials with high photoactivity by a simple and cheap method.

5.2.1.2 Characterizations

For the binary composite, the micromorphology is very important to reflect the compounding between the components. For the most reported 2D-structured TiO_2/graphene composite, the graphene shows a large-scale layer structure and plays the role of support. And the nanosized TiO_2 particles play the role of major active component, which is loading on the surface of graphene. The degree of dispersion of TiO_2 nanoparticles on the graphene relates to the photocatalytic activity of composite. The TEM and SEM are considered as the most intuitive characterizations for the micromorphology of TiO_2/graphene composite.

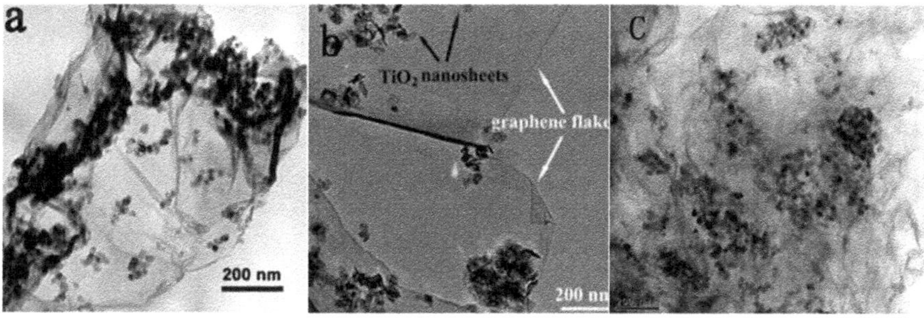

Fig. 5.4 (**a**) Typical TEM image of P25–GR, with P25 loading on the surface of graphene and concentrating along the wrinkles. Reprinted with permission from Ref. [41]. Copyright 2010, American Chemical Society. (**b**) TEM images of TiO2/GO sample [56]. (**c**) TEM image of TiO2/RGO nanocomposites. (Reprinted with permission from Ref. [57]. Copyright 2016, American Chemical Society)

In the early stages, Zhang et al. [41] found that the TiO_2 nanoparticles preferred to grow along the edge and wrinkles of the graphene sheets, according to the TEM image of P25-GR (Fig. 5.4a). Many carboxylic acid functional groups still remained along the edge and wrinkles after a hydrothermal treatment, which are beneficial to react with the surface hydroxyl groups on the surface of TiO_2, resulting in the firmly interaction between TiO_2 and graphene. Xiang et al. [56] have also reported the TiO_2 nanosheets/graphene composite with TEM image shown in Fig. 5.4b. After the reduction, the reduced graphene remained in the 2D-structured sheets with micrometer-long wrinkles after the hydrothermal treatment. The TiO_2 nanosheets with an average side size of 50–80 nm and a thickness of 6–8 nm were dispersed on the graphene sheets with face-to-face orientation, owing to the interaction between carboxylic acid functional groups on graphene and hydroxyl groups on TiO_2. Liu et al. [57] presented a simple and relatively general approach for synthesis of TiO_2/RGO nanocomposites, which used the GO and TiO_2 nanoparticles as starting materials. As seen in Fig. 5.4c, many nanoparticles were on the graphene sheet, which could be contributed to the interaction between TiO_2 nanoparticles and functional groups such as epoxides, hydroxyl, and carboxylic acids on the GO sheets.

The corresponding SEM images also provide similar micromorphology results, which are always consistent with the TEM results. Huang et al. [35] prepared a C–Ti chemically bonded interface TiO_2/graphene composites by a facile solvothermal method using TBOT as the Ti source. The TEM and SEM images of G2.5–TiO_2 (where G2.5 represents that the mass ratio of graphene to titania was 2.5 wt %) were shown in Fig. 5.5, each of which exhibited the loading of TiO_2 on the graphene surfaces.

Additionally, the controlling of size of graphene is also important to the micromorphology of TiO_2/graphene composite. Atomic force microscope (AFM) characterization is always used to detect the size of graphene sheets. Peng et al. [58] reported a simple, surfactant-free, low-temperature process to synthesize

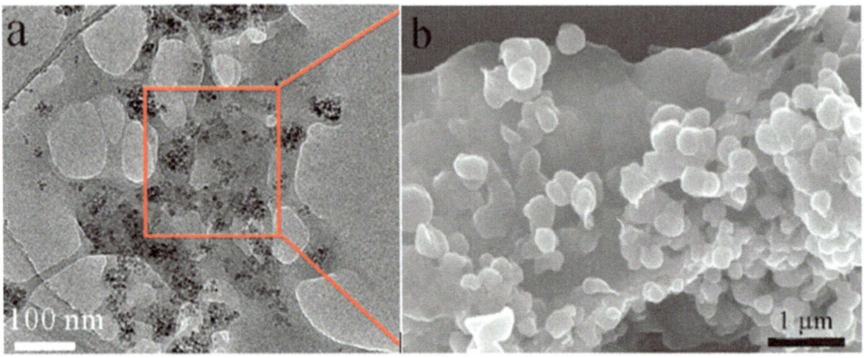

Fig. 5.5 (**a**) TEM image of G2.5–TiO_2 nanocomposite. (**b**) SEM image of G2.5–TiO_2. (Reprinted with permission from Ref. [35]. Copyright 2013, American Chemical Society)

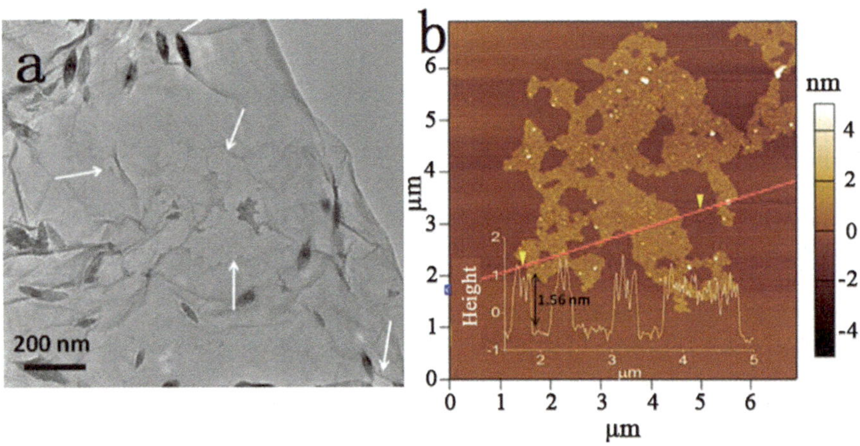

Fig. 5.6 Structural characterization of TiO_2 nanospindles/reduced graphene oxide. (**a**) TEM images of TiO_2/RGO. (**b**) AFM images of TiO_2/RGO. (Reprinted with permission from Ref. [58]. Copyright 2016, American Chemical Society)

mesoporous TiO_2 nanospindles. Then following the same procedure, the TiO_2 nanospindles were chemically anchored onto the graphene oxide (GO) to form a TiO_2/GO composite. As shown in Fig. 5.6, the TiO_2 nanospindles could grow on the RGO, which is confirmed by TEM in Fig. 5.6a. The AFM image of TiO_2/RGO after 8-h photodegradation showed the size and morphology of the composite. The thickness of the RGO sheets was ~ 1.5 nm, corresponding to a single layer. This height is larger than the theoretical value of ~0.8 nm due to the oxygen-containing function groups on both sides.

There are also several characterizations for the investigation of the reduction of graphene oxides such as the XRD, Raman, XPS, and so on. The characteristic peak at 9.8° for the graphene oxides which is ascribed to the (001) interlayer spacing of

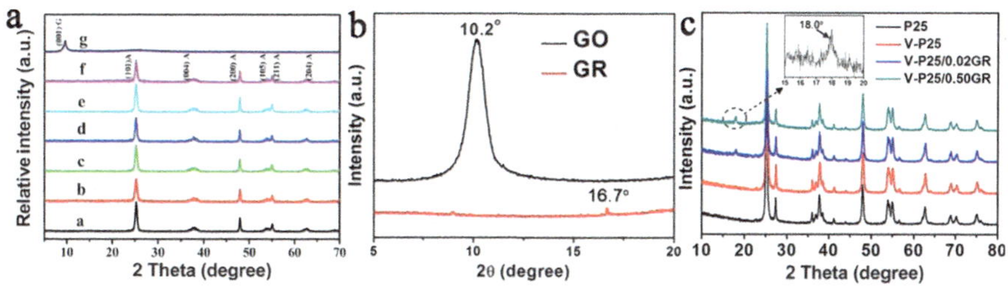

Fig. 5.7 (**a**) XRD patterns of the TiO_2/graphene composites ("g" is the XRD spectrum of graphene oxide, and others are the XRD spectra of the composites). Reprinted with permission from Ref. [56]. Copyright 2016, American Chemical Society. (**b**) XRD spectra of the graphene oxide before and after vacuum activation. (Reprinted with permission from Ref. [53]. Copyright 2015, American Chemical Society. (**c**) XRD spectra of the composites [59])

0.90 nm [56], as shown in Fig. 5.7a. Obviously, the peak at 9.8° disappeared to all the TiO_2/graphene composites after a microwave–hydrothermal treatment, which indicated the successful reduction of graphene oxides by a chemical reduction treatment. The regular stacking of GO was destroyed by the reduction process under the microwave–hydrothermal conditions and could not be resolved by XRD. Our previous work demonstrated that the absence of a peak at 10.2° and the presence of another peak at 16.7° after the vacuum activation treatment also indicated the reduction of graphene oxides (Fig. 5.7b) [53]. The peak at a small angle of 16.7° is indicative of graphene with lower crystallinity and some defects induced by the vacuum activation [59]. Interestingly, the peak assigned to graphene changed from 16.7° to 18.0° after the loading of TiO_2 nanoparticles. This shift may be resulted from the existence of chemical connection between TiO_2 nanoparticles and carbon groups on GR surface. The loading of TiO_2 surface gave a distinct pressure on the graphene, which led to the characteristic peak of GR shifting from 16.7° to 18.0°.

XPS characterization is another measurement to demonstrate the successful reduction of graphene oxides. Xiang et al. [56] found that one of the characteristic peaks of GO located at 287.0 eV in the C1s XPS spectrum of TiO_2/graphene composite was an obvious decrease after the hydrothermal reduction, indicating the damage of the oxygen-containing carbonaceous bands (C–OH) during the hydrothermal process (Fig. 5.8a, b). Our previous work also investigated the reduction of GO through the vacuum activation method [48]. Seen from the C1s XPS of GO, the peaks at 285.7, 286.8, and 288.4 eV were assigned to the C–OH, C–O–C, and C=O bonds (Fig. 5.8c). These oxidation groups in GO were broken after a vacuum activation treatment, inducing the intensity decrease of the peak mentioned above in Fig. 5.9c, which suggested the successful reduction of GO by a vacuum activation method.

In addition to the XRD and XPS, the Raman spectroscopy is another useful technology to detect the reduction of graphene oxides. Raman spectroscopy is a powerful nondestructive tool to characterize the crystalline quality of carbon. Peng et al. [58] employed the Raman to investigate the reduction of graphene oxides in the TiO_2/RGO composites. As shown in Fig. 5.9, the free TiO_2 nanospindles and TiO_2/

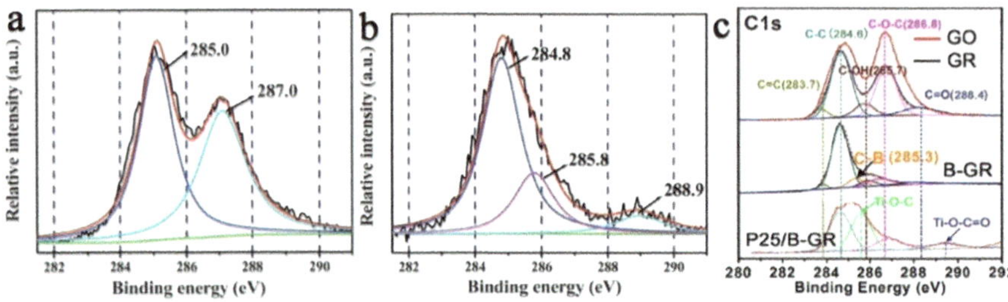

Fig. 5.8 High-resolution XPS spectra of C1s for GO (**a**) and the G1.0 (**b**) sample. (Reprinted with permission from Ref. [56]. Copyright 2016, American Chemical Society. (**c**) C1s XPS spectra for the GO, GR, B–GR, and TiO_2/graphene composite [48])

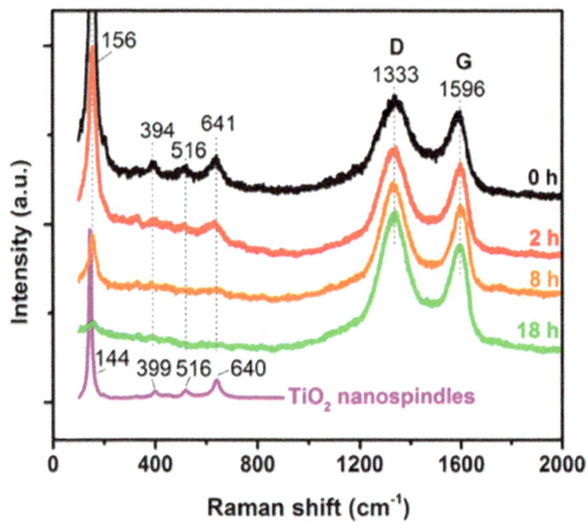

Fig. 5.9 Raman spectra of TiO_2 nanospindles and TiO_2/RGO at different photodegradation stages. (Reprinted with permission from Ref. [58]. Copyright 2016, American Chemical Society)

RGO exhibited characteristic peaks of anatase phase. For the free TiO_2 nanospindles, the main E_g vibration peak is centered at 144 cm^{-1}, while the vibration peaks at 399, 516, and 640 cm^{-1} are assigned to B_{1g}, A_{1g}, and E_g modes of anatase phase, respectively. Compared with those Raman peaks of free TiO_2 nanospindles, the main E_g vibration peak of TiO_2/RGO is blueshifted to 156 cm^{-1}, while the peak at 394 cm^{-1} is redshifted. In conclusion, the Raman peak shifts indicated the strong interaction between the TiO_2 nanospindles and the RGO, which promotes the charge separation and electron transportation.

In order to investigate the electrochemical properties of TiO_2/graphene composite, the photocurrent responses and EIS measurements are always carried out to characterize the electroconductivity of the composite. For instance, Zhang et al. [41] carried out the ESI measurement in the presence of a 2.5 mM $K_3[Fe(CN)_6]/K_4[Fe(CN)_6]$ (1:1) mixture as a redox probe in 0.1 M KCl aqueous solution, and the ESI results are shown in Fig. 5.10a. The typical electrochemical impedance spectra were presented as Nyquist plots. It was noticeable that the semicircle of P25 in the plot

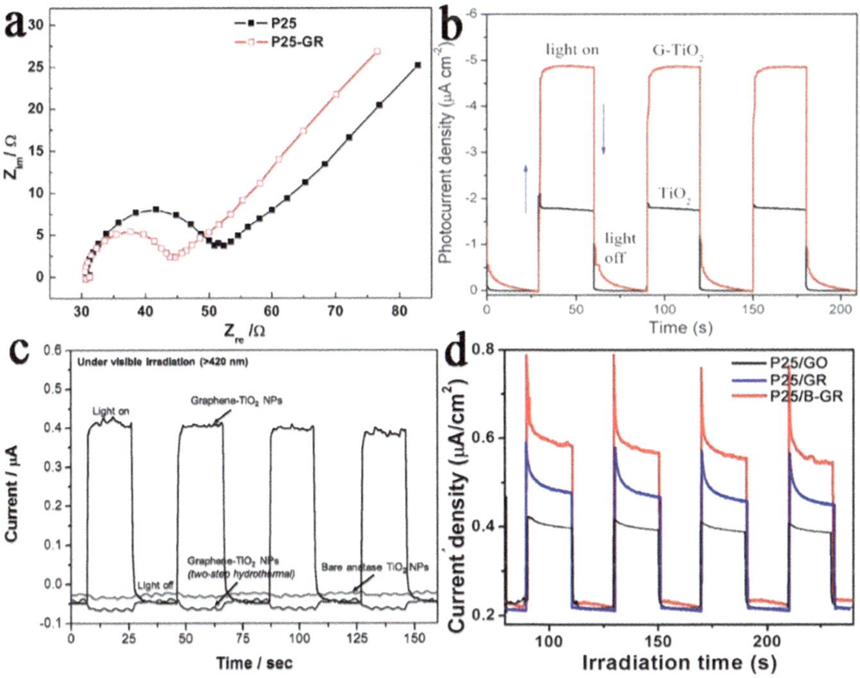

Fig. 5.10 (**a**) EIS changes of P25 and P25–GR electrodes. Reprinted with permission from Ref. [41]. Copyright 2010, American Chemical Society. (**b**) Photocurrents of TiO_2 and G–TiO_2 composites under intermittent irradiation by an ultraviolet lamp at a bias potential of 0.2 V. Reprinted with permission from Ref. [60]. Copyright 2010, Royal Society of Chemistry. (**c**) Photocurrent responses of bare anatase TiO_2 NPs, graphene–TiO_2 NPs, and graphene–TiO_2 NPs under visible light irradiation ($\lambda > 420$ nm). Reprinted with permission from Ref. [43]. Copyright 2012, John Wiley and Sons. (**d**) Transient photocurrent responses of P25/GO, P25/GR, and P25/B–GR in 0.5 M Na_2SO_4 aqueous solution under solar irradiation [48]

became shorter with the introduction of graphene, which indicated a decrease in resistance of the solid-state interface layer and the charge transfer on the surface. Obviously, both the electron-accepting and electron-transporting properties of graphene in the composite could contribute to the suppression of charge recombination.

In order to investigate the light-responsive property of the TiO_2/composite, the photoelectrochemical behaviors of the pure TiO_2 and TiO_2/graphene composites are compared in Fig. 5.10b [60]. Under the UV light irradiation, the TiO_2/graphene composites exhibited a much higher photocurrent density than pure TiO_2. The authors classified that the improved photoresponse benefited from the introduction of graphene. With the extensive 2D p–p conjugation structure, graphene could accept the photo-generated electrons from TiO_2 and transfer them to the external circuit quickly. Lee et al. [43] found that the visible light also could induce a current signal on the TiO_2/graphene composite as shown in Fig. 5.10c. A strong photocurrent response was observed for each switch-on/off event in the case of graphene–TiO_2 NPs, which confirmed high photoactivity of graphene–TiO_2 NPs under visible light irradiation. Additionally, our recent research work demonstrated that the P25/graphene composite could produce the electrons and exhibit a high photocurrent response under the simulated solar light irradiation (Fig. 5.10d) [48].

5.2.2 Three-Dimensional TiO_2/Graphene Composite

In addition to 2D-structured composites, the 3D-structured TiO_2/graphene composites also attracted intense attention in recent years. Compared to the 2D-structured graphene, 3D graphene such as hydrogel or aerogel has some unique properties. For example, it has an ultra-low density, three-dimensional network structure, controllable morphology, controllable mechanical strength, excellent electrical conductivity, and strong adsorption for the gasoline and other organic compounds [61–63].

The unique and outstanding performance of 3D graphene mentioned above determines that the composite of semiconductors compounded with the 3D graphene has a broad application prospect in the field of environment protection and energy storage. For photocatalysis application, the TiO_2-based materials are still the predominant photocatalysts owing to fruitful achievements realized by so many researchers. Thus, in this section, we also would like to give a simple summarization of the 3D-structured TiO_2/graphene composites and applications in photocatalysis.

5.2.2.1 Preparations

Generally, there are three synthesized methods including the evaporation-induced self-assembly, hydrothermal–freeze-drying, and thermal reduction for the preparation of TiO_2/graphene hydrogen or aerogel.

The hydrothermal technology is still the main preparation method for the synthesis of 3D-structured TiO_2/graphene composites. Different from the preparation of 2D-structured TiO_2/graphene by the hydrothermal method, the freeze-drying after treatment plays an important role in the establishment of 3D network structure during the hydrothermal process. A room-temperature or high-temperature drying will destroy the skeleton of 3D graphene, resulting in the failure of the formation of aerogel or hydrogel. Wan et al. [64] developed a facile low-temperature hydrothermal method to prepare anatase titania/cellulose aerogels with strong photocatalytic activities. This involved low-temperature hydrothermal preparation of ATC aerogel. Zhang et al. [65] also reported a new type of multifunctional TiO_2–graphene nanocomposite hydrogel (TGH) by a similar one-pot hydrothermal approach and explored its environmental and energy applications as photocatalyst, reusable adsorbents, and supercapacitor.

Our recently research also found that the hydrothermal–freeze-drying method is a very convenient technology to prepare the TiO_2/GAs aerogel [66]. Very

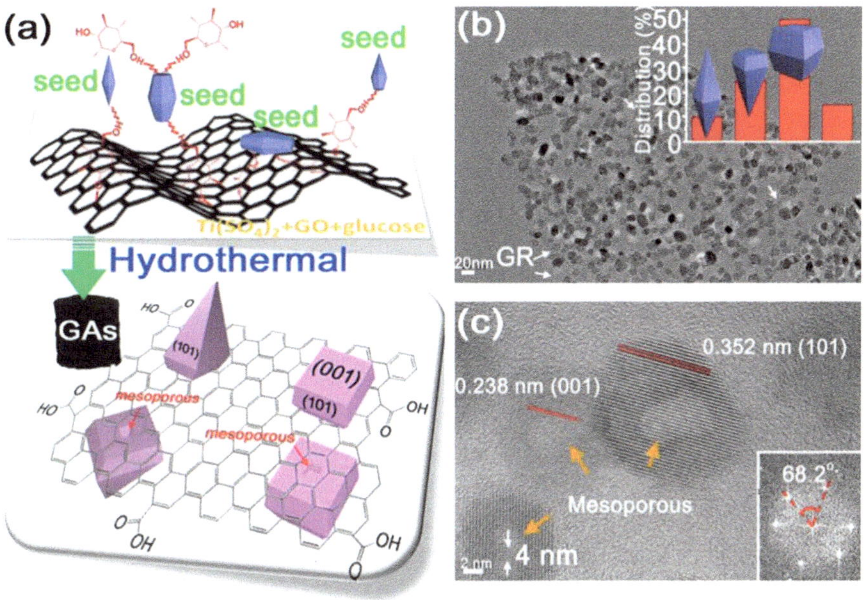

Fig. 5.11 (**a**) Glucose-linked transformation pathway for the in situ growth of TiO_2 nanocrystals with (001) facets on the GA surface. (**b**) TEM image for TiO_2/GAs (67 wt% of TiO_2 in TiO_2/GAs). Inset (*b*) is the corresponding morphology distribution of the TiO_2 nanocrystals derived from 100 of TiO_2 crystals in image (*b*). (**c**) HRTEM image for TiO_2/GAs (67 wt%). Inset (*c*) is the corresponding fast Fourier transform (FFT) pattern. (Reprinted with permission from Ref. [66]. Copyright 2014, American Chemical Society)

interestingly, we found that the addition of glucose during the hydrothermal treatment could achieve the in situ growth of TiO_2 nanocrystals with (001) facets on the surface of aerogel, which results in the highly dispersed TiO_2 nanocrystals on the graphene due to the strong interaction between TiO_2 and graphene. As described in our report (Fig. 5.11), $Ti(SO_4)_2$ was used as the titanium source to be first dissolved in an aqueous solution. The cultivation of TiO_2 crystal seeds was necessary before a known amount of glucose being adsorbed on the seeds. During the hydrothermal process, large numbers of glucose molecules are adsorbed onto the TiO_2 (001) facets, generating nanosized seeds with exposed (001) facets owing to many hydroxyl groups on glucose. The hydroxyl groups at one end of glucose are connected with GR, while the hydroxyl groups at the other end are connected with the TiO_2 facets. Thus, the hydroxyl groups enabled glucose to serve as the linker achieving the in situ growth of TiO_2 nanocrystals on a GA surface.

Evaporation-induced self-assembly technology is another method involving low energy consumption and operability, which has widely been used in the preparation of 3D-structured TiO_2/graphene composites. Hou et al. [67] reported a room-temperature synthesis of chemically bonded TiO_2–graphene composite hydrogels and applied them in the visible light-driven photocatalysis. Graphene oxide, P25, and other precursors were mixed together at room temperature, and then the mixture

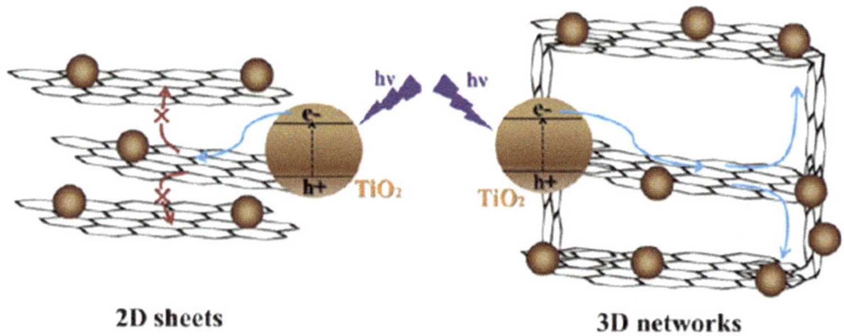

Fig. 5.12 Schematic diagrams for illuminating the charge behavior at interfaces in 2D P25–graphene sheets and 3D P25–graphene networks. (Reprinted with permission from Ref. [67]. Copyright 2012, Elsevier)

was ultra-sonicated for half an hour. After that, the solvent of the 2D composite was slowly evaporated at room temperature for 8 h to obtain hydrogels. With the reduction proceeding of the evaporation, which induced self-assembly process, the π-conjugated structures of reduced GO sheets were growing larger and larger, which would increase the amount of π–π stacking cross-links between graphene sheets. As a result, the partial overlapping or coalescing of flexible graphene sheets is formed into the construction of the 3D structures of hydrogels in Fig. 5.12.

5.2.2.2 Characterizations

Very similar to the 2D graphene, the XRD, Raman, and XPS characterization are also always used to detect the reduction degree of the graphene oxides in the 3D graphene-based composite after a thermal reduction treatment, and the TEM and SEM are employed to detect the micromorphology of the composite.

Zhang et al. [65] developed a new type of 3D-structured TiO_2/graphene composite hydrogels with multifunctions by using a facile one-step hydrothermal method. Seen from Fig. 5.13, the composite had a 3D block appearance (Fig. 5.13a), and the TEM image showed that the TiO_2 nanoparticles are highly dispersed on the surface of graphene sheets (Fig. 5.13b). From the XRD and XPS analysis of 3D-structured TiO_2/graphene composite, the disappearing of the peak at 10.4° in the XRD spectra after a heat treatment indicated the successful reduction of GO (Fig. 5.13c), and the absence of the characteristic peaks of C–O and C=O in the C1s XPS spectra also suggested the reduction of GO during a hydrothermal process (Fig. 5.13d, e).

In our recent work, we used the FTIR characterization to investigate the connection between TiO_2 and graphene in the TiO_2/GA composite [66]. In the FTIR spectra of TiO_2/GAs (Fig. 5.14a), the appearance of –C–O–C– signals resulted from the covalent linkage between the hydroxide radicals of GR and glucose. This reaction led to a new and very broad peak in the range of 600–800 cm^{-1} belonging to the resulting Ti–O–C bond, which indicates the interaction between the GA and TiO_2. It was because of the formation of these chemical bonds that highly dispersed TiO_2

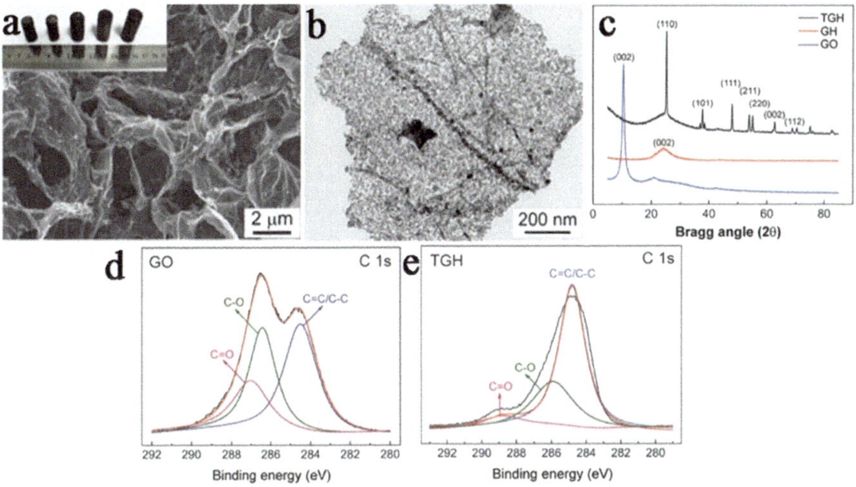

Fig. 5.13 (**a**) Low-magnified SEM images of freeze-dried TGH; inset of (*a*) is the photograph of TGH with different ratios of TiO_2 to graphene at 180 °C for 2 h. (**b**) Low-magnified TEM images of freeze-dried TGH. (**c**) XRD patterns of TGH, GH, and GO; C1s XPS spectra of hydrogels before (**d**) and after (**e**) TiO_2 loading. (Reprinted with permission from Ref. [65]. Copyright 2013, American Chemical Society)

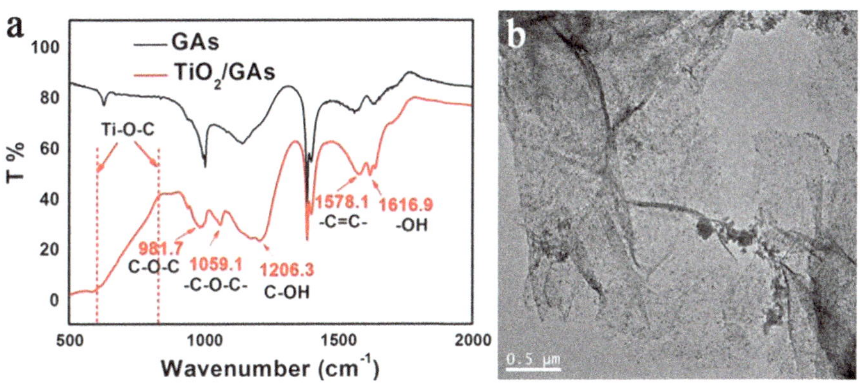

Fig. 5.14 (**a**) FTIR spectra for GAs with and without TiO_2 nanocrystals (67 wt% TiO_2 on GAs); (**b**) TEM image for the TiO_2/GAs composite. (Reprinted with permission from Ref. [66]. Copyright 2014, American Chemical Society)

nanocrystals were achieved by in situ growth on the 3D-structured graphene (Fig. 5.14b).

As we mentioned above, most of the reported graphene aerogels and hydrogels are very fragile, which is difficult to undergo the stirring treatment during the photocatalysis process. Thus, the mechanical strength is another important characteristic for the TiO_2/graphene aerogels or hydrogels. In order to promote the mechanical strength of graphene hydrogels, Cong et al. [63] used the polyacrylamide

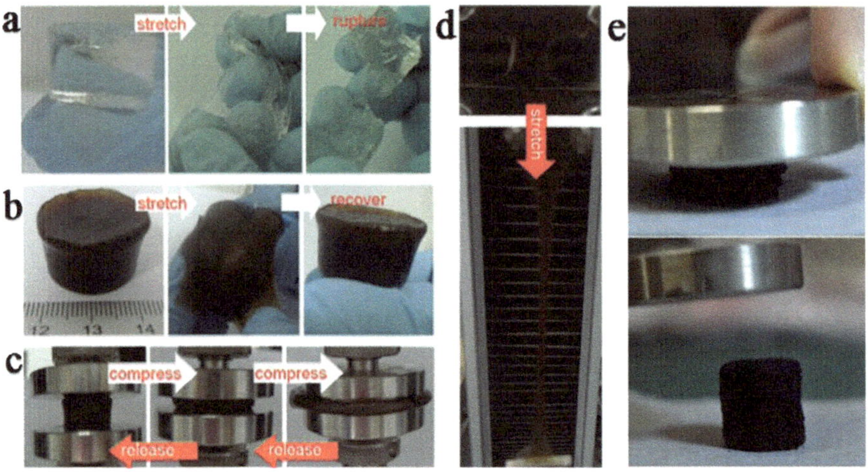

Fig. 5.15 Photographs demonstrating the excellent mechanical behavior of the GO/PAM hydrogels. PAM hydrogel is easily ruptured by stretching (**a**). The GO/PAM columnar hydrogels recover their initial shapes after stretching to an irregular film (**b**). After compression testing by>90% (**c**). The GO/PAM hydrogel fixed to two clamps is stretched to 11 times its initial length in a tensile machine (**d**). Reprinted with permission from Ref. [63]. Copyright 2013, John Wiley and Sons. (**e**) The compressibility of TiO_2/GAs. TiO_2/GAs can be squeezed into a pellet under pressure. Once the external pressure is removed, the TiO_2/GAs unfolds almost completely. (Reprinted with permission from Ref. [66]. Copyright 2014, American Chemical Society)

(PMA) as the additive to prepare a new kind of elastic GO/PAM hydrogels with exceptional mechanical strength by the combination of the characteristics of conventional double-network hydrogel and nanocomposite hydrogel, as shown in Fig. 5.15a. The authors explained that the Ca^{2+} coordination-induced GO nanosheet network and covalently cross-linked PAM network were intertwined by the hydrogen bonds between the GO sheets and PAM chains, owing to the rich hydroxyl and epoxy groups on the GO. The special interaction between ion-induced GO and PMA contributed to the dissipation of crack energy through unzipping and/or sliding of the weak cross-links progressively or deforming the network conformation, which was responsible for the high toughness and good elasticity of the GO/PMA composite. Although the addition of organic polymer is beneficial to the mechanical strength of graphene hydrogels, the nonconducting high-molecular polymer may be harmful to the conductivity of the graphene which will hinder the application to the photocatalysis. Hence, in our recent work, we employed the glucose as the linker to synthesize the mesoporous TiO_2 nanocrystals growing in situ on graphene aerogels [66]. The prepared aerogels also had a good mechanical strength, as shown in Fig. 5.15e. The TiO_2/GAs prepared in the presence of glucose unfolded almost completely after removing of the external pressure.

5.3 Applications in Photocatalysis

TiO_2/graphene composites have been widely used for the photodegradation of organic pollutants in recent years, but they have also been applied in many other fields, including hydrogen production and CO_2 photoreduction, which are related with the protection of the environment.

5.3.1 Photodegradation of Organic Pollutants

Nowadays, water and air are considered as the most valuable human resources which have attracted much more attentions than before. Especially with the speeding up of the industrial process, a large amount of waste water and gasses are directly discharged into the rivers and atmosphere, which will induce the accumulation of large amounts of toxic organic compounds in our daily lives. And in these organic pollutants, some compounds are very difficult to be degraded by the ordinary processing methods. The photocatalytic technology is an ideal method to degrade these organic compounds by using the inexhaustible solar light energy.

In our group, we also chose some organic dyes as the simulated pollutant molecules, which were photodegraded by the TiO_2/graphene composites, such as methylene blue, methyl orange, and rhodamine B solutions. For instance, we successfully prepared the graphene nanoribbons and used it for the photodegradation of rhodamine B under the solar light irradiation [68]. We found that the bandgap of the graphene could be adjusted by changing the morphology of graphene and the nanosized graphene exhibited a semiconductor property and showed a good photocatalytic activity for the first time. The prepared graphene nanoribbons continued to be used as the precursors to synthesize the nanocomposite of TiO_2 and graphene nanoribbons by a facile ultrasonic–mechanical mixing method [48]. As shown in Fig. 5.16, the prepared composite had an excellent performance for the

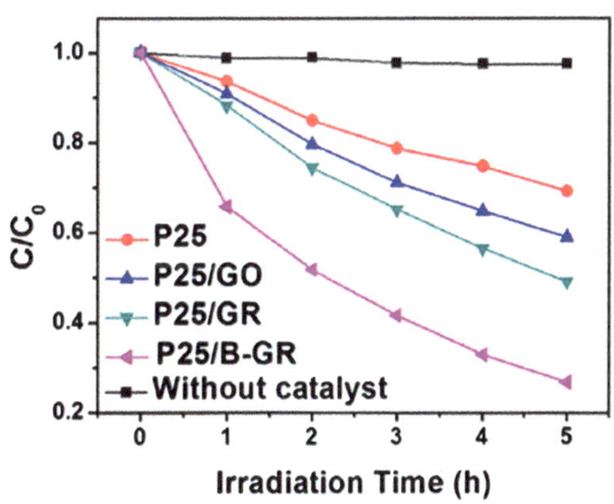

Fig. 5.16 Visible light-induced photocatalytic activities for degradation of MO by P25/B–GR

photodegradation of methyl orange under the solar light irradiation. The outstanding transfer efficiencies of photo-generated carriers induced by the nanosized compounding between TiO_2 and graphene were responsible for the enhancement photocatalytic activity.

In addition to the colored dyes, some colorless organic compounds also can be photodegraded by the graphene-based photocatalysts. In our previous work, the Ti^{3+} -doped TiO_2 nanorods/boron-doped graphene composite was prepared by a hydrothermal method, and the synergistic effect on the visible light activity was investigated by the photodegradation of colorless phenol [20]. The TiO_2/graphene composite showed a higher photodegradation of phenol than other photocatalysts, and the exposed (100) facets on TiO_2, the self-doping of Ti^{3+}, and the boron doped in graphene were the reasons for the excellent photocatalytic activity. Ng et al. prepared the electrodes which were consisted of TiO_2/graphene nanocomposite. The graphene-based electrodes had a significant activity for the complete photocatalytic decomposition of 2,4-dichlorophenoxyacetic acid under the UV light irradiation. Kamegawa et al. [50] designed a composite of TiO_2 nanoparticles supported on a mesoporous silica surface (TiO_2/MCM-41) which were selectively coated with graphene through the formation of surface complexes between TiO_2 nanoparticles and 2,3-dihydroxynaphthalene followed by carbonization under N_2 flow. The selective graphene modification induced the enhanced photocatalytic activities of TiO_2/MCM-41 for the decomposition of 2-propanol in water compared with unmodified samples, owing to the appropriate adsorption properties of organics to transfer to the surface of TiO_2 nanoparticles.

5.3.2 *Water Splitting*

Water splitting is another important application of the TiO_2-based photocatalysis, due to the exhaustion of energy in the world. The hydrogen is regarded as the clean energy which is standing out in recent years owing to its low carbon consuming.

Fan et al. [69] developed a nanocomposite of TiO_2 and reduced graphene oxide as an efficient photocatalysts for the H_2 evolution. As an ideal conductor, graphene exhibited an excellent electron transfer capacity, leading to the enhancement of the lifetime of photo-generated electrons which is the key factor for the high photocatalytic water-splitting activity of the TiO_2/graphene composite. Wang et al. [70] reported the synthesis of graphene@TiO_2 nanocomposites with controlled exposed crystal facets by a simple one-pot hydrothermal process. The prepared TiO_2/graphene composite showed an enhanced photocatalytic H_2 evolution under the simulated solar light irradiation, owing to the exposed high reactive crystal facets and high dispersed TiO_2 nanocrystals on graphene.

Li et al. [56] reported a reduced TiO_2–graphene oxide heterostructure by a facile chemical reduction method, which synthesized different samples named as sample 1 (1 mg GO/50 mg TiO_2), sample 2 (2 mg GO/50 mg TiO_2), sample 3 (3 mg GO/50 mg TiO_2), and sample 4 (4 mg GO/50 mg TiO_2). The photocatalytic activity

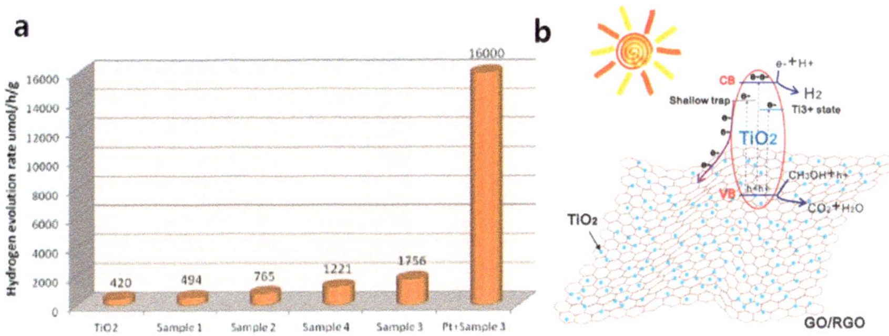

Fig. 5.17 (**a**) Rates of hydrogen evolution for samples under AM 1.5 G irradiation by different samples. (**b**) Schematic illustration of hydrogen evolution mechanism for the strong coupling between TiO_2 and RGO sheets. (Reprinted with permission from Ref. [56]. Copyright 2016, American Chemical Society)

of the samples is performed by measuring the hydrogen evolution experiments. As shown in Fig. 5.17a, they proceeded to optimize the catalyst composition by measuring hydrogen evolution rate for different GO concentrations. As a result, the hydrogen production rate increases first and then decreases with the increasing GO amount. Importantly, they found that a maximum H_2 production rate is measured to be 16 mmol/h/g by using Pt as a cocatalyst under the simulated sunlight irradiation (AM 1.5G, 135 mW/cm^2).

5.3.3 CO_2 Photoreduction

With the intensification of the global greenhouse effect, CO_2 photoreduction becomes a hot research in the field of environment and energy. Unfortunately, there is much less report on the CO_2 photoreduction compared with the reports on the organic pollutant photodegradation and water splitting on the TiO_2/graphene composites, up to now.

Tu et al. [71] fabricated the TiO_2–graphene 2D sandwich-like hybrid nanosheets in a binary ethylenediamine (En)/H_2O solvent by using an in situ simultaneous reduction hydrolysis technology. The high photocatalytic activity of G–TiO_2 hybrid was confirmed by photocatalytic reduction of CO_2 to valuable hydrocarbons (CH_4 and C_2H_6) in the presence of water vapor and without any noble metal cocatalysts (Fig. 5.18a). The enhanced conversion rate could be assigned to the well matching in the energy levels of d orbital of TiO_2 and π orbital of graphene. Moreover, the G–TiO_2 had chemical bond interactions and formed d–π electron orbital overlap. Under the UV light irradiation, the photo-generated electrons are transferred from the TiO_2 to the graphene, which could reduce the CO_2 to generate the hydrocarbons, as shown in Fig. 5.18b. The synergistic effect of the surface–Ti^{3+} sites and graphene favors the

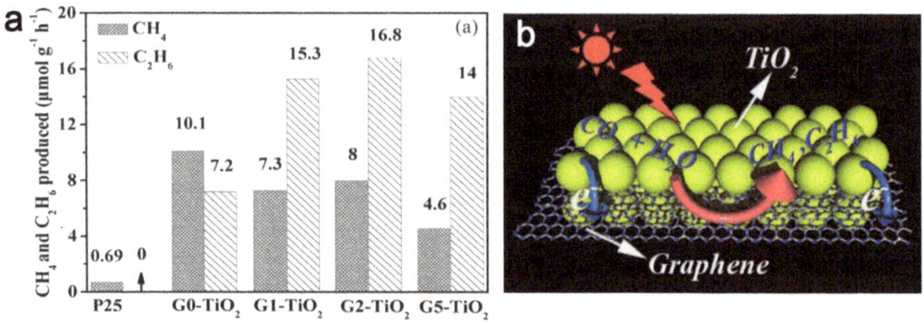

Fig. 5.18 (**a**) Comparison of photocatalytic activity of samples G_x–TiO_2 ($x = 0, 1, 2, 5$) and P25. The molar ratio of C_2H_6 to CH_4 increases from 0.71 (for G_0–TiO_2), 2.09 (G_1–TiO_2), 2.10 (G_2–TiO_2), to 3.04 (G_5–TiO_2). (**b**) Schematic illustration of the charge separation and transfer in the G–TiO_2 system and photoreduction of CO_2 into CH_4 and C_2H_6. (Reprinted with permission from Ref. [71]. Copyright 2012, John Wiley and Sons)

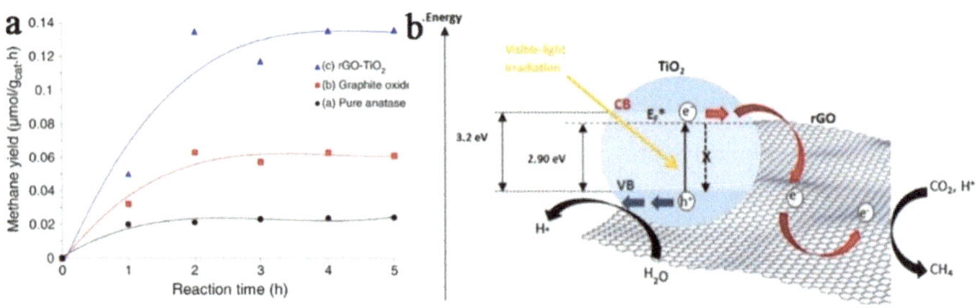

Fig. 5.19 (**a**) Time dependence on the photocatalytic formation rate of CH_4. (**b**) Charge transfer and separation in the rGO–TiO_2 composite. Schematic illustrating the charge transfer and separation in the rGO–TiO_2 composite for the photoreduction of CO_2 under visible light irradiation with the introduction of a new energy level, E*F [44]

generation of C_2H_6, and the yield of the C_2H_6 increases with the content of incorporated graphene.

In addition to the UV light, the visible light also could induce the photoreduction of CO_2 on the catalyst of TiO_2/graphene composite. Tan et al. [44] successfully prepared the reduced graphene oxide (rGO)–TiO_2 hybrid nanocrystals through a novel and simple solvothermal synthetic route. Compared with the pure graphene oxide and the blank TiO_2, the TiO_2/graphene composite had the highest photoreduction efficiency of CO_2 in the visible light irradiation, as shown in Fig. 5.19a. What's more, the authors used the energy band theory to explain the reason of the photocatalytic activity of TiO_2/graphene, and the corresponding schematic illustration was given in Fig. 5.19b. Generally, the overall mechanism of the CO_2 photoreduction process is a sequential combination of H_2O oxidation and CO_2 reduction. Seen from Fig. 5.19b, the d orbital of TiO_2 and the π orbital of rGO matched well in energy levels, which induced a chemical bond interaction to form d–π electron

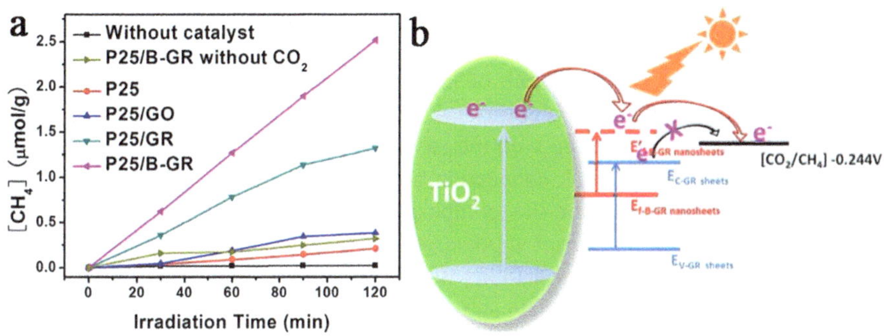

Fig. 5.20 (**a**) Simulated solar light-induced CO_2 reduction (the dark yellow line is the data of the blank photocatalytic test of P25/B–GR in the absence of CO_2). (**b**) Schematic diagram of photogenerated electron transfer between TiO_2 and graphene materials [48]

orbital overlap between TiO_2 and graphene. The CB flat band potential of TiO_2 is more negative than the reduction potential of CO_2/CH_4, indicating that the photogenerated electrons and holes on the irradiated rGO–TiO_2 composites could react with adsorbed CO_2 and H_2O to produce CH_4 via an eight-electron reaction.

Our previous work also developed the application on the solar light-driven CO_2 photoreduction by using modified TiO_2/graphene composite as the photocatalyst. Highly dispersed boron-doped graphene nanosheets loaded with TiO_2 nanoparticles were synthesized in our lab by using a simple mixing method [48]. The prepared P25/B–GR showed higher CO_2 photoreduction capacity than other samples, as shown in Fig. 5.20a. Under the solar light irradiation, the photo-produced electrons of B–GR make its Fermi level ($E'_{\text{f-B-GR nanosheets}}$) being higher than the conduction band ($E_{\text{C-GR sheets}}$) of GR sheets (from $E_{\text{f-B-GR nanosheets}}$ to $E'_{\text{f-B-GR nanosheets}}$ in Fig. 5.20b). The level of $E'_{\text{f-B-GR nanosheets}}$ just falls in between the conduction band of TiO_2 and the relevant redox potentials of CO_2/CH_4, which is beneficial to the transfer of photo-generated electrons. Our results opened the way to further implementation of graphene-based materials as photocatalysts that can be used in the photoreduction of CO_2 and photodegradation of other organic pollutants in gas phases.

5.4 Conclusions and Prospective

In conclusion, graphene has become an ideal support and conductor in the field of photocatalysis, which can capture the photo-generated electrons from the TiO_2 and transfer them to participate in the photocatalytic reactions. In addition, these electrons can also be stored in the π–π network of the composites, which can not only reduce the graphene oxides by themselves but also shuttle to other metal particles deposited on the graphene layers. In that case, the introduction of graphene hinders the electron–hole recombination on TiO_2 and further extends the absorption of light in the visible region.

However, most of the applications of TiO_2/graphene composites in photocatalysis concentrate on the laboratory conditions, and it is really difficult to achieve its industrial application in dealing with the environmental issues. The poor dispersion of TiO_2 on graphene, the weak connection between TiO_2 and graphene, the uncontrollable micromorphology of the composites, and the difficult recycle for the powders all hinder the industrial application of TiO_2/graphene composites. Future works should be focused on the stability of TiO_2/graphene composites as well as on their immobilization on appropriate substrates for the operation of photoreactors in continuous mode, which is close to the industrial application in environmental issues. Compared with the 2D-structured TiO_2/graphene, we believe that the 3D-structured TiO_2/graphene has relative larger potential applications in photocatalysis, owing to its strong adsorption capacity, excellent photocatalytic activity, and stable appearance. Arrangement of a serial of TiO_2/graphene aerogels or hydrogels with strong mechanical strength together will exhibit an outstanding adsorption capacity and photocatalytic activity for the organic pollutants. The block appearance of hydrogels can undergo strong stirring treatment and facilitate the recycle, which will decrease the cost and promote the industrial applications.

References

1. Fujishima A, Honda K (1972) Electrochemical photolysis of water at a semiconductor electrode. Nature 238:37–38
2. Bach U, Lupo D, Comte P, Moser JE, Weissortel F et al (1998) Solid-state dye-sensitized mesoporous TiO_2 solar cells with high photon-to-electron conversion efficiencies. Nature 395:583–585
3. Crossland EJW, Noel N, Sivaram V et al (2013) Mesoporous TiO_2 single crystals delivering enhanced mobility and optoelectronic device performance. Nature 495:215–219
4. O'Regan B, Gratzel M (1991) A low-cost, high-efficiency solar cell based on dye-sensitized colloidal TiO_2 films. Nature 353:737–740
5. Yang H, Sun C, Qiao S, Zou J, Liu G et al (2008) Anatase TiO_2 single crystals with a large percentage of reactive facets. Nature 453:638–641
6. Thompson TL, Yates JT (2006) Surface science studies of the photoactivation of TiO_2 new photochemical processes. Chem Rev 106:4428–4453
7. Linsebigler AL, Lu G, Yates JT (1995) Photocatalysis on TiO_2 surfaces: principles, mechanisms, and selected results. Chem Rev 95:735–758
8. Schneider J, Matsuoka M, Takeuchi M, Zhang J, Horiuchi Y, Anpo M, Bahnemann DW (2014) Understanding TiO_2 photocatalysis: mechanisms and materials. Chem Rev 114:9919–9986
9. Xing M, Zhang J, Chen F (2009) New approaches to prepare nitrogen-doped TiO_2 photocatalysts and study on their photocatalytic activities in visible light. Appl Catal B Environ 89:563–569
10. Xing M, Zhang J, Chen F (2009) Photocatalytic performance of N-doped TiO_2 adsorbed with Fe^{3+} ions under visible light by a redox treatment. J Phys Chem C 113:12848–12853
11. Xing M, Wu Y, Zhang J, Chen F (2010) Effect of synergy on the visible light activity of B, N and Fe Co-doped TiO_2 for the degradation of MO. Nanoscale 2:1233–1239

12. Xing M, Zhang J, Chen F, Tian B (2011) An economic method to prepare vacuum activated photocatalysts with high photo-activities and photo-sensitivities. Chem Commun 47:4947–4949
13. Chen X, Mao S (2007) Titanium dioxide nanomaterials: synthesis, properties, modifications, and applications. Chem Rev 107:2891–2959
14. Qi D, Xing M, Zhang J (2014) Hydrophobic carbon-doped TiO_2/MCF-F composite as a high performance photocatalyst. J Phys Chem C 118:7329–7336
15. Cong Y, Chen F, Zhang J, Anpo M (2006) Carbon and nitrogen-codoped TiO_2 with high visible light photocatalytic activity. Chem Lett 35:800–801
16. Wu Y, Xing M, Tian B, Zhang J, Chen F (2010) Preparation of nitrogen and fluorine Co-doped mesoporous TiO_2 microsphere and photodegradation of acid orange 7 under visible light. Chem Eng J 162:710–717
17. Wu Y, Xing M, Zhang J (2011) Gel-hydrothermal synthesis of carbon and boron Co-doped TiO_2 and evaluating its photocatalytic activity. J Hazard Mater 192:368–373
18. Xing M, Qi D, Zhang J, Chen F (2011) One-step hydrothermal method to prepare carbon and lanthanum Co-doped TiO_2 nanocrystals with exposed {001} facets and their high UV and visible-light photocatalytic activity. Chem Eur J 17:11432–11436
19. Xing M, Fang W, Nasir M, Ma Y, Zhang J, Anpo M (2013) Self-doped Ti^{3+}-enhanced TiO_2 nanoparticles with a high-performance photocatalysis. J Catal 297:236–243
20. Xing M, Li X, Zhang J (2014) Synergistic effect on the visible light activity of Ti^{3+} doped TiO_2 nanorods/boron doped graphene composite. Sci Rep 4:5493
21. Yang P, Lu C, Hua N, Du Y (2002) Titanium dioxide nanoparticles Co-doped with Fe^{3+} and Eu $^{3+}$ ions for photocatalysis. Mater Lett 57:794–801
22. Yu L, Yuan S, Shi LY, Zhao Y, Fang J (2010) Synthesis of Cu^{2+} doped mesoporous titania and investigation of its photocatalytic ability under visible light. Microporous Mesoporous Mater 134:108–114
23. Yuan X, Zhang J, Anpo M, He D (2010) Synthesis of Fe^{3+} doped ordered mesoporous TiO_2 with enhanced visible light photocatalytic activity and highly crystallized anatase wall. Res Chem Intermed 36:83–93
24. Cong Y, Zhang J, Chen F, Anpo M, He D (2007) Preparation, photocatalytic activity, and mechanism of nano-TiO_2 Co-doped with nitrogen and iron (III). J Phys Chem C 111:10618–10623
25. Chen X, Lu D, Lin S (2012) Preparation and properties of sulfur-doped visible-light response S-TiO_2/SiO_2 photocatalyst. Chin J Catal 33:993–999
26. Cheng C, Amini A, Zhu C, Xu Z et al (2014) Enhanced photocatalytic performance of TiO_2-ZnO hybrid nanostructures. Sci Rep 4:4181
27. Dong F, Sun Y, Fu M (2012) Enhanced visible light photocatalytic activity of V_2O_5 cluster modified N-doped TiO_2 for degradation of toluene in air. Int J Photoenergy 2012
28. Dong R, Tian B, Zhang J, Wang T, Tao Q, Bao S et al (2013) AgBr@Ag/TiO_2 core-shell composite with excellent visible light photocatalytic activity and hydrothermal stability. Catal Commun 38:16–20
29. He C, Tian B, Zhang J (2010) Thermally stable SiO_2-doped mesoporous anatase TiO_2 with large surface area and excellent photocatalytic activity. J Colloid Interface Sci 344:382–389
30. Liu Y, Xing M, Zhang J (2014) Ti^{3+} and carbon Co-doped TiO_2 with improved visible light photocatalytic activity. Chin J Catal 35:1511–1519
31. Geim AK (2009) Graphene: status and prospects. Science 324:1530–1534
32. Li Z, Wang J, Liu X et al (2011) Electrostatic layer-by-layer self-assembly multilayer films based on graphene and manganese dioxide sheets as novel electrode materials for supercapacitors. J Mater Chem 21:3397–3403
33. Chen Z, Berciaud S, Nuckolls C, Heinz TF, Brus LE (2010) Energy transfer from individual semiconductor nanocrystals to graphene. ACS Nano 4:2964–2968
34. Rao CNR, Sood AK, Subrahmanyam KS, Govindaraj A (2009) Graphene: the new two-dimensional nanomaterial. Angew Chem Int Ed 48:7752–7777

35. Huang Q, Tian S, Zeng et al (2013) Enhanced photocatalytic activity of chemically bonded TiO_2/graphene composites based on the effective interfacial charge transfer through the C–Ti bond. ACS Catal 3:1477–1485
36. Chen C, Cai W, Long M et al (2010) Synthesis of visible-light responsive graphene oxide/TiO_2 composites with p/n heterojunction. ACS Nano 4:6425–6432
37. Wang D, Choi D, Li J et al (2009) Self-assembled TiO_2/graphene hybrid nanostructures for enhanced Li-ion insertion. ACS Nano 3:907–914
38. Williams G, Seger B, Kamat PV (2008) TiO_2-graphene nanocomposites UV-assisted photocatalytic reduction of graphene oxide. ACS Nano 2:1487–1491
39. Xin X, Zhou X, Wu J et al (2012) Scalable synthesis of TiO_2/graphene nanostructured composite with high-rate performance for lithium ion batteries. ACS Nano 6:11035–11043
40. Zhang Y, Tang ZR, Fu X, Xu Y (2010) TiO_2-graphene nanocomposites for gas-phase photocatalytic degradation of volatile aromatic pollutant: is TiO_2-graphene truly different from other TiO_2-carbon composite materials? ACS Nano 4:7303–7314
41. Zhang H, Lv X, Li Y, Wang Y, Li J (2009) P25-graphene composite as a high performance photocatalyst. ACS Nano 4:380–386
42. Liang Y, Wang H, Chen Z, Dai H (2010) TiO_2 nanocrystals grown on graphene as advanced photocatalytic hybrid materials. Nano Res 3:701–705
43. Lee JS, You KH, Park CB (2012) Highly photoactive, low bandgap TiO_2 nanoparticles wrapped by graphene. Adv Mater 24:1084–1088
44. Tan LL, Ong WJ, Chai SP, Mohamed A (2013) Reduced graphene oxide-TiO_2 nanocomposite as a promising visible-light-active photocatalyst for the conversion of carbon dioxide. Nanoscale Res Lett 8:1–9
45. Sanjaya D, Ruperto G, Khiem V et al (2012) Hydrothermal synthesis of graphene-TiO_2 nanotube composites with enhanced photocatalytic activity. ACS Catal 2:949–956
46. Morales-Torres S, Pastrana-Martínez L, Figueiredo J, Faria J, Silva AT (2012) Design of graphene-based TiO_2 photocatalysts-a review. Environ Sci Pollut Res 19:3676–3687
47. Zhang L, Xi Z, Xing M, Zhang J (2013) Effects of the preparation order of the ternary P25/GO/Pt hybrid photocatalysts on hydrogen production. Int J Hydrog Energy 38:9169–9177
48. Xing M, Shen F, Qiu B, Zhang J (2014) Highly-dispersed boron-doped graphene nanosheets loaded with TiO_2 nanoparticles for enhancing CO_2 photoreduction. Sci Rep 4:6341–6346
49. Guo S, Zhen M, Liu L et al (2016) Facile preparation and lithium storage properties of TiO_2@graphene comppsite electrodes with low carbon content. Chem Eur J 22:11943–11948
50. Kamegawa T, Yamahana D, Yamashita H (2010) Graphene coating of TiO_2 nanoparticles loaded on mesoporous silica for enhancement of photocatalytic activity. J Phys Chem C 114:15049–15053
51. Li W, Wang F, Feng S et al (2013) Sol-gel design strategy for ultradispersed TiO_2 nanoparticles on graphene for high-performance lithium ion batteries. J Am Chem Soc 135:18300–18303
52. Sellappan R, Sun J, Galeckas A, Lindvall N et al (2013) Influence of graphene synthesizing techniques on the photocatalytic performance of graphene-TiO_2 nanocomposites. Phys Chem Chem Phys 15:15528–15537
53. Qiu B, Zhou Y, Ma Y, Yang X, Sheng W, Xing M, Zhang J (2015) Facile synthesis of the Ti^{3+} self-doped TiO_2-graphene nanosheet composites with enhanced photocatalysis. Sci Rep 5:8591–8589
54. Mohamed RM (2012) UV-assisted photocatalytic synthesis of TiO_2-reduced graphene oxide with enhanced photocatalytic activity in decomposition of sarin in gas phase. Desalin Water Treat 50:147–156
55. Pu X, Zhang D, Gao Y, Shao X, Ding G, Li S, Zhao S (2013) One-pot microwave-assisted combustion synthesis of graphene oxide–TiO_2 hybrids for photodegradation of methyl orange. J Alloys Compd 551:382–388
56. Li L, Yu L, Lin Z et al (2016) Reduced TiO_2-graphene oxide heterostructure as broad spectrum-driven efficient water-splitting photocatalysts. ACS Appl Mater Interfaces 8:8536–8545

57. Liu J, Niu Y, He X et al (2016) Photocatalytic reduction of CO_2 using TiO_2-graphene nanocomposites. J Nanomater 2016:1–5
58. Peng G, Ellis J, Xu G et al (2016) In situ grown TiO_2 nanospindles facilitate the formation of holey reduced graphene oxide by photodegradation. ACS Appl Mater Interfaces 8:7403–7410
59. Wu Y, Wang B, Ma Y et al (2010) Efficient and large-scale synthesis of few-layered graphene using an arc-discharge method and conductivity studies of the resulting films. Nano Res 3:661–669
60. Zhou K, Zhu Y, Yang X, Jiang X, Li C (2011) Preparation of graphene-TiO_2 composites with enhanced photocatalytic activity. New J Chem 35:353–359
61. Zhao Y, Hu C, Hu Y, Cheng H, Shi G, Qu L (2012) A versatile, ultralight, nitrogen-doped graphene framework. Angew Chem Int Ed 51:11371–11375
62. Sun H, Xu Z, Gao C (2013) Multifunctional, ultra-flyweight, synergistically assembled carbon aerogels. Adv Mater 25:2554–2560
63. Cong H, Wang P, Yu S (2014) Highly elastic and superstretchable graphene oxide/polyacrylamide hydrogels. Small 10:448–453
64. Wan C, Lu Y, Jin C, Sun Q, Li J (2014) A facile low-temperature hydrothermal method to prepare anatase titania/cellulose aerogels with strong photocatalytic activities for rhodamine B and methyl orange degradations. J Nanomater:717016
65. Zhang Z, Xiao F, Guo Y, Wang S, Liu Y (2013) One-pot self-assembled three-dimensional TiO_2-graphene hydrogel with improved adsorption capacities and photocatalytic and electrochemical activities. ACS Appl Mater Interfaces 5:2227–2233
66. Qiu B, Xing M, Zhang J (2014) Mesoporous TiO_2 nanocrystals grown in situ on graphene aerogels for high photocatalysis and lithium-ion batteries. J Amer Chem Soc 136:5852–5855
67. Hou C, Zhang Q, Li Y, Wang H (2012) P25-graphene hydrogels: room-temperature synthesis and application for removal of methylene blue from aqueous solution. J Hazard Mater 205:229–235
68. Xing M, Fang W, Yang X, Tian B, Zhang J (2014) Highly-dispersed boron-doped graphene nanoribbons with enhancing conductibilities and photocatalysis. Chem Commun 136:5852–5855
69. Fan W, Lai Q, Zhang Q, Wang Y (2011) Nanocomposites of TiO_2 and reduced graphene oxide as efficient photocatalysts for hydrogen evolution. J Phys Chem C 115:10694–10701
70. Wang Z, Huang B, Dai Y et al (2012) Crystal facets controlled synthesis of graphene@TiO_2 nanocomposites by a one-pot hydrothermal process. Cryst Eng Comm 14:1687–1692
71. Tu W, Zhou Y, Liu Q et al (2013) An in situ simultaneous reduction-hydrolysis technique for fabrication of TiO_2-graphene 2D sandwich-like hybrid nanosheets: graphene-promoted selectivity of photocatalytic-driven hydrogenation and coupling of CO_2 into methane and ethane. Adv Funct Mater 23:1743–1749

Chapter 6
Phase Control of TiO_2 Photocatalyst

6.1 Introduction

In recent years, the environmental pollution and energy exhaustion problems are attracting more and more attention of mankind on a global scale. Due to the increase of population and industrial growth, the energy consumption has been accelerating, and a large amount of toxic agents and industrial wastes have been released into the air and waterways, leading to energy crisis, global warming, and pollution-related diseases. Thus, nowadays most scientists face the major challenges to develop the environmentally harmonious, ecologically clean, safe, and sustainable chemical technologies, materials, and processes for addressing energy as well as pollution and climatic change.

Although many different kinds of approaches to solve these issues exist, ever since Honda and Fujishima [1] found that TiO_2 photoelectrode could induce cleavage of water into H_2 and O_2 under the irradiation of ultraviolet light, there has been enormous increasing interest in the use of TiO_2 as well as other extended oxide and chalcogenide semiconductors. The organic toxic materials at low concentrations are photocatalytically converted to harmless oxidation products such as CO_2 and H_2O, achieving the purpose of the environmental remediation [2–29]. TiO_2 photocatalyst with a convenient bandgap of 3.2 eV (ca. 400 nm) has many attractive features including high stability, low cost, nontoxicity, good biocompatibility, and good photocatalytic performance in the oxidation of organic pollutants to CO_2 and H_2O. For these reasons, TiO_2-based photocatalysts have drawn much attention for various applications in the fields of energy regeneration and environmental protection.

As mentioned above, due to its excellent properties, lots of systematic in-depth studies of TiO_2 has been performed by the majority of scientists, promoting the application process in various aspects related to energy and environment in recent years. The related research mainly includes studies on photocatalytic mechanism [3–5], regulation of crystal structure and morphology [6–15], improvement of

J. Zhang et al., *Photocatalysis*, Lecture Notes in Chemistry 100,
https://doi.org/10.1007/978-981-13-2113-9_6

photocatalysis and photoelectric conversion efficiency [16–24], and expansion of the optical response range [25–29]. However, traditional single-phase TiO_2 has defects in the photocatalytic performance, which are mainly on account of two aspects: on one hand, owing to the limits of its bandgap, the absorption of visible light is little, causing the inefficient utilization of the sunlight [3–5]; on the other hand, the high recombination rate of photo-generated electrons and holes greatly limits the photocatalytic performance of TiO_2 [30, 31].

Researches have reported that modification of TiO_2 by organic dye photosensitization [32–36], noble metal deposition [37–41], or doping [42–45] or semiconductor compounding [45–49] can effectively enhance the photon utilization and reduce the recombination rate of photo-generated electrons with holes, thereby improving the photocatalytic efficiency of TiO_2. However, those methods have many disadvantages such as low economic effect, poor reaction stability and controllability, and complicated operation process, among others. In these respects, mixed-phase TiO_2 with relatively simple preparation process, low cost, and adjustable crystal type has attracted much attention. In the mixed-phase photocatalyst, the effective separation of the photo-generated electrons and holes, taken place on the two-phase interface, results in the reduction of their recombination rate. Moreover, adding rutile TiO_2 with a narrower bandgap into pure anatase TiO_2 to form mixed-phase TiO_2 of rutile and anatase can enhance the utilization of visible light to some extent, further significantly improving the photocatalytic activity of TiO_2 [50–53]. These abovementioned advantages of the mixed phase can efficiently enhance the comprehensive properties of TiO_2 photocatalyst. Apart from these researches, other kinds of mixed-phase TiO_2 have also been studied in depth, such as anatase/brookite [54, 55], anatase/brookite/rutile [56], and brookite/TiO_2(B) [57]. The results of these photocatalytic studies have demonstrated that the photocatalytic activity of mixed-phase TiO_2 is higher than that of single-phase TiO_2, which results from the existence of different band positions and the promoted separation rate of photo-generated electron–hole pairs and catalytic "hot spots" at the interface of different phases [51, 56].

In this chapter, first, we briefly introduce three main kinds of TiO_2 phases (anatase, rutile, and brookite), and then we focus on recent advances in the development of mixed-phase TiO_2 photocatalysts, including the synthesis for the mixed-phase catalysts and their applications to various photocatalytic reaction systems such as photocatalytic hydrogen production, photoreduction of CO_2, and photocatalytic degradation of organic pollutants. Following this we discuss the mechanism of enhanced photocatalytic activity of the mixed-phase TiO_2. Finally, the existing problems of the mixed-phase TiO_2 are summarized, and the application prospects of this kind of nanomaterials are outlooked.

6.2 Phases of TiO_2

TiO_2 possesses eight types of crystal phases, which are rutile, anatase, brookite, TiO_2-B, TiO_2-R, TiO_2-H, TiO_2-II, and TiO_2-III [58]. Among them, rutile, anatase, and brookite have been researched mostly for applications because they are naturally occurring oxides of titanium at atmospheric pressure [59, 60]. The other five phases have also been investigated [61–64]; however, because they are usually formed at high pressure, there is minor significance for practical research and applications. Therefore, in this chapter, we mainly focus on the three phases of rutile, anatase, and brookite.

6.2.1 Structure Properties of Rutile, Anatase, and Brookite

The three kinds of TiO_2 phases possess different crystallographic properties, as shown in Table 6.1 [65].

In their structures, TiO_6 octahedron exists in a more or less tortuous configuration formed by the fundamental building block made up of a titanium atom surrounded by six oxygen atoms. In the structures of the three kinds of TiO_2 phase, the stacking of the TiO_6 octahedra causes threefold coordinated oxygen atoms. As shown in Fig. 6.1, the structural units in these three TiO_2 crystals form from TiO_6 octahedron basic units and exist in different lattice configurations. In rutile, the chains were formed by TiO_6 octahedra connected by sharing an edge with the c-axis and then

Table 6.1 Crystallographic properties of anatase, rutile, and brookite TiO_2

Crystal structure	Density (g/cm [3])	System	Space group	Cell parameters (nm)		
				a	b	c
Rutile	4.240	Tetragonal	D_{4b}[14]-$P4_2$/mnm	0.4584	–	0.2953
Anatase	3.830	Tetragonal	D_{4a}[19]-$I4_1$/amd	0.3758	–	0.9514
Brookite	4.170	Rhombohedral	D_{2h}[15]-Pbca	0.9166	0.5436	0.5135

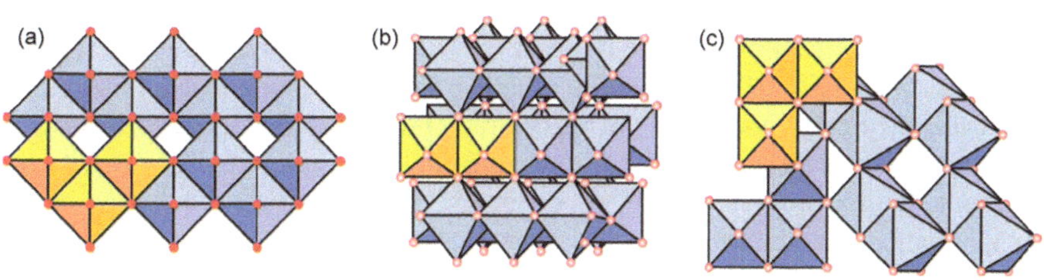

Fig. 6.1 Crystalline structure of (**a**) anatase, (**b**) brookite, and (**c**) rutile [68]. (Reprinted with permission from Ref. [68]. Copyright 2010, Elsevier)

interlinked by sharing corner oxygen atoms to form a three-dimensional lattice. Contrarily in anatase, the three-dimensional lattice is formed only through sharing edge bonding among TiO_6 octahedra. It means that the octahedra in anatase share four edges and are arranged in zigzag chains while the octahedra in brookite share both edges and corners and form an orthorhombic structure [66, 67].

Researchers usually use the X-ray diffraction (XRD) experimental method to determine these crystal structures and estimate the crystal grain size of anatase, rutile, and brookite. Anatase peaks in X-ray diffraction are found at $\theta = 12.65°$, 18.9°, and 24.054°; the rutile peaks occur at $\theta = 13.75°$, 18.1°, and 27.2°, while brookite peaks emerge at $\theta = 12.65°$, 12.85°, 15.4°, and 18.1° (θ represents the X-ray diffraction angle) [69, 70].

6.2.2 *Stability and Phase Transformation*

Compared to the other two types of phases, rutile is the most stable phase. It is almost impossible for rutile to decompose or undergo a phase transformation even at very high temperatures. However, metastable anatase and brookite can be transformed into thermodynamically stable rutile when they are calcined to a certain temperature. There has been extensive research on the phase transfer mechanism of TiO_2 during the calcination process. Shannon [71] proposed that the transformation of anatase to rutile from crystallography includes a nucleation and growth process. At first, rutile nucleates on the surface of anatase and then expands to the bulk. Due to the great diversities between anatase and rutile, the transformation involving the breaking and reforming of bond processes can be reconstructed [72]. During the course of anatase transforming to rutile, the {112} planes in anatase are persisted as the {100} planes in the newly generated rutile. And Ti and O atoms synergistically rearrange in these planes by moving Ti atoms to a new location to form rutile via the breaking of two Ti–O bonds in the TiO_6 octahedron. As a result, the newly formed oxygen vacancies accelerate the transformation; meanwhile, Ti interstitials inhibit the phase transformation. The transformation from anatase to rutile is a nonequilibrium phase transition, which usually occurs at a certain range of temperature (400~1000 °C). During the transformation process, the calcination temperature has a great influence on the impurities, particle size, and surface area of products. Because impurities and processing atmosphere can result in different defect structure, they also strongly influence the temperature and rate of phase transition. Generally, impurities such as the oxides of Li, K, Na, Fe, Ce, and Mn usually promote the phase transformation via increasing the oxygen vacancies; on the contrary, impurities like S, P, and W usually restrain the phase transformation. A reductive atmosphere such as H_2 and Cl_2 can accelerate the transformation, while a conducive atmosphere can inhibit the phase transformation via the formation of Ti interstitials.

Artificial synthesis conduces to the preparation of anatase nanoparticles, especially the synthesis of TiO_2 in aqueous solution [73] for the reason that the energy of the three kinds of TiO_2 phases is quite close. On condition that the nanoparticles are small enough (<13 nm), the small surface free energy will play a decisive role in the phase transformation [74]. For TiO_2 nanocrystals with the size smaller than 11 nm, anatase is the most stable phase, and the rutile phase with nanocrystals size bigger than 35 nm performs the thermodynamic stability. As for brookite phase, its stability lies between the anatase and rutile. In fact, as a metastable phase, the major physical parameters of brookite are between those of anatase and rutile. For example, the bandwidths of anatase and rutile are 3.19 and 3.0 eV, respectively, while that of brookite is 3.11 [75].

6.2.3 *Photocatalytic Activity of Rutile, Anatase, and Brookite*

Among the three types of phases, anatase exhibits the highest photocatalytic activity, which is on account of the following aspects:

1. The bandgap of anatase is 3.19 eV, while that of rutile and brookite are 3.0 eV and 3.11 eV, respectively [62]. Therefore, the electron–hole pair of anatase has more positive or more negative potential, improving the oxidation ability [76].
2. The surface of anatase has a stronger adsorption ability for H_2O, O_2, and OH, which conduces to high photocatalytic activity because the adsorption capacity of the surface has a dramatic influence on the photocatalytic activity during the photocatalytic reaction, and strong adsorption capacity benefits to high activity.
3. Compared to rutile and brookite, anatase usually exhibits smaller grain size and larger specific surface area in the crystallization process, enhancing the photocatalytic activity.

However, due to the crystallization process greatly influencing the photocatalytic activity, the above rules do not suit for all situations. When the amorphous TiO_2 crystallize, rutile usually forms large grains with poor surface properties and thus exhibits low photocatalytic activity. Under the same conditions, if the rutile can have the same grain size and adsorption ability as anatase, it can also exhibit high photocatalytic activity. For example, Lee et al. [77] found that through the treatment of laser exposure, anatase phase of TiO_2 can transfer to rutile phase without the change of the specific surface area and grain size. This resultant rutile TiO_2 exhibited considerable high photocatalytic activity. Tsai et al. [78] fabricated anatase and rutile TiO_2 via different methods and investigated their photocatalytic activity for the degradation of phenol. It is found that the preparation methods as well as the calcination temperature have a remarkable influence on the photocatalytic activity of TiO_2 catalyzers. Under certain conditions, rutile TiO_2 exhibited very high catalytic activity. Therefore, whether anatase or rutile, the photocatalytic activity of materials greatly depends on the grain size and the surface properties. Apart from that, Ohno et al. [79] demonstrated that the photocatalytic activity of different phase TiO_2 relates to the electron acceptors in the system. When the electron acceptor is O_2, the photocatalytic activity of anatase is relatively higher than that of rutile. And when Fe^{3+} is the electron acceptor, rutile exhibits higher catalytic activity than anatase. This is because when O_2 acts as an electron acceptor, it is quite sensitive to properties of the catalyst materials in the process of the photocatalytic reactions. The surface structure and the low-energy band of rutile may make it show low transfer efficiency of electrons to O_2, and thus rutile exhibits low catalytic activity when O_2 is used as the electron acceptor. Therefore, rutile usually exhibits low photocatalytic activity based on the fact that most researchers use O_2 for the studies of photocatalytic reactions.

Because there have been few studies and experiments on brookite and the general products are mixed-phase TiO_2, including anatase/brookite [80], brookite/rutile [81], or anatase/brookite/rutile instead of pure brookite, the photocatalytic activity of brookite TiO_2 is quite controversial [44].

Recently, lots of studies have demonstrated that the mixed-phase TiO_2 crystals in an appropriate composition ratio exhibit higher photocatalytic activity than that of single-phase TiO_2 crystals. Bacsa et al. [82] found that the single-phase TiO_2 such as pure anatase or pure rutile has relatively lower photocatalytic activity, but a mixture of anatase and rutile in various ratios achieved enhanced photocatalytic activity than 100% anatase or 100% rutile, when the mixed phase in the ratio of 30% rutile and 70% anatase had the best photocatalytic activity. Thus the TiO_2 catalyzer with two kinds of phases has a certain synergistic effect for improving photocatalytic activities. And the commercial photocatalyst P25 with considerable high photocatalytic activity is also a kind of mixed-phase TiO_2 instead of single-phase TiO_2. Therefore, great efforts have been made to study in depth on the synthesis and photocatalysis applications of mixed-phase TiO_2.

6.3 Synthesis of Mixed-Phase TiO_2 Photocatalysts

It has been found that adjusting the experimental parameters of the synthesis can change the structure and properties of the three types of crystals, making the preparation of different mixed-phase TiO_2 come true. At present, there is a large number of methods to fabricate mixed-phase TiO_2, such as pulsed laser deposition (PLD), hydrothermal method, hydrolysis method, solvothermal method, microemulsion-mediated solvothermal method, solvent mixing and calcination method (SMC method), high-temperature calcination method, high-temperature vapor decomposition method, and so on [83]. In the synthesis process of mixed-phase TiO_2, with different preparation methods, the same factor has different effects on the mixed-phase crystals. In all kinds of synthesis methods, the mixed-phase ratio, morphology, and surface properties of photocatalysts are controllable via optimizing the experimental conditions, such as calcination temperature, pressure, concentration, and types of reagents. The ratio of different crystals in an excellent mixed-phase photocatalyst should have the ability to be tuned according to the actual demands, and the product should possess controllable shape as well as uniform dispersion and should be difficult to agglomerate. The preparation conditions and influence parameters of mixed-phase TiO_2 will be discussed in the following subsection, combining concrete synthesis methods.

6.3.1 Hydrothermal Method and Solvothermal Method

The hydrothermal method is used most widely in the laboratory fabrication of TiO_2 nanomaterials [25, 84–90]. Using aqueous solution as the reaction system, through heating or other means, the target photocatalysts are often synthesized in a sealed reaction vessel (such as an autoclave), which creates an environment of high temperature and high pressure during the hydrothermal process. In this condition, with the increase of the water vapor pressure, the density and surface tension of the materials decrease, and the ion product increases, leading to the dissolution and the recrystallization of the substances which can be dissolved in water at room temperature. The hydrothermal method is one of the liquid-phase chemical synthesis methods, and its main factors influencing the properties of the product are vessel volume, reaction temperature, heating rate, and reaction time.

Chen and Sum et al. [91] fabricated highly crystalline single-phase TiO_2 brookite and two-phase anatase/brookite TiO_2 nanostructures in NaOH solutions via a simple hydrothermal method, using TiS_2 as the precursors. They successfully controlled the phase composition via varying the solution concentration as well as reaction time. In the reaction system, the precursors TiS_2 underwent a hydrolysis reaction as shown in the following two equations:

$$TiS_2 + 4\,NaOH \rightarrow Ti^{4+} + 4\,OH^- + 2\,Na_2S \qquad (6.1)$$

$$Ti^{4+} + x(OH)^- \rightarrow Ti(OH)_x^{4-x} \qquad (6.2)$$

The reaction began with the hydrolysis of TiS_2, forming Ti^{4+} ion (Eq. 6.1). Then Ti^{4+} reacted with OH^- in the solution during the hydrothermal process (Eq. 6.2). The formation of different crystal phases is related to concentration of OH^-. When the concentration of OH^- is low and thus the hydrolysis rate of Ti^{4+} is slow, the hydrothermal reactions are more likely to form anatase TiO_2 via connecting Ti^{4+} with OH^-; when the OH^- concentration is high, the fast hydrolysis rate of Ti^{4+} leads to the formation of mixed-phase TiO_2 nanomaterials of brookite and anatase. The mass fraction of brookite (B) and anatase (A) can be calculated by the following two formulas [92]:

$$W_B = K_B I_B / (K_A I_A + K_B I_B) \qquad (6.3)$$

$$W_A = K_A I_A / (K_A I_A + K_B I_B) \qquad (6.4)$$

In the formula, $K_A = 0.886$, $K_B = 2.721$, and I_A represent the integral intensity of the strongest diffraction peaks in anatase, and I_B represents that in brookite. The

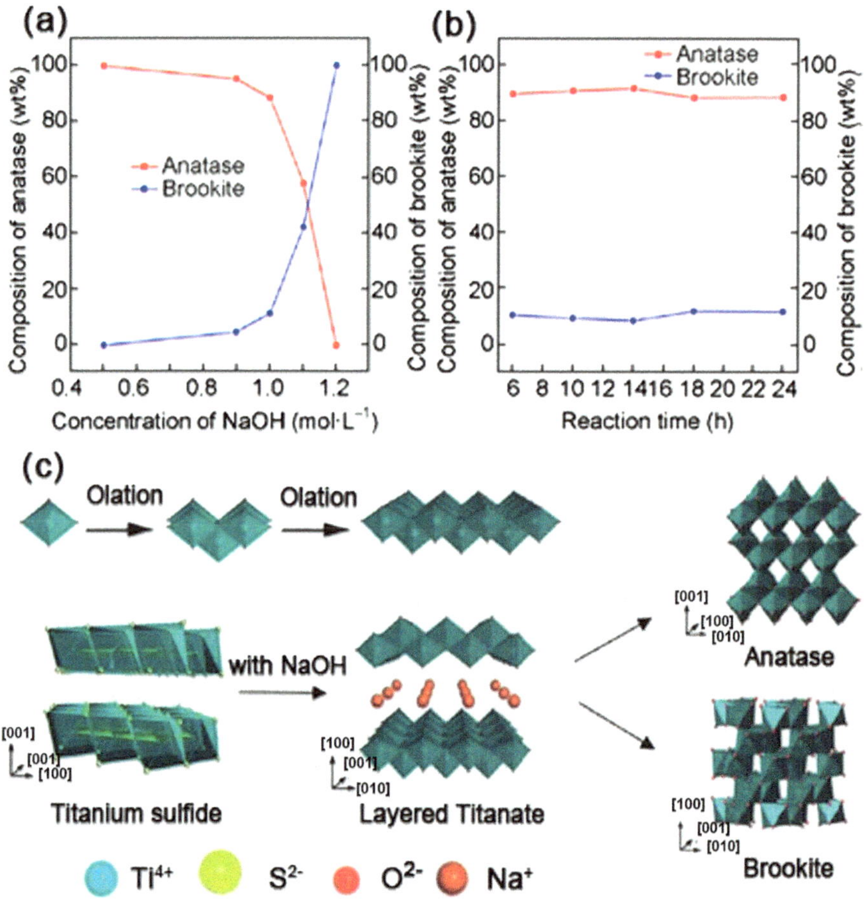

Fig. 6.2 The composition of anatase and brookite in the mixed-phase TiO_2 prepared with (**a**) various NaOH concentrations and (**b**) different reaction times under hydrothermal reaction [91]; (**c**) The scheme of formation of anatase and brookite in mixed-phase TiO_2 [91]. (Reprinted with permission from Ref. [91]. Copyright 2013, American Chemical Society)

influences of different NaOH concentrations and distinct reaction times on TiO_2 crystals are shown in Fig. 6.2a, b. As NaOH concentration increases, the content of brookite increases, and when the reaction time is over 6 h, the crystalline formation of the product does not change significantly. In addition, single-phase brookite TiO_2 can be obtained by replacing TiS_2 with sodium titanate. Since Na^+ can stabilize the layered structure, when there is no Na^+ ion in the reaction system, a layered structure is instable, and the hydrothermal process causes the destruction of the structure, forming anatase TiO_2; when Na^+ is excessive, the stable layered structure does not easily collapse and ultimately form titanate; when Na^+ concentration is proper, the layered structure partially collapses, and part of the layered structure remains

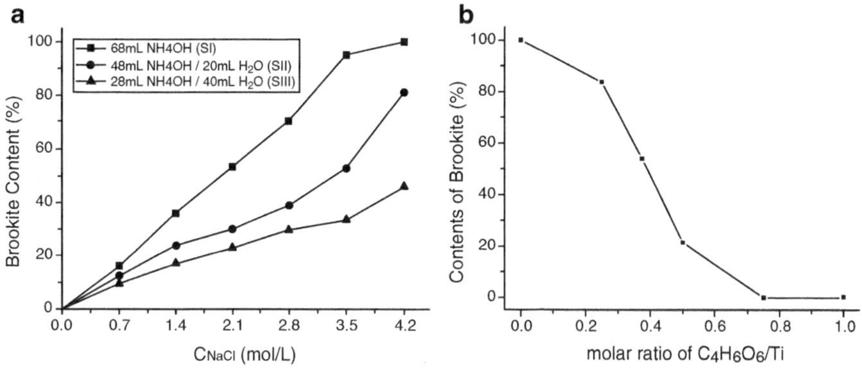

Fig. 6.3 (**a**) Relationship between the contents of brookite and the applied NaCl concentration [93]; Reprinted with permission from Ref. [93]. Copyright 2013, Elsevier. (**b**) Relationship between the contents of brookite in the products and the applied tartaric acid to $TiCl_3$ molar ratio [97]. Reprinted with permission from Ref. [97]. Copyright 2012, Springer

instable, leading to form brookite (Fig. 6.2c). The above results are consistent with the previous literature [87].

Zhang et al. [93] simultaneously introduced NaCl and NH_4OH into the hydrothermal reaction system with $TiCl_3$ as a titanium source. The results showed that changing the concentration of NaCl and the volume ratio of NH_4OH to H_2O can well control the phase composition of the products in the reaction system (Fig. 6.3a). They put forward that Na^+ was able to stabilize the layered structure of titanate, which was formed by the hydrolysis reaction of $TiCl_3$ in aqueous ammonia in the hydrothermal system. In the absence of NaCl, NH^{4+} ions which exist in the interlayer of titanate can balance the negative charges of the layered titanate. However, during the process of hydrothermal treatment with high temperature up to 200 °C, NH^{4+} is hydrolyzed to form NH_4OH and thus is released from the interlayer. Then the H^+ ions generated from the hydrolysis of NH^{4+} ions induce a hydroxyl condensation reaction, forming anatase TiO_2. In the presence of NaCl, Na^+ ions benefit to stabilize the layered structure with NH_4^+. The simultaneous existence of Na^+ and NH_4^+ in the interlayer of titanate helps to balance the negative charges of titanate. After the layered titanate was continuously treated by hydrothermal methods at a high temperature for several hours, the increased hydrolysis of NH_4^+ caused the collapse of layered structures. However, some bonds in the structure were kept by Na^+ ions. Hence the structural transformation was delayed which resulted in forming a brookite-like structure [94–96]. A brookite lattice was formed by the expansion of as-formed brookite-like structure. Increasing the concentration of NaCl would be helpful for the brookite lattice to compete with the anatase lattice. Therefore, with the increase of NaCl concentration, the brookite content in the mixed-phase TiO_2 increased. In addition to this research, Zhang et al. [97] developed another facial hydrothermal method for the synthesis of anatase/brookite TiO_2 using tartaric acid ($C_4H_6O_6$) as the phase content regulator, $TiCl_3$ as the titanium source, and NaOH to adjust the pH of the reaction solution. Changing the molar ratio of $C_4H_6O_6$ to $TiCl_3$

in the hydrothermal reaction system could successfully control the contents of brookite and anatase in the TiO_2 particles (Fig. 6.3b). According to the ligand field theory, they explained the mechanism of the phase evolution between brookite and anatase in which $C_4H_6O_6$ could chelate with Ti to form a stable titanium complex [98]. When the $C_4H_6O_6/TiCl_3$ molar ratio was below 0.75 in the reaction system, there were two forms of Ti species. They were the insoluble Ti-contained species Ti $(OH)_4$ and the soluble Ti-contained complex $[Ti(OH)x(C_4H_6O_6)y]z$-. Under hydrothermal treatment, the amorphous $Ti(OH)_4$ could first transform to layered titanate. Given a proper concentration of Na^+ and OH^-, the layered titanate would be transformed to brookite [94]. Due to the large steric hindrance of carboxylic acidic ligands, the $[Ti(OH)x(C_4H_6O_6)y]z$- complexes were inclined to combine together by sharing equatorial or apical edges and being arranged in zigzag chains, which was beneficial to form anatase crystallites, thus resulting in an mixed-phase TiO_2 of anatase and brookite. They also pointed out that the brookite as the pure phase or the main phase could not be obtained on the conditions that the pH value of the reaction system was lower than 9 or the hydrothermal temperature was below 180 °C. In other words, it was hard to control the contents of anatase and brookite in samples just by changing the $C_4H_6O_6/TiCl_3$ molar ratio while $pH < 9$ or temperature < 180 °C in the hydrothermal reaction system.

A variation of the hydrothermal method is the solvothermal method wherein many kinds of organic solvents such as ethanol, glycol, and toluene can be used to replace an aqueous system. Similar to the hydrothermal method, both the crystal types and morphology of the TiO_2 nanomaterials can be well controlled by regulating parameters in the solvothermal reaction system, including temperature and pressure inside the system, the reaction time, and the titanium source.

Through the solvothermal method, Li et al. [99] synthesized anatase/rutile mixed-phase TiO_2 crystals with different rutile content by hydrolysis of tetraisopropyltitanate in an acid alcoholic solution and studied influence of hydrochloric acid on the rutile content in the mixed-phase crystal. The results showed that high H_2O/Ti mole ratio favored the formation of brookite/anatase mixed-phase TiO_2, and low H_2O/Ti mole ratio benefited to form anatase/rutile mixed-phase TiO_2. Lei et al. [100] prepared anatase/rutile mixed-phase TiO_2 crystal using a low-temperature (80 °C) solvothermal method by pre-oxidizing $TiCl_3$ into Ti^{4+} with HNO_3 followed by diluting with urea, water, and ethanol. They found that the anatase content in the mixed-phase crystal increased by increasing ethanol content in the solution (Fig. 6.4). The average particle size of anatase and rutile in the mixed-phase crystal is both below 10 nm calculated by the Scherrer formula.

The mixed-phase TiO_2 nanomaterials synthesized by hydrothermal and solvothermal methods are usually well crystallized and are not required to be calcinated at a certain high temperature. The size and phase type of TiO_2 can be regulated by simply adjusting the experimental parameters such as the type of base or acid in the reaction system, the reaction temperature, the heating rate, the reaction time, and the autoclave pressure. Besides these advantages, there still remain some drawbacks such as the requirement for the equipment to withstand high pressure as well as high temperature and experimental safety concerns. Moreover, it is difficult

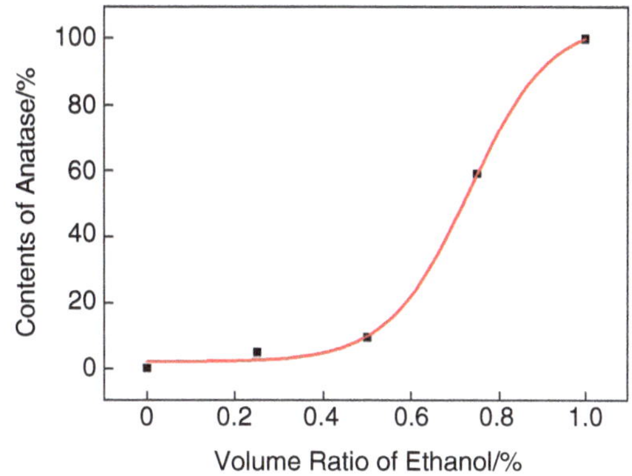

Fig. 6.4 Relationship between the contents of anatase in the product and the volume ratio of ethanol in a solvothermal reaction system [100]. (Reprinted with permission from Ref. [100]. Copyright 2008, Elsevier)

to monitor the real time of the hydrothermal and solvothermal process, and the reaction usually proceeds without stirring, sometimes leading to the incomplete reaction and irregularly arranged product. Due to the fact that a variety of organic solvents can be used in the solvothermal method, it is easier to control the crystalline and morphology of the products than the hydrothermal method. Therefore, we can foresee that the solvothermal method has better application prospects.

6.3.2 Microemulsion-Mediated Solvothermal Method

The method of microemulsion usually contains two processes. Firstly, two immiscible solvents divided into many micro-reactors (microbubbles) with the help of the surfactants form emulsion, in which nanomaterials are obtained through nucleation, coalescence, agglomeration, and heat treatment process in the oil-in-water or water-in-oil microbubbles. The reactants of immiscible solvents are well dispersed, and a uniform nucleation limited in the micro-reactors occurs. Therefore, the particle size and stability of the nanomaterials can be well controlled. Compared with other traditional synthesis methods, the microemulsion method has shown obvious advantages in terms of the preparation of nanomaterials with perfect monodispersity and interfacial properties. Moreover, this method is a very versatile technique which allows the synthesis of a great variety of nanomaterials by combining with other techniques. Recently, the microemulsion-mediated solvothermal method, which is named for the combination of the microemulsion method with solvothermal method, has been used to the preparation of the mixed-phase TiO_2 nanomaterials [66, 101, 102].

Yan et al. [66] fabricated anatase/rutile mixed-phase TiO_2 crystal through the combination of microemulsion and solvothermal method, using tert-octylphenoxypolyethoxyethanol (Triton X-100) as the surfactant, n-hexanol as the co-surfactant, and cyclohexane as the continuous oil phase. Tetrabutyl titanate and $(NH_4)_2SO_4$ were completely dissolved in the hydrochloric acid to serve as the

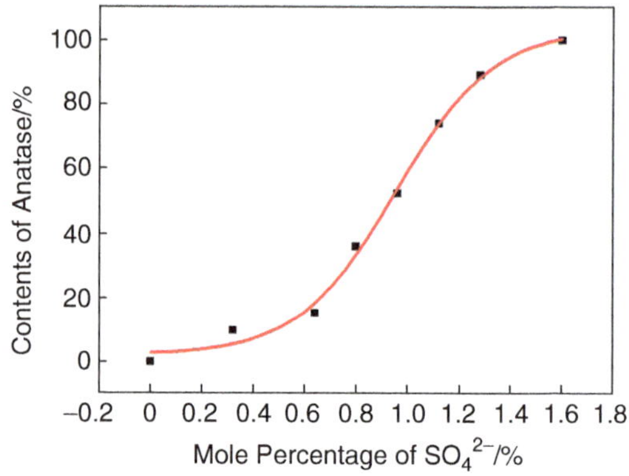

Fig. 6.5 Relationship between the contents of anatase in the product and the mole percentage of SO_4^{2-} in the microemulsion-mediated hydrothermal reaction system [66]. (Reprinted with permission from Ref. [66]. Copyright 2005, American Chemical Society)

aqueous phase. Then the aqueous phase was added dropwise to the oil phase to form a clear microemulsion. After that, the obtained solution was solvothermally treated below 120 °C for 13 h to produce the mixed-phase TiO_2. The surface area of the resultant product is about 86~169 m^2/g. Besides grain size of anatase is 15 nm and that of rutile is about 10 nm. Their results showed that the adjustment of SO_4^{2-} concentration has obvious influence on the content of anatase in the mixed-phase TiO_2 (Fig. 6.5). With the increase of SO_4^{2-} concentration, the anatase content in the mixed-phase TiO_2 increased. Different polymorphs can be achieved by affecting the polycondensation of TiO_6 octahedra with SO_4^{2-} (Fig. 6.6a). During the hydrothermal reaction process, the SO_4^{2-} interacts with the hydroxy groups on the surface of TiO_6 octahedra, forming hydrogen bonds (Fig. 6.6b). Because of the steric effect of SO_4^{2-}, the polycondensation of the octahedron with SO_4^{2-} and another octahedron along the converse direction results in the decrease of the repulsion (Fig. 6.6c), and the orientation of the third octahedron is more conducive to form an anatase nucleus (Fig. 6.6d). When the concentration of SO_4^{2-} is low, a rutile structure can be easily formed for the sake of the lowest free energy in the system. Therefore, controlled adjustment of anatase content in mixed-phase TiO_2 can be achieved by changing the concentration of SO_4^{2-}. In the same oil-phase system, Zhang et al. [102] synthesized mixed-phase TiO_2 crystal in different ratio of anatase and rutile content by the microemulsion-mediated solvothermal method. In the reaction system, a mixture of $TiCl_3$, urea, and hydrochloric acid solution was used as the aqueous phase. The results showed that anatase content in the mixed-phase TiO_2 increased with urea concentration increasing. The hydrochloric acid in the reaction solution served as an inhibitor to the generation of anatase, promoting the formation of rutile phase. Because TiO_6 octahedra polycondensed in chain to form a rutile structure was in accordance with the minimum energy principle, Cl^- and SO_4^{2-} had different influence on the formation of TiO_2 phases. A high concentration of Cl^- with small radius and weak steric hindrance favored the formation of rutile TiO_2, while a high concentration of SO_4^{2-} benefited to form anatase TiO_2.

Among the synthesis methods of mixed-phase TiO_2, the microemulsion-mediated solvothermal method is relatively complicated. However, the grain size of the

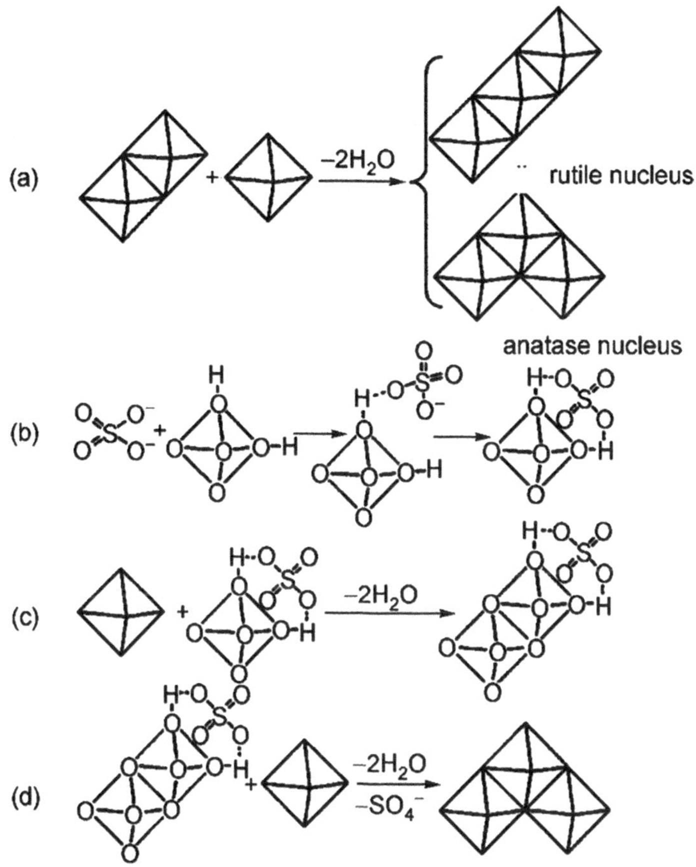

Fig. 6.6 Proposed mechanism: (**a**) the orientation of the third octahedron determines the formation of rutile or an anatase nucleus; (**b**) interaction between SO_4^{2-} anion and TiO_6^{2-} octahedral hydroxyls; (**c**) two TiO_6^{2-} octahedra share edge when SO_4^{2-} is present; (**d**) the formation of anatase in the presence of SO_4^{2-} [66]. (Reprinted with permission from Ref. [66]. Copyright 2005, American Chemical Society)

product is small, and the phases can be well controlled, which makes this method to have good prospects for development.

6.3.3 Sol–Gel Method

Sol–gel method is a commonly used method for synthesis of amorphous and crystalline TiO_2 nanomaterials for the sake of its advantages such as processing at low temperature and the ability to prepare materials in various morphologies. Titanium alkoxides or titanium halides are usually used as the titanium precursor

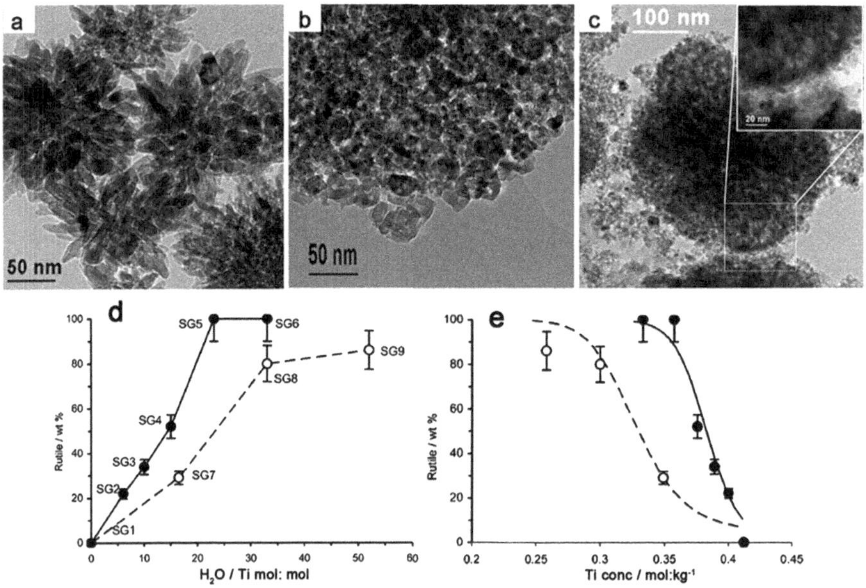

Fig. 6.7 TEM images of (**a**) 100% rutile, (**b**) 100% anatase, and (**c**) mixed-phase TiO_2 with 48 wt % anatase and 52 wt % rutile; plots of rutile content (wt %) in the product (**d**) vs H_2O/Ti molar ratio (**e**) vs titanium concentration [52]. (Reprinted with permission from Ref. [52]. Copyright 2008, American Chemical Society)

in the sol–gel reaction system. Firstly, the hydrolysis of titanium precursor forms a sol. After aging the sol for a certain time, a three-dimensional cross-linked gel is obtained, and amorphous white powder is obtained by grinding the gel. Finally, crystallized TiO_2 is obtained by the treatment of the calcination at a certain high temperature.

Scotti et al. [52] prepared anatase/rutile mixed-phase TiO_2 crystal by a sol–gel method and studied the effects of HCl/Ti and H_2O/Ti molar ratio on the phase type of TiO_2 crystal. At first, $TiCl_4$ and triblock copolymer were dissolved in ethanol. Then water and HCl were added to adjust the pH of the solution, obtaining a sol. After aging the sol for 3–13 days, a gel was formed. After drying and calcination, mixed-phase TiO_2 crystals consisting of anatase and rutile in different content were obtained. In the products, pure rutile-phase TiO_2 (Fig. 6.7a) displayed chestnut burr aggregates of elongated nanocrystals in radial growing shape with average sizes of 10–20 nm in width and 100–200 nm in length. The pure anatase phase (Fig. 6.7b) showed aggregates of almost square-ended nanoparticles whose average sizes were 5–15 nm. The mixed-phase sample (Fig. 6.7c) was observed to possess two types of phases with the chestnut burr aggregates of rutile surrounded by the small anatase particles. Changing the H_2O/Ti (*r*w) and HCl/Ti (*r*a) molar ratios systematically can well control the phase content of the products. Results have proved that, in the titanium alkoxide or acidic titanium halide precursor solution, with the increase of H_2O content, the concentration of Ti^{4+} decreased, leading to the

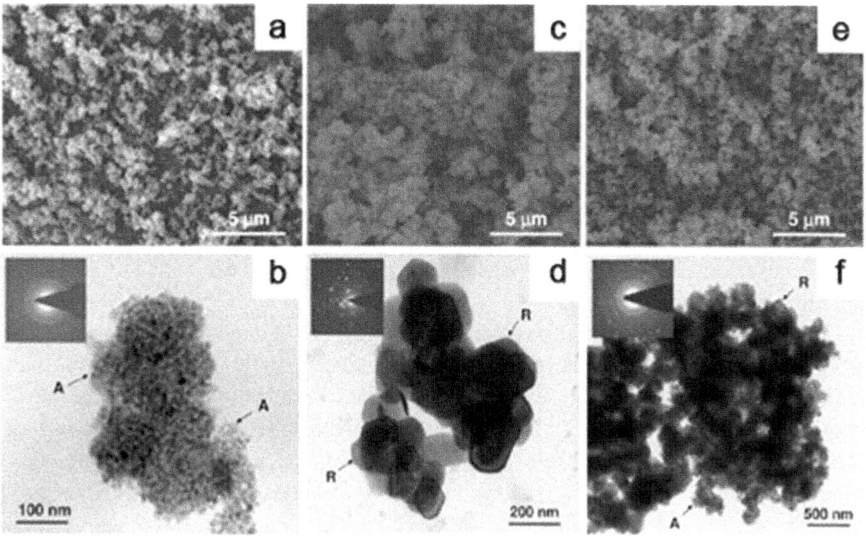

Fig. 6.8 SEM and TEM images of (**a**, **b**) pure anatase, (**c**, **d**) pure rutile, and (**e**, **f**) mixed-phase TiO_2 with an optimum rutile content of 40 wt% prepared by the SMC method [103]. (Reprinted with permission from Ref. [103]. Copyright 2009, American Chemical Society)

decrease of the condensation reaction rate of TiO_6 octahedra to form rutile TiO_2 with better thermodynamically stability (Fig. 6.7d, e).

Mixed-phase TiO_2 crystals with small particles and high purity can be obtained by the sol–gel method, but the experimental procedure usually requires longtime aging, and the agglomeration of particles is easy to occur after high-temperature calcination, which may cause significant decrease of the photocatalytic activity.

6.3.4 Solvent Mixing and Calcination Method

The solvent mixing and calcination method (SMC method) means to mix TiO_2 with different phases in a certain solvent at first and then to evaporate the solvent completely, finally followed by calcinations at high temperature in order to make a close contact among different crystal phases. Zachariah et al. [103] prepared anatase/rutile mixed-phase TiO_2 by the SMC method wherein anhydrous isopropanol was used as the solvent. According to the Scherrer equation, the crystalline calculated size of pure anatase was 10 nm, and the grain size of pure rutile was 40 nm. In the resultant mixed-phase TiO_2 containing 40% rutile, the size of anatase was 29 nm and that of rutile was 47 nm. From Fig. 6.8a, b, it could be observed that pure anatase consisted of small particles whose size was 100~150 nm. And as Fig. 6.8c, d shows, the particle size of pure rutile was 300~500 nm. The morphology of the mixed-phase TiO_2 nanomaterial was a mixture of anatase and rutile particles (Fig. 6.8c, f). From the high-resolution transmission electron microscopy (HRTEM), the phase interface could be seen clearly, indicating a close interaction existing between the two phases. Bojinova et al. [104] synthesized an anatase/rutile mixed-phase TiO_2 by the SMC method using ethanol as the solvent. In the product, the grain size of anatase was 42 nm and that of rutile was 56 nm which are calculated according to the Scherrer formula. Liu et al. [105] used a layer-by-layer self-assembly technique to make the anatase small particles adsorbed onto rutile nanorods with assist of polystyrene sulfonate (PSS) as a medium, and finally the PSS was removed by calcination. The anatase content in the mixed-phase TiO_2 crystal could be controlled by adjusting the number of loading cycles.

The phase content of the mixed-phase TiO_2 can be regulated easily by using the SMC synthesis method. However, because the final calcination process requires high temperature, the TiO_2 is easy to agglomerate. In addition, it is usually difficult to obtain the mixed-phase TiO_2 with anatase and rutile mixing uniformly, causing products with many pure rutile aggregates as well as pure anatase aggregates. Therefore, the photocatalytic performance of the mixed-phase TiO_2 crystals produced by this method is limited.

6.3.5 High-Temperature Calcination Method

The high-temperature calcination method, which is mostly used to research the phase transition of TiO_2, is one of the earliest methods for the synthesis of mixed-phase TiO_2 [25].

Through the high-temperature calcination method, Nair et al. [106] succeeded in preparing the mixed-phase TiO_2 nanomaterials with different anatase and rutile contents. They firstly prepared small anatase particles by a sol–gel method followed by calcinating the anatase particles at elevated temperatures. It was found that the samples which were calcined at low temperatures ($T < 600$ °C) remained pure anatase phase. The phase transformation began at the temperature of 650 °C. When the calcination temperature was raised to 850 °C, the product converted to pure rutile. As calculated by the Scherrer formula, the grain sizes of anatase and rutile in the mixed-phase products were 33.66~51.48 nm and 45.2~60.6 nm, respectively. Figure 6.9 showed that the particle size of anatase was about 100 nm and that of rutile was about 200 nm, and the mixed-phase particle size was between the above two and increased with the increase of calcination temperature. The phase transformation includes nucleation and growth. However, there is some dispute on the concrete process. Gouma et al. [107] considered that the rutile firstly formed nuclear on the surface of anatase and then expanded to the bulk, while Zhang et al. [108] verified the phase transformation of anatase–rutile occurred in the bulk at first and then spread to the surface with the increase of calcination temperature via UV Raman spectroscopy. Figure 6.10 shows the phase's transformation.

The simple high-temperature calcination method can be used to obtain mixed-phase TiO_2 nanomaterials with perfect polymorphs and tunable phase content.

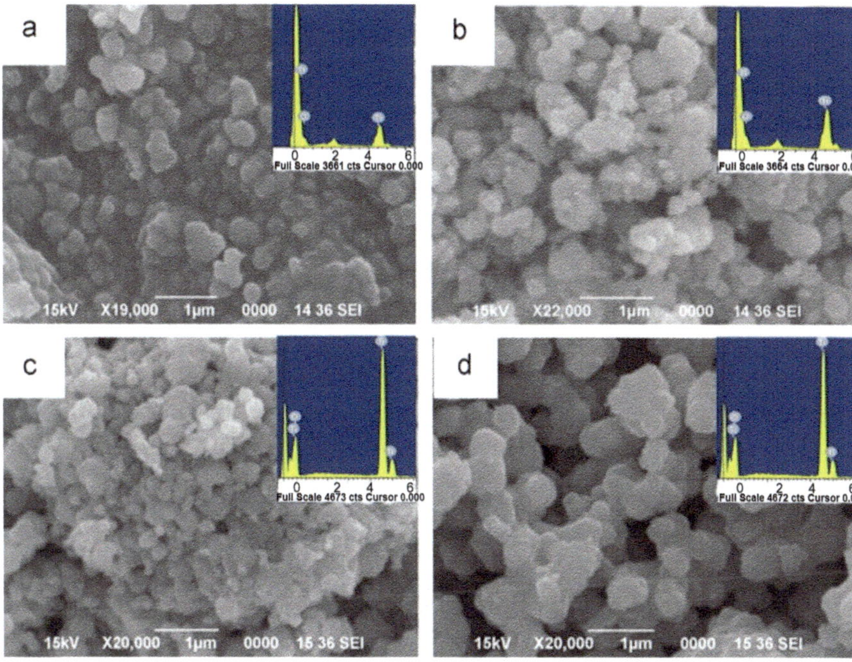

Fig. 6.9 SEM and EDAX images of TiO_2 photocatalysts calcined at different temperature: (**a**) 600 °C, (**b**) 700 °C, (**c**) 750 °C, and (**d**) 800 °C [106]. (Reprinted with permission from Ref. [106]. Copyright 2011, Elsevier)

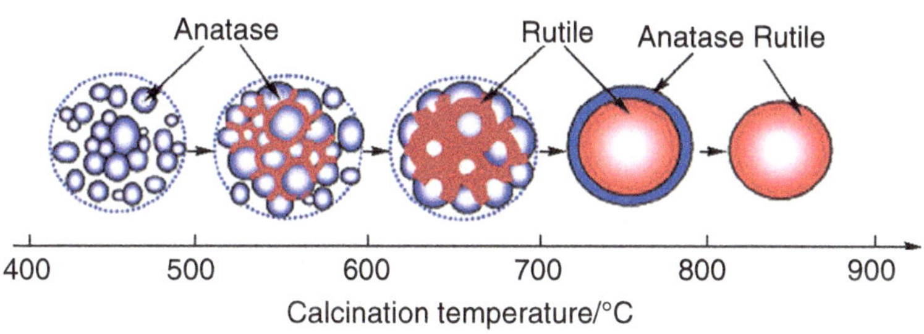

Fig. 6.10 Proposed schemes for the phase transformation of titania with increasing calcination temperature [108]. (Reprinted with permission from Ref. [108]. Copyright 2006, American Chemical Society)

However, it is easy to agglomerate in the process of calcination, and the particles size of product is large, resulting in the decrease of the photocatalytic activity.

Generally speaking, both the SMC method and the high-temperature calcination method require a high temperature to form the mixed-phase TiO_2 or close contact between different crystalline forms. The electron micrographs and XRD results showed that the grain size of crystal after calcination at high temperature is significantly larger than that of mixed-phase TiO_2 crystals obtained by the in situ method.

Table 6.2 The photocatalytic reaction rate constants of mixed-phase TiO_2 synthesized by different methods

Synthesis method	Rate constants/min^{-1}	References
Hydrothermal method	0.019	[53]
Sol–gel method	2.400	[51]
Microemulsion-mediated	0.030	[75]
Solvothermal method		
SMC method	0.023	[77]
Calcination method	0.003	[79]

Besides, the agglomeration of TiO_2 product is also serious. All of the before mentioned can easily decrease the photocatalytic activity of mixed-phase TiO_2. Therefore, in recent years, these methods are used less and less.

The photocatalytic reaction rate constants of mixed-phase TiO_2 crystals synthesized by different methods are summarized in Table 6.2. From this table, we can see that the reaction rate of TiO_2 by prepared in situ methods, such as the sol–gel method and microemulsion-mediated hydrothermal method, is usually higher than the high-temperature calcination method or SMC method. This is because the latter requires high-temperature calcination, leading to larger particle size and serious agglomeration that decrease the contact area with the degradation target and further impact the photocatalytic activity. However, the photocatalytic reaction rate of mixed-phase TiO_2 prepared by Zheng et al. [54] was low, which may be attributed to the high-temperature calcination which leads to a hollow structure and decreases the specific surface area, thus causing low photocatalytic activity. Photocatalytic activities of mixed-phase TiO_2 prepared on different conditions are also influenced by many other factors such as product morphology, particle size, and crystal proportion.

Apart from the commonly used approaches mentioned above, there are some new preparation methods of mixed-phase TiO_2, such as the microwave-heating method [109]. The particle size of TiO_2 prepared by this method is smaller than that synthesized by heating with an oil bath. With the development of new technologies, we can foresee that there will be more new methods for preparing mixed-phase TiO_2.

6.4 Applications of Mixed-Phase TiO_2 in Photocatalysis

6.4.1 Photocatalytic Hydrogen Production

Faced with the increasingly serious energy crisis, hydrogen has been widely recognized as an ideal energy source to replace a significant fraction of fossil fuels. However, nowadays the H_2 source demanded is mostly produced from fossil feedstock, mainly via steam reforming of methane. It is therefore necessary to develop sustainable methods for hydrogen production. Among them, the photocatalytic hydrogen production is one of the most promising technologies, which possesses various advantages, such as sustainability of the primary energy source (the solar light), the renewability of the starting feedstock, as well as the possible production of by-products with a high added value [110]. Due to the merits of TiO_2 such as nontoxicity, relatively cheap cost, easy to prepare, and excellent stability under the reaction conditions, it is the most investigated photocatalyst for H_2 production. However, the present solar-to-hydrogen energy conversion efficiency of TiO_2 is not high enough for the applications in commercial production. The rapid recombination of photo-generated electron–hole pairs of the materials, backward reaction, and the poor utilization of visible light hinder the practical applications. In order to solve the problems mentioned above, many researchers have been conducting studies concentrated on an emphasis to develop effective remediation methods like modification of TiO_2 by noble metal loading, metal ion doping, anion doping, dye sensitization, composite semiconductor, and metal ion implantation. In addition, mixed-phase TiO_2 has also been investigated for hydrogen production [91, 110–114], since mixed-phase TiO_2 nanomaterials demonstrated higher performances than the correspondent monophasic systems in many photocatalytic processes.

Chen et al. [91] successfully prepared pure anatase nanoparticles, pure brookite nanoplates, and two-phase anatase/brookite TiO_2 photocatalysts by a simple hydrothermal method. The photocatalytic activity of the as-synthesized catalysts for hydrogen production was studied in methanol solution. Their experimental results have shown that the photocatalytic activity of the two-phase anatase/brookite TiO_2 was higher compared with that of pure brookite nanoplates and pure anatase nanoparticles, as shown in Fig. 6.11. Moreover, in comparison with the highly active two-phase commercial P25, the synthesized two-phase anatase/brookite TiO_2 was 220% more active when measured by the H_2 yield per unit area of the photocatalyst surface. Similar results were achieved by Montini et al. [111]. They also prepared the

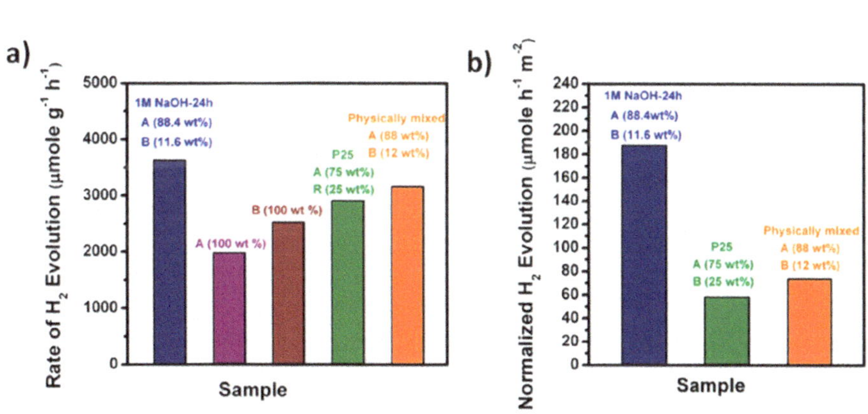

Fig. 6.11 Hydrogen evolved per gram of photocatalyst per hour under UV–vis irradiation in aqueous methanol solution over 0.3 wt % Pt-loaded photocatalysts. A, B, and R stand for anatase, brookite, and rutile, respectively [91]. (Reprinted with permission from Ref. [91]. Copyright 2013, American Chemical Society)

mixed-phase TiO_2 of anatase/brookite via hydrothermal treatments. Compared with a reference TiO_2 prepared by conventional sol–gel synthesis, the anatase/brookite nanocomposites which are obtained by hydrothermal method showed the higher photocatalytic activity for producing H_2. Furthermore, they found that the anatase/brookite ratio in the nanocomposite had a great influence on the photocatalytic activity in H_2 production.

Apart from the research applying mixed-phase anatase/brookite TiO_2 as the photocatalyst to produce hydrogen, a great deal of research on the mixed-phase anatase/rutile TiO_2 has also been done. Li et al. [112] fabricated photocatalysts with a tuned anatase/rutile structure by calcination of commercial P25 at different temperatures and investigated their photocatalytic activity for hydrogen production through photocatalytic biomass reforming. Surprisingly, it was found that compared with P25 without any treatment, the photocatalytic performance of the thermal-treated P25 for hydrogen production is better. The overall photocatalytic activity for hydrogen production on thermal-treated P25 had a dramatic improvement of three to five times. It was proposed that the anatase/rutile junction structure was mainly responsible for the enhanced photocatalytic performance. The research work implied that the photocatalytic ability of TiO_2 could be further improved by elaborately designing the anatase/rutile structure. Amal et al. [113] obtained the similar conclusions. They did a systematical study on photocatalytic H_2 evolution using mixed-phase TiO_2 as a function of anatase and rutile phase compositions with methanol as hole scavengers. The TiO_2 nanomaterials they prepared contain 4–95 mol % anatase, with the remaining being rutile. Synergistic effects on H_2 evolution were observed for a wide range of anatase contents, from 13 to 79 mol %, while due to insufficient physical contact, no synergistic effect was observed for the sample obtained by mixing anatase and rutile particles physically. Recently, in addition to the above research on particulate TiO_2 (Fig. 6.12a, b), Yu et al. [114] prepared a kind of anatase/rutile TiO_2 nanofiber photocatalyst (Fig. 6.12b, c). The enhanced photocatalytic performance for H_2 production was also observed in the synthesized anatase/rutile composite nanofibers. The product with 45 wt.% rutile phase and 55 wt.% anatase phase exhibited the highest photocatalytic activity with the H_2 production rate of 324 mmol h^{-1} and the apparent QE of 20.9% at 365 nm.

Though the research on the mixed-phase TiO_2 of anatase/rutile generally demonstrated that the combination of the two phases favored the photocatalytic hydrogen production efficiency, the low utilization rate of visible light still remains as a big limit for its practical application. The relatively narrow bandgap of rutile extends some visible light absorption to some extent. However, this is not enough for TiO_2 photocatalyst to apply in the practical applications. Researchers have made great efforts on to expand the visible light absorption of the mixed-phase TiO_2 crystals. Keller et al. [115] reported Au-modified anatase/rutile mixed-phase TiO_2 photocatalyst (Fig. 6.13) for hydrogen production. The light absorption of the sample with and without Au deposition was investigated via UV–vis light absorption spectra. Before the Au deposition, the catalyst with rutile phase whose content was higher than that of P25 exhibited more light absorption extended up to 550 nm, but its intensity is relatively low. After Au deposition, the samples showed an obvious

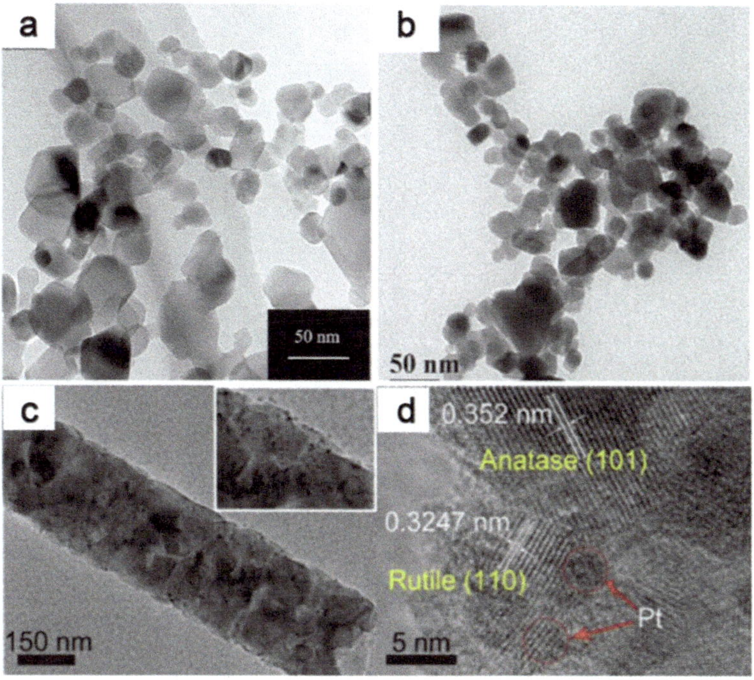

Fig. 6.12 (**a** and **b**) TEM images of particulate anatase and rutile phase TiO_2; (**c** and **d**) TEM and HRTEM images of fibrous anatase and rutile TiO_2 [112]. (Reprinted with permission from Ref. [112]. Copyright 2011, Elsevier)

modification of the light absorption properties, yielding an additional absorption band around 550 nm. It could be attributed to a plasmon resonance phenomenon [116] owing to collective oscillations of the conduction electrons located on the 6s orbital of Au and induced by the incident electromagnetic wave. In addition to this plasmon absorption, Au deposition also benefited to shift the absorption spectrum of TiO_2 photocatalysts deeper into the visible light range of 380–450 nm. This enhanced light absorption in the visible region was demonstrated to contribute to the improved photocatalytic performance of TiO_2 for the hydrogen production under the irradiation of simulated solar light. Furthermore, they pointed out that there were various factors which were crucial to improve H_2 evolution efficiency: (i) the surface and the crystallographic porosity properties of the TiO_2 anatase/rutile photocatalyst, (ii) the anatase/rutile ratio, (iii) the amount and nature of the metallic cocatalyst, (iv) the metal–support interactions, and (v) the relative content of sacrificial reagent. After studying the influence of these different factors in depth, they optimized the experimental conditions, obtaining the important H_2 production efficiency up to120 μmol min^{-1} over days without deactivation and with very low amounts of sacrificial reagent.

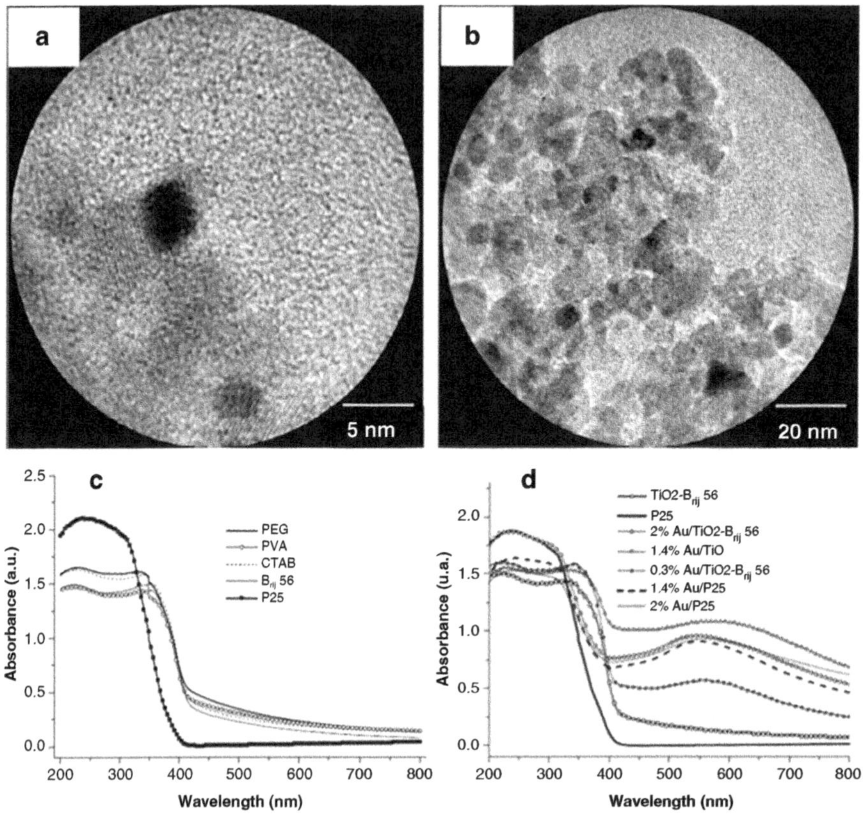

Fig. 6.13 (**a, b**) TEM images of mixed-phase TiO_2 photocatalyst deposited Au; (**c**) Light absorption properties of mixed-phase TiO_2 without and with Au deposition [115]. (Reprinted with permission from Ref. [115]. Copyright, 2010 Elsevier)

6.4.2 Photocatalytic Reduction of CO_2 with Water on Mixed-Phase TiO_2

As we know, CO_2 is the main cause of the greenhouse effect, which leads to global warming. Meanwhile, based on the fact that CO_2 can be converted into various useful chemical compounds and fuels such as CH_4, CH_3OH, and HCOOH, it is also a promising carbon resource. Therefore, in order to reduce the emissions of CO_2 and to obtain a sustainable energy, many kinds of novel materials and new technologies have been developed to convert CO_2. Besides the methods of solar thermochemical conversion as well as electrochemical reduction of CO_2, solar-activated photocatalytic reduction of CO_2 with water by TiO_2 at atmospheric pressure and room temperature has attracted much attention due to its “green chemistry” and relatively low cost.

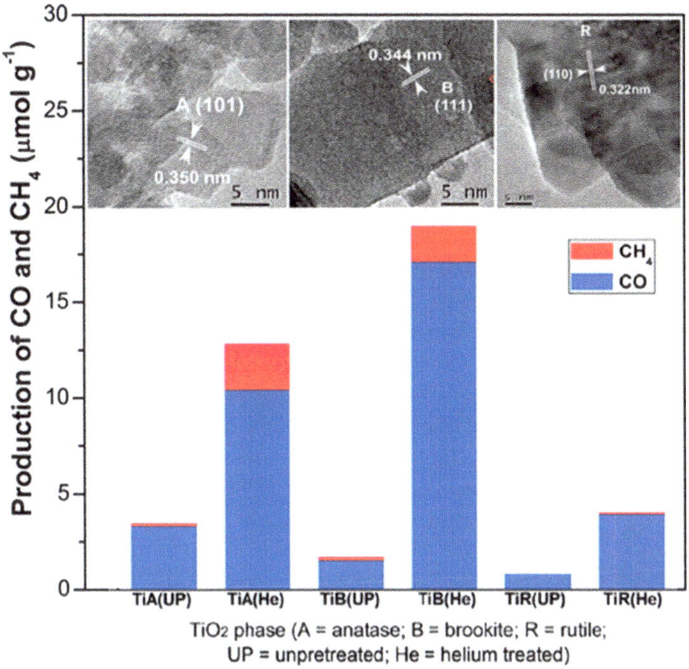

Fig. 6.14 The top figure are TEM images of different TiO_2 crystals: anatase, brookite, and rutile (from left to right). The bottom figure is the production of CO and CH_4 with the three different TiO_2 polymorphs [117]. (Reprinted with permission from Ref. [117]. Copyright 2012, American Chemical Society)

During the process of CO_2 photocatalytic reduction with H_2O by TiO_2, the irradiation of solar light activates the generation of electron–hole pairs in TiO_2 nanomaterials. The excited electrons in the conduction band (CB) of TiO_2 photocatalyst migrate to the surface to reduce CO_2. Meanwhile, the holes left in the valence band (VB) of TiO_2 can oxidize H_2O into O_2.

It has been proved that the above photocatalytic process had a close relation with the phase type of TiO_2. Li et al. [117] investigated the CO_2 photoreduction with water vapor on three types of TiO_2 nanocrystal polymorphs (anatase, rutile, and brookite). The results showed that the photocatalytic reduction activity of different-phase TiO_2 nanomaterials follows the order: anatase > brookite > rutile. Because of the fast e^- and h^+ recombination in rutile, it showed the least photocatalytic activity. Besides, the photoreduction activity of the TiO_2 catalysts with a helium treatment was investigated. The photoreduction data indicated that photocatalysts with a helium treatment were more active than the samples without a helium treatment. And the photocatalytic activity for production of CO and CH_4 from the photoreduction of CO_2 followed the order, brookite > anatase > rutile (Fig. 6.14), where brookite showed the highest photocatalytic activity. Thus their study also implied that the brookite phase is a promising material for the photoreduction of CO_2; they performed further research on this topic, including the study on brookite-containing mixed-phase TiO_2. They prepared bicrystalline anatase/brookite TiO_2 photocatalysts via a hydrothermal method [118]. The as-prepared anatase/brookite TiO_2 was also applied for CO_2 photoreduction with water vapor for the production of CO and CH_4. Compared with those of pure anatase, pure brookite, and a commercial anatase/rutile TiO_2 (P25), the photocatalytic activities of bicrystalline anatase/brookite TiO_2 were better. Wherein, the bicrystalline mixture with a composition of 75% anatase and 25% brookite showed the best photocatalytic performance, whose photocatalytic activity was nearly twice as high as that of 100% anatase (A100) and three times as high as that of pure brookite (B100) TiO_2. Because pure anatase A100 possessed the largest specific surface area and the smallest bandgap among three types of TiO_2 phases, the higher photocatalytic activity of bicrystalline anatase/brookite was very likely ascribed to the interactions between the anatase and brookite nanocrystals. In addition, the result that the anatase-rich bicrystalline anatase/brookite mixtures were superior to anatase/rutile mixture P25 which indicated the interaction between anatase and brookite of as-prepared mixed-phase TiO_2 seems to be more effective than that of P25 in the photoreduction of CO_2.

6.4.3 Photocatalytic Degradation of Organic Pollutants on Mixed-Phase TiO_2

In 1977, S. N. Frank and A. J. Bard reported that TiO_2 could effectively decompose the cyanide in aqueous medium under sunlight for the first time [119]. From then on, the application of TiO_2 for photocatalytic degradation of organic pollutants such as organic dyes has raised wide attention due to its effectiveness in degrading and mineralizing the toxic, bio-refractory, and highly concentrated organic compounds as well as the possibility of utilizing the solar ultraviolet and visible light spectrum. Owing to the superiority that mixed-phase TiO_2 usually exhibit better photocatalytic activity than single-phase TiO_2, TiO_2 nanomaterials consisting of different phases have also been widely used to degrade organic pollutants.

There have been many studies on the photocatalytic degradation of organic dyes using mixed-phase TiO_2 as photocatalyst. For example, under simulated solar light irradiation, the degradation of methyl blue (MB) was carried out in the aqueous solution by anatase/rutile TiO_2 heterojunction nanoflowers [120]. It was found that 72% of MB could be degraded in 120 min in the presence of the mixed-phase photocatalyst prepared in optimized experimental conditions, while only 30% of MB could be degraded in the solution with 100% anatase. Besides, the TiO_2 nanoflower photocatalysts showed excellent stability after nine cycles under the same conditions, suggesting that the mixed-phase anatase/rutile TiO_2 heterojunction nanoflower materials have great potential for the future photodegradation of practical dye wastewater. Since high surface area favors high photocatalytic activity, mesoporous mixed-phase anatase/rutile TiO_2 was also fabricated [121] (Fig. 6.15a) to apply for

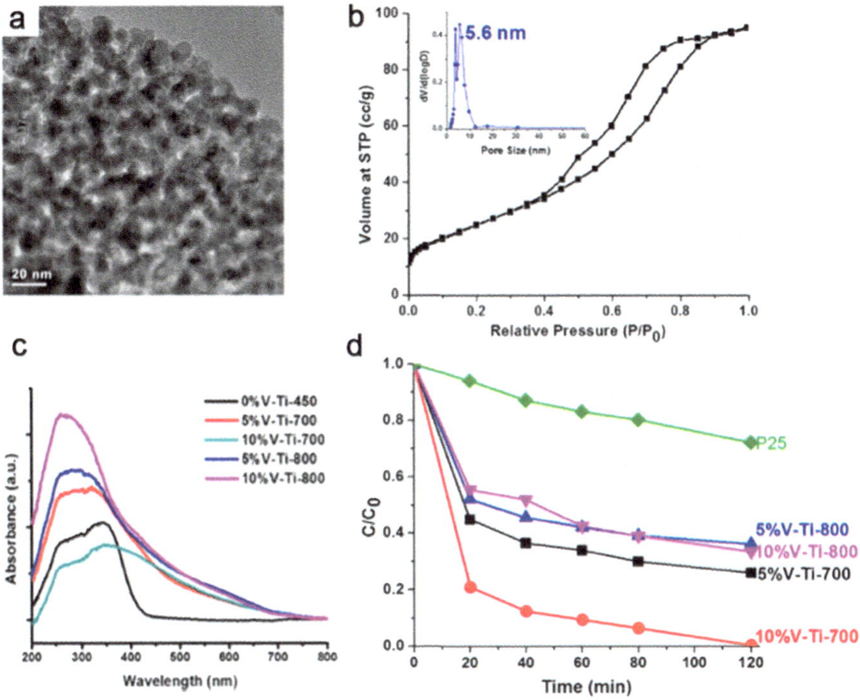

Fig. 6.15 (**a**) TEM image of mesoporous anatase/rutile mixed-phase TiO_2 (10%V-Ti-700); (**b**) N_2 sorption isotherms of 10%V-Ti-700. The inset, the BJH desorption pore-size distributions; (**c**) DR UV−vis spectra for various vanadium-doped mixed-phase TiO_2; (**d**) photocatalytic decomposition experiments of 100 mL, 10^{-4} MB dye under visible light in 2 h by using 100 mg of 5% V-Ti-700, 10%V-Ti-700, 5%V-Ti-800, 10%V-Ti-800, and P25 samples, respectively [121]. (Reprinted with permission from Ref. [121]. Copyright 2014, American Chemical Society)

the photocatalytic degradation of MB. The optimum mesoporous photocatalyst (10%V-Ti-700) in situ doped with vanadium with the surface area up to 94 m^2 g^{-1} and an average pore size of 5.6 nm (Fig. 6.15b) exhibited higher photocatalytic activity under visible light than that of the commercial P25 (Fig. 6.15d), in which the vanadium doping helped to extend the absorption of TiO_2 to the visible light region (>400 nm) (Fig. 6.15c). It was concluded that the high surface area, the mixed-phase effect, and the vanadium doping altogether contributed to the enhanced photocatalytic performance of the mesoporous mixed-phase anatase/rutile TiO_2.

In addition to the anatase/rutile TiO_2, many studies on brookite/rutile TiO_2 for the photodegradation of dyes have been done, such as Zhang et al.'s [81] work. They prepared mixed-phase TiO_2 photocatalysts with a tunable brookite-to-rutile ratio through a facile controllable one-pot hydrothermal method. And the photocatalytic performance of the as-prepared TiO_2 nanocrystals was tested in the degradation of Rhodamine B under the simulated solar light. Compared with the samples with other brookite-to-rutile ratios, the mixed-phase TiO_2 nanocrystals with 38% brookite and

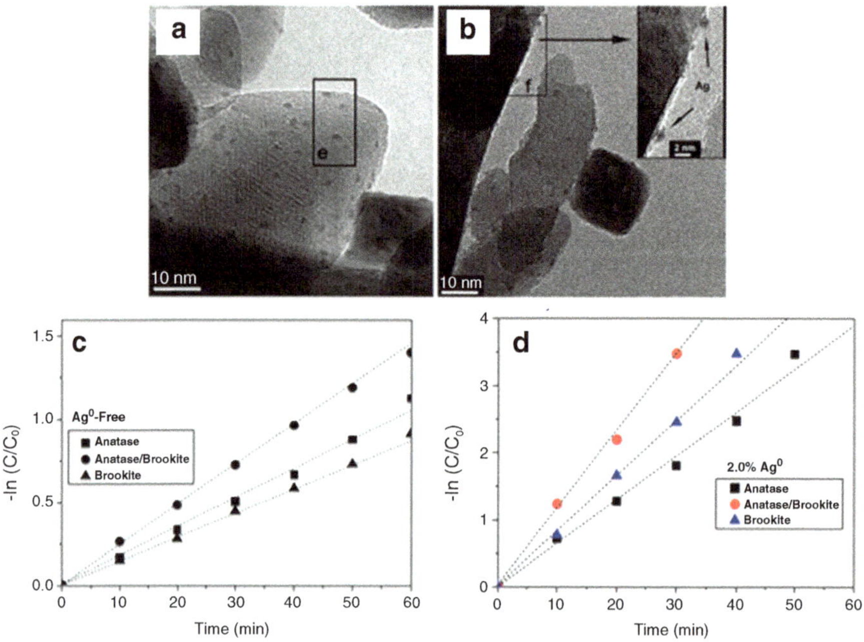

Fig. 6.16 (**a** and **b**) TEM images of 2.0% Ag^0–TiO_2 (brookite/anatase); (**c**) photocatalytic degradation of MO with Ag-free and 2.0% Ag^0-loaded (**d**) TiO_2 photocatalysts with different crystal phase composition under UV irradiation [55]. (Reprinted with permission from Ref. [55]. Copyright 2012, Elsevier)

62% rutile showed the highest photocatalytic activity, which was about six times that of the commercial P25 catalysts. Recently, Chen et al. [55] grew anatase grains on the surface of brookite petals to form a brookite/anatase TiO_2 hybrid, which appeared to have superior photocatalytic activity to the single-phase TiO_2 during the degradation of methyl orange (MO) as well as 2,4-dichlorophenol (2,4-DCP). The hybrid consisting of 60% brookite and 40% anatase exhibited the highest activity, whose degradation rate constants are 2.27 and 1.80 times higher than that of the corresponding physically mixed brookite/anatase TiO_2 for the photodegradation of MO and 2,4-DCP, respectively. Following this study, Ag was utilized to further modify the brookite/anatase composite [122]. Ag^0 clusters, which possessed an average diameter of ca. 1.5 nm, formed on the surface of the mixed-phase TiO_2 (Fig. 6.16a, b). The photocatalytic performance of the as-prepared photocatalysts was examined for the degradation of methyl orange (MO). As shown in Fig. 6.16d, the photodegradation reaction rate constant of Ag^0-loaded brookite/anatase composite was higher than that of the pure anatase and brookite, indicating an improved photocatalytic activity. Besides, it was found that the Ag^0-loaded mixed-phase TiO_2 photocatalysts showed better photocatalytic performance than that of Ag-free sample by comparing the results in Fig. 6.16c, d. The enhanced photocatalytic reactivity could be ascribed to the significant improvement in the

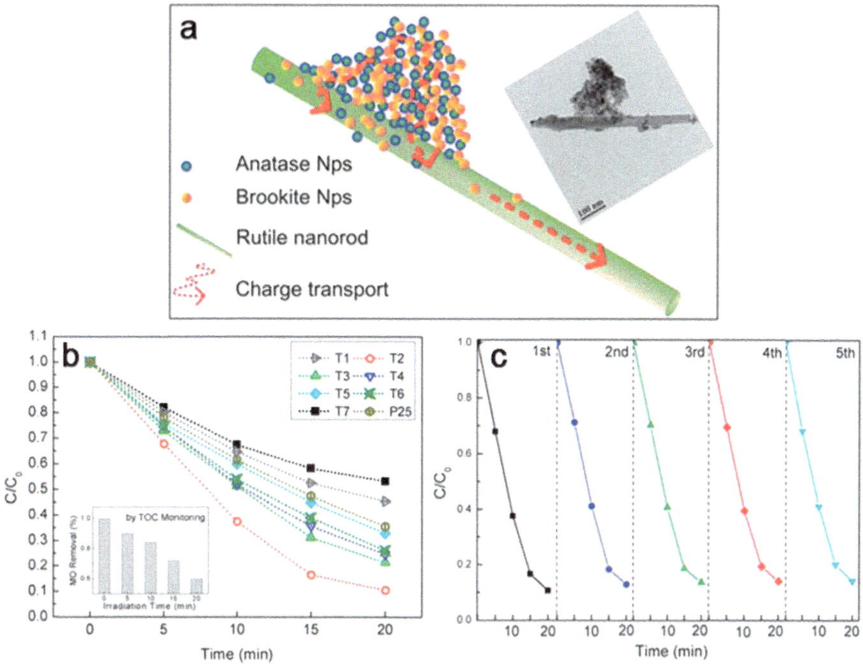

Fig. 6.17 (**a**) Scheme of the brookite/anatase/rutile nanocomposites; inset, TEM image of as-prepared composite; (**b**) photocatalytic degradation of MO solution by using the brookite/anatase/rutile nanocomposites prepared with various reagent ingredients, and the inset shows the removal of MO monitored by TOC with the sample T2; (**c**) cycling degradation curves of the brookite/anatase/rutile triphasic nanocomposites (sample T2) [123]. (Reprinted with permission from ref. [123]. Copyright 2012, Royal Society of Chemistry)

separation of the photo-generated electrons and holes, which was caused by two aspects: (1) synergistic effect of anatase and brookite in the composite and (2) the Schottky barrier at the interface of TiO_2 and Ag^0.

While most research focused on photocatalytic degradation properties of biphasic TiO_2 photocatalysts, Que. et al. [123] prepared brookite/anatase/rutile triphasic TiO_2 composite as the photocatalysts for degradation of MO. Their experimental results indicated that in the brookite/anatase/rutile coexisting nanomaterials, the brookite and anatase phases were crystallized into irregular nanoparticles with diameter sizes of less than 20 nm, whereas the rutile phase was crystallized into single-crystalline nanorods about 20 nm in diameter and 100–500 nm in length, as Fig. 6.17a shows. Under irradiation of UV light, the as-prepared triphasic TiO_2 photocatalyst was demonstrated to possess better photocatalytic activity during degradation of MO than the biphasic commercial P25. The catalyst with 29.9% anatase, 27.9% brookite, and 42.2% rutile (referred as T2) had the highest photocatalytic activity, removing 90% of MO in 20 min (Fig. 6.17b). The photodegradation rate constant k of this sample was 0.10180 min^{-1}, which is nearly twice higher than that of P25

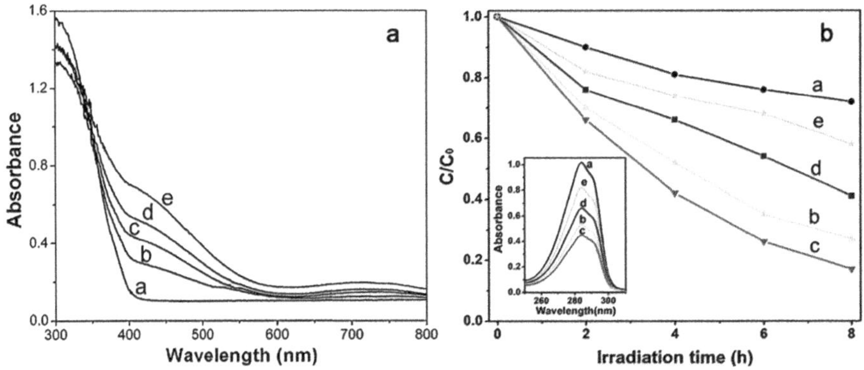

Fig. 6.18 (**a**) UV–vis DRS and (**b**) photocatalytic degradation curves of 2,4-DCP over TiO_2 without doping and different Cr-TiO_2 samples under visible light irradiation: (a) TiO_2 without doping, (b) 0.5% Cr-TiO_2, (c) 1% Cr-TiO_2, (d) 2% Cr-TiO_2, and (e) 5% Cr-TiO_2. The inset is the absorption spectra of 2,4-DCP over different as-prepared photocatalysts after visible light irradiation for 8 h [126]. (Reprinted with permission from ref. [126]. Copyright 2012, Elsevier

($k = 0.05397\ min^{-1}$). In addition, as shown in Fig. 6.17c, the photocatalytic activity of the best photocatalyst T2 showed only a slight loss after five recycles, indicating a good stability of the triphasic TiO_2 photocatalyst in the degradation process. Similarly, Shao et al. also developed brookite/anatase/rutile coexisting TiO_2, and the as-prepared photocatalyst was demonstrated to show excellent photocatalytic performance for the degradation of methylene blue solutions.

Apart from dyes, some other organic compounds such as phenol and benzyl alcohol have also been used as the target pollutants in the mixed-phase TiO_2 photocatalysis system. Because phenol as well as its derivatives, as one type of primary pollutants, is able to do harm to organisms at low concentrations, many studies on the photodegradation of phenol have been widely carried out [124]. Lu et al. [125] synthesized a series of mixed-phase TiO_2 samples with different anatase-to-rutile ratios and studied the effects of crystal phase in photocatalytic oxidation of phenol aqueous solution. They found that the photocatalysts with higher anatase-to-rutile ratios showed higher photocatalytic activities for phenol degradation. Tian et al. [126] reported Cr-doped anatase/rutile bicrystalline phase TiO_2 (Cr-TiO_2) nanoparticles for the photocatalytic degradation of 2,4-dichlorophenol (2,4-DCP). Their experimental results showed that Cr^{3+} doping could not only effectively extend the visible light response of TiO_2 nanomaterials (Fig. 6.18a) but also enhance the anatase-to-rutile transformation. The photocatalytic activities of different Cr-TiO_2 photocatalysts for the photocatalytic degradation of 2,4-dichlorophenol (2,4-DCP) were evaluated under the irradiation of visible light (Fig. 6.18b). It is found that appropriate Cr^{3+} doping can remarkably enhance the visible light photocatalytic activity of TiO_2, which is ascribed to improving the response of visible light as well as suitable anatase-to-rutile ratio. In addition, excess Cr^{3+} doping is unfavorable for improving visible light photocatalytic activity, because of the formation of Cr_2O_3 clusters and the excessive rutile content.

6.5 Mechanism of the Enhanced Photocatalytic Activities by the Mixed-Phase TiO_2 Photocatalysis

Generally, most of the photo-generated electrons and holes in the single-phase TiO_2 will recombine, and only a small amount of the excitons can migrate to the surface to participate in the oxidation–reduction reactions with absorbed molecules, causing relatively low photocatalytic efficiency. In 1991, Bickley et al. [76] first put forward the mechanism of enhanced photocatalytic activity of P25 consisting of 80% anatase and 20% rutile. They demonstrated that the mixed-phase structure of P25 is an anatase nanostructure coated with a layer of rutile film via TEM. Compared with any other single-phase TiO_2 photocatalyst, the evaluation results showed that the photocatalytic activity of the mixed-phase TiO_2 was significantly improved. Because of the different bandgaps, a bending band was formed on the interface of anatase and rutile, as shown in Fig. 6.19. Under light irradiation, the photo-generated electrons migrated from rutile phase to anatase phase, while the holes migrated from anatase phase to rutile phase; thus the electrons and holes could be separated effectively, leading to a high photocatalytic activity of the mixed-phase P25. Meanwhile, they proposed that the mechanism of the improved photocatalytic performance of mixed-phase TiO_2 was actually more complex and required further study. Subsequently, much research work on the mobility direction of the photo-generated carriers, the molecular dynamic characterizations of the mixed-phase interface [76], the band structure of mixed-phase TiO_2 photocatalysts [127, 128], and the suitable phase proportion in the mixed-phase crystal TiO_2 has been done to study the mechanism in depth.

In 1995, through observing the mixed-phase crystal structure in P25 via X-ray diffraction (XRD) and high-resolution transmission electron microscopy (HRTEM)

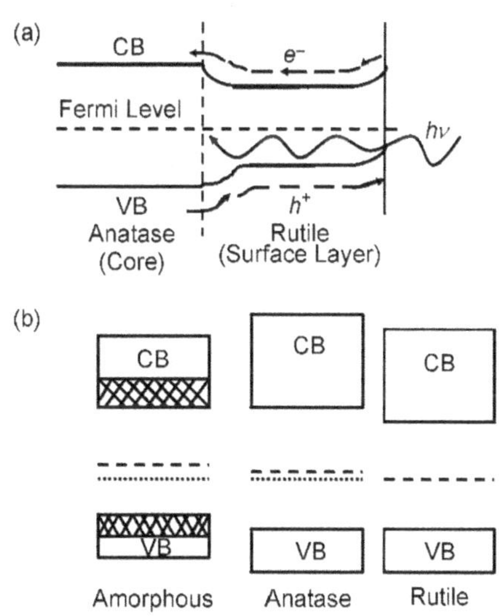

Fig. 6.19 Schematic diagram of P25 band structure [76]. (Reprinted with permission from Ref. [76]. Copyright 1991, Elsevier)

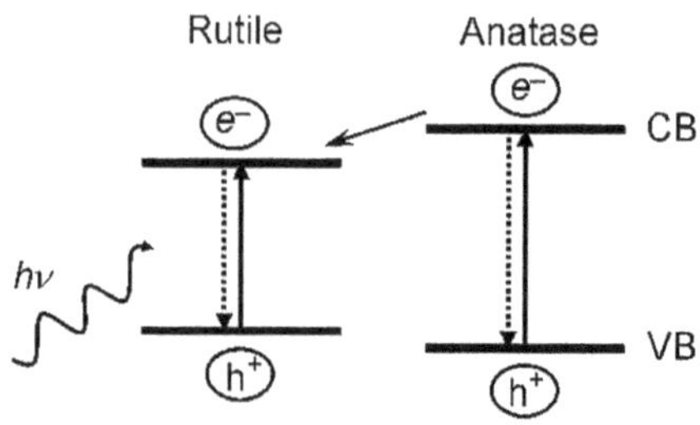

Fig. 6.20 A proposed schematic diagram showing the migration behavior of the electron transferring from anatase to rutile [132, 133]. (Reprinted with permission from Ref. [132]. Copyright 2002, Wiley Online Library. Reprinted with permission from Ref. [133]. Copyright 2007, Elsevier)

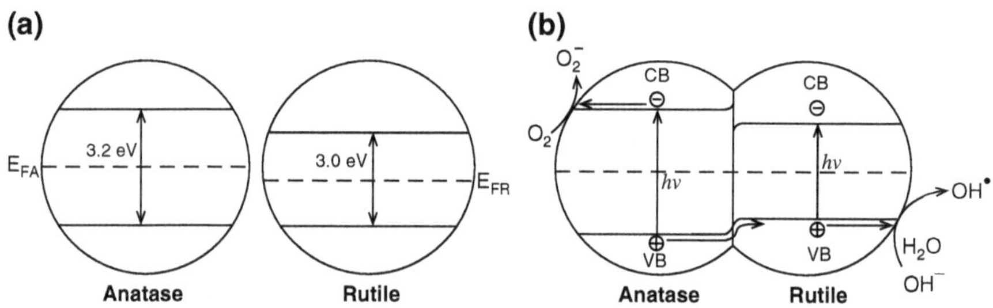

Fig. 6.21 (**a**) Band structure of anatase and rutile before they contact each other; (**b**) band structure and the separation of photo-generated excitons after anatase and rutile contact each other [134]. (Reprinted with permission from Ref. [134]. Copyright 2003 American Chemical Society)

techniques, Datye et al. [129] put forth distinguishing views against Bickley et al. [76]. The characterization results showed that the mixed-phase P25 was composed of individual single-crystal particles of anatase and rutile rather than anatase particles covered by a layer of rutile. Subsequently, their view was confirmed by Zhang et al. [130] and Ohno et al. [131]. But they did not further study the mechanism of the enhanced photocatalytic performance of the mixed-phase TiO_2.

In 2002, Kawahara et al. [132] designed a film model to confirm the migration direction of the photo-generated excitons in anatase/rutile mixed-phase crystal TiO_2 under light irradiation. The mixed-phase crystal TiO_2 film was immersed into an $AgNO_3$ solution with the exposure to light over a period of time at the atmosphere of argon. It could be observed that there were a large number of Ag particles appearing on the surface of the rutile phase. According to the phenomenon, they inferred that Ag^+ caused a reduction reaction on the rutile surface and the photo-generated electrons migrated from the conduction band of anatase to rutile, as shown in Fig. 6.20.

In 2003, Sun et al. [134, 135] deposited Pt on P25 and applied the catalyst to the photocatalytic degradation of phenol. They found that loading P25 with Pt did not show the increase of photocatalytic activity for phenol decomposition as well as the total carbon removal rates. Besides, they found that Pt appeared on the surface of anatase. Therefore, as shown in Fig. 6.21, they proposed a charge separation

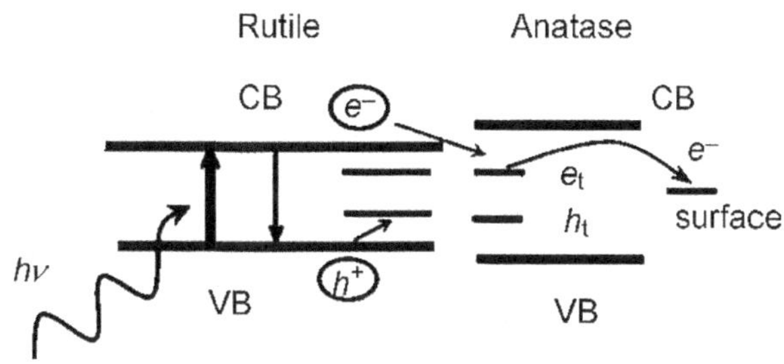

Fig. 6.22 A proposed schematic illustration showing the separation of photo-generated holes and electrons using EPR [51]. (Reprinted with permission from Ref. [51]. Copyright 2003, American Chemical Society)

mechanism for P25 that the contact of the two phases led to the bending of the conduction band, resulting in the holes in anatase transferring to rutile, while the electrons in anatase not transferring to rutile, further causing the results that holes were concentrated in rutile and electrons stayed in the anatase particles. It was to say that the oxidation reaction mostly happened on rutile and the reduction reactions mostly took place on anatase; thus the Pt deposition on anatase was not able to increase the photocatalytic efficiency for phenol oxidation in water.

Hurum et al. [51] characterized the charge separation process of P25 via electron paramagnetic resonance (EPR) in the same year. Due to the wide bandgap of anatase, no exciton was generated under the visible light irradiation. However, the electron trapping sites were found on anatase surface in electron paramagnetic resonance spectroscopy. It could be concluded that the electrons in rutile excited by visible light transferred to the lower-energy anatase lattice trapping sites, improving the separation charges and thus enhancing the photocatalytic activity of P25 (Fig. 6.22). Thereafter, Liu et al. [136] confirmed this mechanism via the photocatalytic activity experiments of anatase/rutile mixed-phase TiO_2 nanotubes.

EPR spectroscopy is a new characterization method which has been developed in recent years. It can be used for the detection of the unpaired electrons in the sample. Many researchers have started to characterize the migration behaviors of photo-generated electrons and holes in the mixed-phase TiO_2 via the EPR technique [137].

An "antenna mechanism" was proposed to explain the mechanism of improved photocatalytic performance in mixed crystal phases by Wang et al. [138] in 2006. They thought that during the photocatalytic degradation process, TiO_2 absorbed light to produce photo-generated excitons, which then take part in the oxidation–reduction reactions with the target molecules adsorbed on the catalyst's surface; however, due to the lack of light in the depth, the particles deeper in the liquid could only receive the photon energy via the particles irradiated by light. The energy transferred from the shallow particles induced the oxidation–reduction reaction of the particles which light cannot reach, enhancing the photocatalytic activity. The long-chain particles served as an antenna to transfer the photon energy from the location of light absorption to the location of reaction in the process of photon energy being transferred to the photocatalyst deeper in the liquid; thus this is called an "antenna mechanism" (Fig. 6.23).

However, the theoretical models in the research objects mentioned above are all P25 consisting of a certain phase ratio of anatase and rutile. Because mixed-phase

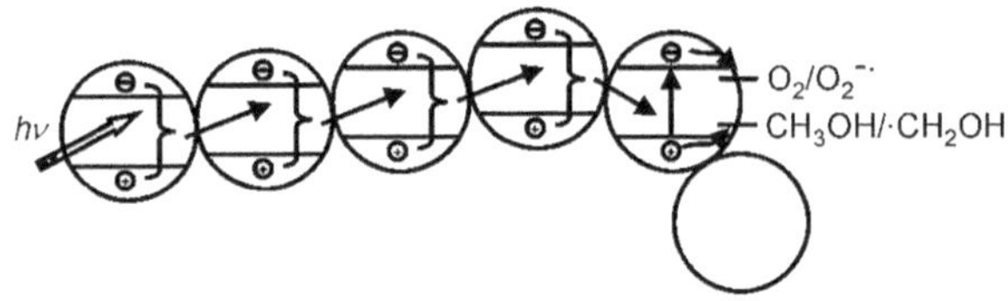

Fig. 6.23 Antenna effect by network structure leading to the enhancement of photocatalytic activity [138]. (Reprinted with permission from Ref. [138]. Copyright, 2006 Elsevier)

TiO_2 with different phase ratios exhibit different photocatalytic activities, it is difficult to adequately explain the physical phenomenon responsible for the existence of an optimum phase ratio in mixed-phase TiO_2 with the maximum photocatalytic activity using these existing models. Therefore, further research is essential to find out the photocatalytic mechanism of mixed-phase TiO_2 in different phase contents.

In 2008, Zachariah et al. [103] took the factors of crystallite size distribution as well as phase composition into account, and proposed a new mechanism model, which overcame the limitations of the existing models in terms of explaining the abovementioned phenomenon. The mechanism is based on the charge separation mechanism proposed by Sun et al. [134]. Zachariah et al. [103] thought that when the size of nano-TiO_2 was identical (Fig. 6.24a), the bandgap of adjacent anatase crystallites was the same, so there was no driving force to migrate the photo-generated holes, resulting in the high recombination rate of photo-generated charges and the low photocatalytic activity. If the sizes of anatase crystallite were different and were below the critical size (Fig. 6.24b, c), the bandgaps of the connected crystallites would depend on their size. Thus, photo-generated holes in one crystallite could easily escape into the other, leading to the effective separation of photo-generated excitons. When a small amount of rutile was mixed with anatase (Fig. 6.24d), owing to the different bandgap, photo-generated holes would transfer from anatase to rutile, so that the photocatalytic activity could be further enhanced. However, the excessive rutile content limited the migration of photo-generated holes in mixed-phase crystal TiO_2 (Fig. 6.24e), and the photocatalytic activity of photocatalysts began to decline. On the condition that the crystallite sizes of pure rutile are bigger than the critical size (Fig. 6.24f), their bandgap values kept the same, and then the migration of photo-generated holes could not occur, and the photocatalytic activity was minimum. The mechanism showed that whether the TiO_2 was mixed-phase crystalline or single crystalline, the photo-generated electron–hole pairs could be separated and the photocatalytic activity of photocatalysts could be enhanced as long as the bandgap values were different.

In summary, the mechanisms mentioned above for the enhanced photocatalytic activity of mixed-phase TiO_2 all involve the migration behaviors of photo-generated excitons. In these mechanisms, researchers all agree that the mixed-phase crystal structure favors the effective separation of photo-generated electrons and holes. However, the specific migration paths of excitons are still controversial. The

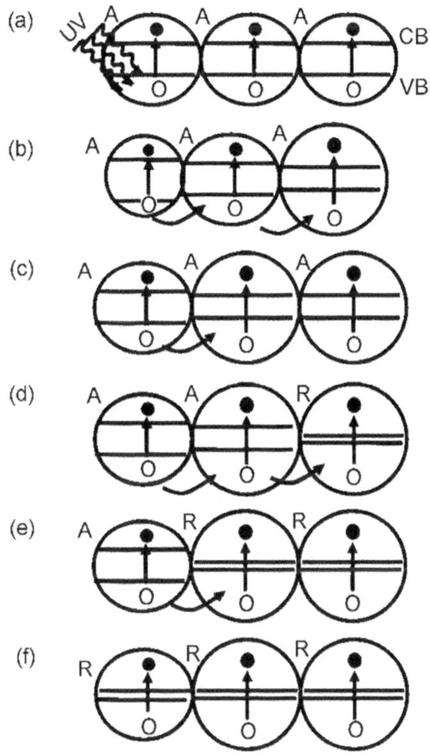

Fig. 6.24 A schematic illustration of the newly proposed mechanism in the basis of the bandgap variation in the connected nanocrystallites as a function of the phase and the size distribution involved [103]. A, R, CB, and VB stand for anatase TiO_2, rutile TiO_2, the conduction band, and the valence band, respectively [103]. (Reprinted with permission from Ref. [103]. Copyright 2009, American Chemical Society)

determining factors for the photocatalytic activity of mixed-phase crystal TiO_2 contain grain size, structural forms (core–shell structure, cladding structure, random composite, etc.), phase composition, and so on. These factors cause the emergence of bending band, trapping sites, and other phenomena in the mixed-phase TiO_2. Therefore, the obtained mixed-phase TiO_2 differ in photocatalytic mechanism on different experimental conditions, so further research is still required.

6.6 Conclusion and Outlook

In the field of photocatalysis, since it was discovered that TiO_2 nanomaterials could be used for photocatalytic water splitting and degradation of organic pollutants in the 1970s, the synthetic methods and applications of mixed-phase TiO_2 have achieved great progress. The superior photocatalytic performance of mixed-phase TiO_2 to single-phase TiO_2 has been well recognized, and the mechanism of the improved photocatalytic activities of the mixed-phase TiO_2 photocatalysts has also been widely studied. However, there are still many problems remaining to be resolved. For instance, (1) particulate mixed-phase TiO_2 nanomaterials are easy to agglomerate, which will greatly inhibit the photocatalytic activity and impede the further development of the materials; (2) though the relatively narrow bandgap of rutile is able to extend the absorption of the mixed-phase TiO_2 to some of the visible light range, it is still unable to completely utilize all visible light from sunlight, limiting its practical application in photocatalysis; and (3) although researchers all approve that the mixed-phase crystal structure favors the effective separation of photo-generated electrons and holes, thereby improving the photocatalytic activity, there is still much controversy about the mobility direction of photo-generated excitons. It is necessary for researchers to further study on the mechanism of the enhanced photocatalytic performance of mixed-phase TiO_2. Therefore, developing the mixed-phase TiO_2 nanomaterials, combining other advantageous structures (such as hierarchical structure) to enhance the utilization of the visible light, and exploring the mechanism of photocatalytic reactions by newly developed characterization techniques will retain as the challenges and hot research themes in the future.

Although mixed-phase TiO_2 photocatalysts have been studied for decades, it is still a hot research theme owing to its excellent photocatalytic activity. With the increase of serious energy crisis and environmental problems, the applications of mixed-phase crystal TiO_2 are predicted to attract lasting and considerable attention.

References

1. Fujishima A, Honda K (1972) Electrochemical photolysis of water at a semiconductor electrode. Nature 238(5358):37–38
2. Carey JH, Lawrence J, Tosine HM (1976) Photodechlorination of PCB's in the presence of titanium dioxide in aqueous suspensions. Bull Environ Contam Toxicol 16(6):697–701
3. Habisreutinger SN, Schmidt-Mende L, Stolarczyk JK (2013) Photocatalytic reduction of CO_2 on TiO_2 and other semiconductors. Angew Chem Int Ed 52(29):7372–7408
4. Hu K, Robson KC, Johansson PG et al (2012) Intramolecular hole transfer at sensitized TiO_2 interfaces. J Am Chem Soc 134(20):8352–8355
5. Guo Q, Xu C, Ren Z et al (2012) Stepwise photocatalytic dissociation of methanol and water on TiO_2 (110). J Am Chem Soc 134(32):13366–13373
6. Tian B, Chen F, Zhang J et al (2006) Influences of acids and salts on the crystalline phase and morphology of TiO_2 prepared under ultrasound irradiation. J Colloid Interface Sci 303 (1):142–148
7. Tomkiewicz M, Dagan G, Zhu Z (1994) Morphology and photocatalytic activity of TiO_2 aerogels. Res Chem Intermed 20(7):701–710
8. Zhu S, Xie G, Yang X et al (2013) A thick hierarchical rutile TiO_2 nanomaterial with multilayered structure. Mater Res Bull 48(5):1961–1966
9. Beuvier T, Richard-Plouet M, Mancini-Le Granvalet M et al (2010) TiO_2 (B) nanoribbons as negative electrode material for lithium ion batteries with high rate performance. Inorg Chem 49(18):8457–8464
10. Xin X, Scheiner M, Ye M et al (2011) Surface-treated TiO_2 nanoparticles for dye-sensitized solar cells with remarkably enhanced performance. Langmuir 27(23):14594–14598
11. Hosono E, Fujihara S, Imai H et al (2007) One-step synthesis of nano–micro chestnut TiO_2 with rutile nanopins on the microanatase octahedron. ACS Nano 1(4):273–278
12. Sinha AK, Jana S, Pande S et al (2009) New hydrothermal process for hierarchical TiO_2 nanostructures. Cryst Eng Comm 11(7):1210–1212

13. Cheng QQ, Cao Y, Yang L et al (2011) Synthesis and photocatalytic activity of titania microspheres with hierarchical structures. Mater Res Bull 46(3):372–377
14. Wang Y, Zhang L, Deng K et al (2007) Low temperature synthesis and photocatalytic activity of rutile TiO_2 nanorod superstructures. J Phys Chem C 111(6):2709–2714
15. Wei J, Yao J, Zhang X et al (2007) Hydrothermal growth of titania nanostructures with tunable phase and shape. Mater Lett 61(23):4610–4613
16. Hu YH (2012) A highly efficient photocatalyst—hydrogenated black TiO_2 for the photocatalytic splitting of water. Angew Chem Int Ed 51(50):12410–12412
17. Oh JK, Lee JK, Kim HS et al (2010) TiO_2 branched nanostructure electrodes synthesized by seeding method for dye-sensitized solar cells. Chem Mater 22(3):1114–1118
18. Ye M, Liu HY, Lin C et al (2013) Hierarchical rutile TiO_2 flower cluster-based high efficiency dye-sensitized solar cells via direct hydrothermal growth on conducting substrates. Small 9 (2):312–321
19. Si P, Ding S, Yuan J et al (2011) Hierarchically structured one-dimensional TiO_2 for protein immobilization, direct electrochemistry, and mediator-free glucose sensing. ACS Nano 5 (9):7617–7626
20. Etgar L, Gao P, Xue Z et al (2012) Mesoscopic CH_3 NH_3 PbI_3/TiO_2 heterojunction solar cells. J Am Chem Soc 134(42):17396–17399
21. Guo W, Xu C, Wang X et al (2012) Rectangular bunched rutile TiO_2 nanorod arrays grown on carbon fiber for dye-sensitized solar cells. J Am Chem Soc 134(9):4437–4441
22. Wang YQ, Gu L, Guo YG et al (2012) Rutile-TiO_2 nanocoating for a high-rate $Li_4Ti_5O_{12}$ anode of a lithium-ion battery. J Am Chem Soc 134(18):7874–7879
23. So S, Lee K, Schmuki P (2012) Ultrafast growth of highly ordered anodic TiO_2 nanotubes in lactic acid electrolytes. J Am Chem Soc 134(28):11316–11318
24. Wang WN, An WJ, Ramalingam B et al (2012) Size and structure matter: enhanced CO_2 photoreduction efficiency by size-resolved ultrafine Pt nanoparticles on TiO_2 single crystals. J Am Chem Soc 134(27):11276–11281
25. Leghari SAK, Sajjad S, Chen F et al (2011) WO_3/TiO_2 composite with morphology change via hydrothermal template-free route as an efficient visible light photocatalyst. Chem Eng J 166 (3):906–915
26. Chen F, Zou W, Qu W et al (2009) Photocatalytic performance of a visible light TiO_2 photocatalyst prepared by a surface chemical modification process. Catal Commun 10 (11):1510–1513
27. Xing M, Zhang J, Chen F (2009) New approaches to prepare nitrogen-doped TiO_2 photocatalysts and study on their photocatalytic activities in visible light. Appl Catal B Environ 89(3):563–569
28. Kim TH, Gómez-Solís C, Moctezuma E et al (2014) Sonochemical synthesis of Fe–TiO_2–SiC composite for degradation of rhodamine B under solar simulator. Res Chem Intermed 40 (4):1595–1605
29. Zuo F, Wang L, Wu T et al (2010) Self-doped Ti^{3+} enhanced photocatalyst for hydrogen production under visible light. J Am Chem Soc 132(34):11856–11857
30. Kim YJ, Lee MH, Kim HJ et al (2009) Formation of highly efficient dye-sensitized solar cells by hierarchical pore generation with nanoporous TiO_2 spheres. Adv Mater 21(36):3668–3673
31. Liu S, Li Q, Hou C et al (2013) Hierarchical nitrogen and cobalt co-doped TiO_2 prepared by an interface-controlled self-aggregation process. J Alloys Compd 575:128–136
32. Cai M, Pan X, Liu W et al (2013) Multiple adsorption of tributyl phosphate molecule at the dyed-TiO_2/electrolyte interface to suppress the charge recombination in dye-sensitized solar cell. J Mater Chem A 1(15):4885–4892
33. Liang MS, Khaw CC, Liu CC et al (2013) Synthesis and characterisation of thin-film TiO_2 dye-sensitised solar cell. Ceram Int 39(2):1519–1523
34. Kim H, Hwang YH, Cho G et al (2011) Partially dyed-TiO_2 dispersions for adaptation to the continuous fabrication of photoanodes. Electrochim Acta 56(25):9476–9481

35. Zhang C, Huang Y, Chen S et al (2012) Photoelectrochemical analysis of the dyed TiO_2/electrolyte interface in long-term stability of dye-sensitized solar cells. J Phys Chem C 116 (37):19807–19813
36. Bae EG, Kim H, Hwang YH et al (2012) Genetic algorithm-assisted optimization of partially dyed-TiO_2 for room-temperature printable photoanodes of dye-sensitized solar cells. J Mater Chem 22(2):551–556
37. Ashkarran AA, Ghavamipour M, Hamidinezhad H et al (2015) Enhanced visible light-induced hydrophilicity in sol–gel-derived Ag–TiO_2 hybrid nanolayers. Res Chem Intermed 41 (10):7299–7311
38. Haruta M, Uphade BS, Tsubota S et al (1998) Selective oxidation of propylene over gold deposited on titanium-based oxides. Res Chem Intermed 24(3):329–336
39. Tian B, Zhang J, Tong T et al (2008) Preparation of Au/TiO_2 catalysts from Au (I)–thiosulfate complex and study of their photocatalytic activity for the degradation of methyl orange. Appl Catal B Environ 79(4):394–401
40. Wu Y, Liu H, Zhang J et al (2009) Enhanced photocatalytic activity of nitrogen-doped titania by deposited with gold. J Phys Chem C 113(33):14689–14695
41. Wang W, Zhang J, Chen F et al (2010) Catalysis of redox reactions by Ag@TiO_2 and Fe^{3+}-doped Ag@TiO_2 core–shell type nanoparticles. Res Chem Intermed 36(2):163–172
42. Wang Y, Feng C, Zhang M et al (2010) Enhanced visible light photocatalytic activity of N-doped TiO_2 in relation to single-electron-trapped oxygen vacancy and doped-nitrogen. Appl Catal B Environ 100(1):84–90
43. Feng C, Wang Y, Zhang J et al (2012) The effect of infrared light on visible light photocatalytic activity: an intensive contrast between Pt-doped TiO_2 and N-doped TiO_2. Appl Catal B Environ 113:61–71
44. Charanpahari A, Umare SS, Gokhale SP et al (2012) Enhanced photocatalytic activity of multi-doped TiO_2 for the degradation of methyl orange. Appl Catal A Gen 443:96–102
45. Tian B, Li C, Gu F et al (2009) Synergetic effects of nitrogen doping and Au loading on enhancing the visible-light photocatalytic activity of nano-TiO_2. Catal Commun 10 (6):925–929
46. Zhang P, Shao C, Li X et al (2012) In situ assembly of well-dispersed Au nanoparticles on TiO_2/ZnO nanofibers: a three-way synergistic heterostructure with enhanced photocatalytic activity. J Hazard Mater 237:331–338
47. Zhang Z, Yuan Y, Liang L et al (2008) Preparation and photoelectrocatalytic activity of ZnO nanorods embedded in highly ordered TiO_2 nanotube arrays electrode for azo dye degradation. J Hazard Mater 158(2):517–522
48. Chattopadhyaya G, Macdonald DG, Bakhshi NN et al (2006) Removal of nitric oxide over Saskatchewan lignite and its derivatives. Catal Lett 108(1):1–5
49. Yang M, Men Y, Li S et al (2012) Enhancement of catalytic activity over TiO_2-modified Al_2O_3 and ZnO–Cr_2O_3 composite catalyst for hydrogen production via dimethyl ether steam reforming. Appl Catal A Gen 433:26–34
50. Su R, Bechstein R, Sø L et al (2011) How the anatase-to-rutile ratio influences the photoreactivity of TiO_2. J Phys Chem C 115(49):24287–24292
51. Hurum DC, Agrios AG, Gray KA et al (2003) Explaining the enhanced photocatalytic activity of Degussa P25 mixed-phase TiO_2 using EPR. J Phys Chem B 107(19):4545–4549
52. Scotti R, Bellobono IR, Canevali C et al (2008) Sol−gel pure and mixed-phase titanium dioxide for photocatalytic purposes: relations between phase composition, catalytic activity, and charge-trapped sites. Chem Mater 20(12):4051–4061
53. Puddu V, Choi H, Dionysiou DD et al (2010) TiO_2 photocatalyst for indoor air remediation: influence of crystallinity, crystal phase, and UV radiation intensity on trichloroethylene degradation. Appl Catal B Environ 94(3):211–218
54. Zheng R, Meng X, Tang F (2009) Synthesis, characterization and photodegradation study of mixed-phase titania hollow submicrospheres with rough surface. Appl Surf Sci 255 (11):5989–5994

55. Jiao Y, Chen F, Zhao B et al (2012) Anatase grain loaded brookite nanoflower hybrid with superior photocatalytic activity for organic degradation. Colloids Surf A Physicochem Eng Asp 402:66–71
56. Di Paola A, Cufalo G, Addamo M et al (2008) Photocatalytic activity of nanocrystalline TiO_2 (brookite, rutile and brookite-based) powders prepared by thermohydrolysis of $TiCl_4$ in aqueous chloride solutions. Colloids Surf A Physicochem Eng Asp 317(1):366–376
57. Li W, Liu C, Zhou Y et al (2008) Enhanced photocatalytic activity in anatase/TiO_2 (B) core−shell nanofiber. J Phys Chem C 112(51):20539–20545
58. Hanaor DA, Sorrell CC (2011) Review of the anatase to rutile phase transformation. J Mater Sci 46(4):855–874
59. Gamboa JA, Pasquevich DM (1992) Effect of chlorine atmosphere on the anatase-rutile transformation. J Am Ceram Soc 75(11):2934–2938
60. Ding XZ, He YZ (1996) Study of the room temperature ageing effect on structural evolution of gel-derived nanocrystalline titania powders. J Mater Sci Lett 15(4):320–322
61. Muscat J, Swamy V, Harrison NM (2002) First-principles calculations of the phase stability of TiO_2. Phys Rev B 65(22):224–112
62. Arlt T, Bermejo M, Blanco MA et al (2000) High-pressure polymorphs of anatase TiO_2. Phys Rev B 61(21):14414–14419
63. Ren R, Yang Z, Shaw LL (2000) Polymorphic transformation and powder characteristics of TiO_2 during high energy milling. J Mater Sci 35(23):6015–6026
64. Dubrovinskaia NA, Dubrovinsky LS, Ahuja R et al (2001) Experimental and theoretical identification of a new high-pressure TiO_2 polymorph. Phys Rev Lett 87(27):455–475
65. Diebold U (2003) The surface science of titanium dioxide. Surf Sci Rep 48(5):53–229
66. Yan M, Chen F, Zhang J et al (2005) Preparation of controllable crystalline titania and study on the photocatalytic properties. J Phys Chem B 109(18):8673–8678
67. Li Y, White TJ, Lim SH (2004) Low-temperature synthesis and microstructural control of titania nano-particles. J Solid State Chem 177(4):1372–1381
68. Khataee AR, Kasiri MB (2010) Photocatalytic degradation of organic dyes in the presence of nanostructured titanium dioxide: influence of the chemical structure of dyes. J Mol Catal A Chem 328(1):8–26
69. Daneshvar N, Rasoulifard MH, Khataee AR et al (2007) Removal of CI acid orange 7 from aqueous solution by UV irradiation in the presence of ZnO nanopowder. J Hazard Mater 143 (1):95–101
70. Moret MP, Zallen R, Vijay DP et al (2000) Brookite-rich titania films made by pulsed laser deposition. Thin Solid Films 366(1):8–10
71. Shannon RD, Pask JA (1965) Kinetics of the anatase-rutile transformation. J Am Ceram Soc 48(8):391–398
72. Batzill M, Morales EH, Diebold U (2006) Influence of nitrogen doping on the defect formation and surface properties of TiO_2 rutile and anatase. Phys Rev Lett 96(2):026–103
73. Oskam G, Nellore A, Penn RL et al (2003) The growth kinetics of TiO_2 nanoparticles from titanium (IV) alkoxide at high water/titanium ratio. J Phys Chem B 107(8):1734–1738
74. Banfield J (1998) Thermodynamic analysis of phase stability of nanocrystalline titania. J Mater Chem 8(9):2073–2076
75. Li JG, Ishigaki T, Sun X (2007) Anatase, brookite, and rutile nanocrystals via redox reactions under mild hydrothermal conditions: phase-selective synthesis and physicochemical properties. J Phys Chem C 111(13):4969–4976
76. Bickley RI, Gonzalez-Carreno T, Lees JS et al (1991) A structural investigation of titanium dioxide photocatalysts. J Solid State Chem 92(1):178–190
77. Lee SK, Robertson PK, Mills A et al (1999) Modification and enhanced photocatalytic activity of TiO_2 following exposure to non-linear irradiation sources. J Photochem Photobiol A Chem 122(1):69–71
78. Tsai SJ, Cheng S (1997) Effect of TiO_2 crystalline structure in photocatalytic degradation of phenolic contaminants. Catal Today 33(1–3):227–237

79. Ohno T, Sarukawa K, Matsumura M (2001) Photocatalytic activities of pure rutile particles isolated from TiO_2 powder by dissolving the anatase component in HF solution. J Phys Chem B 105(12):2417–2420
80. Ozawa T, Iwasaki M, Tada H et al (2005) Low-temperature synthesis of anatase–brookite composite nanocrystals: the junction effect on photocatalytic activity. J Colloid Interface Sci 281(2):510–513
81. Xu H, Zhang L (2009) Controllable one-pot synthesis and enhanced photocatalytic activity of mixed-phase TiO_2 nanocrystals with tunable brookite/rutile ratios. J Phys Chem C 113 (5):1785–1790
82. Bacsa RR, Kiwi J (1998) Effect of rutile phase on the photocatalytic properties of nanocrystalline titania during the degradation of p-coumaric acid. Appl Catal B Environ 16(1):19–29
83. Jung KY, Park SB, Jang HD (2004) Phase control and photocatalytic properties of nano-sized titania particles by gas-phase pyrolysis of $TiCl_4$. Catal Commun 5(9):491–497
84. Zhu J, Zheng W, He B et al (2004) Characterization of Fe–TiO_2 photocatalysts synthesized by hydrothermal method and their photocatalytic reactivity for photodegradation of XRG dye diluted in water. J Mol Catal A Chem 216(1):35–43
85. Wu Y, Xing M, Zhang J (2011) Gel-hydrothermal synthesis of carbon and boron co-doped TiO_2 and evaluating its photocatalytic activity. J Hazard Mater 192(1):368–373
86. Ng J, Wang X, Sun DD (2011) One-pot hydrothermal synthesis of a hierarchical nanofungus-like anatase TiO_2 thin film for photocatalytic oxidation of bisphenol A. Appl Catal B Environ 110:260–272
87. Ovenstone J, Yanagisawa K (1999) Effect of hydrothermal treatment of amorphous titania on the phase change from anatase to rutile during calcination. Chem Mater 11(10):2770–2774
88. Li G, Ciston S, Saponjic ZV et al (2008) Synthesizing mixed-phase TiO_2 nanocomposites using a hydrothermal method for photo-oxidation and photoreduction applications. J Catal 253 (1):105–110
89. Fehse M, Fischer F, Tessier C et al (2013) Tailoring of phase composition and morphology of TiO_2-based electrode materials for lithium-ion batteries. J Power Sources 231:23–28
90. Zhang Y, Chen J, Li X (2010) Preparation and photocatalytic performance of anatase/rutile mixed-phase TiO_2 nanotubes. Catal Lett 139(3–4):129–133
91. Tay Q, Liu X, Tang Y et al (2013) Enhanced photocatalytic hydrogen production with synergistic two-phase anatase/brookite TiO_2 nanostructures. J Phys Chem C 117 (29):14973–14982
92. Zhang H, Banfield JF (2000) Understanding polymorphic phase transformation behavior during growth of nanocrystalline aggregates: insights from TiO_2. J Phys Chem B 104 (15):3481–3487
93. Shen X, Tian B, Zhang J (2013) Tailored preparation of titania with controllable phases of anatase and brookite by an alkalescent hydrothermal route. Catal Today 201:151–158
94. Zhao B, Chen F, Huang Q et al (2009) Brookite TiO_2 nanoflowers. Chem Commun 34:5115–5117
95. Zhao LM, Zhang ZJ, Zhang SY et al (2011) Metal–organic frameworks based on transition-metal carboxylate clusters as secondary building units: synthesis, structures and properties. CrystEngComm 13(3):907–913
96. Zhao B, Chen F, Jiao Y et al (2010) Phase transition and morphological evolution of titania/titanate nanomaterials under alkalescent hydrothermal treatment. J Mater Chem 20 (37):7990–7997
97. Shen X, Zhang J, Tian B (2012) Tartaric acid-assisted preparation and photocatalytic performance of titania nanoparticles with controllable phases of anatase and brookite. J Mater Sci 47 (15):5743–5751
98. Cheng H, Ma J, Zhao Z et al (1995) Hydrothermal preparation of uniform nanosize rutile and anatase particles. Chem Mater 7(4):663–671
99. Li G, Gray KA (2007) Preparation of mixed-phase titanium dioxide nanocomposites via solvothermal processing. Chem Mater 19(5):1143–1146

100. Lei SHI, Duan WENG (2008) Highly active mixed-phase TiO_2 photocatalysts fabricated at low temperature and the correlation between phase composition and photocatalytic activity. J Environ Sci 20(10):1263–1267
101. Wu M, Long J, Huang A et al (1999) Microemulsion-mediated hydrothermal synthesis and characterization of nanosize rutile and anatase particles. Langmuir 15(26):8822–8825
102. Shen X, Zhang J, Tian B (2011) Microemulsion-mediated solvothermal synthesis and photocatalytic properties of crystalline titania with controllable phases of anatase and rutile. J Hazard Mater 192(2):651–657
103. Zachariah A, Priya R, Baiju KV, Shukla S et al (2008) Synergistic effect in photocatalysis as observed for mixed-phase nanocrystalline titania processed via sol-gel solvent mixing and calcination. J Phys Chem C 112:11345–11356
104. Bojinova A, Kralchevska R, Poulios I et al (2007) Anatase/rutile TiO_2 composites: influence of the mixing ratio on the photocatalytic degradation of malachite green and orange II in slurry. Mater Chem Phys 106(2):187–192
105. Liu Z, Zhang X, Nishimoto S et al (2007) Anatase TiO_2 nanoparticles on rutile TiO_2 nanorods: a heterogeneous nanostructure via layer-by-layer assembly. Langmuir 23(22):10916–10919
106. Nair RG, Paul S, Samdarshi SK (2011) High UV/visible light activity of mixed phase titania: a generic mechanism. Sol Energy Mater Sol Cells 95(7):1901–1907
107. Gouma PI, Mills MJ (2001) Anatase-to-rutile transformation in titania powders. J Am Ceram Soc 84(3):619–622
108. Zhang J, Li M, Feng Z et al (2006) UV Raman spectroscopic study on TiO_2 I. Phase transformation at the surface and in the bulk. J Phys Chem B 110(2):927–935
109. Hsu YC, Lin HC, Chen CH et al (2010) Nonaqueous seeded growth of flower-like mixed-phase titania nanostructures for photocatalytic applications. J Solid State Chem 183 (9):1917–1924
110. Ni M, Leung MK, Leung DY et al (2007) A review and recent developments in photocatalytic water-splitting using TiO_2 for hydrogen production. Renew Sust Energ Rev 11(3):401–425
111. Lei J, Li H, Zhang J et al (2016) Mixed-phase TiO_2 nanomaterials as efficient photocatalysts. In: Low-dimensional and nanostructured materials and devices. Springer International Publishing, Cham, pp 423–460
112. Xu Q, Ma Y, Zhang J et al (2011) Enhancing hydrogen production activity and suppressing CO formation from photocatalytic biomass reforming on Pt/TiO_2 by optimizing anatase–rutile phase structure. J Catal 278(2):329–335
113. Kho YK, Iwase A, Teoh WY et al (2010) Photocatalytic H_2 evolution over TiO_2 nanoparticles. The synergistic effect of anatase and rutile. J Phys Chem C 114(6):2821–2829
114. Xu F, Xiao W, Cheng B et al (2014) Direct Z-scheme anatase/rutile bi-phase nanocomposite TiO_2 nanofiber photocatalyst with enhanced photocatalytic H_2-production activity. Int J Hydrog Energy 39(28):15394–15402
115. Rosseler O, Shankar MV, Karkmaz-Le Du M et al (2010) Solar light photocatalytic hydrogen production from water over Pt and Au/TiO_2 (anatase/rutile) photocatalysts: influence of noble metal and porogen promotion. J Catal 269(1):179–190
116. Marcı G, Addamo M, Augugliaro V et al (2003) Photocatalytic oxidation of toluene on irradiated TiO_2: comparison of degradation performance in humidified air, in water and in water containing a zwitterionic surfactant. J Photochem Photobiol A Chem 160(1):105–114
117. Liu L, Zhao H, Andino JM et al (2012) Photocatalytic CO_2 reduction with H_2O on TiO_2 nanocrystals: comparison of anatase, rutile, and brookite polymorphs and exploration of surface chemistry. ACS Catal 2(8):1817–1828
118. Zhao H, Liu L, Andino JM et al (2013) Bicrystalline TiO_2 with controllable anatase–brookite phase content for enhanced CO_2 photoreduction to fuels. J Mater Chem A 1(28):8209–8216
119. Frank NS, Bard AJ (1977) Heterogeneous photocatalytic oxidation of cyanide ion in aqueous solutions at titanium dioxide powder. J Am Chem Soc 99(1):303–304
120. Xu H, Li G, Zhu G et al (2015) Enhanced photocatalytic degradation of rutile/anatase TiO_2 heterojunction nanoflowers. Catal Commun 62:52–56

121. Luo Z, Poyraz AS, Kuo CH et al (2015) Crystalline mixed phase (anatase/rutile) mesoporous titanium dioxides for visible light photocatalytic activity. Chem Mater 27(1):6–17
122. Zhao B, Chen F, Jiao Y et al (2011) Ag 0-loaded brookite/anatase composite with enhanced photocatalytic performance towards the degradation of methyl orange. J Mol Catal A Chem 348(1):114–119
123. Liao Y, Que W, Jia Q et al (2012) Controllable synthesis of brookite/anatase/rutile TiO_2 nanocomposites and single-crystalline rutile nanorods array. J Mater Chem 22(16):7937–7944
124. Grabowska E, Reszczyńska J, Zaleska A (2012) Mechanism of phenol photodegradation in the presence of pure and modified-TiO_2: a review. Water Res 46(17):5453–5471
125. Ding Z, Lu GQ, Greenfield PF (2000) Role of the crystallite phase of TiO_2 in heterogeneous photocatalysis for phenol oxidation in water. J Phys Chem B 104(19):4815–4820
126. Tian B, Li C, Zhang J (2012) One-step preparation, characterization and visible-light photocatalytic activity of Cr-doped TiO_2 with anatase and rutile bicrystalline phases. Chem Eng J 191:402–409
127. Deák P, Aradi B, Frauenheim T (2011) Band lineup and charge carrier separation in mixed rutile-anatase systems. J Phys Chem C 115(8):3443–3446
128. Scanlon DO, Dunnill CW, Buckeridge J et al (2013) Band alignment of rutile and anatase TiO_2. Nat Mater 12(9):798–801
129. Datye AK, Riegel G, Bolton JR et al (1995) Microstructural characterization of a fumed titanium dioxide photocatalyst. J Solid State Chem 115(1):236–239
130. Zhang Z, Wang CC, Zakaria R et al (1998) Role of particle size in nanocrystalline TiO_2-based photocatalysts. J Phys Chem B 102(52):10871–10878
131. Ohno T, Sarukawa K, Tokieda K et al (2001) Morphology of a TiO_2 photocatalyst (Degussa, P-25) consisting of anatase and rutile crystalline phases. J Catal 203(1):82–86
132. Kawahara T, Konishi Y, Tada H et al (2002) A patterned TiO_2 (anatase)/TiO_2 (rutile) bilayer-type photocatalyst: effect of the anatase/rutile junction on the photocatalytic activity. Angew Chem 114(15):2935–2937
133. Li G, Gray KA (2007) The solid–solid interface: explaining the high and unique photocatalytic reactivity of TiO_2-based nanocomposite materials. Chem Phys 339(1):173–187
134. Sun B, Vorontsov AV, Smirniotis PG (2003) Role of platinum deposited on TiO_2 in phenol photocatalytic oxidation. Langmuir 19(8):3151–3156
135. Sun B, Smirniotis PG (2003) Interaction of anatase and rutile TiO_2 particles in aqueous photooxidation. Catal Today 88(1):49–59
136. Liu B, Peng L (2013) Facile formation of mixed phase porous TiO_2 nanotubes and enhanced visible-light photocatalytic activity. J Alloys Compd 571:145–152
137. Li G, Chen L, Graham ME et al (2007) A comparison of mixed phase titania photocatalysts prepared by physical and chemical methods: the importance of the solid–solid interface. J Mol Catal A Chem 275(1):30–35
138. Wang CY, Pagel R, Dohrmann JK et al (2006) Antenna mechanism and deaggregation concept: novel mechanistic principles for photocatalysis. C R Chim 9(5):761–773

Chapter 7
The Preparation and Applications of g-C_3N_4/TiO_2 Heterojunction Catalysts

7.1 Introduction

Nowadays, there have been two global problems all the society is facing, environmental pollution and energy shortage, which have caused great harm to human health and life. To solve these problems, photocatalysis as an effective approach has attracted widespread concern of researchers. In this approach, by utilizing photocatalysts, light as a clean excitation power can be used to induce a series of catalytic reactions, with regard to environment and energy, such as photocatalytic degradation of pollutants in water [1–4], removal of indoor harmful gases [5], reduction of CO_2 [6–8], as well as splitting of water to produce H_2 and O_2 [9].

In many of the photocatalysts studied, TiO_2 has been widely recognized as the most potential one due to its merits of low cost, good stability, nontoxicity, and so on [10–13]. However, the conventional TiO_2 has shortcomings in the following two aspects: (1) the large bandgap (~3.2 eV) can only absorb UV light with $\lambda \leq 387$ nm, and the absorption of visible light is almost zero, which leads to the low utilization efficiency of sunlight, and (2) the recombination rate of photo-generated electron–hole pairs is high, greatly limiting the photocatalytic performance of TiO_2. At present, various methods for the improvement of the photocatalytic activity of TiO_2 have been reported, such as metal and nonmetal oxide loading [14, 15], noble metal deposition [16, 17], nonmetal element doping [18, 19], modifications of morphology, and so on [20–23]. These methods can broaden the absorption wavelength of TiO_2 and enhance the absorption efficiency of solar light in some extent. But the recombination of photo-generated electrons and holes results in a lower quantum yield, further seriously affecting the catalytic activity of the catalysts.

Heterojunction catalyst is one of the hot spots in catalytic field in recent years, which usually consists of two different semiconductors contacting with each other to form the structure of heterojunction. The heterojunction structure can promote the transfer of the photo-generated electrons and holes in opposite direction, greatly improving the effective utilization rate of the excitons. Therefore, heterojunction

J. Zhang et al., *Photocatalysis*, Lecture Notes in Chemistry 100,
https://doi.org/10.1007/978-981-13-2113-9_7

catalyst possesses better catalytic effect than that of single-component catalyst. Nowadays, there have been a great number of materials which can be used to modify TiO_2 by forming the heterojunction structure, such as ZnO [24, 25], SnO_2 [26, 27], $g\text{-}C_3N_4$, etc. [28–31]. Among them, $g\text{-}C_3N_4$ as a stable nonmetal semiconductor has attracted much attention in the catalytic field. Due to its special triazine structure, $g\text{-}C_3N_4$ exhibits many special properties including low density, high chemical stability, enhanced biological compatibility, good abrasion resistance, and so on. In addition, the relatively narrow bandgap of $g\text{-}C_3N_4$ (about 2.7 eV) extends its light response to visible region (about 450 nm). Zhang et al. reported that the $g\text{-}C_3N_4$ exhibited high photocatalytic performance for water splitting under visible light irradiation [32]. Dong and coworkers facilely synthesized polymeric $g\text{-}C_3N_4$-layered catalyst by directly heating urea or thiourea [33, 34]. In the research on heterojunction catalysts, $g\text{-}C_3N_4$ has gained the majority of researchers' attention owing to its simple preparation method, abundant kinds of precursors, as well as the advantages of low cost, becoming the first choice to form heterojunction with TiO_2.

In this chapter, the recent developments of the research on the heterojunction photocatalysts formed by $g\text{-}C_3N_4$ and TiO_2 were introduced, including its synthesis methods and applications. Firstly, the synthesis methods were summarized based on the synthesis order of each component during the preparation process and divided into three categories. In each category, the preparation procedures as well as their advantages and drawbacks were introduced in detail. Through these synthesis methods, an efficient heterojunction structure can be obtained between $g\text{-}C_3N_4$ and TiO_2. The photocatalytic activity of the photocatalysts can be greatly enhanced due to the formation of the heterojunction structure, which can effectively promote the separation of photo-generated charge carriers [35, 36]. The excellent photocatalytic activity of the $g\text{-}C_3N_4/TiO_2$ heterojunction photocatalysts enables them to be applied in many aspects. Therefore, the chapter also introduced the applications of $g\text{-}C_3N_4/TiO_2$ heterojunction photocatalysts in the field of photocatalysis, containing depollution of environment, hydrogen generation, photofixation of carbon dioxide, bacteria disinfection, and so on. In the end of the chapter, a short summary and outlook on the development of $g\text{-}C_3N_4/TiO_2$ heterojunction photocatalysts were provided.

7.2 The Preparation Methods of $g\text{-}C_3N_4/TiO_2$ Heterojunction Catalyst

Because the $g\text{-}C_3N_4/TiO_2$ heterojunction photocatalyst consists of two single-component $g\text{-}C_3N_4$ and TiO_2, the synthesis methods of $g\text{-}C_3N_4/TiO_2$ heterojunction catalyst can be classified into three categories according to the order of synthesis of each component: (1) physically mixing TiO_2 and $g\text{-}C_3N_4$, (2) growing $g\text{-}C_3N_4$ on TiO_2 catalyst, and (3) loading TiO_2 on $g\text{-}C_3N_4$ catalyst. No matter what the method is, the final aim is to make g-C_3N_4 and TiO_2 contact with each other, further forming a heterojunction structure between g-C_3N_4 and TiO_2.

7.2.1 Physically Mixing g-C_3N_4 and TiO_2

This preparation method refers to firstly synthesizing g-C_3N_4 and TiO_2, respectively, and then physically mixing the two components together by ball milling or evaporation of dispersion solution. Hongjian Yan et al. prepared TiO_2-C_3N_4 by mixing TiO_2 and g-C_3N_4 powder using a ball milling method with different contents of g-C_3N_4. The TiO_2 was synthesized by the hydrolysis of $TiCl_4$ in ammonia, and the g-C_3N_4 was prepared by directly heating melamine at an atmosphere of Ar [37]. Yongfa Zhu et al. also fabricated g-C_3N_4/TiO_2 hybrid photocatalysts by a facile ball milling method. In their experiments, g-C_3N_4 was synthesized by directly heating melamine, and g-C_3N_4/TiO_2 photocatalysts were obtained by mixing g-C_3N_4 and TiO_2 powder in a ball mill. Their results showed that a layered structure of g-C_3N_4 was formed on the surface of TiO_2 [38]. Interestingly, they found that the as-prepared catalyst showed highly enhanced photocatalytic performance and the photocatalytic efficiency increased gradually with the increase of milling rate.

Apart from the ball milling, solvent evaporation is also a commonly used physical mixing method for the synthesis of g-C_3N_4/TiO_2 heterojunction photocatalyst. In this preparation process, g-C_3N_4 and TiO_2 are homogeneously dispersed in a solvent such as methanol, and then the solvent is evaporated to make g-C_3N_4 and TiO_2 contact with each other and thus form the heterojunction structure. Jingyu Wang et al. hybridized anatase TiO_2 nanosheets with dominant (001) facets with g-C_3N_4 via this facile solvent evaporation method. The polymeric g-C_3N_4 was synthesized by directly calcinating urea, and the anatase TiO_2 nanosheets with dominant (001) facets were prepared by a solvothermal reaction of tetrabutyl titanate (TBT). After the solvothermal treatment, the well-washed precipitate was dispersed into methanol and mixed with g-C_3N_4, followed by sonication for 30 min to completely disperse the g-C_3N_4. After that, the above sample was stirred in a fume hood for 12 h to evaporate the methanol, and the rest powder was dried at 100 °C for 4 h [39]. Dongjiang Yang et al. synthesized g-C_3N_4/TiO_2(B) nanofibers with exposed (001) plane with the enhanced visible light photoactivity through a facile solvent evaporation operation to the methanol solution of g-C_3N_4 and TiO_2 (B). The g-C_3N_4 was prepared by directly heating melamine in air at the temperature of 550 °C for 4 h, and TiO_2 (B) nanofibers were synthesized using a hydrothermal method combined with a subsequent calcination treatment [40]. Hong Huang et al. prepared heterostructured g-C_3N_4/Ag/TiO_2 microspheres with improved photocatalytic performance under visible light irradiation. As shown in Fig. 7.1, the protonated g-C_3N_4 sheets were synthesized by calcinating melamine and followed by the protonation in HCl solution, and TiO_2 nanomaterial was prepared by a typical hydrothermal method of $Ti(OC_4H_9)_4$, and then Ag/TiO_2 microspheres were obtained by depositing

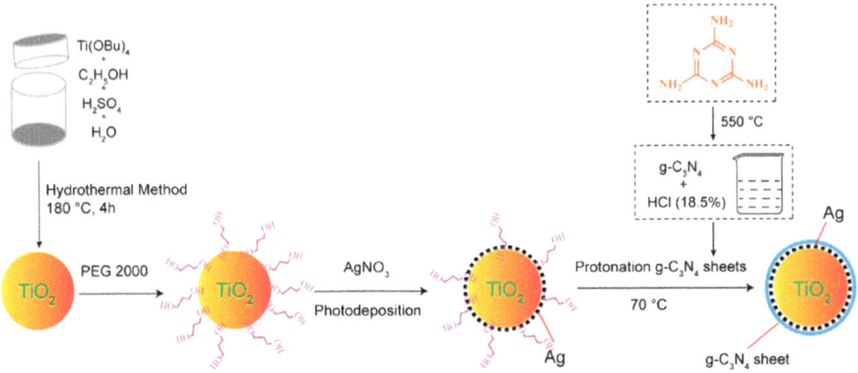

Fig. 7.1 Scheme for the synthesis of g-C_3N_4/Ag/TiO_2 microspheres [41] (Reprinted with permission from Ref. [41]. Copyright 2014, American Chemical Society)

Ag on the surface of TiO_2 microspheres, which was then mixed with g-C_3N_4 by forming suspension at 70 °C [41].

Guangshe Li et al. reported an effective visible light-driven photocatalyst of brookite TiO_2 (br-TiO_2) hybridized with g-C_3N_4 for the first time via a facile calcination of br-TiO_2 and g-C_3N_4 in air. The optimum photocatalytic activity of the as-prepared samples was higher than that of other phase types of TiO_2 (anatase and rutile) hybridized with g-C_3N_4 [42]. Tianyou Peng et al. synthesized porous g-C_3N_4 by a simple pyrolysis of urea, and Pt-TiO_2 was fabricated by photodepositing Pt on the TiO_2. Then g-C_3N_4–Pt-TiO_2 nanocomposite was synthesized via a facile chemical adsorption followed by a calcination treatment [30].

The physical mixing method for preparing g-C_3N_4/TiO_2 heterojunction catalyst is easy to operate and beneficial for scale-up, which provides a potential for mass production. However, uniform mixing may not be easy to achieve. In the aspect of designing catalyst, morphology control is widely considered as an effective way to improve the catalytic activity, while physical mixing method is difficult to achieve this goal. Moreover, the close contact between g-C_3N_4 and TiO_2 may be not easy to form, causing the poor stability of the heterojunction catalysts.

7.2.2 *Growing TiO_2 on g-C_3N_4*

In this method, g-C_3N_4 is firstly synthesized by one-step calcinations of precursors, and then the prepared g-C_3N_4 reacts with the precursor of TiO_2 to achieve the in situ growth of TiO_2 on the surface of g-C_3N_4. For example, Deliang Cui et al. fabricated g-C_3N_4/TiO_2 composite through this method. The g-C_3N_4 was synthesized by polymerization of dicyandiamide at the temperature of 600 °C for 5 h under N_2 atmosphere. Then the as-synthesized g-C_3N_4 was taken into the hydrolysis of Ti ($OC_4H_9n)_4$. After hydrothermal reaction, the hybrid composite of g-C_3N_4/TiO_2 was obtained, which showed better photocatalytic activity than hybrid composite of g-C_3N_4/TiO_2 and the pure TiO_2 for degradation of rhodamine B (RhB) under the UV and visible light irradiation [28]. Similarly, Hongtao Yu et al. prepared g-C_3N_4/TiO_2 hybrid photocatalyst with wide absorption wavelength range from 300 nm to 450 nm by taking g-C_3N_4 into the hydrolysis reaction of $TiCl_4$. TEM images showed that TiO_2 nanoparticles were dispersed well on the surface of g-C_3N_4 sheet, and the average size of particles was much smaller than that of TiO_2 samples without g-C_3N_4 sheet. The synthesized g-C_3N_4/TiO_2 exhibited much better photocatalytic activity for the degradation of phenol than pristine g-C_3N_4 and TiO_2 [43]. Qianhong Shen et al. developed a novel and facile template-free method to synthesize a network structure of mesoporous g-C_3N_4/TiO_2 nanocomposite with enhanced visible light photocatalytic activity. Firstly, they synthesized g-C_3N_4 by directly heating melamine, and then g-C_3N_4/TiO_2 was obtained by adding g-C_3N_4 into the solution of titanium sulfate $Ti(SO_4)_2$ and followed by hydrothermal reaction [44].

In recent years, more and more attention has been paid in the research on nitrogen-doped titanium dioxide (N-TiO_2), due to its promising extension for environmental application [2, 45]. Many groups grew the N-TiO_2 on the surface of g-C_3N_4 to form heterojunction. Fatang Li et al. reported an in situ microwave-assisted synthesis method to fabricate N-TiO_2/g-C_3N_4 composites by using H_2TiO_3 as the reactant and $NH_3 \cdot H_2O$ as the N-doping source. In their experiments, they firstly took g-C_3N_4 into the H_2TiO_3 solution then followed by a microwave-assisted reaction. The preparation process was as shown in Fig. 7.2. The catalyst had a porous structure and large surface area, which increased the contact area of the catalyst with pollutants. The photocatalytic degradation of rhodamine B (RhB) and methylene blue (MB) with the as-prepared samples was carried out under visible light irradiation to evaluate the photocatalytic activity. Among them, N-TiO_2/g-C_3N_4 composite with 40 wt % of N-TiO_2 showed the highest photocatalytic activity [46].

Through heating the mixture of the hydrolysis product of $TiCl_4$ and g-C_3N_4 at different weight ratios, W. F. Zhang et al. successfully prepared N-doped TiO_2/C_3N_4 composite samples. Due to the introduction of g-C_3N_4, the composite samples showed slight visible light absorption. XPS result revealed that some nitrogen was doped into TiO_2, and g-C_3N_4 existed in the composite sample [29]. Similarly, as shown in Fig. 7.3, g-C_3N_4 nanosheets (g-C_3N_4 NSs) hybridized nitrogen-doped TiO_2 (N-TiO_2) nanofibers (GCN/NT NFs) have been synthesized in situ through a

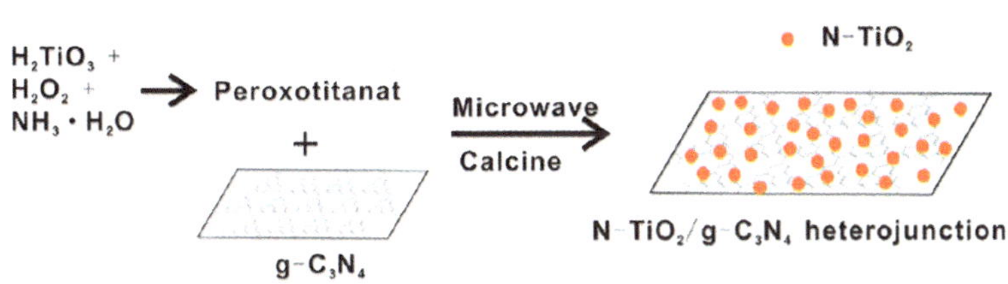

Fig. 7.2 Diagrammatic sketch for the in situ deposition of N-TiO_2 nanoparticles on g-C_3N_4 sheets [46]. (Reprinted with permission from Ref. [46]. Copyright 2013, American Chemical Society)

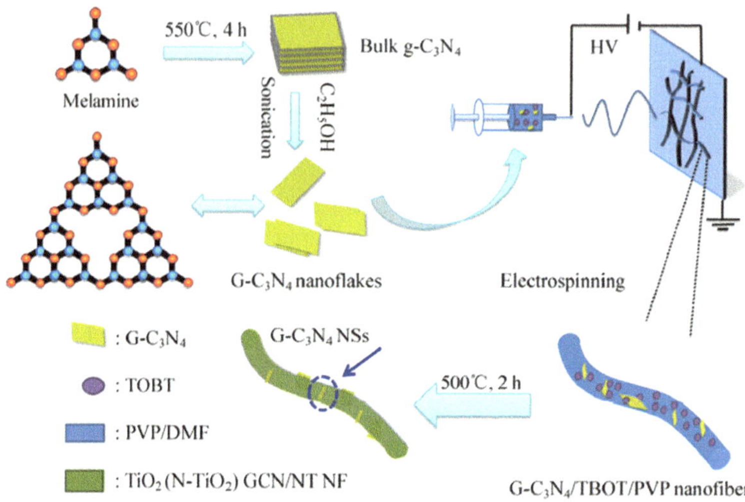

Fig. 7.3 Schematic illustration of the fabrication of GCN/NT NFs [47]. (Reprinted with permission from Ref. [47]. Copyright 2013, Springer)

simple electrospinning process combined with a modified heat-etching method by Cheng Han et al. [47]. The melamine was thermal polymerized to form g-C_3N_4, which was dispersed into acetic acid solution including poly(vinylpyrrolidone) (PVP) and titanium (IV) n-butoxide (TNBT). Doping nitrogen into TiO_2 narrowed its energy bandgap, and the catalyst could be activated under visible irradiations, leading to higher photocatalytic efficiency.

In addition, most of TiO_2 nanoparticles grown on the surface of g-C_3N_4 were present as crystals. Solvothermal reaction can control the exposure of high-energy surfaces. For example, Kangle Lv et al. grew TiO_2 hollow nanobox (TiO_2-HNB) assembled from high-energy TiO_2 nanosheets (TiO_2-NS) on g-C_3N_4 to form the g-C_3N_4/TiO_2 hybrid and investigated the effect of contact interfaces of high-energy TiO_2, (101) and (001) facets on the photocatalytic activity. The catalyst was fabricated through a solvothermal strategy using TBA as the solvent [48]. In our previous work, well-dispersed TiO_2 nanocrystals with (001) facets were successfully grown in situ on g-C_3N_4 through a facial solvothermal method, as shown in Fig. 7.4. During the solvothermal process, the ammonium acetate (AMAT) serving as a catalyst for the hydrolysis of tetrabutyl titanate (TBOT) was added into the nonaqueous system. In addition, because carboxylic acid is easy to adsorb on the surface of anatase (001), part of the acetic acid produced by the decomposition of AMAT serves as face-growth inhibitors, slowing the growth of the (001) facet of TiO_2 in the TiO_2 nanoparticles, leading to the exposure of high-energy facets. The characterization results showed an enhanced separation efficiency of photo-generated charge carriers compared with that of pure g-C_3N_4, and well-matched energy levels between TiO_2 and g-C_3N_4 altogether led to the enhancement of photocatalytic activity [49].

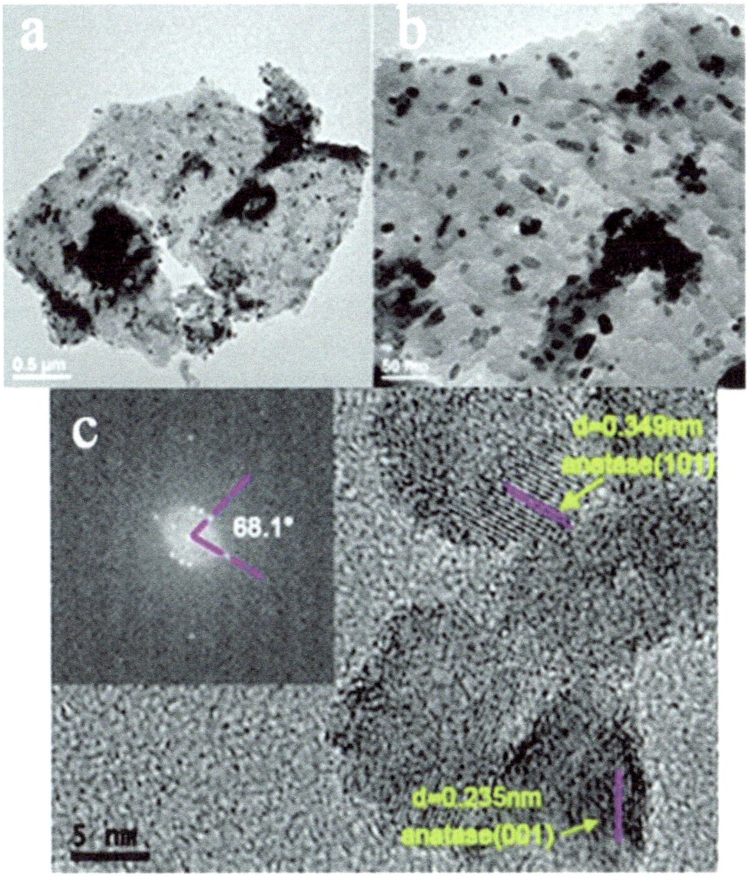

Fig. 7.4 (**a** and **b**) TEM images of TiO_2/g-C_3N_4(1.5) catalyst, (**c**) HRTEM image of TiO_2/g-C_3N_4(1.5) catalyst. Inset: the corresponding fast Fourier transform (FFT) pattern [49]. (Reprinted with permission from Ref. [49]. Copyright 2015, Royal Society of Chemistry)

The strategy of growing TiO_2 on g-C_3N_4 is very effective to form the heterojunction structure. The heterojunction which formed in the growing process possesses chemical stability during the multiple cycle experiments. However, due to the rapid hydrolysis process of the titanium precursor, it is difficult to achieve ultradispersed TiO_2 nanocrystals on the surface of g-C_3N_4 by this method [50]. Besides, a great challenge for controlling the microstructures of coupled TiO_2 with desired size distribution and dispersity still retains.

7.2.3 Loading g-C_3N_4 on TiO_2

In this synthesis method, TiO_2 is firstly obtained by hydrolysis, hydrothermal, microwave method, or directly using the commercial P25, and then TiO_2 is impregnated in the precursor solution of g-C_3N_4 to obtain the g-C_3N_4/TiO_2 heterojunction catalyst after drying and calcination. Weide Zhang et al. modified TiO_2 nanorod arrays with g-C_3N_4 via chemical vapor deposition using melamine as a precursor. The rutile TiO_2 nanorod arrays were firstly synthesized by hydrothermal process, and then the TiO_2/FTO was loaded with melamine and followed by heating process in a muffle furnace to obtain g-C_3N_4/TiO_2/FTO. The g-C_3N_4/TiO_2/FTO electrode exhibited high photoelectrocatalytic activity for degradation of RhB. Under visible light irradiation, the photocurrent response of the g-C_3N_4/TiO_2/FTO electrode is about 10 times as that of the TiO_2/FTO electrode, making it a promising nanomaterial for future applications in solar cells, water treatment, as well as photoelectric devices [51]. Min Fu et al. prepared a kind of novel visible light photocatalyst g-C_3N_4/TiO_2 composite by calcinating the mixtures of melamine and commercial TiO_2 at different weight ratios. In their work, the samples at the optimized precursor weight ratio ($M_{melamine}$: $M_{titania} = 2.5$) exhibited highest adsorption ability and visible light photocatalytic activity, evaluated by photocatalytic degradation of methylene blue (MB) [52]. Furthermore, Min Fu et al. also synthesized novel g-C_3N_4-coated TiO_2 nanocomposites by a facile and cost-effective solid-state method through thermal treatment of the mixture of urea and commercial TiO_2. The as-prepared g-C_3N_4-coated TiO_2 nanocomposites showed efficient visible light photocatalytic activity for degradation of aqueous MB owing to the increased visible light absorption and enhanced MB adsorption [31]. Burapat Inceesungvorn et al. fabricated g-C_3N_4/TiO_2 films by directly heating the mixture of melamine and pre-synthesized TiO_2 nanoparticles at the atmosphere of Ar. The TiO_2 was prepared by hydrolysis of titanium tetraisopropoxide (TTIP) and calcination. The obtained samples showed enhanced photocatalytic degradation of MB. In addition, as Fig. 7.5

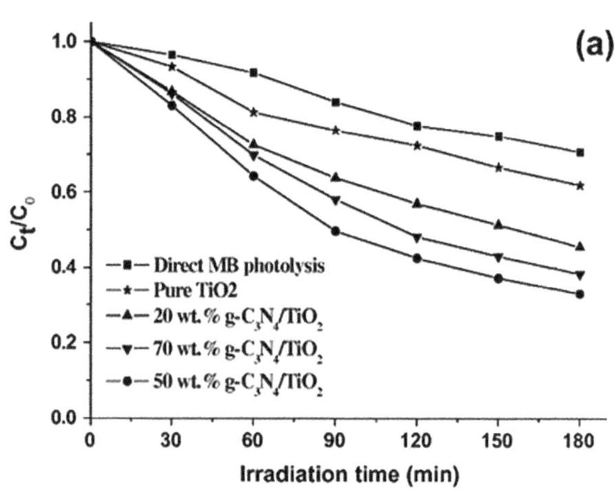

Fig. 7.5 MB photolysis and photocatalytic degradation using pure TiO_2, pure g-C_3N_4, and g-C_3N_4/TiO_2 composite films as photocatalysts [53]. (Reprinted with permission from Ref. [53]. Copyright 2014, Elsevier)

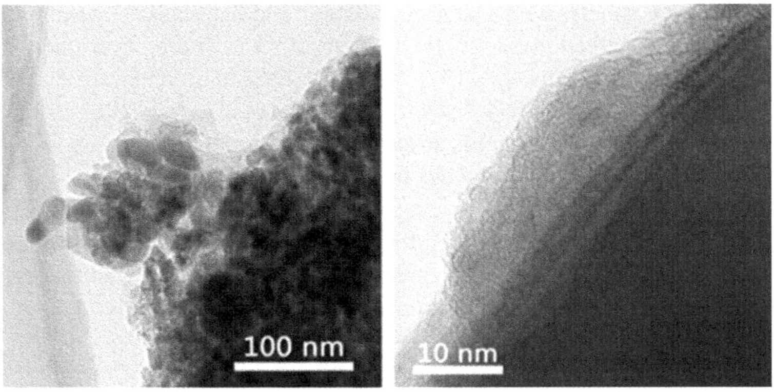

Fig. 7.6 TEM image of g-C_3N_4-modified TiO_2 composites [55]. (Reprinted with permission from Ref. [55]. Copyright 2015, Elsevier)

shows, the 50 wt% g-C_3N_4/TiO_2 composite with the best loading content exhibited the best performance [53].

Honglei Zhu et al. synthesized a series of g-C_3N_4-P25 composite photocatalysts with different mass ratios using an in situ preparation method. In the method, g-C_3N_4-P25 nanocomposites were obtained by calcinating mixtures of the P25 and dicyandiamide. The optimal g-C_3N_4 content was determined to be 84%. The sample in the optimal weight ratio exhibited almost 3.3 times higher photocatalytic activity than that of individual g-C_3N_4 under visible light irradiation [54]. Our group also prepared g-C_3N_4-modified TiO_2 composites through a simple calcination process of anatase and cyanamide. TEM images of as-synthesized catalyst, presented in Fig. 7.6, show TiO_2 is covered by a thin shell of g-C_3N_4, and the polymer shell on the surface is around 5–10 nm thick. The photocatalytic activities of the composites were evaluated by photocatalytic degradation of Acid Orange 7 (AO7). The photocatalyst showed excellent activity under both visible and UV light. In addition, no nitrogen doping was found in TiO_2 lattice, demonstrating the g-C_3N_4 was surface attached on TiO_2 and ascribing all improvement of photocatalytic activity of g-C_3N_4/TiO_2 composite to the synergy between TiO_2 and g-C_3N_4 [55]. After that, we reported a highly condensed g-C_3N_4-modified TiO_2 photocatalyst prepared by a vacuum calcination method. A close-to-theoretical C/N ratio was detected in the catalyst by element analysis. The results indicated a complete and neat polymerization of the g-C_3N_4 on TiO_2. Excellent photocatalytic activities of as-prepared catalysts have been achieved under both visible and UV light irradiation. The heterojunction can be easily obtained during the calcination process, and the preparation procedures are easy to operate, but the amount of loading g-C_3N_4 is influenced by numerous factors, such as gas condition, flow rate, heating temperature, and heating rate [56].

Photochemical and electrochemical methods were also developed to load g-C_3N_4 on TiO_2. These methods are hard to control, but this in situ growth strategy has drawn more and more attention in recent years. Xiaoxin Zou et al. synthesized mesoporous TiO_2 spheres with a large surface area and rich surface hydroxyl groups by a light-driven synthetic strategy. It can be used for activating urea under a mild condition to form $g\text{-}C_3N_4$ material [57]. Xiaosong Zhou et al. synthesized a $g\text{-}C_3N_4/TiO_2$ nanotube array (CN/TNT) heterojunction photocatalyst with visible light response via a simple electrochemical method. $g\text{-}C_3N_4$ polymer was deposited into the crystallized TiO_2 nanotubes by electrodeposition [58].

In this method, because TiO_2 is prepared firstly, it is allowed for selection or structure design of TiO_2, but the high-temperature calcination for the formation of $g\text{-}C_3N_4$ is prone to resulting in the aggregation of TiO_2 and may lead to a negative impact on the improvement of photocatalytic activity.

7.3 The Applications of $g\text{-}C_3N_4/TiO_2$ Heterojunction Catalyst

Compared with single-component catalysts, the $g\text{-}C_3N_4/TiO_2$ heterojunction catalysts formed by the combination of $g\text{-}C_3N_4$ and TiO_2 show greatly enhanced photocatalytic activity. Therefore, fabricating the $g\text{-}C_3N_4/TiO_2$ has many promising applications in various fields of photocatalysis. Currently, the researches on the applications of $g\text{-}C_3N_4/TiO_2$ mainly focus on the degradation of organic pollutants, hydrogen generation from water, photocatalytic reduction of CO_2, treatment of heavy metal ion, and inactivation of bacteria.

7.3.1 Degradation of Organic Pollutants

With the rapid development of economy, environmental pollution problems have greatly affected our daily lives, among which the most serious problems are water pollution and air pollution. The majority in the source of pollution is organic pollutants. Therefore, the degradation of organic pollutants is a hot research topic in recent decades. Various kinds of $g\text{-}C_3N_4/TiO_2$ heterojunction catalysts have also been developed and applied to solve these pollution problems.

7.3.1.1 Degradation of Pollutants in Liquid Phase

Many research works have been carried out to examine the photocatalytic degradation of organic dyes such as RhB and AO7 in aqueous solution in the presence of $g\text{-}C_3N_4/TiO_2$ heterojunction catalyst. For instance, methyl blue (MB) was degraded by $g\text{-}C_3N_4/TiO_2$ catalyst which was synthesized by directly heating the mixture of urea and commercial TiO_2 [31]. The catalyst exhibited efficient photocatalytic degradation of MB under visible light irradiation. The degradation efficiency can

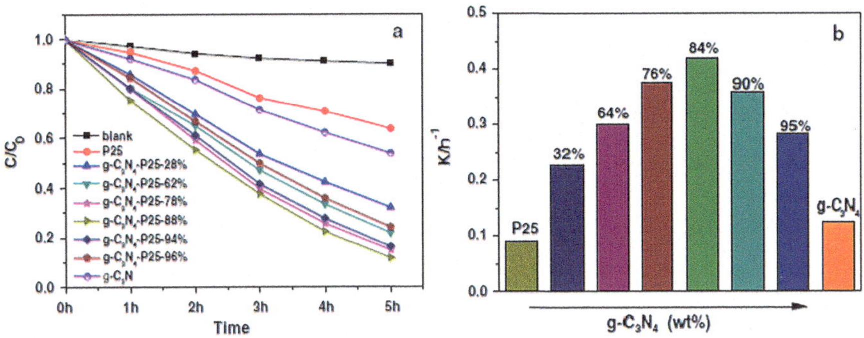

Fig. 7.7 (**a**) Photolysis and photocatalytic degradation of MB with TiO_2, g-C_3N_4, and g-C_3N_4-P25 photocatalysts. (**b**) Degradation rate constants of MB over TiO_2, g-C_3N_4, and g-C_3N_4-P25 photocatalysts [54]. (Reprinted with permission from Ref. [54]. Copyright 2015, Springer)

be adjusted by tuning the treatment temperature in the synthesis process of the composite catalyst, and the g-C_3N_4/TiO_2 nanocomposite prepared at 450 °C exhibited the best photocatalytic performance, which was much higher than the pure TiO_2. In addition to the preparation temperature, the mass ratio between g-C_3N_4 and TiO_2 also had a great influence on the degradation efficiency. Honglei Zhu et al. fabricated g-C_3N_4-P25 composite catalysts with different mass ratios and examined their photocatalytic activity toward the degradation of MB [54]. As shown in Fig. 7.7, the degradation efficiency varied with different g-C_3N_4 and P25 mass ratios. The sample with an optimal g-C_3N_4 content of 88% exhibited the highest photocatalytic activity which was almost 3.3 times higher than that of pure g-C_3N_4 under visible light irradiation.

In addition to MB, other dyes were also degraded by the g-C_3N_4/TiO_2 heterojunction photocatalysts. Photocatalytic degradation of RhB and MB was carried out by Fatang Li et al. to test the visible light photocatalytic activity of N-TiO_2/g-C_3N_4. As Fig. 7.8 shows, N-TiO_2/g-C_3N_4 composite with 40 wt% of N-TiO_2 showed the highest photocatalytic activity. The efficient separation of photo-generated electrons and holes, which resulted from the formation of N-TiO_2/g-C_3N_4 heterostructure, led to the excellent photocatalytic performance [46]. Guohong Wang et al. also degraded RhB using a novel macro-/mesoporous g-C_3N_4/TiO_2 heterojunction photocatalyst. The good photocatalytic activity of this kind of product ascribed to the fact that the sample possessed a large specific surface area and an excellent heterostructure [59]. Xiaosong Zhou et al. synthesized a carbon nitride/TiO_2 nanotube array (CN/TNT), and the catalyst exhibited high photocatalytic activity toward the degradation of methyl orange (MO) [58]. They prepared the photocatalysts denoted as CT_x (x represents the deposition time) by electrodeposition of g-C_3N_4 into the crystallized TiO_2 nanotubes. Their experimental results showed that the photocatalytic activities of CN/TNTs increased as the deposition time increased at the first, then decreased, and the $CT_{5.0}$ exhibited the highest photocatalytic activity.

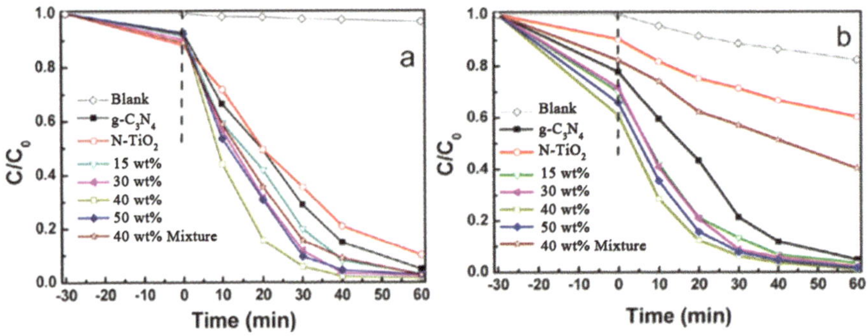

Fig. 7.8 Photocatalytic activities of N-TiO_2, g-C_3N_4, and N-TiO_2/g-C_3N_4 composites on the photodegradation of (**a**) RhB and (**b**) MB driven by visible light irradiation [46]. (Reprinted with permission from Ref. [46]. Copyright 2013, American Chemical Society)

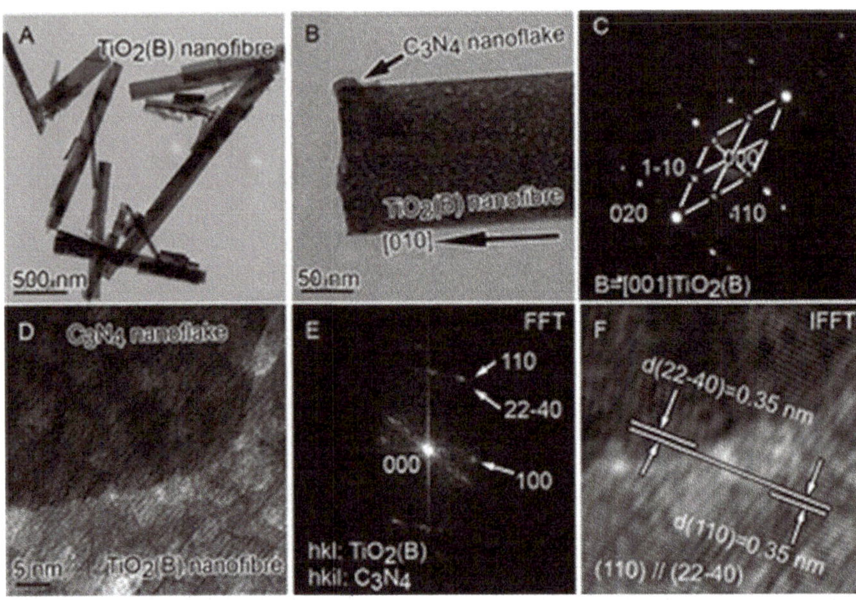

Fig. 7.9 (**a** and **b**) TEM images of the g-C_3N_4/TiO_2(B)-1 catalyst; (**c**) EDP of the g-C_3N_4/TiO_2(B)-1 catalyst; (**d**) HRTEM image of a g-C_3N_4 nanoflake deposited on the TiO_2(B) nanofiber; (**e**) fast Fourier transformation (FFT) image of the joint area between TiO_2(B) and g-C_3N_4 in image; (D), (**f**) inverse fast Fourier transformation (IFFT) image of g-C_3N_4/TiO_2(B)-1 sample [40]. (Reprinted with permission from Ref. [40]. Copyright 2014, Royal Society of Chemistry)

Dongjiang Yang et al. synthesized g-C_3N_4/TiO_2(B) nanofibers with selective exposure of high-energy (001) plane and applied it to the degradation of sulforhodamine B (SRB) dye. Figure 7.9 revealed that the diffraction spots of electron diffraction pattern (EDP) of samples could be indexed as (110), (1–10),

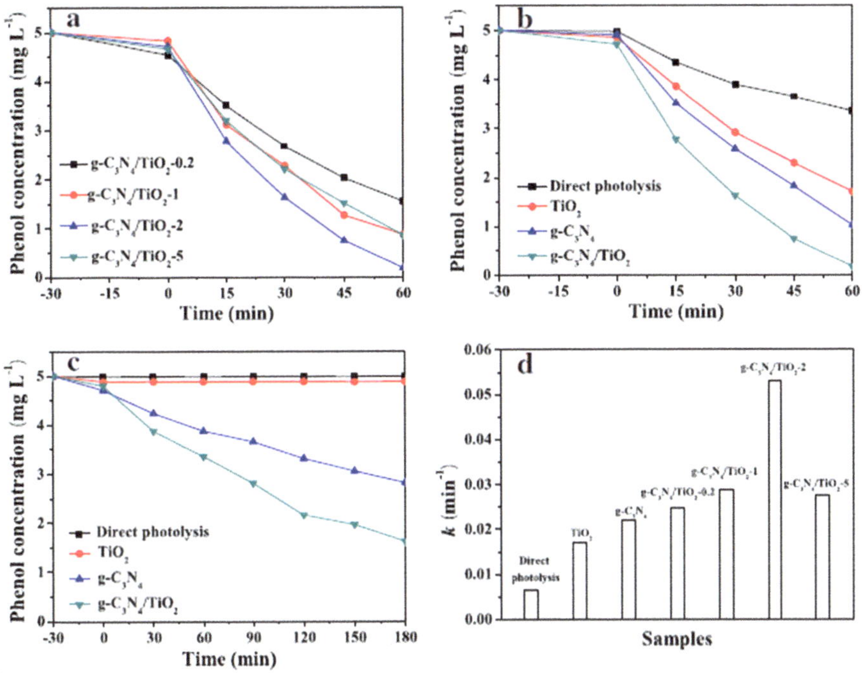

Fig. 7.10 (**a**) Photocatalytic degradation of phenol using $g\text{-}C_3N_4/TiO_2$ catalysts with various weight ratios under full-spectrum irradiation; (**b**) photocatalytic degradation of phenol by $g\text{-}C_3N_4$, TiO_2, and $g\text{-}C_3N_4/TiO_2$ under full-spectrum irradiation and (**c**) under visible light irradiation; (**d**) the kinetic constants of phenol degradation under full-spectrum irradiation [43]. (Reprinted with permission from Ref. [43]. Copyright 2012, Elsevier)

and (020). The $g\text{-}C_3N_4/TiO_2(B)$ system showed better photocatalytic degradation ability than the $g\text{-}C_3N_4$/anatase system, although the photocatalytic activity of the anatase nanofibers was much better than that of the $TiO_2(B)$ nanofibers [40].

In addition to dyes, some other organic compounds such as phenol have also been served as the target pollutants in the $g\text{-}C_3N_4/TiO_2$ photocatalysis system. For example, Hongtao Yu et al. investigated the photocatalytic activity of $g\text{-}C_3N_4/TiO_2$ for the photocatalytic degradation of phenol under visible and UV light. The $g\text{-}C_3N_4/TiO_2$ exhibited higher photocatalytic activity than pure TiO_2 and $g\text{-}C_3N_4$, as shown in Fig. 7.10, and the $g\text{-}C_3N_4/TiO_2\text{–}2$ with the mass ratio of $g\text{-}C_3N_4/TiO_2 = 2$ possessed the best photocatalytic activity [43].

In addition, a kind of $TiO_2/g\text{-}C_3N_4$ catalyst with highly dispersed TiO_2 nanocrystals on $g\text{-}C_3N_4$ has also been used for the photocatalytic degradation of phenol by Jinlong Zhang et al. It was found that high dispersion of TiO_2 with high-energy (001) facet was beneficial for the enhancement of the photocatalytic activity. As Fig. 7.11a shows, the photocatalytic activity of $TiO_2/g\text{-}C_3N_4$ catalysts showed an obvious increase for phenol decomposition compared with the pure TiO_2 and $g\text{-}C_3N_4$. The optimal catalyst $TiO_2/g\text{-}C_3N_4(1.5)$ successfully degraded 100% phenol

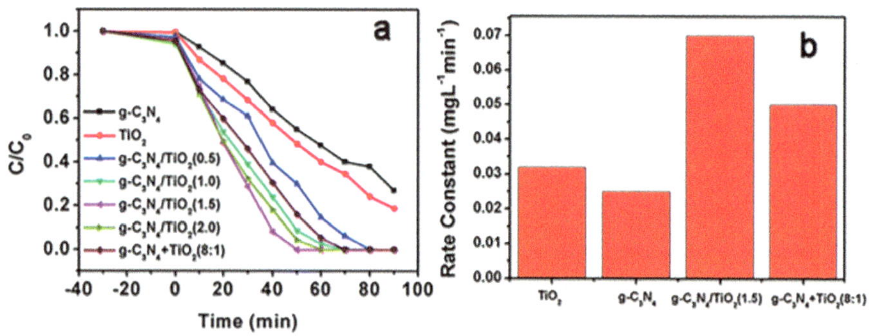

Fig. 7.11 Photocatalytic degradation results of 10 mg L^{-1}phenol with different catalysts under simulated sunlight irradiation of 300 W xenon lamp coupled with AM 1.5 [49]. (Reprinted with permission from Ref. [49]. Copyright 2015, Royal Society of Chemistry)

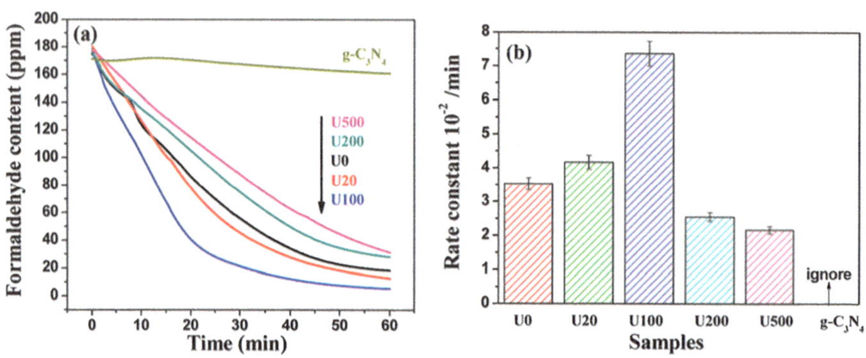

Fig. 7.12 The photocatalytic decomposition results of HCHO in air using U0, U20, U100, U200, U500, and g-C_3N_4 samples, respectively. (**a**) The corresponding concentration–time plots of HCHO; (**b**) the decomposition rate constant for all samples [5]. (Reprinted with permission from Ref. [5]. Copyright 2013, Royal Society of Chemistry)

within 50 min, whose rate constant was about 2.2 times that of pure TiO_2 and 2.8 times that of pure g-C_3N_4 (Fig. 7.11b) [49].

7.3.1.2 Degradation of Pollutants in Gas Phase

The status of air quality has great influence on people's health. After interior decoration, the concentration of formaldehyde (HCHO) in the air will rise sharply, causing a great impair to people's health. Jiaguo Yu et al. synthesized g-C_3N_4-TiO_2 photocatalyst and applied it for the photocatalytic oxidation decomposition of HCHO in air. As shown in Fig. 7.12, the weight percentage ratio of urea against P25 in the precursors was tuned to be 0, 20, 100, 200, and 500 (wt%), and the resulting catalysts were labeled as Ux (x represented the urea to P25 weight ratio),

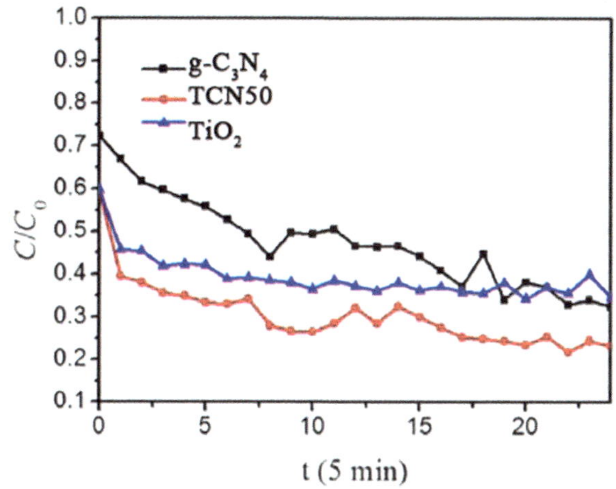

Fig. 7.13 Photocatalytic degradation results of gaseous acetone under the simulated solar light irradiation [60]. (Reprinted with permission from Ref. [60]. Copyright 2016, Elsevier)

and its value was equal to 0, 20, 100, 200, and 500, respectively. Their experimental results showed that the pure g-C_3N_4 exhibited very low photocatalytic activity for HCHO oxidation decomposition, while pure TiO_2 was active for decomposition of HCHO, and the photocatalytic activity of g-C_3N_4-TiO_2 was highly dependent on the amount of incorporated g-C_3N_4. The U100 sample with g-C_3N_4 content of 94% exhibited the highest photocatalytic activity for HCHO decomposition [5].

In addition to formaldehyde, gaseous acetone has also been degraded by the g-C_3N_4/TiO_2 photocatalyst. Xiangli Li fabricated microspherical g-C_3N_4/TiO_2 with high percentage of TiO_2 (001) facets through a solvothermal method and evaluated its photocatalytic activity for the degradation of gaseous acetone [60]. As Fig. 7.13 shows, the g-C_3N_4/TiO_2 catalyst (TCN50) could degrade more than 70% acetone within 120 min under simulated solar light irradiation. The photocatalytic efficiency of g-C_3N_4/TiO_2 for degrading acetone was much higher than that of pure g-C_3N_4 and TiO_2. Moreover, their experimental results also proved that acetone was oxidized by the highly active $O_2\cdot$ into CO_2 and H_2O in their reaction system.

In the process of the degradation of the gaseous pollutants, the adsorption capacity of the catalyst was the main factor, which greatly affects the photocatalytic activity. Tailoring the performance of materials via adjusting the morphologies and structures of the catalysts has emerged as a new and important direction of the research on g-C_3N_4/TiO_2 heterojunction catalyst for the photocatalytic degradation of gaseous organic pollutants.

7.3.2 Hydrogen Generation from Water

Due to the fact of the global energy depletion, the development and production of new sources of energy especially the clean energy have attracted more and more experimental interests. Hydrogen is widely considered as a highly effective

Fig. 7.14 The average hydrogen production rates as a function of wt% of g-C_3N_4 in the TiO_2-g-C_3N_4 composite from the first 3 h of the reaction duration [37]. (Reprinted with permission from Ref. [37]. Copyright 2011, Elsevier)

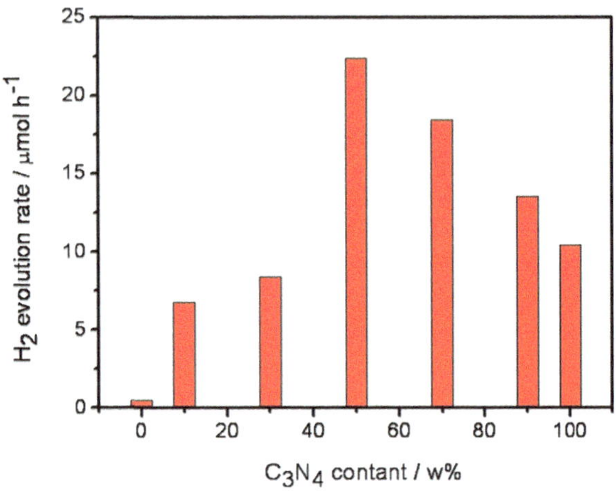

environmental and green energy, whose production methods have been studied and explored a lot. Among the numerous production methods, water splitting with g-C_3N_4/TiO_2 as photocatalyst has been favored by many researchers due to its merit of environment friendly.

Hongjian Yan et al. fabricated TiO_2-g-C_3N_4 composite catalysts with varying the wt% of g-C_3N_4 and used the samples in photocatalytic H_2 generation. The visible light-induced H_2 evolution rate was remarkably improved by coupling TiO_2 with g-C_3N_4, and the sample TiO_2–50 wt% C_3N_4 showed the highest activity, as shown in Fig. 7.14 [37].

Tianyou Peng et al. prepared porous g-C_3N_4-Pt-TiO_2, and their experimental results showed that coupling TiO_2 with g-C_3N_4 could remarkably enhance the visible light-induced photocatalytic hydrogen evolution rate. Besides, the g-C_3N_4-Pt-TiO_2 composite with a mass ratio of 70:30 exhibited the maximum photocatalytic activity as well as excellent photostability for hydrogen production under visible light irradiation (Fig. 7.15) [30].

Zhenyi Zhang et al. synthesized ternary heterostructured nanofibers (NFs) consisting of g-C_3N_4 nanosheets (NSs), plasmonic noble metal nanoparticles (Au, Ag, or Pt NPs), and TiO_2 NPs. The ternary composite photocatalyst exhibited improved charge-carrier migration efficiency and achieved highly efficient photocatalytic H_2 evolution [61]. Yanping Hong et al. prepared an anatase boron-doped TiO_2 (B-TiO_2) with exposed (001) facets and composited it with the g-C_3N_4 to form B-TiO_2–001/g-C_3N_4 heterojunctions. The heterojunction photocatalyst had the greatest photocatalytic activity for H_2 production as shown in Fig. 7.16, which was ascribed to the broad range of visible light absorption, the efficiently reduced charge recombination, and relatively higher catalytic activity of (001) facets compared to the (101) facets [62].

In addition, Yan-Yan Song et al. modified the g-C_3N_4/TiO_2 nanotube arrays with Pt nanoparticles. Compared with g-C_3N_4-free aligned TiO_2 nanotube layers, the obtained sample exhibited a strong enhancement for photoelectron–chemical and

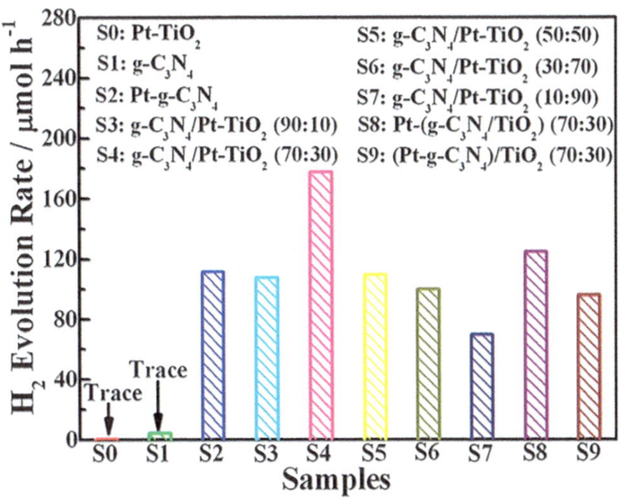

Fig. 7.15 Photocatalytic H_2 evolution rates of Pt-TiO_2, g-C_3N_4, Pt-g-C_3N_4, and g-C_3N_4–Pt-TiO_2 composites in different mass ratios as well as Pt-(g-C_3N_4–TiO_2) and (Pt-g-C_3N_4)/TiO_2 composites with a mass ratio of 70:30 under visible light irradiation [30]. (Reprinted with permission from Ref. [30]. Copyright 2012, Royal Society of Chemistry)

Fig. 7.16 Photocatalytic H_2 evolution of the as-synthesized samples: (*a*) g-C_3N_4, (*b*) TiO_2–001, (*c*) B-TiO_2–001, (*d*) TiO_2–001/g-C_3N_4 (5.0%), (*e*) P25/g-C_3N_4 (5.0%), (*f*) B-TiO_2–001/g-C_3N_4 (1.0%), (*g*) B-TiO_2–001/g-C_3N_4 (2.0%), (*h*) B-TiO_2–001/g-C_3N_4 (5.0%), (*i*) B-TiO_2–001/g-C_3N_4 (10%) [62]. (Reprinted with permission from Ref. [62]. Copyright 2016, Elsevier)

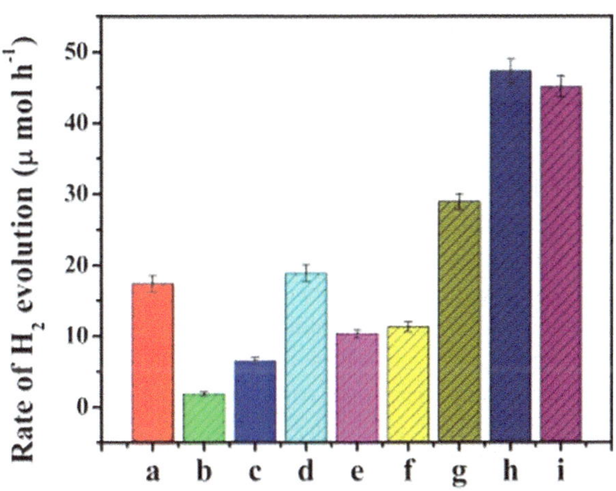

bias-free H_2 evolution of 15.62 μL h^{-1} cm^{-2}, which was almost a 98-fold increase in the H_2 production rate of aligned TiO_2 nanotube layers (0.16 μL h^{-1} cm^{-2}) [63]. Jian-guo Wang et al. designed and fabricated a novel g-C_3N_4/TiO_2 nanobelt heterostructure material. As shown in Fig. 7.17, the g-C_3N_4/TiO_2 nanobelt heterostructure with a mass ratio of 3:1 showed the highest H_2 production rate of 46.6 μmol h^{-1} [64].

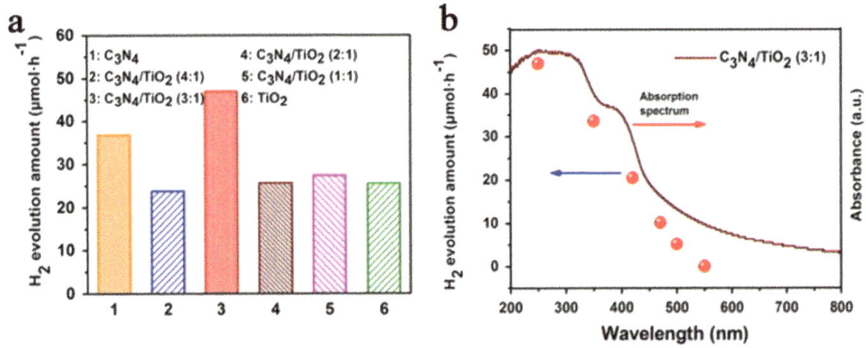

Fig. 7.17 (**a**) Photocatalytic H_2 evolution rates of the pure TiO_2 NBs, g-C_3N_4, and g-C_3N_4/TiO_2 NB heterostructure with different mass ratios; (**b**) wavelength dependence of H_2 evolution rate for g-C_3N_4/TiO_2 NBs (3:1) [64]. (Reprinted with permission from Ref. [64]. Copyright 2014, Elsevier)

7.3.3 Other Applications

Apart from the above applications, other applications of the g-C_3N_4/TiO_2 heterojunction catalysts have also been explored, such as the photocatalytic reduction of CO_2, treating heavy metal ion, as well as the inactivation of bacteria.

Guiyuan Jiang et al. fabricated a series of composites of g-C_3N_4 and in situ N-doped TiO_2 and then applied them to photocatalytic reduction of CO_2 under simulated light irradiation with water vapor at room temperature. Their research results showed efficient photocatalytic conversion of CO_2 to CO, and CH_4 was achieved. In addition, the photocatalytic activity and product selectivity were easy to adjust through simply varying the ratios of the precursor for g-C_3N_4 to the precursor for TiO_2 during the synthesis process of the catalyst. Moreover, as shown in Fig. 7.18, compared with g-C_3N_4 and commercial P25, the as-prepared g-C_3N_4-N-TiO_2 heterojunction photocatalysts showed improved photocatalytic performance for the reduction of CO_2, indicating the g-C_3N_4/TiO_2 heterojunction catalysts have good application prospects for mitigating the greenhouse effect and producing hydrocarbon and chemical compounds [65].

Guangshe Li et al. synthesized br-TiO_2/g-C_3N_4 by a facile calcination in air of brookite TiO_2 (br-TiO_2) hybridized with g-C_3N_4. The obtained samples were used for oxidation of toxic As^{3+} [42]. The intimately contacted hybrid br-TiO_2/g-C_3N_4 showed excellent photocatalytic activity in oxidation of As^{3+} to As^{5+}, which is less harmful than As^{3+}. Figure 7.19 indicated that the br-TiO_2/g-C_3N_4 catalyst with 35% weight ratio of the g-C_3N_4 exhibited much higher efficiency than pure br-TiO_2 and g-C_3N_4 for the application of As^{3+} oxidization.

Additionally, Taicheng An et al. investigated effective removal of biohazards from water using g-C_3N_4/TiO_2 hybrid photocatalyst [66]. The photocatalyst they synthesized was composed of micron-sized TiO_2 spheres wrapped with lamellar g-C_3N_4. A significantly improved visible light absorption and effectively reduced recombination of photo-generated electron–hole pairs were achieved by the

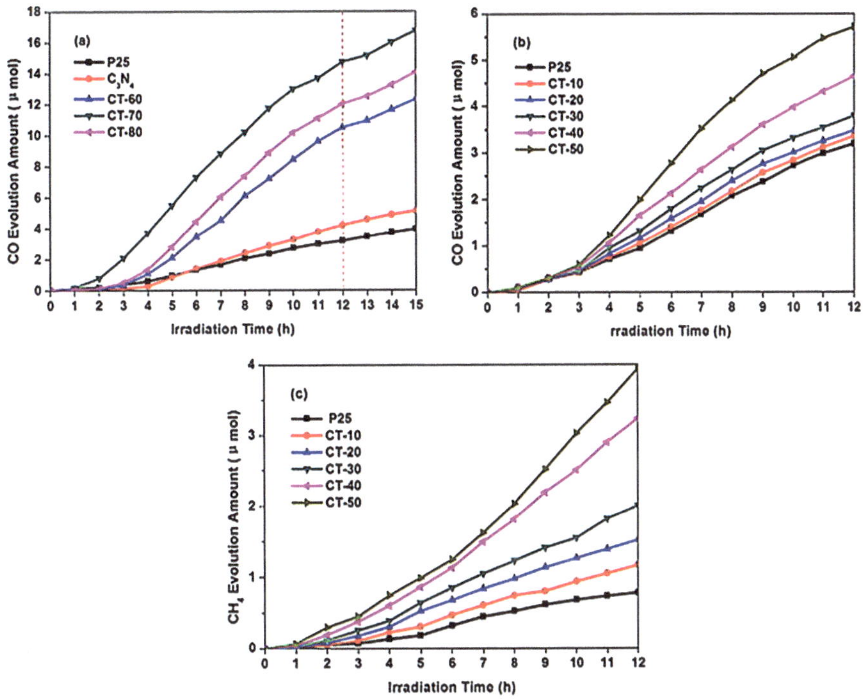

Fig. 7.18 (**a**) CO generation using different samples as a function of UV–vis light irradiation time. (**b**) CO generation using different samples as a function of UV–vis light irradiation time. (**c**) CH_4 generation using different samples as a function of UV–vis light irradiation time [65]. (Reprinted with permission from Ref. [65]. Copyright 2014, Elsevier)

combination of these two components. Using this hybrid photocatalyst, 107 cfu mL^{-1} of *Escherichia coli* K-12 could be completely inactivated within 180 min under visible light irradiation. Figure 7.20 showed that bacterial cells were seriously damaged during the photocatalytic inactivation processes, resulting in a severe leakage of intracellular components. Their research revealed that, through this kind of g-C_3N_4/TiO_2 heterojunction catalyst, bacterial cell destruction and water disinfection could be achieved easily. Besides, their results showed that substantial interaction between TiO_2 and g-C_3N_4 in the hybrid photocatalyst was a vital prerequisite for the enhancement of photocatalytic activity, which subsequently increased the trapping of the photoinduced charge carriers, benefiting for the production of reactive species. Furthermore, besides h^+, other reactive species such as subsequently generated $\cdot O_2^-$ and H_2O_2 also attacked biohazards, causing efficient photocatalytic inactivation and completely decomposition of bacteria.

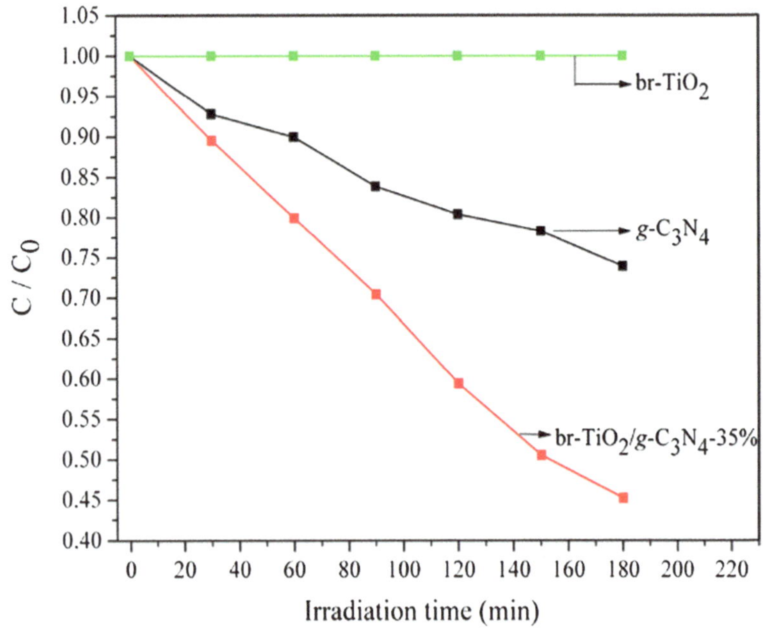

Fig. 7.19 Oxidation efficiency of As^{3+} over the hybrid br-TiO_2/g-C_3N_4–35%, pure g-C_3N_4, and 100% br-TiO_2 under visible light irradiation [42]. (Reprinted with permission from Ref. [42]. Copyright 2014, Royal Society of Chemistry)

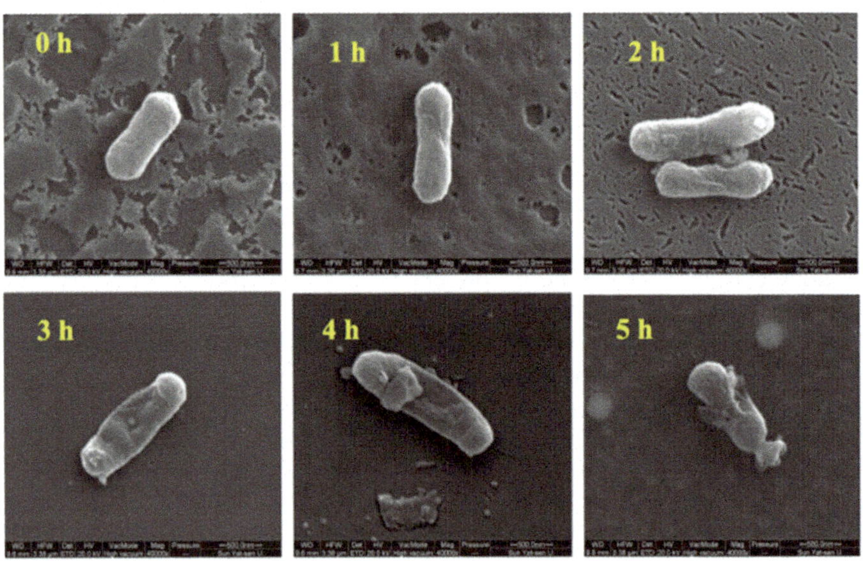

Fig. 7.20 SEM images of *E. coli* K-12 after the treatment of g-C_3N_4/TiO_2 hybrid photocatalysts at different times under visible light irradiation [66]. (Reprinted with permission from Ref. [66]. Copyright 2015, Elsevier)

7.4 Conclusions

This chapter summarized the recent progress of the research on the preparation methods and catalytic applications of g-C_3N_4/TiO_2 heterojunction catalysts. The synthesis methods of g-C_3N_4/TiO_2 heterojunction catalysts can be simply divided into three classes according to the order of each component prepared in the preparation process, which includes physically mixing g-C_3N_4 and TiO_2, growing TiO_2 on g-C_3N_4, and loading g-C_3N_4 on TiO_2. In spite of the different advantages and disadvantages exciting in the three different methods, every one of them can effectively form the heterojunctions between g-C_3N_4 and TiO_2, resulting in enhanced photocatalytic activity of the catalysts. The g-C_3N_4/TiO_2 heterojunction catalysts with excellent photocatalytic performance have been mainly applied in the photocatalytic degradation of organic pollutants, photolysis of water for producing H_2, photocatalytic reduction of CO_2, as well as the treatment of heavy metal ion and inactivation of bacteria.

Apart from the above research, g-C_3N_4/TiO_2 heterojunction catalysts are still worthy of exploration. Some groups found that there exists the phenomenon that the electron can transfer from dye to TiO_2, which implies that dye self-sensitized degradation also exists in this kind of composite system during the degradation process, providing the possibility for g-C_3N_4/TiO_2 heterojunction catalysts being applied to the dye-sensitized solar cells (DSSC). Moreover, some researchers have tried to design the ternary heterojunction catalysts such as g-C_3N_4/Ag/TiO_2 [41]. The studies on the structure design, the morphology control, and the expansion of applications related to g-C_3N_4/TiO_2 heterojunction catalysts are still significant. The heterojunction catalysts contain multiple components; therefore the stability of the heterojunction is not so satisfactory, and the interaction force between different components is still unknown. The preparation method of heterojunction catalysts looks more complex than the synthesis of other catalyst, and it still remains a great challenge for the development of a simplified synthesis method of the heterojunction catalysts.

References

1. Yang L, Wang L, Xing M et al (2016) Silica nanocrystal/graphene composite with improved photoelectric and photocatalytic performance. Appl Catal B Environ 180:106–112
2. Cheng C, Tan X, Lu D et al (2015) Carbon-dot-sensitized, nitrogen-doped TiO_2 in mesoporous silica for water decontamination through nonhydrophobic enrichment–degradation mode. Chem Eur J 21(49):17944–17950
3. Cheng C, Lu D, Shen B et al (2016) Mesoporous silica-based carbon dot/TiO_2 photocatalyst for efficient organic pollutant degradation. Microporous Mesoporous Mater 226:79–87
4. Liu F, Yu J, Tu G et al (2017) Carbon nitride coupled Ti-SBA15 catalyst for visible-light-driven photocatalytic reduction of Cr (VI) and the synergistic oxidation of phenol. Appl Catal B Environ 201:1–11

5. Yu J, Wang S, Low J et al (2013) Enhanced photocatalytic performance of direct Z-scheme g-C_3N_4–TiO_2 photocatalysts for the decomposition of formaldehyde in air. Phys Chem Chem Phys 15(39):16883–16890
6. Xing M, Shen F, Qiu B et al (2014) Highly-dispersed boron-doped graphene nanosheets loaded with TiO_2 nanoparticles for enhancing CO_2 photoreduction. Sci Rep 4:6341
7. Ansari MB, Min BH, Mo YH et al (2011) CO_2 activation and promotional effect in the oxidation of cyclic olefins over mesoporous carbon nitrides. Green Chem 13(6):1416–1421
8. Dong C, Xing M, Zhang J (2016) Economic hydrophobicity triggering of CO_2 photoreduction for selective CH_4 generation on noble-metal-free TiO_2–SiO_2. J Phys Chem Lett 7(15):2962
9. Xing M, Zhang J, Chen F et al (2011) An economic method to prepare vacuum activated photocatalysts with high photo-activities and photosensitivities. Chem Commun 47:4947
10. Dong C, Song H, Zhou Y et al (2016) Sulfur nanoparticles in situ growth on TiO_2 mesoporous single crystals with enhanced solar light photocatalytic performance. RSC Adv 6 (81):77863–77869
11. Qiu B, Xing M, Yi Q, Zhang J (2015) Angew Chem 127(36):10667
12. Ashkarran AA, Ghavamipour M, Hamidinezhad H et al (2015) Enhanced visible light-induced hydrophilicity in sol–gel-derived ag–TiO_2 hybrid nanolayers. Res Chem Intermed 41 (10):7299–7311
13. An L, Wang G, Cheng Y et al (2015) Ultrasonic-assisted synthesis of visible-light-driven TiO2/Bi2O3 nanocomposite photocatalysts: characterization, properties and azo dye removal application. Res Chem Intermed 41(10):7449
14. Qiu B, Xing M, Zhang J (2014) Mesoporous TiO_2 nanocrystals grown in situ on graphene aerogels for high photocatalysis and lithium-ion batteries. J Am Chem Soc 136(16):5852–5855
15. Kitano S, Murakami N, Ohno T et al (2013) Bifunctionality of Rh^{3+} modifier on TiO_2 and working mechanism of Rh^{3+}/TiO_2 photocatalyst under irradiation of visible light. J Phys Chem C 117(21):11008–11016
16. Shiraishi Y, Takeda Y, Sugano Y et al (2011) Highly efficient photocatalytic dehalogenation of organic halides on TiO_2 loaded with bimetallic Pd–Pt alloy nanoparticles. Chem Commun 47 (27):7863–7865
17. Tsukamoto D, Shiro A, Shiraishi Y et al (2012) Photocatalytic H_2O_2 production from ethanol/O_2 system using TiO_2 loaded with au–ag bimetallic alloy nanoparticles. ACS Catal 2 (4):599–603
18. Qi D, Xing M, Zhang J (2014) Hydrophobic carbon-doped TiO_2/MCF-F composite as a high performance photocatalyst. J Phys Chem C 118(14):7329–7336
19. Xie Y, Kum J, Zhao X et al (2011) Enhanced photocatalytic activity of mesoporous SN-codoped TiO_2 loaded with ag nanoparticles. Semicond Sci Technol 26(8):085–037
20. Wang P, Lei J, Xing M et al (2015) J Environ Chem Eng 3(2):961
21. Li H, Shen X, Liu Y, Wang L, Lei J, Zhang J (2016) J Alloys Compd 687
22. Ünlü H, Horing NJ, Dabowski J (eds) (2015) Low-dimensional and nanostructured materials and devices: properties, synthesis, characterization, modelling and applications. Springer
23. Li H, Shen X, Liu Y et al (2015) Facile phase control for hydrothermal synthesis of anatase-rutile TiO_2 with enhanced photocatalytic activity. J Alloys Compd 646:380–386
24. Zhang D, Gu X, Jing F et al (2015) High performance ultraviolet detector based on TiO_2/ZnO heterojunction. J Alloys Compd 618:551–554
25. Lin L, Yang Y, Men L et al (2013) A highly efficient TiO_2@ZnO n–p–n heterojunction nanorod photocatalyst. Nanoscale 5(2):588–593
26. Malik R, Tomer VK, Chaudhary V et al (2016) Facile synthesis of hybridized mesoporous au@TiO_2/SnO_2 as efficient photocatalyst and selective VOC sensor. ChemistrySelect 1 (12):3247–3258
27. Zhang ZL, Wang ML, Mao YL (2015) Mater Technol 30(1):2
28. Lu X, Wang Q, Cui D (2010) Preparation and photocatalytic properties of g-C_3N_4/TiO_2 hybrid composite. J Mater Sci Technol 26(10):925–930

29. Yang N, Li G, Wang W et al (2011) Photophysical and enhanced daylight photocatalytic properties of N-doped $TiO_2/g\text{-}C_3N_4$ composites. J Phys Chem Solids 72(11):1319–1324
30. Chai B, Peng T, Mao J et al (2012) Graphitic carbon nitride ($g\text{-}C_3N_4$)–Pt-TiO_2 nanocomposite as an efficient photocatalyst for hydrogen production under visible light irradiation. Phys Chem Chem Phys 14(48):16745–16752
31. Fu M, Pi J, Dong F et al (2013) A cost-effective solid-state approach to synthesize $g\text{-}C_3N_4$ coated TiO_2 nanocomposites with enhanced visible light photocatalytic activity. Int J Photoenergy 2013
32. Zhang J, Zhang M, Sun RQ et al (2012) A facile band alignment of polymeric carbon nitride semiconductors to construct isotype heterojunctions. Angew Chem 124(40):10292–10296
33. Dong F, Wu L, Sun Y et al (2011) Efficient synthesis of polymeric $g\text{-}C_3N_4$ layered materials as novel efficient visible light driven photocatalysts. J Mater Chem 21(39):15171–15174
34. Dong F, Sun Y, Wu L et al (2012) Facile transformation of low cost thiourea into nitrogen-rich graphitic carbon nitride nanocatalyst with high visible light photocatalytic performance. Cat Sci Technol 2(7):1332–1335
35. Chu VB, Cho JW, Park SJ et al (2014) Use of a precursor solution to fill the gaps between indium tin oxide nanorods, for preparation of three-dimensional $CuInGaS_2$ thin-film solar cells. Res Chem Intermed 40(1):49–56
36. Chen J, Qin S, Liu Y et al (2014) Preparation of a visible light-driven Bi_2O_3–TiO_2 composite photocatalyst by an ethylene glycol-assisted sol–gel method, and its photocatalytic properties. Res Chem Intermed 40(2):637–648
37. Yan H, Yang H (2011) TiO_2-$g\text{-}C_3N_4$ composite materials for photocatalytic H_2 evolution under visible light irradiation. J Alloys Compd 509(4):26–29
38. Zhou J, Zhang M, Zhu Y (2015) Photocatalytic enhancement of hybrid C_3N_4/TiO_2 prepared via ball milling method. Phys Chem Chem Phys 17(5):3647–3652
39. Gu L, Wang J, Zou Z et al (2014) Graphitic-C_3N_4-hybridized TiO_2 nanosheets with reactive {001} facets to enhance the UV-and visible-light photocatalytic activity. J Hazard Mater 268:216–223
40. Zhang L, Jing D, She X et al (2014) Heterojunctions in $g\text{-}C_3N_4/TiO_2$ (B) nanofibres with exposed (001) plane and enhanced visible-light photoactivity. J Mater Chem A 2(7):2071–2078
41. Chen Y, Huang W, He D et al (2014) Construction of heterostructured $g\text{-}C_3N_4/ag/TiO_2$ microspheres with enhanced photocatalysis performance under visible-light irradiation. ACS Appl Mater Interfaces 6(16):14405–14414
42. Zang Y, Li L, Xu Y et al (2014) Hybridization of brookite TiO_2 with $g\text{-}C_3N_4$: a visible-light-driven photocatalyst for As^{3+} oxidation, MO degradation and water splitting for hydrogen evolution. J Mater Chem A 2(38):15774–15780
43. Zhao S, Chen S, Yu H et al (2012) $G\text{-}C_3N_4/TiO_2$ hybrid photocatalyst with wide absorption wavelength range and effective photogenerated charge separation. Sep Purif Technol 99:50–54
44. Shen J, Yang H, Shen Q et al (2014) Template-free preparation and properties of mesoporous $g\text{-}C_3N_4/TiO_2$ nanocomposite photocatalyst. CrystEngComm 16(10):1868–1872
45. Li X, Liu P, Mao Y et al (2015) Preparation of homogeneous nitrogen-doped mesoporous TiO_2 spheres with enhanced visible-light photocatalysis. Appl Catal B Environ 164:352–359
46. Wang X, Yang W, Li F et al (2013) In situ microwave-assisted synthesis of porous N-$TiO_2/g\text{-}C_3N_4$ heterojunctions with enhanced visible-light photocatalytic properties. Ind Eng Chem Res 52(48):17140
47. Han C, Wang Y, Lei Y et al (2015) In situ synthesis of graphitic-C_3N_4 nanosheet hybridized N-doped TiO_2 nanofibers for efficient photocatalytic H_2 production and degradation. Nano Res 8(4):1199–1209
48. Sun Q, Lv K, Zhang Z et al (2015) Effect of contact interface between TiO_2 and $g\text{-}C_3N_4$ on the photoreactivity of $g\text{-}C_3N_4/TiO_2$ photocatalyst:(001) vs (101) facets of TiO_2. Appl Catal B Environ 164:420–427
49. Li H, Zhou L, Wang L (2015) In situ growth of TiO_2 nanocrystals on $g\text{-}C_3N_4$ for enhanced photocatalytic performance. Phys Chem Chem Phys 17(26):17406–17412

50. Li Y, Wang J, Yang Y et al (2015) Seed-induced growing various TiO $_2$ nanostructures on $g\text{-}C_3N_4$ nanosheets with much enhanced photocatalytic activity under visible light. J Hazard Mater 292:79–89
51. Wang J, Zhang WD (2012) Modification of TiO_2 nanorod arrays by graphite-like C_3N_4 with high visible light photoelectrochemical activity. Electrochim Acta 71:10–16
52. Fu M, Liao J, Dong F et al (2014) Growth of $g\text{-}C_3N_4$ layer on commercial TiO_2 for enhanced visible light photocatalytic activity. J Nanomater 2014:1
53. Boonprakob N, Wetchakun N, Phanichphant S et al (2014) Enhanced visible-light photocatalytic activity of $g\text{-}C_3N_4/TiO_2$ films. J Colloid Interface Sci 417:402–409
54. Zhu H, Chen D, Yue D et al (2014) In-situ synthesis of $g\text{-}C_3N_4$-P25 TiO_2 composite with enhanced visible light photoactivity. J Nanopart Res 16(10):2632
55. Lei J, Chen Y, Shen F et al (2015) Surface modification of TiO_2 with $g\text{-}C_3N_4$ for enhanced UV and visible photocatalytic activity. J Alloys Compd 631:328–334
56. Lei J, Chen Y, Wang L et al (2015) Surface modification of TiO_2 with $g\text{-}C_3N_4$ for enhanced UV and visible photocatalytic activity. J Mater Sci 50(9):3467
57. Zou XX, Li GD, Wang YN et al (2011) Direct conversion of urea into graphitic carbon nitride over mesoporous TiO_2 spheres under mild condition. Chem Commun 47(3):1066–1068
58. Zhou X, Jin B, Li L et al (2012) A carbon nitride/TiO_2 nanotube array heterojunction visible-light photocatalyst: synthesis, characterization, and photoelectrochemical properties. J Mater Chem 22(34):17900–17905
59. Li K, Gao S, Wang Q et al (2015) In-situ-reduced synthesis of Ti^{3+} self-doped $TiO_2/g\text{-}C_3N_4$ heterojunctions with high photocatalytic performance under LED light irradiation. ACS Appl Mater Interfaces 7(17):9023–9030
60. Ma J, Tan X, Yu T et al (2016) Fabrication of $g\text{-}C_3N_4/TiO_2$ hierarchical spheres with reactive {001} TiO_2 crystal facets and its visible-light photocatalytic activity. Int J Hydrog Energy 41 (6):3877–3887
61. Wei X, Shao C, Li X et al (2016) Facile in situ synthesis of plasmonic nanoparticles-decorated $g\text{-}C_3N_4/TiO_2$ heterojunction nanofibers and comparison study of their photosynergistic effects for efficient photocatalytic H 2 evolution. Nanoscale 8(21):11034–11043
62. Chen L, Zhou X, Jin B et al (2016) Heterojunctions in $g\text{-}C_3N_4$/B-TiO_2 nanosheets with exposed {001} plane and enhanced visible-light photocatalytic activities. Int J Hydrog Energy 41 (18):7292–7300
63. Gao ZD, Qu YF, Zhou X et al (2016) Pt-decorated $g\text{-}C_3N_4/TiO_2$ nanotube arrays with enhanced visible-light photocatalytic activity for H_2 evolution. ChemistryOpen 5(3):197–200
64. Zhong X, Jin M, Dong H et al (2014) TiO_2 nanobelts with a uniform coating of $g\text{-}C_3N_4$ as a highly effective heterostructure for enhanced photocatalytic activities. J Solid State Chem 220:54
65. Zhou S, Liu Y, Li J et al (2014) Facile in situ synthesis of graphitic carbon nitride ($g\text{-}C_3N_4$)-N-TiO_2 heterojunction as an efficient photocatalyst for the selective photoreduction of CO_2 to CO. Appl Catal B Environ 158:20–29
66. Li G, Nie X, Chen J et al (2015) Enhanced visible-light-driven photocatalytic inactivation of Escherichia coli using $g\text{-}C_3N_4/TiO_2$ hybrid photocatalyst synthesized using a hydrothermal-calcination approach. Water Res 86:17–24

Chapter 8
Modifications of Photocatalysts by Doping Methods

8.1 Preparation of Visible Light-Responsive TiO_2 Photocatalysts by Chemical Doping Modification Methods

In order to improve the visible light activity of TiO_2, many modification methods have been developed in recent years, like doping impurities, coupling semiconductors, dye sensitization, and so on. After the modification, the visible light-driven TiO_2 can use the solar energy in dealing with the environmental pollution and new energy development. For instance, the dye-sensitized TiO_2 has been widely used in preparing solar cells, owing to its strong visible light absorption ability. In 1991, Michael Grätzel and Brian O'Regan described a photovoltaic cell, created from low- to medium-purity materials through low-cost processes, which exhibited a commercially realistic energy-conversion efficiency [1]. The device is based on a 10-μm-thick, optically transparent film made with TiO_2 particles having a few nanometers in size, coated with a monolayer of dye to sensitize the film for light harvesting. The overall light-to-electric energy conversion is 7.1–7.9% for simulated solar light and 12% for diffuse daylight. The large current densities and high stability of the solar cells based on dye-sensitized colloidal TiO_2 films are making practical applications feasible. Since then, the record of energy conversion for the dye-sensitized TiO_2-based solar cells is constantly improved year by year [2–8]. In 1998, the solar cell based on dye-sensitized mesoporous TiO_2 films converts photons to electric current with a high yield of 33%, which is also achieved by Prof. Michael Grätzel [9]. At present, the photoelectric conversion efficiency of TiO_2-based solar cells has stabilized to be more than 10%, but its cost of manufacturing is still high.

Some other modified visible light-responsible TiO_2 especially the doping-modified TiO_2 also attracted much attention and have been applied to organic pollutant photodegradation and water splitting reaction. Hence, in this section, we would like to introduce the modification methods related to dope trace impurity into TiO_2, including chemical synthesis like high-temperature sintering in an atmosphere,

J. Zhang et al., *Photocatalysis*, Lecture Notes in Chemistry 100,
https://doi.org/10.1007/978-981-13-2113-9_8

wet chemical methods such as sol–gel processes and hydrothermal treatment, spray pyrolysis, and supercritical methods.

These chemical synthesis methods used in doing TiO_2 are summarized in Table 8.1. As can be seen in Table 8.1, all the doped TiO_2 synthesized by different chemical preparation methods have a high photocatalytic activity for the degradation of organic dye pollutants and water splitting. In addition to be divided in different synthesis method, the chemical doping modification on TiO_2 can also be divided into metal doping and nonmetal doping. In metal doping method, a certain amount of metal ions such as Fe^{3+} [10–14], Cr^{3+} [15, 16], Ru^{2+} [17, 18], Ce^{4+} [19, 20], La^{3+} [21–23], V^{5+} [24–26], are introduced into TiO_2 (Table 8.1), as the active "small oxide islands", affecting the lifetime of the photo-formed electrons and holes their-transfer processes, thus affecting the photocatalytic activity of TiO_2. Furthermore, the impurity induced by metal doping into the TiO_2 could efficiently narrow its bandgap and extend the absorption edge into the visible light range. Many studies have demonstrated that metal doping could effectively improve the photocatalytic activity of TiO_2 under the visible light irradiation [27–30]. However, metal doping also showed several drawbacks: thermal instability of doped TiO_2, electron trapping by the metal centers, introduction of the electrons, and hole recombination centers [31]. It is worth mentioning that, in addition to the above traditional impurity metal ion doping, more and more research has focused on the investigation of Ti^{3+} self-doped TiO_2 in recent years [32–36]. Recent research work has found that excessive Ti^{3+} would not easily introduce the electron and hole recombination centers in TiO_2 [33].

On the other hand, nonmetal doping is another technology to modify TiO_2, which could achieve the substitution of lattice oxygen by nonmetal elements [31, 37–39]. Since a work was investigated by Asahi et al. [40] in 2001 in which they reported that nitrogen doping could enhance the photocatalytic activity of TiO_2 for the photodegradation of methylene blue and gaseous acetaldehyde in the visible light irradiation, though the photocatalytic activity in UV light regions decreases. Since then, many researchers have reported about various nonmetal-doped TiO_2, such as N [41, 42], B [43–46], C [21, 47–49], F [50–54], S [55–59], and P [60]. Although nonmetal doping modification could change the band structure of TiO_2 and affect the transfer of electrons and holes, the origin of its visible light photoactivity is still in debate [31], especially the photocatalytic mechanism of nitrogen doping. Recent experimental and theoretical studies suggest that the N doping does not cause the narrowing of the band gap of the TiO_2 but the formation of localized midgap states above the valance band of TiO_2 which is the reason for its enhancing visible light responsiveness and photoactivity [31, 61].

In this section, we mainly introduce and highlight the chemical-synthesized visible light-responsible TiO_2 photocatalysis with doping modification, including metal doping, nonmetal doping, and co-doping modification. The influencing factors on doping modification, the research development of doping and photocatalytic mechanism, and novel investigation of synergistic effect between different elements are also discussed in this section.

Table 8.1 Doping-modified TiO_2 photocatalysts prepared by different chemical methods

Chemical doping method	Doping elements	Precursors	Visible light-driven photocatalysis	Ref.
Sol–gel	N	Ti source: TBOT/TTIP, N source: urea/NH_4Cl/thiourea/ethylmethylamine	Photocatalytic degradation of methylene blue	[62–66]
			Effective agents against both bacteria and stearic acid using a white light source	
	C	Ti source: TTIP/$TiCl_4$	Photodegradation of gaseous toluene or 4-chlorophenol	[67, 68]
		C source: carbon particle ethanol/melamine borate		
	Fe	Ti source: TTIP/TBOT	Photodegradation of active yellow XRG or methyl orange	[10, 29]
		Fe source: $Fe(NO_3)_3 \cdot 9H_2O$		
	B and N	Ti source: TTIP/TBOT	Hydrogen evolution from splitting water	[69–71]
		N source: NH_4OH/urea		
		B source: H_3BO_3		
Hydrothermal	N	Ti source: TiN/$TiCl_4$	Hydrogen evolution from splitting water; photocatalytic degradation of methyl orange	[72, 73]
		N source: TiN/L-lysine		
	Ti^{3+}	Ti source: titanium powder and hydrochloric acid/TBOT and $NaBH_4$	Evolved H_2 under visible light; photodegradation of methyl orange	[33, 74]
	C	Ti source: TBOT/TTIP	Photocatalytic degradation of methyl orange	[75]
		C source: ethanol		
	B and N	Ti source: TTIP/TBOT	Photodegradation for rhodamine B or methyl orange	[76, 77]
		N source: NH_4OH/NH_3/ urea		
		B source: H_3BO_3		
	Ga^{3+}, Y^{3+}, or Nb^{5+}	Ti source: TTIP	Dye-sensitized solar cells (DSSCs).	[78]
		Ga^{3+} source: gallium nitrate		
		Y^{3+} source: yttrium chloride		
		Nb^{5+} source: niobium pentachloride		
	Fe and N	Ti source: TBOT	Photodegradation of rhodamine B	[79]
		N source: NH_4Cl		
		Fe source: $Fe(NO_3)_3$		
	La and C	Ti source: $Ti(SO_4)_2$/TBOT/TTIP	Photodegradation of methyl orange	[21]
		La^{3+} source: lanthanum nitrate		
		C source: glucose		

(continued)

Table 8.1 (continued)

Chemical doping method	Doping elements	Precursors	Visible light-driven photocatalysis	Ref.
Microemulsion	N	Ti source: TBOT, N source: triethylamine, urea, thiourea, and hydrazine hydrate	Photodegradation of rhodamine B and 2,4-dichlorophenol	[41]
	Ag	Ti source: TTIP	Decomposition of phenol	[80]
		Ag source: silver nitrate		
Chemical precipitation	C	Ti source: $TiCl_4$	Photodegradation of 4-chlorophenol	[81]
		C source: tetrabutylammonium hydroxide		
	N	Ti source: TBOT/$Ti(SO_4)_2$	Photodegradation rate of 2,4-DCP or toluene	[42, 64, 82, 83]
		N source: NH_4OH, NH_4NO_3		
	F	Ti source: TTIP	Photocatalytic oxidation of acetone	[84]
		F source: NH_4F		
	Fe	Ti source: K_2TiF_6	Photocatalytic decomposition of bromocresol green	[85]
		Fe source: $FeSO_4{\cdot}7H_2O$		
Chemical deposition	Oxygen vacancies	Ti source: TTIP (CVD method)	Photodegradation of methylene blue	[86]
	Sn^{4+}	Ti source: $TiCl_4$	Photodegradation of phenol	[87]
		Sn^{4+} source: $SnCl_4$		
		Plasma-enhanced chemical vapor deposition (PCVD)		
	Nb, Ta, or F	Ti source: titanium alkoxides/titanium sheets	Photodegradation for methyl orange	[88, 89]
		Nb source: niobium ethoxide		
		Ta source: tantalum ethoxide		
		F source: *t*-butyl fluoride/NH_4F		
		Chemical vapor deposition (PCVD)		
Impregnation	Fe	Ti source: P25/titanate nanotubes	Photodegradation of oxalic acid or photocatalytic oxidation of acetone	[90–92]
		Fe source: $Fe(acac)_3$/$(Fe(NO_3)_3)$/$Fe_2(SO_4)_3$/$FeCl_3$	Photocatalytic degradation of acetophenone	

(continued)

Table 8.1 (continued)

Chemical doping method	Doping elements	Precursors	Visible light-driven photocatalysis	Ref.
	N and Fe	Ti source: $Ti(SO_4)_2$	Photodegradation of methylene blue	[83]
		N source: NH_4OH		
		Fe source: ferric chloride		
	Co, Cr, Cu, Fe, Mo, V, or W	Ti source: titanium trichloride	Photooxidation of 4-nitrophenol	[93]
		Co source: Co $(NO_3)_2{\cdot}6H_2O$		
		Cr source: Cr $(NO_3)_3{\cdot}9H_2O$,		
		Cu source: Cu $(NO_3)_2{\cdot}3H_2O$		
		Fe source: Fe $(NO_3)_3{\cdot}9H_2O$		
		Mo source: $(NH_4)_6Mo_7O_{24}{\cdot}4H_2O$		
		V source: NH_4VO_3		
		W source: $(NH_4)_6W_{12}O_{39}.xH_2O$		
High-temperature calcination	H	TTIP/titanium butoxide, H_2 atmosphere at 200/450 °C	Methylene blue decomposition and hydrogen generation	[94–97]
	Ti^{3+}	TTIP, 2-ethylimidazole calcination at 500 °C	High visible light photocatalytic activity for the generation of hydrogen gas from water	[32]
	Ti^{3+} and N	Rutile TiO_2 and NH_4 flow at 500 °C	Extension of the active spectrum and the superior visible light water photooxidation activity	[35]
	C	Ti source: Ti metal sheet	Water splitting with a total conversion efficiency of 11% and a maximum photo conversion efficiency of 8.35% when illuminated at 40 milliwatts per square centimeter	[98]
		C source: CO_2 and steam (H_2O)		
	B	Ti source: $TiCl_4$	Photocatalytic decomposition of methyl tertiary butyl ether	[99]
		B source: BH_3/THF-solution		
Vacuum activation	Ti^{3+}	TTIP/P25 in vacuum	Photodegradation of methyl orange, phenol, or methylene blue	[36, 100]
			Hydrogen evolution from splitting water	

(continued)

Table 8.1 (continued)

Chemical doping method	Doping elements	Precursors	Visible light-driven photocatalysis	Ref.
	Ti^{3+} and F	Ti source: $Ti(SO_4)_2$	Photodegradation of rhodamine B	[50]
		F source: NH_4F		
Thermal plasma	H	Ti source: commercial amorphous TiO_2	Photocatalytic methyl orange decomposition and photocatalytic H_2 generation	[101]

8.1.1 *Metal Doping Modification*

As can be seen from the reaction mechanism of TiO_2 photocatalysis, it can be concluded that an increase in the number of photo-formed electrons and holes participating in the photoredox reaction is the key factor to improve the visible light-responsive TiO_2 photocatalyst's performance. In order to achieve the substantive enhancement of the quantum yield of photocatalytic reaction for TiO_2 in the visible light irradiation, two issues must be overcome. One issue is how to produce much more photo-formed electrons and holes; another issue is how to improve the separation efficiency of electrons and holes. It is well known that appropriate amount of transition metal ions doped into TiO_2 can introduce electron capture centers and change the crystallinity of TiO_2 and then produce some defects, which results in an decrease in photo-formed electron and hole recombination centers [11, 27–30, 33, 34, 102]. Thus, metallic ion doping is recognized as an effective modification method for improving the reactivity of the visible light-responsive TiO_2.

The methods using metallic ions selected for appropriate doping into TiO_2 can be divided into three types. The first type is the transition metal ions. Most early, Choi et al. [103] studied the photoactivity of TiO_2 doped with 21 transition metal ions by using the model reaction of photocatalytic oxidation of chloroform and photocatalytic reduction of carbon tetrachloride. The researchers discovered that the Fe^{3+}, Mo^{5+}, Ru^{3+}, Os^{3+}, Re^{5+}, V^{4+}, and Rn^{3+} ion doping modification is beneficial to the enhancement of photodegradation of chloroform for TiO_2. Recently, Yan et al. [104] synthesized the TiO_2 nanoparticles doped with different content of cerium ion by a sol–gel method. The Ce-doped TiO_2 act as the capture of photo-formed holes and decrease the recombination of photo-generated electrons and holes, leading to better visible light absorption and photocatalytic degradation of methylene blue than pure TiO_2 (Fig. 8.1). Zhang and coworkers have also done many works on the transition of metal-doped TiO_2 and its application to various photocatalytic reactions [14, 15, 19, 20, 24, 29, 105–110]. Iron ion-doped anatase TiO_2 were prepared by hydrothermal hydrolysis and crystallized in octanol–water solution [105]. The results of photodegradation of active yellow XRG dye indicated that the amount of doped iron ion plays a significant role in affecting the photocatalytic activity and iron doped with optimum content enhances the photocatalytic activity, especially under visible light irradiation. When the Fe^{3+}-

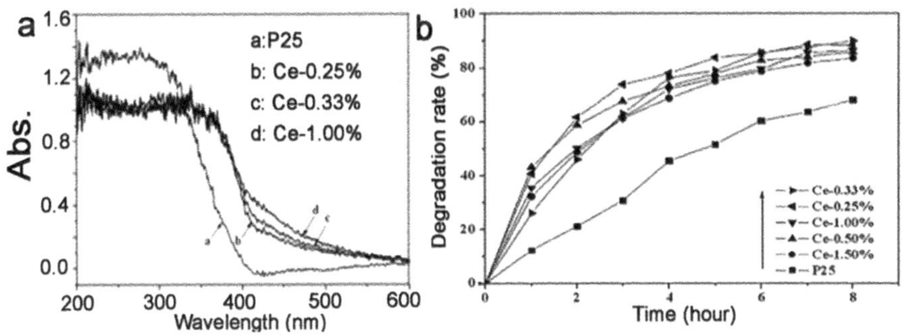

Fig. 8.1 UV–vis absorption spectra of Ce–TiO_2 samples with different Ce ion doping concentrations (**a**). Curves of methylene blue degradation constant by Ce–TiO_2 samples with different Ce ion doping concentrations (**b**) (Reprinted with permission from ref. [104]. Copyright 2015, Elsevier)

doped TiO_2 photocatalysts were prepared by combining sol–gel method with hydrothermal treatment [106], Fe was found to exist in trivalent ionic state and substitute Ti^{4+} in TiO_2, and its concentration was found to decrease from the surface to the deep bulk of TiO_2. As a result, the synthetic method of Fe doping in TiO_2 has a significant influence on the doping structures and concentration. In addition to iron, chromium and vanadium are also widely used as dopant elements into TiO_2. Due to the excitation of 3d electron of Cr^{3+} to the conduction band (CB) of TiO_2, Cr–TiO_2 always shows a good ability for absorbing the visible light to induce the photodegradation of XRG [15]. V^{4+} ions are also successfully incorporated into TiO_2 by flame spray pyrolysis (FSP) technique [24], sol–gel [26, 111], and other chemical methods [25]. V-doping into TiO_2 leads to change the bandgap of TiO_2, leading to an extension of the absorption regions to visible light region, resulting in the improvement of the visible light-driven photocatalytic activity of TiO_2 [26].

The second type is the rare-earth metal doping, and the lanthanide-doped TiO_2 have been accounted for the majority [21, 22, 112–114]. Sun et al. [112] investigated the effects of substitutional La doped on the electronic structures and photocatalytic activity of TiO_2 by the density function theory (DFT) calculation method. Their calculation results indicated that the enhanced absorption in the visible light region for La–TiO_2 was attributed to the adsorptive of La doping rather than the substitutional La. Differently, Anandan et al. [115] believed that the rapid mineralization of monocrotophos over La-doped TiO_2 under the light irradiation could be associated with the suppression of the electrons and holes recombination by La^{3+} doped into TiO_2 and generation of more number of •OH radicals by oxidation of holes. Recently, Zhang and coworkers have studied some other lanthanide metals such as Eu-, Yb-, and Sm-doped TiO_2 and their photocatalytic activities under the visible light irradiation [116–118]. Samarium-doped TiO_2 (Sm–TiO_2) was successfully prepared via a chemical coprecipitation method. The curve in the Sm_{3d} XPS spectrum was found to fit into two peaks [118]. The peak at 1084.3 eV corresponds to the bond of Sm–O. And another peak at 1082.2 eV corresponds to the bond of Sm–O–Ti. Although the ionic radius of Sm^{3+} (1.08 Å) is bigger than the ionic radius of Ti^{4+} (0.68 Å) and the Sm^{3+} ions cannot enter into the lattice of TiO_2, however, Ti^{4+} ions may enter into the lattice of the Sm_2O_3 leading a change in the electronic field of Sm^{3+} and an increase in the electron density and decrease in the binding energy of Sm^{3+}. Well-ordered mesoporous TiO_2 doped with ytterbium was also successfully synthesized by an evaporation-induced self-assembly process [116]. The Yb dopant was beneficial in stabilizing the mesoporous structure and reduces the recombination of photo-generated electrons and holes, being beneficial to its visible light-driven photoactivity. Europium-doped TiO_2 was synthesized by the precipitation–peptization method and used as the photocatalyst to degrade the salicylic acid [117]. The results showed that the doping of Eu was beneficial to the adsorption of salicylic acid and the separation of photo-formed holes and electrons.

In addition to the transition metal ion doping and rare-earth metal ion doping, the studies on some other metal ion-doped TiO_2 can be seen. Stannum doping TiO_2 and Ti (III) self-doping TiO_2 were also reported. The dopant Sn^{4+} substituted Ti^{4+} in the lattice of TiO_2, which was reflected in the lattice expansion in both *a*- and *c*-direction and change in the binding energy [119]. Different from other metal ion doping, the doping of Sn^{4+} in anatase TiO_2 would result in a blue shift of absorption edge and enhance the amount of surface hydroxyl and oxygen vacancies in the UV light region. On the other hand, Sn doping has also been demonstrated as an effective modification method to enhance the visible light response of TiO_2 [120]. Tin would improve the photocatalytic activity of TiO_2 by enhancing the separation rate of photo-generated electrons and holes on the surface of TiO_2. Due to the Fermi level of SnO_2 lower than that of TiO_2, the photo-generated electrons easily transfer from TiO_2 to SnO_2, resulting in a reduction of the number of photo-generated on the surface of TiO_2. In recent years, Ti^{3+}-doped TiO_2 has attracted much interest, since it has been demonstrated to exhibit visible light absorption [121, 122]. Sasikala et al. [122] have found that the surface Ti^{3+} and oxygen vacancies may be responsible for the enhanced visible light absorption of the TiO_2–SnO_2 composite. However, the surface Ti^{3+} and oxygen defects on the TiO_2 are usually not stable enough in air, since the surface Ti^{3+} is easily oxidized into Ti^{4+} by the dissolved oxygen in water [32, 123]. Most research has focused on Ti^{3+} self-doped TiO_2, which exhibits better chemical stability and is active for photocatalytic activity. Zuo et al. [32] have reported a one-step calcination method to synthesize Ti^{3+} self-doped TiO_2 having high stability and found that it exhibits improved visible light absorption and efficient photocatalytic hydrogen production capacity from water under visible light irradiation (Fig. 8.2, top). It has been also reported a vacuum activation method for modifying P25 to obtain Ti^{3+} self-doped TiO_2 with high stability, visible light absorption, and photocatalytic activity under visible light irradiation (Fig. 8.2, bottom) [36]. In addition to the vacuum activation method, a simple one-step solvothermal method with low-cost $NaBH_4$ added as a reductant was also reported to successfully synthesize a series of TiO_2 catalysts self-doped with Ti^{3+} which also exhibited strong visible light absorption and enhanced photocatalytic activity [33].

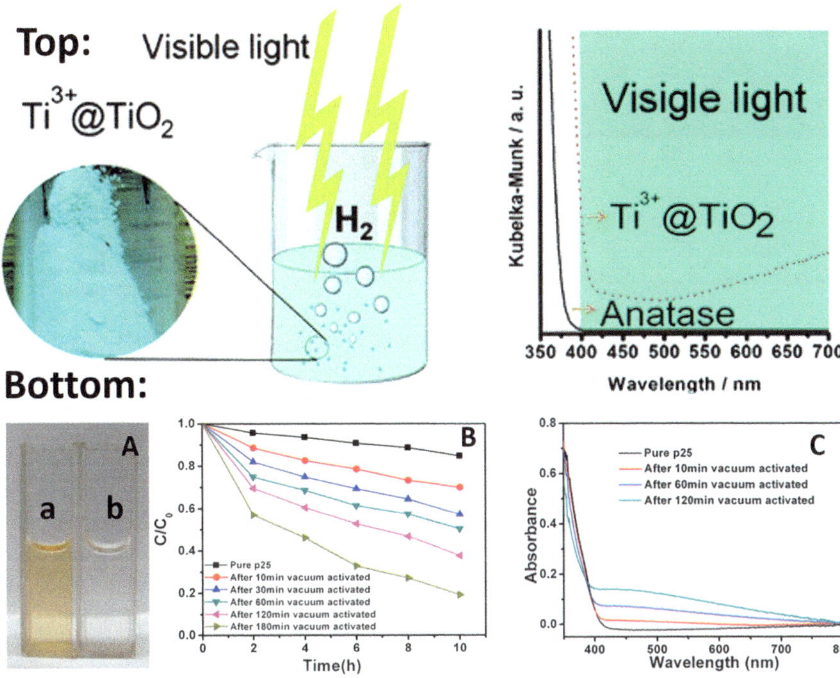

Fig. 8.2 The top figure is the time course of evolved H_2 under visible light (>400 nm) irradiation and the UV–vis diffuse reflectance spectra for commercial anatase TiO_2 (solid line) and Ti^{3+} self-doped TiO_2 (dash line) (the top figures). (Reprinted with permission from ref. [32]. Copyright 2010, American Chemical Society) The bottom figure is (**A**) photooxidation of 5 mg L^{-1} MO before (a) and after (b) visible light (>420 nm) irradiation for 3 h by the sample after vacuum activation for 180 min. (**B**) Photooxidation of 20 mg·L^{-1} phenol under visible light (>420 nm) irradiation for 10 h. (**C**) UV–vis diffuse reflectance spectra for pure P25 and the vacuum-activated samples (Reprinted with permission from ref. [36]. Copyright 2011, Royal Society of Chemistry)

8.1.2 Nonmetal Doping Modification

Although the metal doping modification could promote the absorption of TiO_2 for the visible light, the metallic ions also would induce the poor thermostability of TiO_2 and introduce some recombination centers with an excess doping concentration. Since in 2001 Asahi et al. [50] found that the nitrogen-doped TiO_2 exhibited the visible light absorption and photocatalytic activity, the study on nonmetal doping modification in TiO_2 has been a research hotspot.

Taking into consideration of the poor thermostability caused by the metal doping and the high cost in modification, the increasing number of nonmetal elements is used as the dopant to modify the bandgap of TiO_2 in recent years. Nonmetal doping mainly consists of the N, C, F, B, and other elements having the similar atomic radius with O atom. Among them, nitrogen has attracted much attention and been

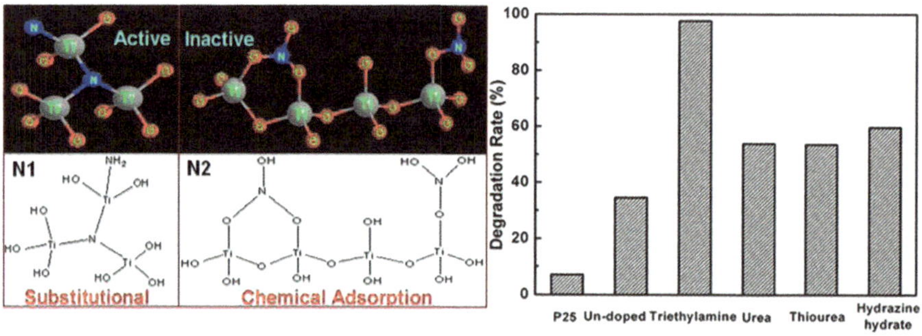

Fig. 8.3 The left figure is the nitrogen forms doped in TiO_2 (Reprinted with permission from ref. [42]. Copyright 2009, Elsevier); the right figure is the photocatalytic activities of TiO_2 prepared from different nitrogen sources with the optimal doping value (Reprinted with permission from ref. [41]. Copyright 2007, American Chemical Society)

widely studied. In some of recent work [51, 52], a new approach to synthesize the N-doped TiO_2 nanocrystals was developed, and the relationship of doped nitrogen species to visible light photoactivity was investigated. It was found that the nitrogen introduced into TiO_2 is beneficial to the visible light photodegradation of 2,4-dichlorophen and the nitrogen species chemically adsorbed on catalyst surface are harmful to the photoactivity, as shown in Fig. 8.3 [42]. Moreover, the nitrogen source also plays an important role in the N doping effect. Urea, triethylamine, thiourea, and hydrazine hydrate all could be used as the nitrogen source to prepare the nitrogen-doped TiO_2 by a microemulsion–hydrothermal method [41]. Among the above nitrogen sources, triethylamine was a more effective nitrogen source, which was used to synthesize the N-doped TiO_2 with the highest visible light photocatalytic degradation of rhodamine B (Fig. 8.3). Yates et al. [124] used NH_3 gas as the nitrogen source to prepare the N-doped TiO_2 with enhanced visible light photoactivity by a calcination method. However, the TiO_2 treated by the N_2 had decreased photoactivity due to the broadening of the bandgap of TiO_2.

In addition to nitrogen, carbon is also demonstrated to be an effective doping element to modify the visible light absorption and photoactivity of TiO_2. Nagaveni et al. [125] successfully prepared the C-doped TiO_2 by using a sol–gel method, which exhibited a high photodegradation of methylene blue under the visible and UV light irradiation. Kamisaka et al. [126] investigated the affection of C doping on the structure and optical property of TiO_2 by the density functional theory (DFT) calculation method. They assumed the carbon atom could substitute four sites of titanium and oxygen to obtain four corresponding C-doping structures. The DFT calculation results indicated that the substitution of C for Ti could not cause the visible light response of TiO_2 because of the formation of titanate anion. On the contrary, the substitution of C for O was beneficial to the visible light absorption of TiO_2 and did not change its crystal structure. Recently, Bai et al. [127] prepared monodisperse, carbon-doped rutile TiO_2 single crystal with exposed (110) facets, which possessed hierarchical structure and highly efficient H_2 generation activity. Yu et al. [128] also fabricated novel carbon self-doped TiO_2 sheets with exposed

(001) facets by a hydrothermal treatment. The C-doped TiO_2 sheets had enhanced absorption in the whole visible light region and a significant redshift at the absorption edges. It was also showing a high photocatalytic degradation of methylene blue under the visible light irradiation. Lin et al. [129] prepared a visible light-driven C-doped mesoporous TiO_2 films by a sol–gel method combined with a hydrothermal treatment. The C-doped TiO_2 film had high surface area and excellent photodegradation of dye reactive brilliant Red X-3B in the UV and visible light irradiation.

Fluorine-doped TiO_2 as one of the widespread modified methods has attracted much more attention in recent years, but conventional F-doped TiO_2 is hard to effectively achieve the enhanced UV and visible light photoactivity, and its photocatalytic mechanism still remains controversial [130–132]. In previous work, NH_4F was used as a hydrophobic modifier, and isopropanol was used as the solvent to prepare super-hydrophobic mesoporous MCF loaded with fluorinated TiO_2 nanoparticles [50], through a simple one-step solvothermal method (Fig. 8.4, top). The prepared catalyst exhibited permanent and excellent super-hydrophobic property, high adsorption capacity, and photocatalytic activity for rhodamine B degradation (Fig. 8.4, bottom). However, through the solvothermal method, the F ion could only adsorb on the surface of the catalyst but not be introduced into TiO_2 lattice. Many researchers also have reported the F-doped TiO_2 having high photocatalytic activity [50, 132]. And the standard substitutional of TiO_2 with F is similar to F-doped SnO_2, with generation of impurity levels close to the conduction band [133–135]. However, its electronic structures and photocatalytic mechanism are still unclear. It has been demonstrated that F substitution for lattice O could not introduce impurity level inside TiO_2 band gap, as well as shift its absorption edge into visible region by the first-principles calculation [136]. That is because of the absence consideration of F substitution for oxygen vacancy during the calculation process. Some studies have reported that F substitution for oxygen vacancy could introduce acceptor impurity level inside ZnO or SnO_2 bandgap [137, 138]. Hence, it was concluded that the achievement of abundant fluorine substitution for lattice oxygen vacancy played a very important role in the diminishing of vacancy-induced recombination sites and the introduction of impurity level inside the TiO_2 bandgap [53].

Compared with the abovementioned nonmetal elements, the study on boron doping in TiO_2 is relatively rare in recent years. In some studies, it was suggested that the boron doping in TiO_2 could lead to a redshift of the absorption band of TiO_2 to the visible light region, because of the overlapping of the impurity levels caused by boron with the 2p electronic states of oxygen [139, 140]. By contrast, some other studies reported that the boron incorporation into TiO_2 lattice could induce a blue shift rather than redshift due to the decrease of the crystal size [141, 142]. The quantization effect of the crystal size would result into a blue shift of absorption band to the UV light region. Even so, the boron has still been used as an important co-dopant together with other nonmetals for modification of TiO_2. Hence, more and more researchers began to study the co-doping modification of boron with other elements in TiO_2 [70, 77, 143, 144].

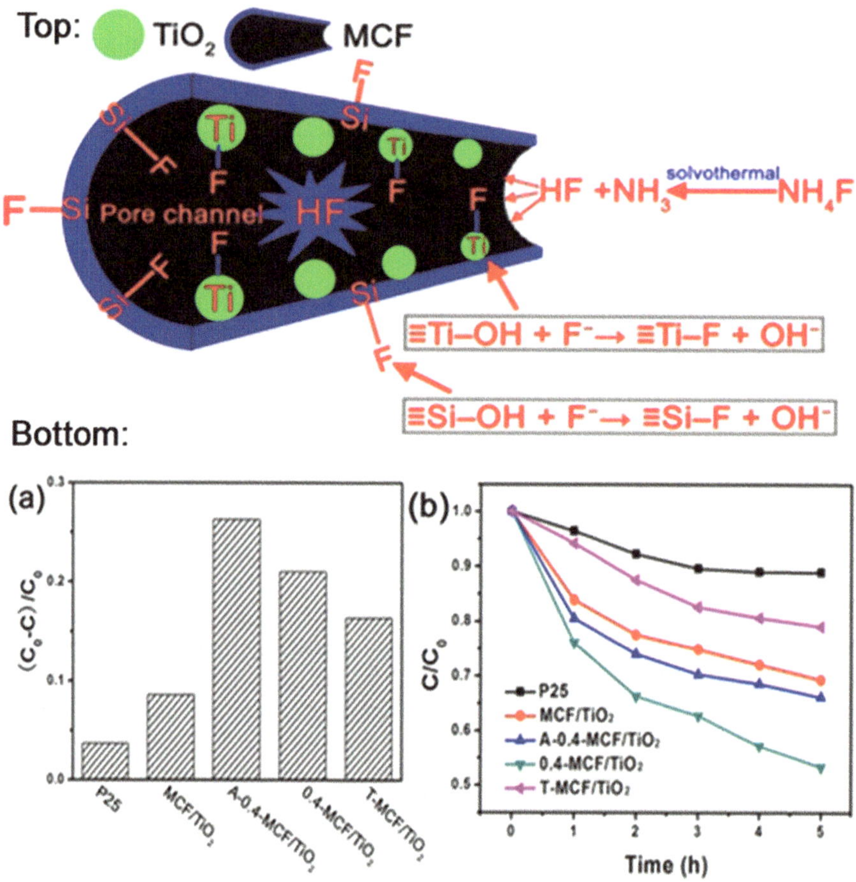

Fig. 8.4 The top figure is the schematic diagram of the fluorination reaction that occurred in the pore channels of MCF; the bottom figure is adsorption capacities of RhB (20 mg/L, catalyst concentration: 0.25 g/L) on different samples (**a**); visible light photocatalytic activities of different samples (**b**) (Reprinted with permission from ref. [50]. Copyright 2012, Elsevier)

8.1.3 Co-doping Modification

In order to further improve the photocatalytic activity of TiO_2 and make effective use of sunlight in visible light region, multicomponent modified TiO_2 such as co-doped with different elements has attracted much more attention in the photocatalysis fields. Some studies found that co-doped TiO_2 with appropriate elements could exhibit a much higher photocatalytic activity than single-doped catalyst, because of the existence of synergistic effect of doped elements which could promote the visible light absorption and facilitate the separation efficiency of photo-formed electrons and holes [11, 25, 35, 76, 77, 112, 116, 145, 146]. Herein, we will focus on summary of the research status of the co-doping modification on TiO_2

photocatalyst in recent years, the modification mechanism of co-doping, and the synergistic effect existing in the co-doped TiO_2.

According to the band theory of semiconductors, the conduction band of TiO_2 semiconductors is mainly determined by the Ti3d orbital energy level, and the valence band mainly depends on the energy level of O2p orbital. Compared with the O2p orbital, nonmetal elements such as N, C, S, and P have the 2p orbital with relatively high energy levels; hence, it is possible to enhance the visible light photocatalytic activity of TiO_2 by the doping N, C, S, P, and other nonmetal elements, due to the increase of electric potential of valence band by the partial substitution of impurity dopants for lattice oxygen.

Compared with other nonmetallic element co-doping modification, nitrogen co-doped with other nonmetal modification is very important, owing to the significant synergistic effect for visible light response. Cong et al. [147] proposed that the energy level of N doping could connect with the states of C doping and facilitate the overlap of C1s and N1s with the VB states of TiO_2, as shown in Fig. 8.5. The co-doping of N and C further narrowed the band gap of TiO_2 and improved the visible light photocatalytic activity of TiO_2 [147]. Meanwhile, boron has also been used as an important co-dopant together with nitrogen for co-doping modification of TiO_2 [69, 76, 77, 99, 143]. In et al. [99] and Liu et al. [76] proposed that B and N co-doped TiO_2 show a high UV and visible light photocatalytic activity, probably due to the existence of a synergistic effect between boron and nitrogen by the formation of Ti–B–N structure at the catalyst surface. However, there is still no detailed illustration of the B–N synergistic effect and its effect on the photocatalytic activity of TiO_2. Xing et al. [77] have illustrated the exact role of synergistic effect in optical absorbance and photocatalytic activity of B and N co-doped TiO_2. Various co-doped TiO_2 was systematically prepared by using the double hydrothermal

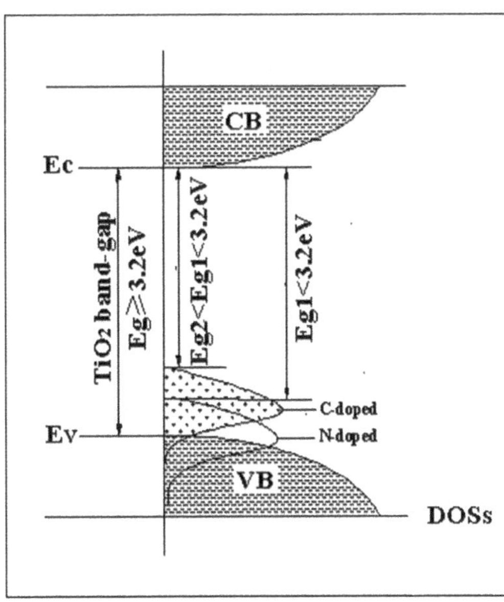

Fig. 8.5 Mechanism for photocatalytic degradation of organic pollutants over C–N–TiO_2 photocatalyst under visible light irradiation (Reprinted with permission from ref. [31]. Copyright 2010, Royal Society of Chemistry; and reprinted with permission from ref. [147]. Copyright 2006, Chemical Society of Japan)

method. Different new bonds were determined to form on the surface of TiO_2 when the order of boron and nitrogen addition was changed, and they could significantly affect the photoactivities of the materials. The abovementioned experimental results are further illustrated by the DFT calculation. Gombac et al. [148] found that the surface N doping did not appreciably modify the TiO_2 structures and texture, and boron incorporation in TiO_2 indeed inhibited the TiO_2 crystallite growth and increased the surface area of TiO_2. Only when B was present in excess with respect to N, a remarkable photoactivity improvement could be obtained. DFT calculation method was used to interpret the observed behavior. And the B in molar excess with respect to N led to the generation of Ti^{3+} sites, which might further induce the generation of reactive superoxide species. Different from the surface bond structures and the reactive Ti^{3+} sites induced by B and N co-doping, Czoska et al. [70] considered that the lattice center (labeled [NOB]$^{\cdot}$) based on the presence of interstitial N and B atoms both bound to the same lattice oxygen ion could introduce an energy level near the edge of VB of TiO_2. [NOB]$^{\cdot}$ can easily trap one electron to produce a diamagnetic center at about 0.4 eV above the top of the VB (Fig. 8.6), which can contribute to the visible light photoactivity.

Recently, nitrogen and sulfur co-doped TiO_2 was successfully immobilized on the surface of nitride Ti substrate, which exhibited high photodegradation of methylene blue in the visible light irradiation [149]. It was estimated that N and S co-doping in the anodic TiO_2 narrowed the band gap of TiO_2 and enhanced its visible light absorption and photocatalytic activity. In order to eliminate the recombination centers induced by the nonmetal doping in TiO_2 nanoparticles, Yang et al. [150] synthesized the fluorine and sulfur co-doped mesoporous TiO_2. The ability to control the morphology and chemical composition of the mesoporous TiO_2 could be beneficial to improve the light-harvesting capacity and decrease the recombination centers. The F and S co-doping in TiO_2 can redshift the threshold of the TiO_2 absorption into the visible light region and improve the photocatalytic efficiency for the degradation of organic pollutants. As well as our previous work, some nonmetal co-doped TiO_2 are successfully synthesized for the photodegradation of organic dyes under the visible light irradiation. N- and F co-doped TiO_2 microspheres were prepared by ethanol solvothermal method [151]. It was found that the co-doped catalyst with mesoporous structure exhibited a significant synergistic

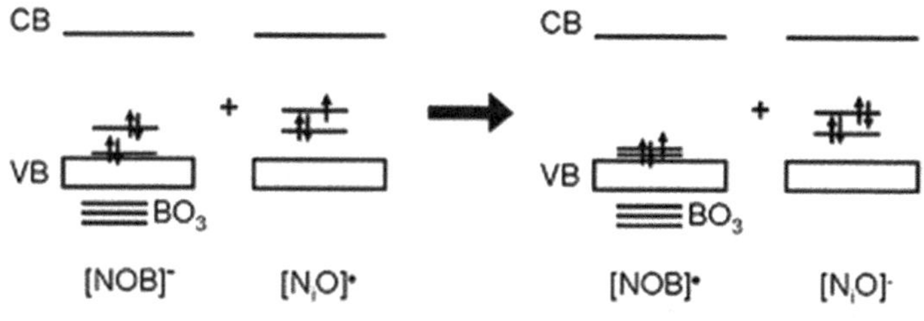

Fig. 8.6 Schematic representation of interplay between the [NOB]$^{\cdot}$ and [NiO]$^{\cdot}$ centers in N–B co-doped TiO_2 (Reprinted with permission from ref. [70]. Copyright 2011, Royal Society of Chemistry)

effect of N and F doping, which led to the high photocatalytic activity of the degradation of AO7 in the visible light irradiation. Moreover, carbon and boron co-doped TiO_2 were also synthesized firstly by the gel–hydrothermal method [144], that is, prepared through sol–gel process followed by hydrothermal in the glucose solution. The experimental results indicated that the coke carbon generated on the co-doped catalyst surface acted as a photosensitizer and had a photosensitization effect under the visible light. And the boron doping could effectively narrow the band gap of TiO_2, which induced the easier transition of photo-formed electrons from the boron dopant level to the Ti^{3+} level. The synergistic effect of B and C is responsible for its excellent visible light photocatalytic activity.

Either metal doping or nonmetal doping, they both will change the electric structure of TiO_2 and create a new doping level inside the band gap of TiO_2. Because of the different position of doping level in TiO_2, it is generally considered that the doping level induced by the substitution of metallic ions for Ti^{4+} is close to the CB of TiO_2. And the impurity level caused by the doping of nonmetallic ions into the oxygen vacancy sites is nearby the VB of TiO_2. These doping levels are located inside the band gap of TiO_2, which can accept the photo-formed electrons from the VB or absorb the photos with longer wavelength and extend the range of absorption spectrum of TiO_2. The synergistic effect between the metal and nonmetal could promote the separation of electrons and holes, resulting into the improvement of the visible light photocatalytic activity of TiO_2.

Generally speaking, the synergistic effect between metal and nonmetal is mainly shown as follows: the nonmetallic ion doping can enhance the absorption of TiO_2 in the visible light region, and the metallic ion doping can introduce traps for electrons and decrease the recombination of electrons and holes. Vanadium and nitrogen co-doped TiO_2 was synthesized by the sol–gel method, and the catalyst showed a high visible light photocatalytic activity for the degradation of RhB [111]. The visible light absorption efficiency of V–N co-doped TiO_2 was better than the V or N single-doped TiO_2, because of the effective narrowing of the band gap induced by the simultaneous incorporation of V and N in TiO_2 lattice, as shown in Fig. 8.7. The energy levels inside the TiO_2 band gap can act as traps for photo-formed holes and electrons thus decrease the recombination between photo-formed charges. The narrowed band gap and enhanced charge separation exhibit synergistic effect to improve the visible light photoactivity of the co-doped TiO_2. Wei et al. [152] considered that the synergistic effect between nitrogen and lanthanum in La and N co-doped TiO_2 was responsible for the high photocatalytic activity. The N doping decreased the band gap of TiO_2 and increased the absorption intensity of TiO_2 in visible light region. And the La^{3+} doping could not only increase the surface area of TiO_2 but also restrain the recombination of electrons and holes, due to the electron capturing capacity of La^{3+}. In order to minimize the role of metallic ions as recombination centers, Kim and coworkers synthesized boron and iron co-doped TiO_2 using a modified sol–gel method, and the presence of boron and iron caused a redshift in the absorption band of TiO_2 [153]. Cong et al. [79] also have successfully prepared the nanosized TiO_2 catalyst co-doped with nitrogen and iron, which exhibited a higher photocatalytic activity than the single doping catalyst under the

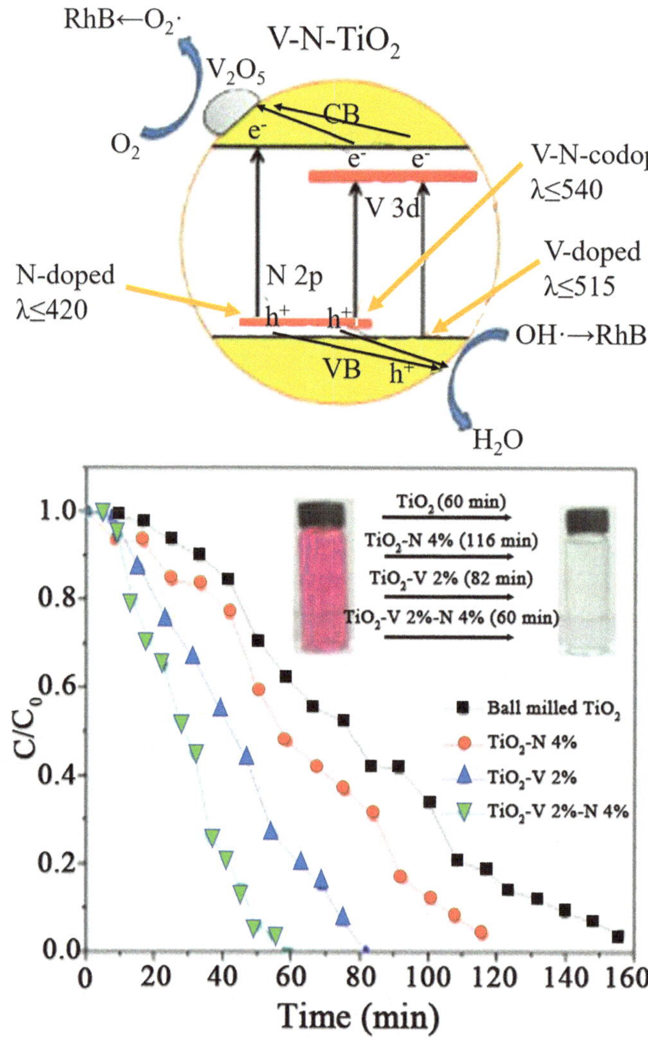

Fig. 8.7 Schematic diagram for photocatalytic degradation of RhB dye over V2–N4-co-doped TiO_2 photocatalyst under light irradiation. And the comparison of photocatalytic degradation of RhB under visible light source in the presence of undoped, V2-doped, N4-doped, and V2–N4-co-doped TiO_2 powders (Reprinted with permission from ref. [111]. Copyright 2012, Elsevier)

visible light irradiation. Xing et al. [21] prepared the highly dispersed carbon and lanthanum co-doped TiO_2 crystals with exposed (001) facets by using glucose as the carbon-doping source and (001)-facet-controlling agent through a simple one-step hydrothermal method. The C doping was responsible for the strong absorption in the visible light region, and the La doping acted as the electron capture center. The synergistic effect of C and La is the reason for the high visible light photocatalytic activity of the co-doped TiO_2.

However, in addition to the abovementioned synergistic effect, there are also some other forms of synergistic effects between metal and nonmetal, such as the increase of the surface active species including the hydroxyl groups and hydrogen peroxide in TiO_2. The role of surface hydroxyl groups and hydrogen peroxide is mainly including two aspects: one is involving in the reaction with the photo-formed holes to generate the hydroxyl radicals; another function is changing the adsorption forms of reactant and acting as the active center to influence the photocatalytic reaction of the reactant molecules. Wei and coworkers found that the content of hydroxyl groups on the surface of sulfur and iron co-doped TiO_2 was increasing, as can be seen in the XPS characterization [154], which is beneficial to its visible light photocatalytic activity. The difference of surface –OH between pure TiO_2 and single-doped TiO_2 is very small, while the amount of surface –OH on Fe and S co-doped TiO_2 is much higher than single-doped catalyst, resulting from the synergistic effect between Fe and S. The increase of surface –OH is in favor of the enhancing of the photoactivity. Gomathi Devi et al. [155] prepared the Ag and nitrogen co-doped TiO_2 by grinding sol–gel titania with urea followed by a photoreduction process. The as-prepared Ag–$TiO_{2-x}N_x$ exhibited much higher visible light photocatalytic activity than the single nitrogen doped TiO_2, which can be accounted to the synergistic effect of Ag loading with N doping. Strongly interacting electron accepting species of hydrogen peroxide at the catalyst surface are acting as the surface states enabling inelastic transfer of electrons from the CB to the oxidizing species. Additionally, the synergistic effect also can be shown as the electrons transfer between the co-dopant states. Hoang et al. [35] reported a synergistic effect involving Ti^{3+} and nitrogen in TiO_2 nanowire arrays, which exhibited an enhanced water photooxidation performance in the visible light irradiation. The authors proposed a reversible electron transfer between the paramagnetic bulk species of N (N_b) and Ti^{3+} centers forming the diamagnetic bulk species of N_b^- and Ti^{4+}. That means the existence of interaction between Ti^{3+} and N in TiO_2 (Fig. 8.8). The lower oxidation states of the substitutional N in the co-doped TiO_2 might be resulted from the electron transfer form Ti^{3+}. Because of the Columbic repulsion, the lower oxidation states of N in co-doped TiO_2 have higher energy than that of the N single-doped TiO_2, thus enabling excitation with photons of longer wavelengths.

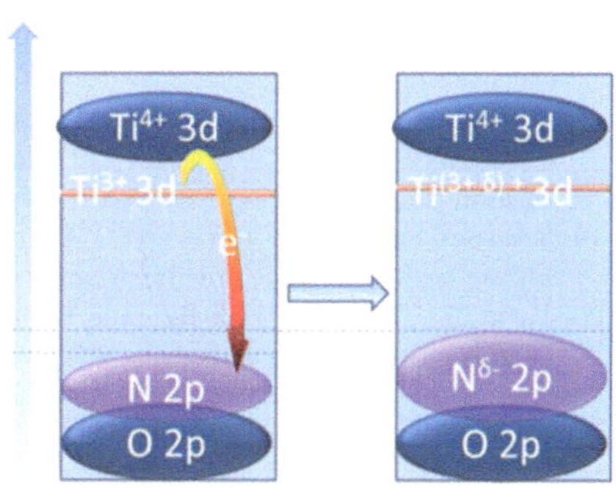

Fig. 8.8 Proposed mechanism for the interaction between Ti^{3+} and substitutional N (Reprinted with permission from ref. [35]. Copyright 2012, American Chemical Society)

References

1. O'Regan B, Gratzel M (1991) A low-cost high-efficiency solar cell based on dye-sensitized colloidal TiO_2 films. Nature 353:737–740
2. Ito S, Chen P, Comte P et al (2007) Fabrication of screen-printing pastes from TiO_2 powders for dye-sensitised solar cells. Prog Photovolt Res Appl 15(7):603–612
3. Ito S, Zakeeruddin SM, Humphry-Baker R et al (2006) High-efficiency organic-dye-sensitized solar cells controlled by Nanocrystalline-TiO_2 electrode thickness. Adv Mater 18 (9):1202–1205
4. Kuang D, Brillet J, Chen P et al (2008) Application of highly ordered TiO_2 nanotube arrays in flexible dye-sensitized solar cells. ACS Nano 2(6):1113–1116
5. Nazeeruddin MK, Humphry-Baker R, Liska P et al (2003) Investigation of sensitizer adsorption and the influence of protons on current and voltage of a dye-sensitized nanocrystalline TiO_2 solar cell. J Phys Chem B 107(34):8981–8987
6. Nazeeruddin MK, Pechy P, Renouard T et al (2001) Engineering of efficient panchromatic sensitizers for nanocrystalline TiO_2-based solar cells. J Am Chem Soc 123(8):1613–1624
7. Wang P, Zakeeruddin SM, Comte P et al (2003) Enhance the performance of dye-sensitized solar cells by co-grafting amphiphilic sensitizer and hexadecylmalonic acid on TiO_2 nanocrystals. J Phys Chem B 107(51):4336–14341
8. Zukalova M, Zukal A, Kavan L et al (2005) Organized mesoporous TiO_2 films exhibiting greatly enhanced performance in dye-sensitized solar cells. Nano Lett 5(9):1789–1792
9. Bach U, Lupo D, Comte P et al (1998) Solid-state dye-sensitized mesoporous TiO_2 solar cells with high photon-to-electron conversion efficiencies. Nature 395(6702):583–585
10. Zhu JF, Chen F, Zhang J et al (2006) Fe^{3+}-TiO_2 photocatalysts prepared by combining sol-gel method with hydrothermal treatment and their characterization. J Photochem Photobiol A Chem 180(1):196–204
11. Yang Y, Tian CX (2012) Effects of calcining temperature on photocatalytic activity of Fe-doped sulfated Titania. Photochem Photobiol 88(4):816–823
12. Shi JW, Zheng JT, Hu Y et al (2007) Influence of Fe^{3+} and Ho^{3+} co-doping on the photocatalytic activity of TiO_2. Meter Chem Phys 106(2):247–249
13. Zhu J, Zheng W, He B et al (2004) Characterization of Fe-TiO_2 photocatalysts synthesized by hydrothermal method and their photocatalytic reactivity for photodegradation of XRG dye diluted in water. J Mol Catal A Chem 216(1):35–43
14. Tong T, Zhang J, Tian B et al (2008) Preparation of Fe3+-doped TiO_2 catalysts by controlled hydrolysis of titanium alkoxide and study on their photocatalytic activity for methyl orange degradation. J Hazard Mater 155(3):572–579
15. Zhu JF, Deng ZG, Chen F et al (2006) Hydrothermal doping method for preparation of Cr^{3+}-TiO_2 photocatalysts with concentration gradient distribution of Cr3+. Appl Catal B Environ 62 (3):329–335
16. Anpo M, Takeuchi MJ (2003) The design and development of highly reactive titanium oxide photocatalysts operating under visible light irradiation. J Catal 216(1):505–516
17. Hamzah N, Nordin NM, Nadzri AHA et al (2012) Enhanced activity of Ru/TiO_2 catalyst using bisupport bentonite-TiO_2 for hydrogenolysis of glycerol in aqueous media. Appl Catal A: General 419:133–141
18. Panagiotopoulou P, Kondarides DI, Verykios XEJ (2010) Mechanistic study of the selective methanation of CO OVER ru/TiO_2 catalyst: identification of active surface species and reaction pathways. J Phys Chem C 115(4):1220–1230
19. Yuan S, Chen Y, Shi LY et al (2007) Synthesis and characterization of Ce-doped mesoporous anatase with long-range ordered mesostructure. Mater Lett 61(21):4283–4286
20. Tong TZ, Zhang JL, Tian BZ et al (2007) Preparation of Ce-TiO_2 catalysts by controlled hydrolysis of titanium alkoxide based on esterification reaction and study on its photocatalytic activity. J Colloid Interface Sci 315:382–388

21. Xing MY, Qi DY, Zhang JL et al (2011) One-step hydrothermal method to prepare carbon and lanthanum co-doped TiO_2 nanocrystals with exposed {001} facets and their high UV and visible-light photocatalytic activity. Chem Eur J 17(41):11432–11436
22. Yuan S, Sheng QR, Zhang JL et al (2005) Synthesis of La^{3+} doped mesoporous titania with highly crystallized walls. Microporous Mesoporous Mater 79(1):93–99
23. Gao HT, Liu WC, Lu B et al (2012) Photocatalytic activity of La, Y co-doped TiO_2 nanoparticles synthesized by ultrasonic assisted sol–gel method. J Nanosci Nanotechnol 12 (5):3959–3965
24. Tian B, Li C, Gu F et al (2009) Flame sprayed V-doped TiO_2 nanoparticles with enhanced photocatalytic activity under visible light irradiation. Chem Eng J 151(1):220–227
25. Liu H, Wu Y, Zhang J (2011) A new approach toward carbon-modified vanadium-doped titanium dioxide photocatalysts. ACS Appl Mater Interfaces 3(5):1757–1764
26. Lin WC, Lin YJ (2012) Effect of vanadium (IV)-doping on the visible light-induced catalytic activity of titanium dioxide catalysts for methylene blue degradation. Environ Eng Sci 29 (6):447–452
27. Sajjad S, Leghari SAK, Chen F et al (2010) Bismuth-doped ordered mesoporous TiO_2: visible-light catalyst for simultaneous degradation of phenol and chromium. Chem Eur J 16:13795–13804
28. Wang WJ, Zhang JL, Chen F et al (2010) Catalysis of redox reactions by Ag@ TiO_2 and Fe^{3+}-doped Ag@ TiO_2 core–shell type nanoparticles. Res Chem Intermed 36(2):163–172
29. Yuan XL, Zhang JL, Anpo M et al (2010) Synthesis of Fe3+- doped ordered mesoporous TiO_2 with enhanced visible light photocatalytic activity and highly crystallized anatase wall. Res Chem Intermed 36(1):83–93
30. Cong Y, Tian BZ, Zhang JL (2011) Improving the thermal stability and photocatalytic activity of nanosized titanium dioxide via La^{3+} and N co-doping. Appl Catal B Environ 101 (3):376–381
31. Zhang J, Wu Y, Xing M et al (2010) Development of modified N doped TiO_2 photocatalyst with metals, nonmetals and metal oxides. Energy Environ Sci 3(6):715–726
32. Zuo F, Wang L, Wu T et al (2010) Self-doped Ti^{3+} enhanced photocatalyst for hydrogen production under visible light. J Am Chem Soc 132(34):11856–11857
33. Xing M, Fang W, Nasir M et al (2013) Self-doped Ti3+-enhanced TiO_2 nanoparticles with a high-performance photocatalysis. J Catal 297:236–243
34. Zheng Z, Huang B, Meng X et al (2013) Metallic zinc-assisted synthesis of Ti^{3+} self-doped TiO_2 with tunable phase composition and visible-light photocatalytic activity. Chem Commun 49(9):868–870
35. Hoang S, Berglund SP, Hahn NT et al (2012) Enhancing visible light photo-oxidation of water with TiO_2 nanowire arrays via cotreatment with H_2 and NH_3: synergistic effects between Ti^{3+} and N. J Am Chem Soc 134(8):3659–3662
36. Xing M, Zhang J, Chen F et al (2011) An economic method to prepare vacuum activated photocatalysts with high photo-activities and photosensitivities. Chem Commun 47 (17):4947–4949
37. Liu GL, Han C, Pelaez M, Zhu DW, Liao SJ, Likodimos V, Ioannidis N, Kontos AG, Falaras P, Dunlop PSM, Byrne JA, Dionysiou DD (2012) Synthesis, characterization and photocatalytic evaluation of visible light activated C-doped TiO_2 nanoparticles. Nanotechnology 23(29):294003
38. Selvam K, Swaminathan M (2012) Nano N-TiO_2 mediated selective photocatalytic synthesis of quinaldines from nitrobenzenes. RSC Adv 2(7):2848–2855
39. Zhang W, Yang B, Chen J (2012) Effects of calcination temperature on preparation of boron-doped TiO_2 by sol-gel method. Int J Photoenergy 2012:1
40. Asahi R, Morikawa T, Ohwaki T et al (2001) Visible-light photocatalysis in nitrogen-doped titanium oxides. Science 293(5528):269–271
41. Cong Y, Zhang JL, Chen F et al (2007) Synthesis and characterization of nitrogen-doped TiO_2 nanophotocatalyst with high visible light activity. J Phys Chem C 111(19):6976–6982

42. Xing M, Zhang J, Chen F et al (2009) New approaches to prepare nitrogen-doped TiO_2 photocatalysts and study on their photocatalytic activities in visible light. Appl Catal B Environ 89(3):563–569
43. Lu XN, Tian BZ, Chen F et al (2010) Preparation of boron-doped TiO_2 films by autoclaved-sol method at low temperature and study on their photocatalytic activity. Thin Solid Films 519 (1):111–116
44. Wu YM, Xing MY, Zhang JL et al (2010) Effective visible light-active boron and carbon modified TiO_2 photocatalyst for degradation of organic pollutant. Appl Catal B Environ 97 (1):182–189
45. Tang YB, Yin LC, Yang Y et al (2012) Tunable band gaps and p-type transport properties of boron-doped graphenes by controllable ion doping using reactive microwave plasma. ACS Nano 6(3):1970–1978
46. Wang XD, Blackford M, Prince K et al (2012) Preparation of boron-doped porous titania networks containing gold nanoparticles with enhanced visible-light photocatalytic activity. ACS Appl Mater Interfaces 4(1):476–482
47. Wu YM, Zhang JL, Xiao L et al (2010) Properties of carbon and iron modified TiO_2 photocatalyst synthesized at low temperature and photodegradation of acid orange 7 under visible light. Appl Surf Sci 256(13):4260–4268
48. Parayil SK, Kibombo HS, Wu CM et al (2012) Enhanced photocatalytic water splitting activity of carbon-modified TiO_2 composite materials synthesized by a green synthetic approach. Int J Hydrog Energy 37(10):8257–8267
49. Zhong J, Chen F, Zhang JL (2009) Carbon-deposited TiO2: synthesis, characterization, and visible photocatalytic performance. J Phys Chem C 114(2):933–939
50. Xing MY, Qi DY, Zhang JL et al (2012) Super-hydrophobic fluorination mesoporous MCF/TiO_2 composite as a high-performance photocatalyst. J Catal 294:37–46
51. Tosoni S, Fernandez Hevia D, González Díaz Ó et al (2012) Origin of optical excitations in fluorine-doped titania from response function theory: relevance to photocatalysis. J Phys Chem Lett 3(16):2269–2274
52. Liu SW, Yu JG, Cheng B et al (2012) Fluorinated semiconductor photocatalysts: tunable synthesis and unique properties. Adv Colloid Interf Sci 173:35–53
53. Seo H, Baker LR, Hervier A et al (2010) Generation of highly n-type titanium oxide using plasma fluorine insertion. Nano Lett 11(2):751–756
54. Kuwahara Y, Maki K, Matsumura Y et al (2009) Hydrophobic modification of a mesoporous silica surface using a fluorine-containing silylation agent and its application as an advantageous host material for the TiO_2 photocatalyst. J Phys Chem C 113(4):1552–1559
55. Xu P, Xu T, Lu J et al (2010) Visible-light-driven photocatalytic S-and C-codoped meso/nanoporous TiO_2. Energy Environ Sci 3(8):1128–1134
56. Niu Y, Xing M, Tian B et al (2012) Improving the visible light photocatalytic activity of nano-sized titanium dioxide via the synergistic effects between sulfur doping and sulfation. Appl Catal B Environ 115–116:253–260
57. Bidaye P, Khushalani D, Fernandes JB (2010) A simple method for synthesis of S-doped TiO_2 of high photocatalytic activity. Catal Lett 134(1–2):169–174
58. Dozzi MV, Livraghi S, Giamello E et al (2011) Photocatalytic activity of S-and F-doped TiO_2 in formic acid mineralization. Photochem Photobiol Sci 10(3):343–349
59. He HY (2010) Solvothermal synthesis and photocatalytic activity of S-doped TiO_2 and TiS_2 powders. Res Chem Intermed 36(2):155–161
60. Yang K, Dai Y, Huang BJ (2007) Understanding photocatalytic activity of S-and P-doped TiO_2 under visible light from first-principles. J Phys Chem C 111(51):18985–18994
61. Di Valentin C, Pacchioni G, Selloni A et al (2005) Characterization of paramagnetic species in N-doped TiO_2 powders by EPR spectroscopy and DFT calculations. J Phys Chem B 109 (23):11414–11419
62. Livraghi S, Paganini MC, Giamello E et al (2006) Origin of photoactivity of nitrogen-doped titanium dioxide under visible light. J Am Chem Soc 128(49):15666–15671

63. Caratto V, Setti L, Campodonico S et al (2012) Synthesis and characterization of nitrogen-doped TiO_2 nanoparticles prepared by sol–gel method. J Sol-Gel Sci Technol 63(1):16–22
64. Dong F, Zhao W, Wu Z et al (2009) Band structure and visible light photocatalytic activity of multi-type nitrogen doped TiO_2 nanoparticles prepared by thermal decomposition. J Hazard Mater 162(2):763–770
65. Hao H, Zhang J (2009) The study of iron (III) and nitrogen co-doped mesoporous TiO_2 photocatalysts: synthesis, characterization and activity. Microporous Mesoporous Mater 121 (1):52–57
66. Jagadale TC, Takale SP, Sonawane RS et al (2008) N-doped TiO_2 nanoparticle based visible light photocatalyst by modified peroxide sol−gel method. J Phys Chem C 112 (37):14595–14602
67. Kim MS, Liu G, Nam WK et al (2011) Preparation of porous carbon-doped TiO_2 film by sol–gel method and its application for the removal of gaseous toluene in the optical fiber reactor. J Ind Eng Chem 17(2):223–228
68. Neville EM, Mattle MJ, Loughrey D et al (2012) Carbon-doped TiO_2 and carbon, tungsten-codoped TiO_2 through sol–gel processes in the presence of melamine borate: reflections through photocatalysis. J Phys Chem C 116(31):16511–16521
69. Gopal NO, Lo HH, Ke SC (2008) Chemical state and environment of boron dopant in B, N-codoped anatase TiO_2 nanoparticles: an avenue for probing diamagnetic dopants in TiO_2 by electron paramagnetic resonance spectroscopy. J Am Chem Soc 130(9):2760–2761
70. Czoska AM, Livraghi S, Paganini MC et al (2011) The nitrogen–boron paramagnetic center in visible light sensitized N–B co-doped TiO_2. Experimental and theoretical characterization. Phys Chem Chem Phys 13(1):136–143
71. Li Y, Ma G, Peng S et al (2008) Boron and nitrogen co-doped titania with enhanced visible-light photocatalytic activity for hydrogen evolution. Appl Surf Sci 254(21):6831–6836
72. Liu G, Yang HG, Wang X et al (2009) Visible light responsive nitrogen doped anatase TiO_2sheets with dominant {001} facets derived from TiN. J Am Chem Soc 131 (36):12868–12869
73. Wang DH, Jia L, Wu XL et al (2012) One-step hydrothermal synthesis of N-doped TiO_2/C nanocomposites with high visible light photocatalytic activity. Nanoscale 4(2):576–584
74. Zuo F, Bozhilov K, Dillon RJ et al (2012) Active facets on titanium (III)-doped TiO_2: an effective strategy to improve the visible-light photocatalytic activity. Angew Chem 124 (25):6327–6330
75. Zhao L, Chen X, Wang X et al (2010) One-step solvothermal synthesis of a carbon@TiO_2 dyade structure effectively promoting visible-light photocatalysis. Adv Mater 22 (30):3317–3321
76. Liu G, Zhao Y, Sun C et al (2008) Synergistic effects of B/N doping on the visible-light photocatalytic activity of mesoporous TiO_2. Angew Chem Int Ed 47(24):4516–4520
77. Xing MY, Li WK, Wu YM et al (2011) Formation of new structures and their synergistic effects in boron and nitrogen codoped TiO_2 for enhancement of photocatalytic performance. J Phys Chem C 115(16):7858–7865
78. Hopper EM, Sauvage F, Chandiran AK et al (2012) Electrical properties of Nb-, Ga-, and Y-substituted Nanocrystalline Anatase TiO_2 prepared by hydrothermal synthesis. J Am Ceram Soc 95(10):3192–3196
79. Cong Y, Zhang J, Chen F et al (2007) Preparation, photocatalytic activity, and mechanism of nano-TiO_2 co-doped with nitrogen and iron (III). J Phys Chem C 111(28):10618–10623
80. Zielińska A, Kowalska E, Sobczak JW et al (2010) Silver-doped TiO_2prepared by microemulsion method: surface properties, bio-and photoactivity. Sep Purif Technol 72 (3):309–318
81. Sakthivel S, Kisch H (2003) Daylight photocatalysis by carbon-modified titanium dioxide. Angew Chem Int Ed 42(40):4908–4911
82. Wu Y, Liu H, Zhang J et al (2009) Enhanced photocatalytic activity of nitrogen-doped titania by deposited with gold. J Phys Chem C 113(33):14689–14695

83. Xing M, Zhang J, Chen F (2009) Photocatalytic performance of N-doped TiO_2 adsorbed with Fe^{3+} ions under visible light by a redox treatment. J Phys Chem C 113(29):12848–12853
84. Yu JC, Yu J, Ho W et al (2002) Effects of F-doping on the photocatalytic activity and microstructures of nanocrystalline TiO_2 powders. Chem Mater 14(9):3808–3816
85. Goswami P, Ganguli JN (2012) Evaluating the potential of a new titania precursor for the synthesis of mesoporous Fe-doped titania with enhanced photocatalytic activity. Mater Res Bull 47(8):2077–2084
86. Justicia I, Ordejón P, Canto G et al (2002) Designed self-doped titanium oxide thin films for efficient visible-light Photocatalysis. Adv Mater 14(19):1399–1402
87. Cao Y, Yang W, Zhang W et al (2004) Improved photocatalytic activity of Sn^{4+} doped TiO_2 nanoparticulate films prepared by plasma-enhanced chemical vapor deposition. New J Chem 28(2):218–222
88. Kurtz SR, Gordon RG (1987) Chemical vapor deposition of doped TiO_2 thin films. Thin Solid Films 147(2):167–176
89. Su Y, Zhang X, Han S et al (2007) F–B-codoping of anodized TiO_2 nanotubes using chemical vapor deposition. Electrochem Commun 9(9):2291–2298
90. Navío J, Colón G, Litter MI et al (1996) Synthesis, characterization and photocatalytic properties of iron-doped titania semiconductors prepared from TiO_2 and iron (III) acetylacetonate. J Mol Catal A Chem 106(3):267–276
91. Yu J, Xiang Q, Zhou M (2009) Preparation, characterization and visible-light-driven photocatalytic activity of Fe-doped titania nanorods and first-principles study for electronic structures. Appl Catal B Environ 90(3):595–602
92. Surolia PK, Tayade RJ, Jasra RV (2007) Effect of anions on the photocatalytic activity of Fe (III) salts impregnated TiO_2. Ind Eng Chem Res 46(19):6196–6203
93. Di Paola A, Marci G, Palmisano L et al (2002) Preparation of polycrystalline TiO_2 photocatalysts impregnated with various transition metal ions: characterization and photocatalytic activity for the degradation of 4-nitrophenol. J Phys Chem B 106(3):637–645
94. Chen X, Liu L, Peter YY et al (2011) Increasing solar absorption for photocatalysis with black hydrogenated titanium dioxide nanocrystals. Science 331(6018):746–750
95. Naldoni A, Allieta M, Santangelo S et al (2012) Effect of nature and location of defects on bandgap narrowing in black TiO_2 nanoparticles. J Am Chem Soc 134(18):7600–7603
96. Xia T, Zhang C, Oyler NA et al (2013) Hydrogenated TiO_2 nanocrystals: a novel microwave absorbing material. Adv Mater 25(47):6905–6910
97. Chen X, Liu L, Liu Z et al (2013) Properties of disorder-engineered black titanium dioxide nanoparticles through hydrogenation. Sci Rep 3
98. Khan SU, Al-Shahry M, Ingler WB (2002) Efficient photochemical water splitting by a chemically modified n-TiO_2. Science 297(5590):2243–2245
99. In S, Orlov A, Berg R et al (2007) Effective visible light-activated B-doped and B, N-codoped TiO_2photocatalysts. J Am Chem Soc 129(45):13790–13791
100. Xia T, Zhang W, Murowchick JB et al (2013) A facile method to improve the photocatalytic and lithium-ion rechargeable battery performance of TiO_2 nanocrystals. Adv Energy Mater 3 (11):1516–1523
101. Wang Z, Yang C, Lin T et al (2013) H-doped black Titania with very high solar absorption and excellent photocatalysis enhanced by localized surface plasmon resonance. Adv Funct Mater 23(43):5444–5450
102. Kurosu H, Yamanobe T (2012) A specialist periodical report on nuclear magnetic resonance (2011/8) synthetic macromolecules. Specialist Periodical Reports- Nucl Magn Res 41:386
103. Choi W, Termin A, Hoffmann MR (1994) The role of metal ion dopants in quantum-sized TiO_2: correlation between photoreactivity and charge carrier recombination dynamics. J Phys Chem 98(51):13669–13679
104. Zhang J, Xu LJ, Zhu ZQ et al (2015) Synthesis and properties of (Yb, N)-TiO_2 photocatalyst for degradation of methylene blue (MB) under visible light irradiation. Mater Res Bull 70:358–364

105. Zhu J, Zheng W, He B et al (2004) Characterization of Fe–TiO_2 photocatalysts synthesized by hydrothermal method and their photocatalytic reactivity for photodegradation of XRG dye diluted in water. J Mol Catal A Chem 216(1):35–43
106. Zhu J, Chen F, Zhang J et al (2006) Fe^{3+}-TiO_2 photocatalysts prepared by combining sol–gel method with hydrothermal treatment and their characterization. J Photochem Photobiol A Chem 180(1):196–204
107. Xiao L, Zhang J, Cong Y et al (2006) Synergistic effects of doped Fe^{3+} and deposited Au on improving the photocatalytic activity of TiO_2. Catal Lett 111(3):207–211
108. Wu Y, Zhang J, Xiao L et al (2009) Preparation and characterization of TiO_2photocatalysts by Fe^{3+} doping together with Au deposition for the degradation of organic pollutants. Appl Catal B Environ 88(3):525–532
109. You X, Chen F, Zhang J et al (2005) A novel deposition precipitation method for preparation of Ag-loaded titanium dioxide. Catal Lett 102(3):247–250
110. Wang W, Zhang J, Chen F et al (2008) Preparation and photocatalytic properties of Fe^{3+}-doped Ag@TiO_2 core–shell nanoparticles. J Colloid Interface Sci 323(1):182–186
111. Jaiswal R, Patel N, Kothari DC et al (2012) Improved visible light photocatalytic activity of TiO_2 co-doped with Vanadium and Nitrogen. Appl Catal B Environ 126:47–54
112. Sun L, Zhao X, Cheng X et al (2012) Synergistic effects in La/N co-doped TiO_2 anatase (101) surface correlated with enhanced visible-light photocatalytic activity. Langmuir 28 (13):5882–5891
113. Yuan S, Sheng Q, Zhang J et al (2006) Synthesis of Pd nanoparticles in La-doped mesoporous titania with polycrystalline framework. Catal Lett 107(1):19–24
114. Chen QL, Wang Y, Zhong CY et al (2011) Effect of Co-doped La^{3+}/halogen on visible light photocatalytic activity of TiO_2. Trans Tech Publ 239:1923–1928
115. Anandan S, Ikuma Y, Murugesan V (2012) Highly active rare-earth-metal La-doped photocatalysts: fabrication, characterization, and their photocatalytic activity. Int J Photoenerg 2012:1
116. Ma Y, Xing M, Zhang J et al (2012) Synthesis of well ordered mesoporous Yb, N co-doped TiO_2 with superior visible photocatalytic activity. Microporous Mesoporous Mater 156:145–152
117. Ma Y, Zhang J, Tian B et al (2012) Synthesis of visible light-driven Eu, N co-doped TiO_2 and the mechanism of the degradation of salicylic acid. Res Chem Intermed 38(8):1947–1960
118. Ma Y, Zhang J, Tian B et al (2010) Synthesis and characterization of thermally stable Sm, N co-doped $TiO_2$with highly visible light activity. J Hazard Mater 182(1):386–393
119. Zhao Y, Liu J, Shi L et al (2011) Solvothermal preparation of Sn4+ doped anatase TiO_2nanocrystals from peroxo-metal-complex and their photocatalytic activity. Appl Catal B Environ 103(3):436–443
120. Xiufeng Z, Juan L, Lianghai L et al (2011) Preparation of crystalline Sn-doped $TiO_2$and its application in visible-light photocatalysis. J Nanomater 2011:47
121. Lu G, Linsebigler A, Jr Y et al (1994) Ti3+ defect sites on TiO_2(110): production and chemical detection of active sites. J Phys Chem 98(45):11733–11738
122. Sasikala R, Shirole A, Sudarsan V et al (2009) Highly dispersed phase of SnO2 on TiO_2nanoparticles synthesized by polyol-mediated route: photocatalytic activity for hydrogen generation. Int J Hydrog Energy 34(9):3621–3630
123. Nakamura I, Negishi N, Kutsuna S et al (2000) Role of oxygen vacancy in the plasma-treated TiO_2photocatalyst with visible light activity for NO removal. J Mol Catal A Chem 161 (1):205–212
124. Linsebigler AL, Lu G, Jr Y (1995) Photocatalysis on TiO_2surfaces: principles, mechanisms, and selected results. Chem Rev 95(3):735–758
125. Nagaveni K, Hegde MS, Ravishankar N et al (2004) Synthesis and structure of nanocrystalline $TiO_2$with lower band gap showing high photocatalytic activity. Langmuir 20(7):2900–2907
126. Kamisaka H, Adachi T, Yamashita K (2005) Theoretical study of the structure and optical properties of carbon-doped rutile and anatase titanium oxides. J Chem Phys 123(8):084704

127. Bai H, Liu Z, Sun DD (2012) Facile preparation of monodisperse, carbon doped single crystal rutile TiO_2nanorod spheres with a large percentage of reactive (110) facet exposure for highly efficient H2 generation. J Mater Chem 22(36):18801–18807
128. Yu J, Dai G, Xiang Q et al (2011) Fabrication and enhanced visible-light photocatalytic activity of carbon self-doped TiO_2sheets with exposed {001} facets. J Mater Chem 21 (4):1049–1057
129. Lin X, Rong F, Ji X et al (2011) Carbon-doped mesoporous TiO_2film and its photocatalytic activity. Microporous Mesoporous Mater 142(1):276–281
130. Czoska AM, Livraghi S, Chiesa M et al (2008) The nature of defects in fluorine-doped TiO2. J Phys Chem C 112(24):8951–8956
131. Fattori A, Peter LM, Wang H et al (2010) Fast hole surface conduction observed for indoline sensitizer dyes immobilized at fluorine-doped tin oxide– TiO_2 surfaces. J Phys Chem C 114 (27):11822–11828
132. Shifu C, Yunguang Y, Wei L (2011) Preparation, characterization and activity evaluation of TiN/F-TiO_2 photocatalyst. J Hazard Mater 186(2):1560–1567
133. Ma HL, Zhang DH, Win SZ et al (1996) Electrical and optical properties of F-doped textured SnO_2 films deposited by APCVD. Sol Energy Mater Sol Cells 40(4):371–380
134. Minami T (2000) New n-type transparent conducting oxides. MRS Bull 25(8):38–44
135. Rakhshani AE, Makdisi Y, Ramazaniyan HA (1998) Electronic and optical properties of fluorine-doped tin oxide films. J Appl Phys 83(2):1049–1057
136. Cui Y, Du H, Wen LS et al (2009) Investigation of electronic structures of F-doped TiO_2 by first-principles calculation. Trans Tech Publ 620:647–650
137. Liu B, Gu M, Liu X et al (2010) First-principles study of fluorine-doped zinc oxide. Appl Phys Lett 97(12):122101
138. Gonzalez-Hernandez R, Martinez AI, Falcony C et al (2010) Study of the properties of undoped and fluorine doped zinc oxide nanoparticles. Mater Lett 64(13):1493–1495
139. Zhao W, Ma W, Chen C et al (2004) Efficient degradation of toxic organic pollutants with Ni_2O_3/TiO_2-x B x under visible irradiation. J Am Chem Soc 126(15):4782–4783
140. Moon SC, Mametsuka H, Suzuki E et al (1998) Characterization of titanium-boron binary oxides and their photocatalytic activity for stoichiometric decomposition of water. Catal Today 45(1):79–84
141. Chen D, Yang D, Wang Q et al (2006) Effects of boron doping on photocatalytic activity and microstructure of titanium dioxide nanoparticles. Ind Eng Chem Res 45(12):4110–4116
142. Jung KY, Park SB, Ihm SK (2004) Local structure and photocatalytic activity of B_2O_3–SiO_2/TiO_2 ternary mixed oxides prepared by sol–gel method. Appl Catal B Environ 51(4):239–245
143. Xing M, Wu Y, Zhang J et al (2010) Effect of synergy on the visible light activity of B, N and Fe co-doped $TiO_2$for the degradation of MO. Nanoscale 2(7):1233–1239
144. Wu Y, Xing M, Zhang J (2011) Gel-hydrothermal synthesis of carbon and boron co-doped $TiO_2$and evaluating its photocatalytic activity. J Hazard Mater 192(1):368–373
145. Naik B, Parida KM (2010) Solar light active photodegradation of phenol over a Fe x Ti1−xO_2−yNy Nanophotocatalyst. Ind Eng Chem Res 49(18):8339–8346
146. Wang W, Lu C, Ni Y et al (2012) Preparation and characterization of visible-light-driven N–F–Ta tri-doped TiO_2 photocatalysts. Appl Surf Sci 258(22):8696–8703
147. Cong Y, Chen F, Zhang J et al (2006) Carbon and nitrogen-codoped TiO_2 with high visible light photocatalytic activity. Chem Lett 35(7):800–801
148. Gombac V, De Rogatis L, Gasparotto A et al (2007) TiO_2 nanopowders doped with boron and nitrogen for photocatalytic applications. Chem Phys 339(1):111–123
149. Komai Y, Okitsu K, Nishimura R et al (2011) Visible light response of nitrogen and sulfur co-doped TiO_2 photocatalysts fabricated by anodic oxidation. Catal Today 164(1):399–403
150. Yang G, Xiao T, Sloan J et al (2011) Low-temperature synthesis of visible-light active fluorine/sulfur Co-doped mesoporous TiO_2 microspheres. Chem Eur J 17(4):1096–1100

151. Wu Y, Xing M, Tian B et al (2010) Preparation of nitrogen and fluorine co-doped mesoporous TiO_2 microsphere and photodegradation of acid orange 7 under visible light. Chem Eng J 162 (2):710–717
152. Wei H, Wu Y, Lun N et al (2004) Preparation and photocatalysis of TiO_2 nanoparticles co-doped with nitrogen and lanthanum. J Mater Sci 39(4):1305–1308
153. Khan R, Kim SW, Kim TJ et al (2008) Comparative study of the photocatalytic performance of boron–iron Co-doped and boron-doped TiO_2 nanoparticles. Mater Chem Phys 112 (1):167–172
154. Wei F, Zhu T (2007) Preparation and photocatalytic property of S and Fe co-doped TiO_2 nanoparticles. Appl Chem Ind 36:421–424
155. Devi LG, Nagaraj B, Rajashekhar KE (2012) Synergistic effect of Ag deposition and nitrogen doping in TiO_2 for the degradation of phenol under solar irradiation in presence of electron acceptor. Chem Eng J 181:259–266

Chapter 9
Hollow or Yolk–Shell-Type Photocatalyst

The photocatalytic efficiency of bulk semiconductor particles generally suffers from poor diffusion efficiency of photo-generated carriers from the inside to the surface and low absorption efficiency to the incident light. The introduction of cavity inside the semiconductor particle has been proven effective to form accessible active sites throughout the bulk particle and enhance the light absorption efficiency through multiple scattering effects. Moreover, through reasonably spatial arrangement of different components in a single particle, the hole–electron separation degree can be enhanced, and functions such as plasmon resonance absorption and magnetic separation can be extra assembled. In this chapter, we mainly discuss the synthesis and characteristics-tuning of hollow-, yolk–shell-, hierarchical-, and tubular-type photocatalyst and the structural effect on the photocatalytic performance.

9.1 Synthesis

The generation of void inside the semiconductor can be achieved mainly through non-template (Kirkendall, Ostwald ripening) and hard-template (silica, carbon, polymer colloids) routes [1–3]. Both methods can form cavitory semiconductors with versatile structures and highly tunable dimensions concerning core, shell, and void.

9.1.1 Non-template Route

Ostwald ripening is the most commonly used strategy for non-template synthesis of cavitory semiconductor [2, 4, 5]. A typical example is the formation of hollow anatase TiO_2 nanospheres using TiF_4 as the precursor in aqueous system and under hydrothermal conditions (Fig. 9.1) [2]. Inner nanospace and highly organized

J. Zhang et al., *Photocatalysis*, Lecture Notes in Chemistry 100,
https://doi.org/10.1007/978-981-13-2113-9_9

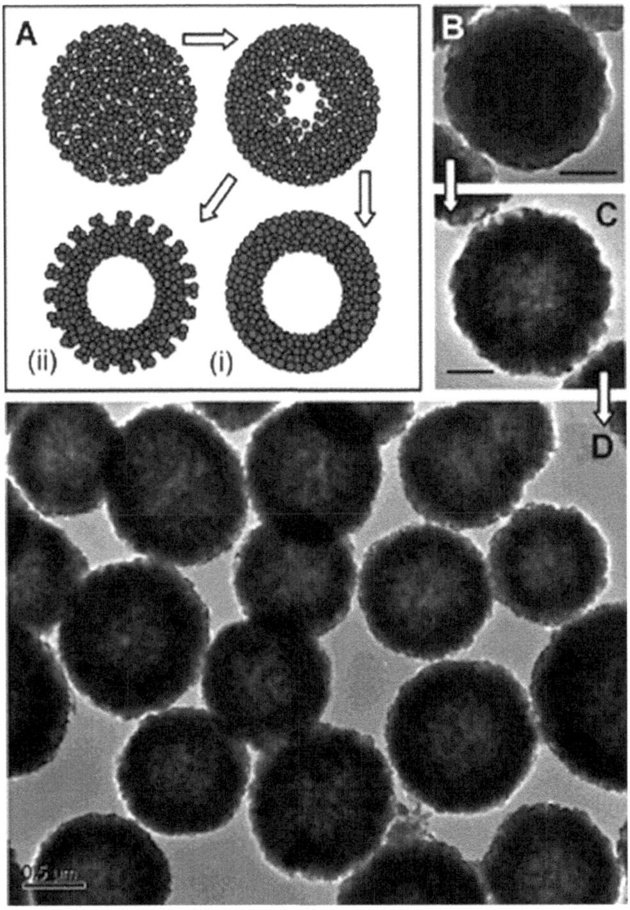

Fig. 9.1 (**a**) Schematic illustration (cross-sectional views) of the ripening process and two types (i & ii) of hollow structures. Evolution (TEM images) of TiO_2 nanospheres synthesized with 30 mL of TiF_4 (1.33 mM) at 180 °C with different reaction times: (**b**) 2 h (scale bar = 200 nm), (**c**) 20 h (scale bar = 200 nm), and (**d**) 50 h (scale bar = 500 nm) (Reprinted with the permission from ref. [2]. Copyright 2004 American Chemical Society)

crystallites in the shell structure and surface regions can be created with a wide range of controlling parameters. The interior void enlarges with the prolonging ripening process.

Bian et al. prepared mesoporous titania spheres with tunable chamber structure, where $TiOSO_4$ as the titania precursor was solvothermally reacted in glycerol, alcohol, and ethyl ether [6]. Judicious choice of the alcohol molecules (e.g., methanol, ethanol, and propanol) and reaction time affords the synthesis of spheres with adjustable morphology, size, and interior structure that is tunable from solid, sphere-in-sphere, to hollow. The formation of such spheres may involve aggregation of titania building clusters into spheres and their subsequent reaction, dissolution, and

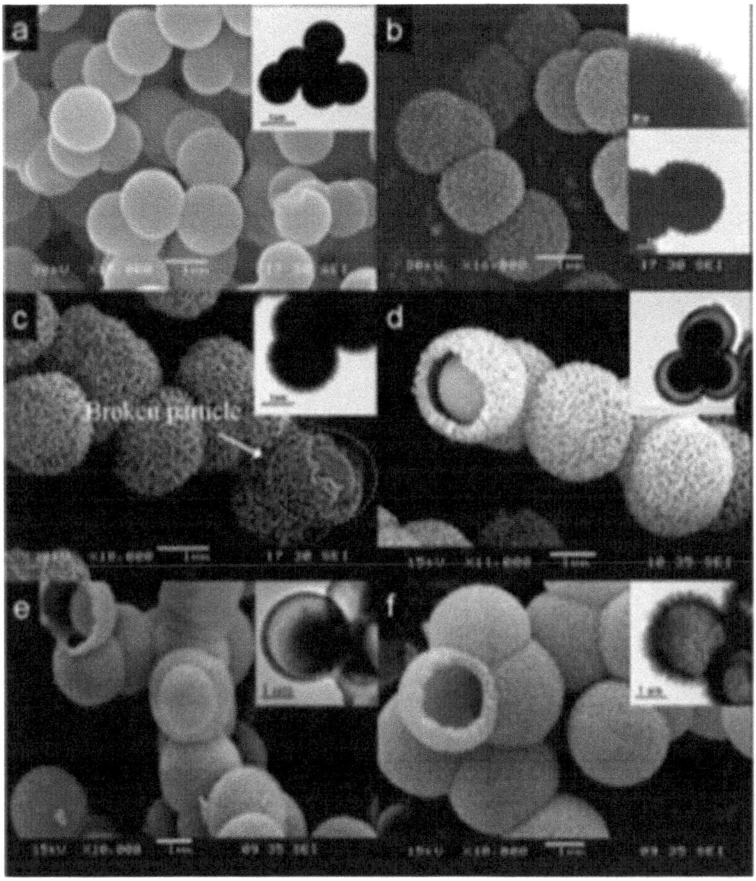

Fig. 9.2 SEM and TEM (insets) images of the titania spheres synthesized for (**a**) 1/24, (**b**) 0.5, (**c**) 1, (**d**) 2, (**e**) 7, and (**f**) 14 days, showing transiting interior structure from dense, to sphere-in-sphere, to hollow and surface morphology from smooth to prickly (Reprinted with the permission from ref. [6]. Copyright 2007 American Chemical Society)

redeposition process. Specially, during the solvothermal condition, etherifying reactions between alcohol and glycerol produce water continuously. Titania building clusters may be generated through alcoholysis reaction or hydrolysis–condensation reactions of $TiOSO_4$. Solid spheres first formed from the clusters aggregation, which contain a large number of hydrolyzable ligands due to slow reaction kinetics. Water continuously produced from the etherifying reactions further drives the hydrolysis–condensation reaction forward, leading to the dissolution and rearrangement of the surface building clusters (Fig. 9.2).

Hierarchical macro-/mesoporous titania can be prepared without the addition of templates or auxiliary additives at room temperature by the simple dropwise addition of tetrabutyl titanate (TBOT) to pure water and then calcined at various temperatures (Fig. 9.3) [7]. Firstly, the contact between the TBOT droplets and distilled water

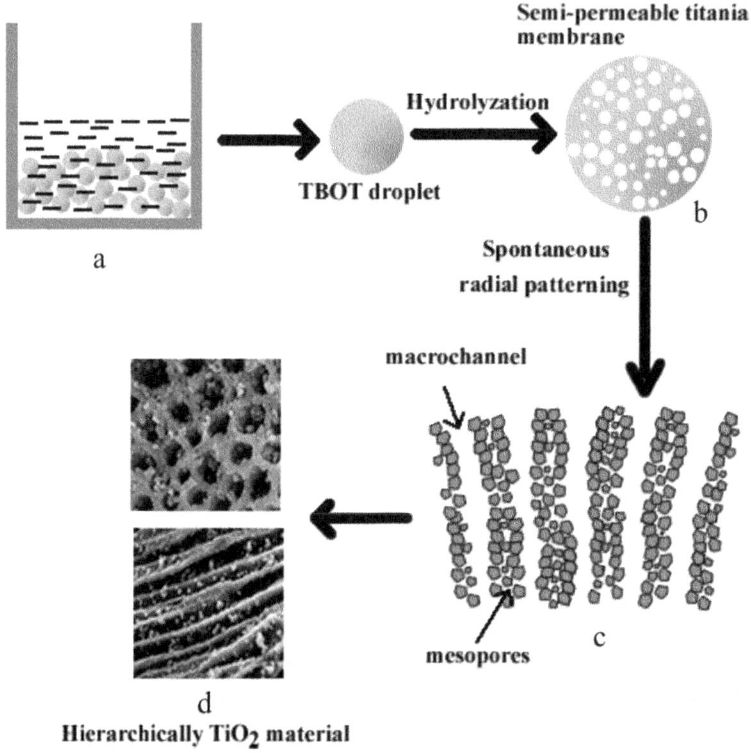

Fig. 9.3 Schematic procedure for the formation mechanism of the hierarchical macro-/mesoporous titania (Reproduced from ref. [7] by permission of John Wiley & Sons Ltd)

immediately produces a thin, dense semipermeable titania membrane at the droplet interface. Then, the distilled water diffuses through the spherical outer membrane, and the reaction proceeding inwards is approximately perpendicular to the external surface of the particles. The water/alcohol channels within the TBOT droplets undergo spontaneous radial patterning caused by the hydrodynamic flow of the solvent. The calcination temperature has a strong effect on the structures and photocatalytic activity of the prepared titania. At 300 °C, the calcined sample shows the highest photocatalytic activity. The calcination temperature higher than 500 °C caused the destruction of the hierarchical macro-/mesoporous structure, thus leading to the decrease of the photocatalytic activity.

9.1.2 Hard-Template Route

Compared with template-free method, hard-template route allows higher tunability over the pore volume, layer numbers, and composition. SiO_2, carbon, and polymer

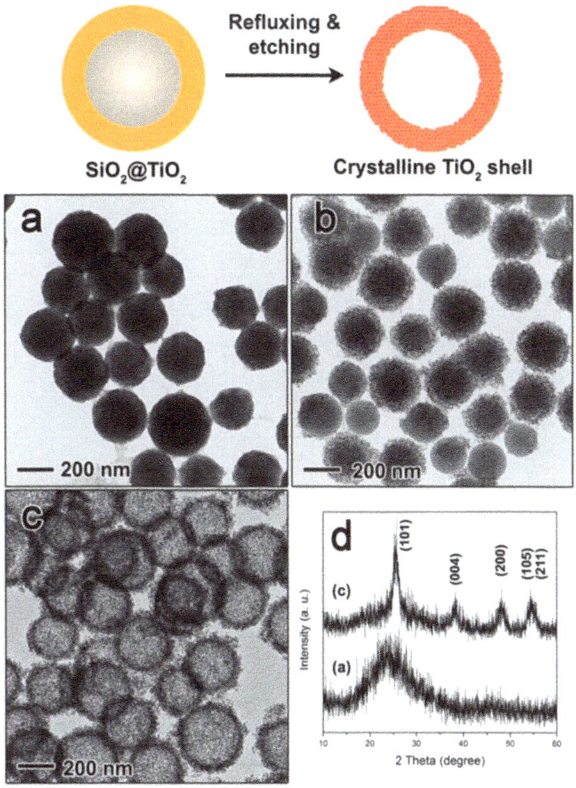

Fig. 9.4 Top: schematic illustration of the water-assisted crystallization strategy for converting an amorphous TiO_2 layer to mesoporous crystalline shells. Bottom, typical TEM images of the samples at each preparation step: (**a**) $SiO_2@TiO_2$ core–shell structures prepared by sol–gel coating; (**b**) $SiO_2@TiO_2$ core–shell structures after water-assisted crystallization; and (**c**) mesoporous TiO_2 hollow nanostructures after removing SiO_2 cores. (**d**) XRD patterns of samples (**a**) and (**c**), showing the transition from the amorphous to the anatase phase after water refluxing (Reproduced from ref. [8] by permission of John Wiley & Sons Ltd)

are the commonly used hard templates, which can be removed through acid–base etching, calcination, or dissolution. Yin et al. produced hollow TiO_2 using SiO_2 as the hard template [8], where amorphous TiO_2 was first coated around SiO_2 and then crystallized into anatase phase. SiO_2 was finally etched through NaOH to form interior void (Fig. 9.4).

Hierarchical macro-/mesoporous silica materials co-incorporated with Cr and Ti (MM–Si–Cr–Ti) were directly synthesized by adopting close-packed array of polystyrene (PS) microsphere as hard template for macropore and triblock copolymer Pluronic P123 as a soft template through a simple soaking-calcination way [9], where the Si/Ti ratio was fixed at 200 and Si/Cr ratio varied from 200 to 10. Ti specie is highly dispersed in the porous matrix, and Cr specie mainly exists as tetra-coordinated CrO_3 when Si/Cr $\leq$ 50 and transforms to a mixture of CrO_3 and crystallized hexa-coordinated Cr_2O_3 when the Si/Cr ratio is higher than 50 (Fig. 9.5).

Hierarchically ordered macro-/mesoporous TiO_2 films (denoted as H–TiO_2) with high thermal stability and crystallinity were synthesized using a confined evaporation-induced self-assembly (EISA) method [10]. P123 is used as a soft template to create the mesopores, and 3D periodic colloidal crystal PS are used as a hard template to create macropores (Scheme 9.1). The surfactant sulfuric acid

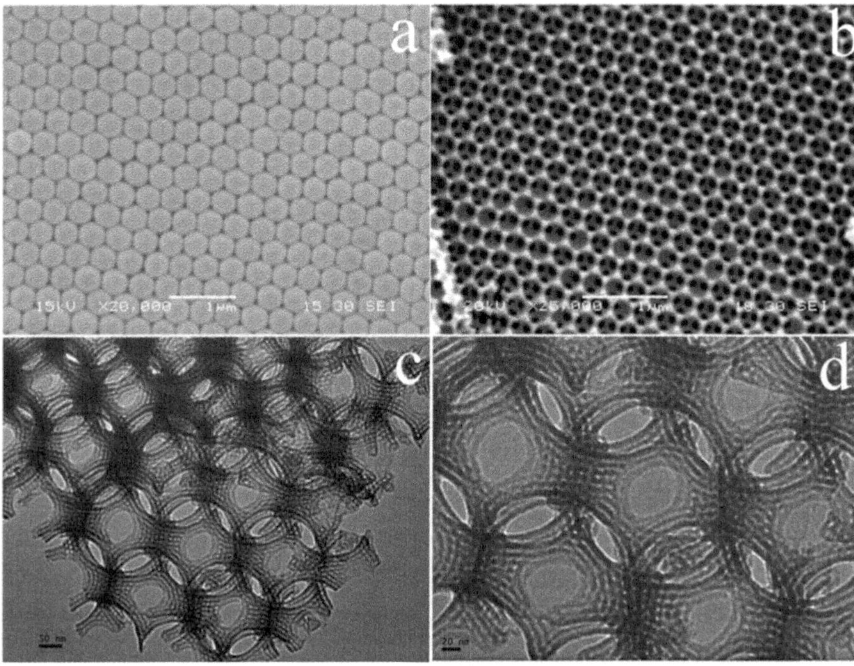

Fig. 9.5 SEM (**a**, **b**) and TEM (**c**, **d**) images of MM–Si–Cr–Ti prepared with molar ratio of Si–Cr–Ti = 200/10/1 (Reprinted from ref. [9], Copyright 2015, with permission from Elsevier)

carbonization method is applied to retain the ordered mesostructure during the high-temperature crystallization (550, 650 °C), while the PS spheres can generate amorphous carbon by heating treatment under an inert environment, which can support the inverse opal macrostructure from collapse. The hierarchically porous TiO_2 films present high surface areas of up to 240 $m^2\ g^{-1}$ and huge pore volume of ca. 1.2 $cm^3\ g^{-1}$. The photoelectrocatalytic water splitting performance of the hierarchical porous TiO_2 films is excellent and is much higher than that of pristine mesoporous TiO_2 films. The photoconversion efficiency is up to 5.23% for the N-doped hierarchically porous TiO_2 films.

9.2 Spatial Arrangement of Different Functions

The fabrication of yolk–shell structure allows the spatial arrangement of different functions. The Ostwald ripening process depicted in Fig. 9.1 can be used to encapsulate the metal core into the interior cavity. For example, Au/TiO_2 was fabricated through the following steps [11]: (1) formation of metallic cores, (2) aggregation of TiO_2 nanocrystallites around a single metallic core, (3) evacuation of central TiO_2 crystallites by Ostwald ripening, and (4) manipulation of the metal

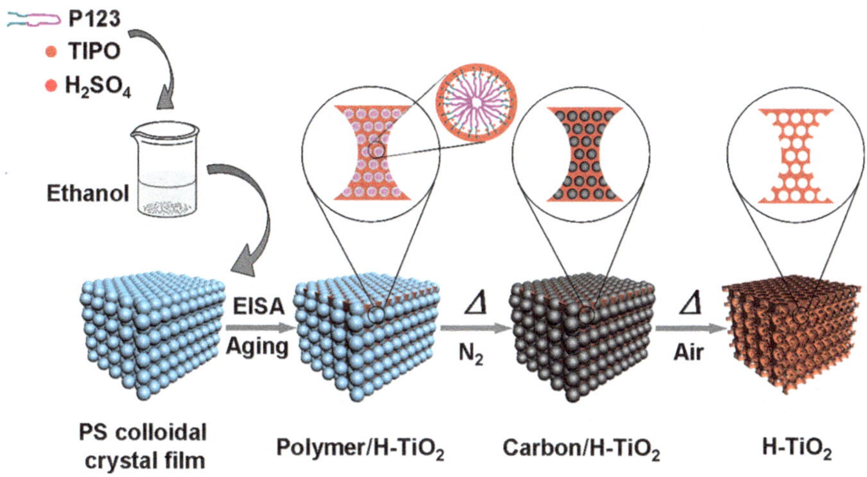

Scheme 9.1 Scheme of the synthesis process of the hierarchically ordered macro-/mesoporous TiO_2 films. The Ti precursor solution containing triblock copolymer Pluronic P123 and H_2SO_4 was impregnated into the interspace of the PS colloidal crystal films. After an EISA and aging process, the as-formed polymer–H–TiO_2 composite films were calcined under N_2 atmosphere at a high temperature (400, 550, 650 °C), and the in situ carbon that came from the carbonization of PS spheres and Pluronic P123 was generated both in macropores and mesopores to form the carbon–H–TiO_2 composite films. After the calcination in air to remove the carbon, the hierarchically ordered macro-/mesoporous TiO_2 films (H–TiO_2) with high crystallinity could be obtained (Reproduced from ref. [10] by permission of John Wiley & Sons Ltd)

core to a desired size via controlling the growth of the inside metal. For the size manipulation, $HAuCl_4$ was first introduced into the vacant space of the nanoreactor, and the $HAuCl_4$-soaked Au/TiO_2 nanoreactors were then placed in a solution of cetyltrimethylammonium bromide (CTAB) and ascorbic acid. It is believed that a slow interdiffusion between the $HAuCl_4$ inside the nanoreactor and the CTAB/ascorbic acid solution outside the nanoreactor would ensure a slow growth of Au cores while preventing TiO_2 shells from direct metal deposition. The size of Au core can be tuned in the range of 150–250 nm (Figs. 9.6 and 9.7).

Li et al. reported a facile "hydrothermal etching assisted crystallization" route to synthesize Fe_3O_4@titanate yolk–shell microspheres with ultrathin nanosheets-assembled double-shell structure [12]. The as-prepared microspheres possess a uniform size, tailored shell structure (Fig. 9.8), good structural stability, versatile ion-exchange capability, high surface area, and large magnetization and exhibit a remarkable photocatalytic performance.

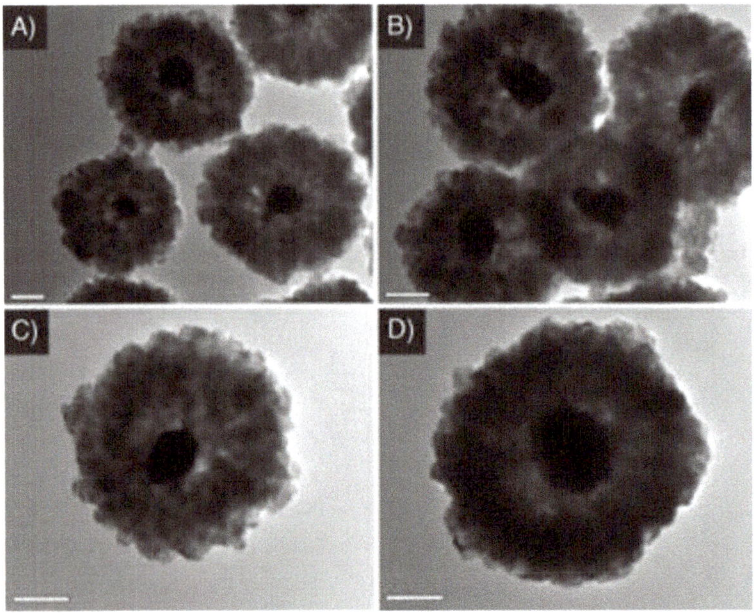

Fig. 9.6 (**a**)–(**d**) TEM images of Au/TiO_2 core–shell nanoreactors after enlargement of Au cores. All bar scales: 200 nm (Reprinted with the permission from ref. [11]. Copyright 2007 American Chemical Society)

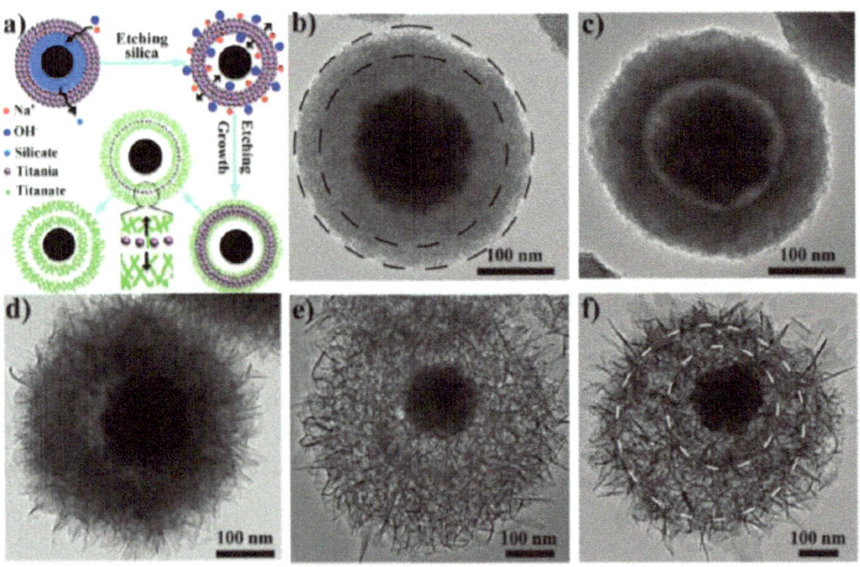

Fig. 9.7 (**a**) Schematic illustration of the "hydrothermal etching assisted crystallization" strategy for the formation of the Fe_3O_4@titanate double-shelled yolk–shell microspheres. TEM images of a single sphere synthesized at 150 C with hydrothermal treatment time: (**b**) 0 min, (**c**) 20 min, (**d**) 2 h, (**e**) 12 h, and (**f**) 24 h (Reprinted with the permission from ref. [12]. Copyright 2011 American Chemical Society)

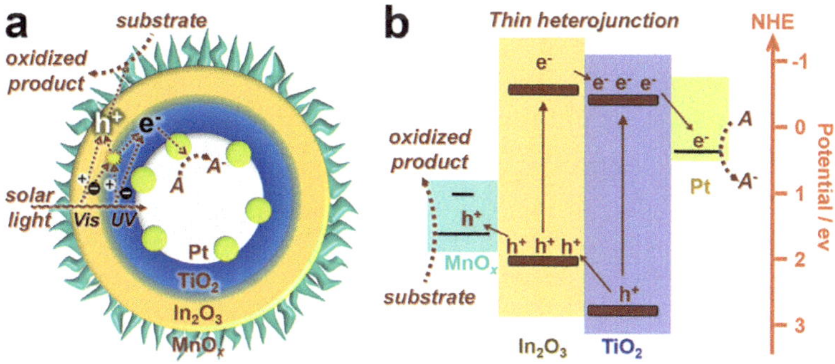

Fig. 9.8 The PTIM-MS structure and the mechanism for photocatalytic oxidation. Pt and MnOx are spatially separated by the TiO_2–In_2O_3 heterogeneous double-layered shell. (**a**) The reaction process. A represents an electron acceptor; here, $NaIO_3$ was used. (**b**) Simplified band structure of the catalyst. The CB positions of In_2O_3 and TiO_2 are 0.63 and 0.40 eV vs NHE, and the VB positions of In_2O_3 and TiO_2 were calculated to be 2.17 and 2.80 eV vs NHE according to the corresponding bandgaps (Reproduced from ref. [14] by permission of John Wiley & Sons Ltd)

9.3 Acceleration of Photo-Carrier Separation

Submicrometer-sized yolk–shell-structured Au@TiO_2 was synthesized by controllably hydrolyzing TiF_4 in Au nanoparticle solution under hydrothermal condition [13]. Both the size and the shell thickness could be easily adjusted by changing the amount of TiF4 in reaction mixture. Notably, when these Au@TiO2 hollow submicrospheres were used as the working electrodes for DSSCs, an obvious improvement in conversion efficiency is achieved compared to those solar cells based on TiO2 hollow spheres or TiO2 nanoparticles. DSSCs fabricated with Au–P25 composite materials and Au/TiO_2 core–shell nanowires can achieve an efficiency of 3.3% and 4.53%, respectively. As comparison, Au@TiO_2 hollow submicrospheres achieve a remarkably higher efficiency of 8.13%. Such a large enhancement should be ascribed to the fact that the sealed Au particles in the TiO_2 shells can effectively suppress both back electron transfer and energy transfer from dye to Au particles (Scheme 9.2).

Pt@TiO_2@In_2O_3@MnO_x mesoporous hollow spheres (PTIM-MSs) was designed and synthesized as shown in Fig. 9.8, which combine the advantages of spatially separated cocatalysts (Pt and MnO_x) and thin heterojunctions (TiO_2@In_2O_3 shell) to simultaneously reduce bulk and surface recombination [14]. Spatially separated cocatalysts drive electrons and holes near the surface to flow in opposite directions, reducing their recombination. Thin heterojunctions can effectively separate charges in the bulk phase and enable their transfer to the surface–subsurface region where they can be easily trapped by cocatalysts for surface reactions. Furthermore, In_2O_3 serves as a sensitizer to enhance light absorption. In combination with other advantages, such as a large surface area, long light-scattering path, and surface reaction kinetics promoted by cocatalysts, the PTIM-MS system is an

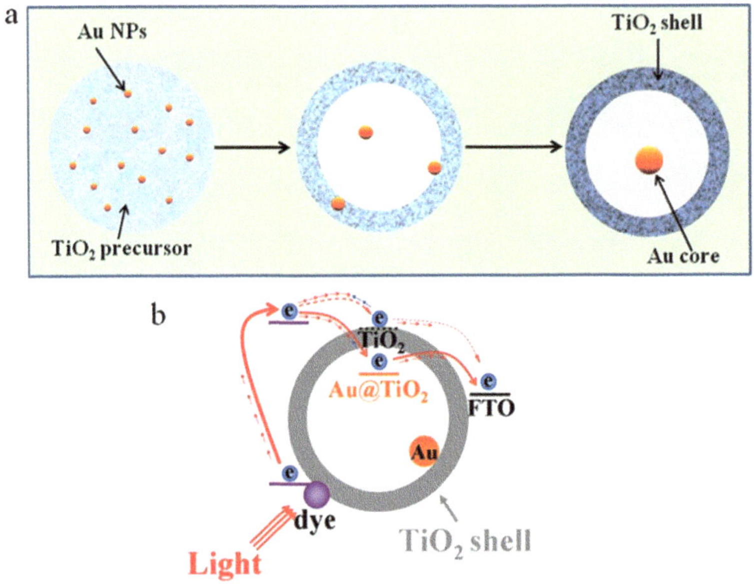

Scheme 9.2 Illustrations of (**a**) formation process of Au@TiO_2 hollow submicrospheres and (**b**) the charge separation process in the DSSCs with the photoanode of Au@TiO_2 hollow submicrospheres (Reproduced from ref. [13] by permission of John Wiley & Sons Ltd)

excellent photocatalyst of both water oxidation and selective benzyl alcohol oxidation.

A new strategy for design and synthesis of Co_9S_8 hollow cubes decorated by CdS QDs was developed by Zhang et al. [15]. The hybrid Z-scheme system is obtained by a simple hydrothermal method containing dimethyl sulfoxide (DMSO) as a solvent and hollow $Co(OH)_2$ cubes as a template. DMSO solvent served as a sulfur source to react with Cd^{2+}, producing CdS QDs, and was also used as the anion exchange reagent to react with easily prepared $Co(OH)_2$ cubes. The hollow CdS–Co_9S_8 cubes exhibit efficient solar light harvesting, as well as impressively enhanced hydrogen evolution reaction (HER) activity and stability under solar light irradiation compared to that of the pure Co_9S_8 and CdS catalysts. An efficient Z-scheme building block and the multiple reflections of solar light within the cavity of hollow cubes are responsible for substantially enhanced HER activity and stability (Fig. 9.9).

MnO_x@CdS/CoP hollow spheres with spatially separated surfaces were fabricated by loading the dual cocatalysts (inside MnO_x and outside CoP) for enhanced photocatalytic H_2 evolution activities [16]. CdS shells (hollow spheres) can be realized by using SiO_2 spheres as sacrificial templates, where the MnO_x and CoP NPs are selectively anchored on the inner and outer surface of CdS shells, respectively. That is, the photo-generated electrons and holes can directionally migrate to the locations of dual cocatalysts for reduction and oxidation reaction, respectively.

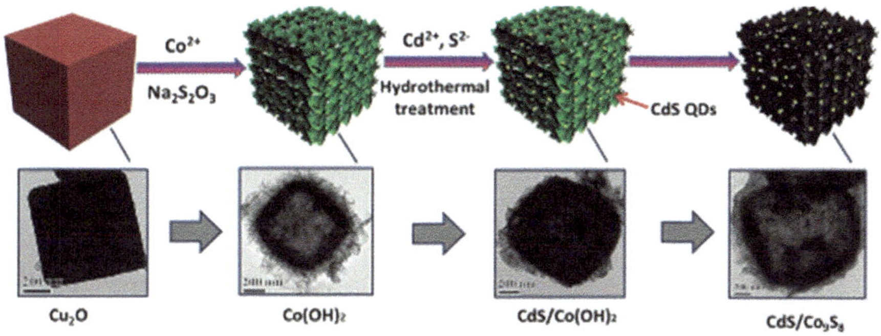

Fig. 9.9 The preparative process for CdS–Co_9S_8 and corresponding TEM images; scale bar = 200 nm (Reproduced from ref. [15] by permission of John Wiley & Sons Ltd)

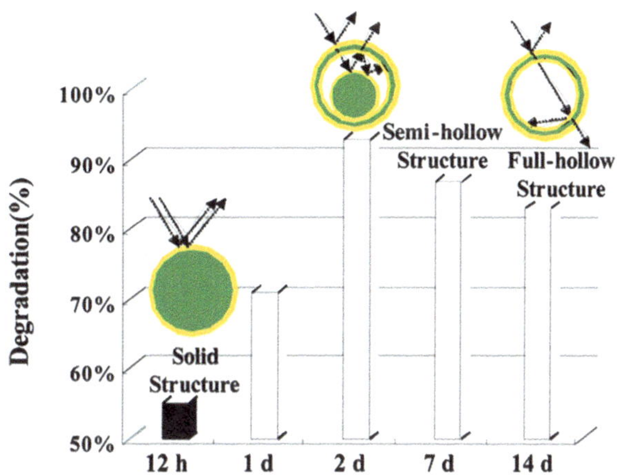

Fig. 9.10 Comparison of photocatalytic activities of the titania spheres with solid, sphere-in-sphere, and hollow structure. Inset shows a schematic illustration of multireflections within the sphere-in-sphere structure (Reprinted with the permission from ref. [6]. Copyright 2007 American Chemical Society)

As expected, in this case, the recommendation of electrons and holes has been significantly restrained, and the MnO_x@CdS/CoP catalyst shows a reduction surface and an enhanced photocatalytic activity for H_2 evolution (Fig. 9.10).

9.4 Multiple Light Scattering

The hollow or yolk–shell structure allows the multiple reflections of incident light within the interior cavity, thus enhancing the photocatalytic activity due to the improved light absorption efficiency (Fig. 9.11) [6]. The destroying of sphere-in-sphere structure by grinding the spheres causes a dramatic decrease of photocatalytic activity that is similar to that of the solid spheres. As schematically illustrated in the inset, a sphere-in-sphere structure with an appropriate inner sphere diameter allows

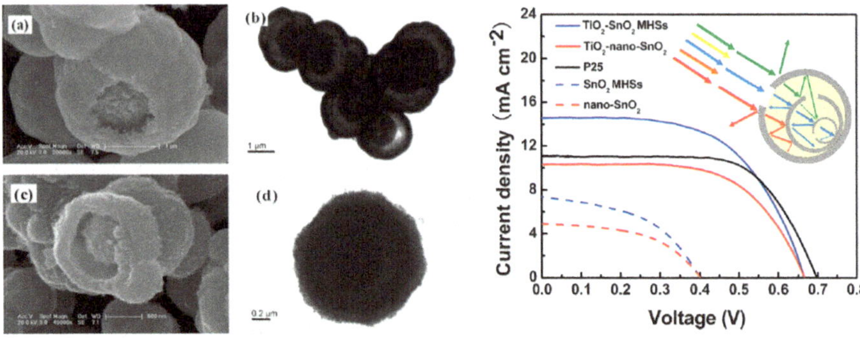

Fig. 9.11 (Left) SEM and (b) HRTEM images of SnO_2 MHSs. (c) SEM and (d) HRTEM images of TiO_2–SnO_2 MHSs. (Right) I–V characteristics of DSSCs with the photoelectrode films of TiO_2–SnO_2 MHSs, TiO_2–nano-SnO_2, SnO_2 MHSs, nano-SnO_2, and TiO_2 (P25) nanoparticles. The inset illustrates the multiple reflecting and scattering of light in the multilayered hollow spheres (Reproduced from ref. [16] by permission of John Wiley & Sons Ltd)

more efficient use of the light source and therefore offers an improved catalytic activity. The hollow and yolk–shell structures with smaller inner spheres show lower activity due to the reduced light reflection efficiency.

Qian et al. designed TiO_2-coated multilayered SnO_2 hollow microspheres (MHS) for dye-sensitized solar cells [16]. Multilayered spherical SnO_2–C composite was first formed through a condensation polymerization and carbonization of sucrose accompanied by hydrolysis of $SnCl_4$ in the hydrothermal reaction. Then, hollow SnO_2 were obtained by removal of carbon via calcination. Compared to TiO_2, SnO_2 has higher electron mobility and larger bandgap. However, SnO_2-based DSSCs generally have lower conversion efficiencies than TiO_2, which are attributed to a faster interfacial electron recombination and lower trapping density. On the other hand, SnO_2 has a lower isoelectric point (pH = 4–5) than anatase TiO2 (pH = 6–7), which leads to less adsorption of the dye with acidic carboxyl groups. To solve these problems, coating a thin layer of an isolating oxide, such as TiO_2, ZnO, Al2O3, or MgO, has been adopted to improve the conversion efficiency of SnO_2 photoelectrodes. TiO_2–SnO_2 multilayered hollow microspheres (SnO_2 MHSs) were synthesized by two steps (Fig. 9.8). First, SnO_2 MHSs were prepared by a chemically induced self-assembly reaction of aqueous sucrose–$SnCl_4$ solution under hydrothermal condition. The second step was to coat TiO_2 nanocrystallites onto the SnO_2 MHSs by impregnating in TiCl4 and then hydrolyzing $TiCl_4$ to form a surface layer of TiO_2.

The distinct photovoltaic behavior of the TiO_2–SnO_2 MHSs is its large short-circuit current (JSC, 14.6 mA cm^{-2}) compared with TiO_2–nano-SnO_2 (J_{SC}, 10.3 mA cm^{-2}) and TiO_2 (J_{SC}, 11.1 mA cm^{-2}) nanoparticles. This enhanced photocurrent could be attributed to better dye adsorption, due to increased active surface area, or better light-harvesting efficiency, due to the hollow spherical structure. Since the amount of dye adsorption on the TiO_2–SnO_2 MHSs is very similar to those of TiO_2–nano-SnO_2 and TiO_2 nanoparticles, the larger JSC value for the TiO_2–SnO_2 MHSs is most likely given rise by an enhanced light harvesting due to multiple light reflecting and scattering in between the hierarchical spherical shells of the TiO_2–SnO_2 MHSs, rather than an increased surface area. The JSC value for the SnO_2 MHSs is 50% higher than that of nano-SnO_2, suggesting that the enhanced JSC value for the TiO_2–SnO_2 MHSs compared with the TiO_2–nano-SnO_2 is due to the hierarchical spherical structure rather than the $TiCl_4$ treatment. The inset in Fig. 9.11 (right) illustrates the reflecting and scattering of light in a TiO_2–SnO_2 MHS. It is believed that the shells with the different sizes in a microsphere cannot only multireflect but also scatter the incident light of different wavelengths in the range of visible light. The improved photocurrent for the TiO_2–SnO_2MHSs is mainly contributed to its multilayered hollow spherical structure, which provides an effective way to enhance light-harvesting efficiency.

Photonic-crystal-based optical coupling offers a unique way of light–matter interaction to increase light harvesting, especially around the absorption edge of a semiconductor. A photonic crystal is a periodic dielectric structure that can forbid the propagation of light in a certain crystal direction within a certain spectrum regime, called a photonic stop-band. The light in a photonic crystal undergoes strong coherent multiple scattering and travels with very low group velocity near the photonic stop-band edges, referred to as slow light. Such a slow-light effect can considerably increase the effective optical path length, therefore leading to a delay and storage of light in photonic materials. 3D photonic crystal design was utilized to enhance incident photon-to-electron conversion efficiency (IPCE) of WO3 photoanodes. Large-area and high-quality WO3 photonic crystal photoanodes with inverse opal structure were prepared (Fig. 9.12). The photonic stop-bands of these WO3 photoanodes were tuned experimentally by variation of the pore size of inverse opal structures. It was found that when the red edge of the photonic stop band of WO3 inverse opals overlapped with the WO3 electronic absorption edge at Eg = 2.6–2.8 eV, a maximum of 100% increase in photocurrent intensity was observed under visible light irradiation ($\lambda > 400$ nm) in comparison with a disordered porous WO3 photoanode (Fig. 9.13). When the red edge of the stop band was tuned well within the electronic absorption range of WO3, noticeable but less amplitude of enhancement in the photocurrent intensity was observed. It was further shown that the spectral region with a selective IPCE enhancement of the WO3 inverse opals exhibited a blueshift in wavelength under off-normal incidence of light, in agreement with the calculated stop band edge locations. The enhancement could be attributed to a longer photon– matter interaction length as a result of the slow-light effect at the photonic stop-band edge, thus leading to a remarkable improvement in the light-harvesting efficiency. The present method can provide a potential and promising approach to effectively utilize solar energy in visible light-responsive photoanodes [17].

A range of TiO_2 inverse opals with tunable macroporous size were synthesized using different sized PS arrays as hard templates [18]. After a simple heating treatment in vacuum, Ti^{3+}-doped TiO_2 inverse opals were obtained. The optical responses of TiO_2- and Ti^{3+}-doped TiO_2 inverse opals could be enhanced by choosing PS arrays with appropriate size as hard templates due to the slow light

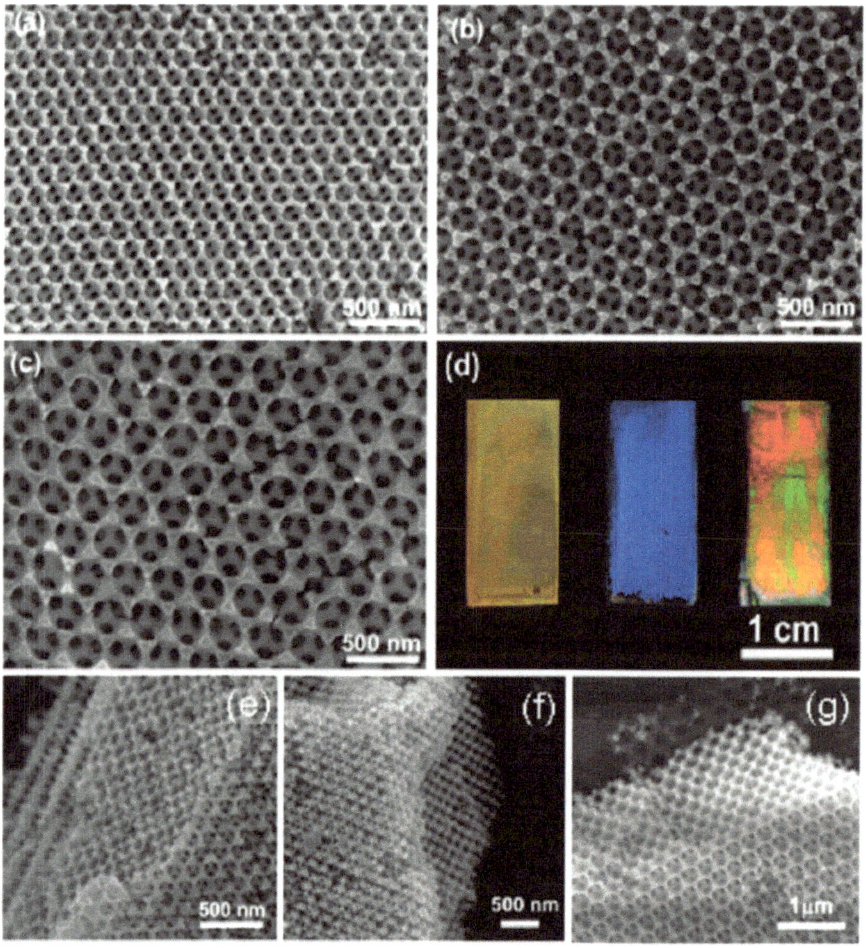

Fig. 9.12 SEM images of WO_3 inverse opals: (**a**) WO_3-200, (**b**) WO_3-260, and (**c**) WO_3-360. (**d**) Photograph of the inverse opal WO_3 photoanodes under white light illumination, WO_3-200, WO_3-260, and WO_3-360, from left to right. (**e–g**) SEM images of cross-sectional view of each inverse opal. The thickness is estimated to be about 2.5 μm (17 layers) for WO_3-200 (**e**), 2.6 μm (14 layers) for WO_3-260 (**f**), and 2.7 μm (11 layers) for WO_3-360 (**g**) (Reprinted with the permission from ref. [17]. Copyright 2011 American Chemical Society)

effect of inverse opal structure, thus coupling the physical and chemical enhancement for the light absorption. The photocatalytic efficiency was evaluated by the photodegradation of AO7, and it is proved that the cooperation of slow light effect and Ti^{3+} doping is an effective way to improve visible light-driven photocatalytic performance of TiO_2 photocatalyst (Scheme 9.3).

To achieve the coupling between the slow photon effect and plasmon resonance absorption, three-dimensional ordered assembly of TiO_2 hollow nanospheres deposited with Au nanoparticles (Au/TiO_2–3DHNSs) was designed (Fig. 9.14) [19]. A

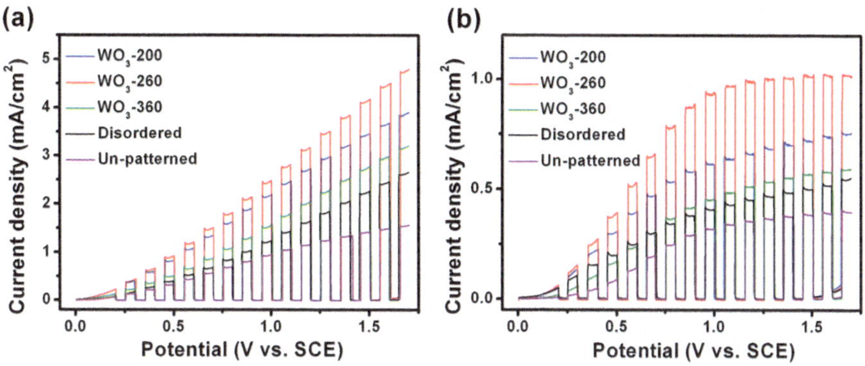

Fig. 9.13 Photocurrent potential curves of different WO_3 photoanodes measured under (**a**) UV–visible light irradiation ($\lambda > 300$ nm) and (**b**) visible light irradiation ($\lambda > 400$ nm) (Reprinted with the permission from ref. [17]. Copyright 2011 American Chemical Society)

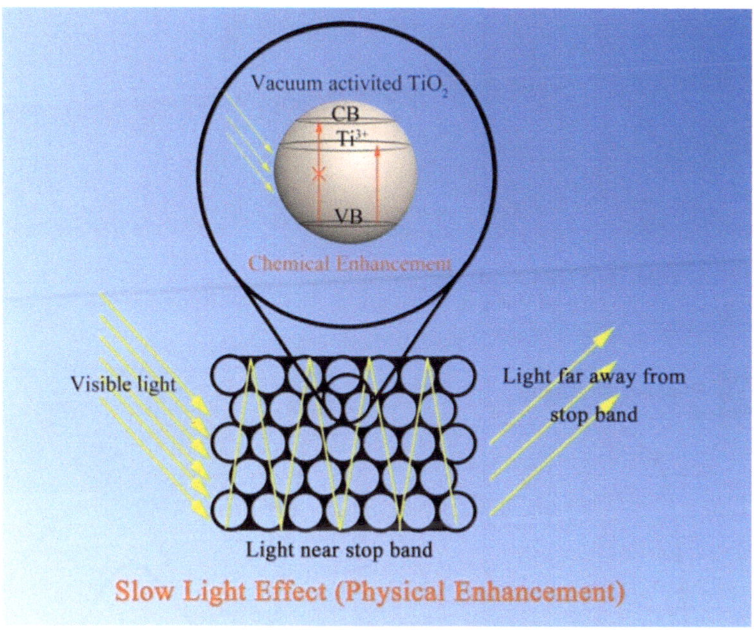

Scheme 9.3 Scheme of slow light effect (physical enhancement) and vacuum activation (chemical enhancement) (Reprinted from ref. [18], Copyright 2014, with permission from Elsevier)

photonic stop band centered at 500 nm was observed from the reflection spectrum (Fig. 9.15). As such, the slow photon effect occurring at the edges of the stop band is expected to appear in the range of 550–620 nm, which matches well with the surface plasmon resonance (SPR) absorption of Au nanoparticles. The wavelength matching of SPR absorption and photonic band edge would be expected to increase the SPR

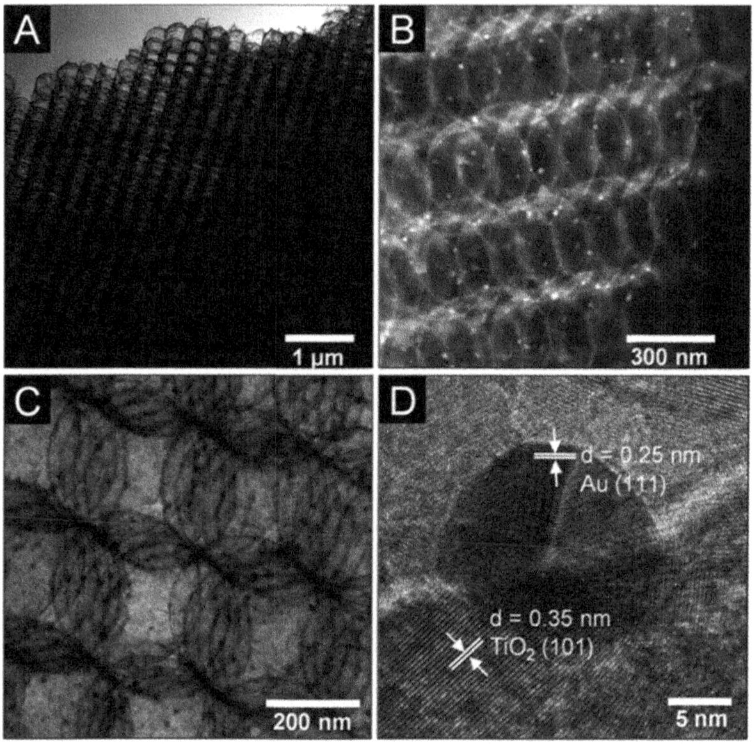

Fig. 9.14 (**a**) TEM; (**b**) STEM; (**c**, **d**) HRTEM images of Au/TiO_2–3DHNSs (Reproduced from ref. [19] by permission of John Wiley & Sons Ltd)

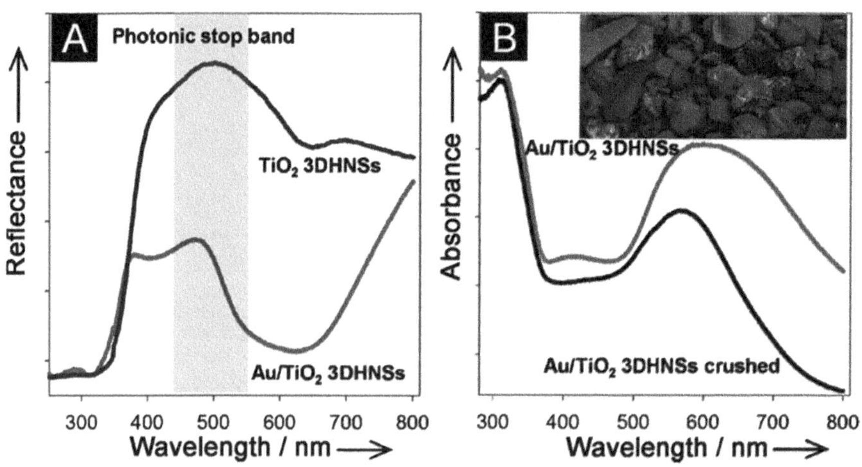

Fig. 9.15 (**a**) UV–vis diffuse reflectance spectra for Au/TiO_2–3DHNSs and the TiO_2–3DHNSs reference sample. The shaded region in (A) shows the presence of photonic stop band. (**b**) UV–vis absorption spectra of Au/TiO_2–3DHNSs before and after being crushed. Inset in (B) shows a digital photo of Au/TiO_2–3DHNSs (Reproduced from ref. [19] by permission of John Wiley & Sons Ltd)

intensity since an increase of the effective path length of light in the photonic band edge regions would result in a significant enhancement of the interaction between photons and Au NPs, which was verified by the decreased SPR absorption from crushed microarray structures (Fig. 9.14).

The photocatalytic activity of Au/TiO_2–3DHNSs for the decomposition of isopropanol to CO_2 under visible light illumination ($\lambda = 420$ nm) was investigated using Au/TiO_2–P25, crushed Au/TiO_2–3DHNSs, and Au/TiO_2–HNSs without the 3D ordered structure as the reference samples. Among all of samples, CO_2 produced from Au/TiO_2–3DHNSs is much higher than those from other samples. Therefore, the 3D ordered assembly of HNSs with periodic voids exhibits the slow photon effect which enhances the visible light absorption and, as a consequence, increases the photocatalytic activity of Au/TiO_2–3DHNSs. Even under the UV light irradiation, Au/TiO_2–3DHNSs still show the highest activity, which can be mainly attributed to the multiple light scattering resulting from the unique 3DHNS structure.

References

1. Yin YD, Rioux RM, Erdonmez CK et al (2004) Formation of hollow nanocrystals through the nanoscale Kirkendall effect. Science 304:711–714
2. Yang HG, Zeng HC (2004) Preparation of hollow anatase TiO_2 nanospheres via Ostwald ripening. J Phys Chem B 108:3492–3495
3. Wang X, Feng J, Bai Y et al (2016) Synthesis, properties, and applications of hollow micro-/nanostructures. Chem Rev 116:10983–11060
4. Li J, Zeng HC (2005) Size tuning, functionalization, and reactivation of Au in TiO_2 nanoreactors. Angew Chem Int Ed 44:4342–4345
5. Wang PH, Yang LG, Wang LZ et al (2016) Template-free synthesis of hollow anatase TiO_2 microspheres through stepwise water-releasing strategy. Mater Lett 164:405–408
6. Li H, Bian Z, Zhu J et al (2007) Mesoporous Titania spheres with tunable Chamber structure and enhanced photocatalytic activity. J Am Chem Soc 129:8406–8407
7. Yu JG, Su YR, Cheng B (2007) Template-free fabrication and enhanced photocatalytic activity of hierarchical macro-/mesoporous Titania. Adv Funct Mater 17:1984–1990
8. Joo JB, Lee I, Dahl M et al (2013) Controllable synthesis of mesoporous TiO_2 hollow shells: toward an efficient photocatalyst. Adv Funct Mater 23:4246–4254
9. Lu LJ, Teng F, Sen T et al (2015) Synthesis of visible-light driven Cr_xO_y–TiO_2 binary photocatalyst based on hierarchical macro–mesoporous silica. Appl Cata B Environ 163:9–15
10. Zhang RY, Shen DK, Xu M et al (2014) Ordered macro-/mesoporous anatase films with high thermal stability and crystallinity for photoelectrocatalytic water-splitting. Adv Energy Mater 4:1301725
11. Li J, Zeng HC (2007) Hollowing Sn-Doped TiO_2 nanospheres via Ostwald Ripening. J Am Chem Soc 129:15839–15847
12. Li W, Deng Y, Wu ZX et al (2011) Hydrothermal etching assisted crystallization: a facile route to functional yolk-shell titanate microspheres with ultrathin nanosheets-assembled double shells. J Am Chem Soc 133:15830–15833
13. Li A, Chang X, Huang Z et al (2016) Thin heterojunctions and spatially separated cocatalysts to simultaneously reduce bulk and surface recombination in photocatalysts. Angew Chem Int Ed 55:13734–13738
14. Qiu B, Zhu Q, Du M et al (2017) Efficient solar light harvesting CdS/Co_9S_8 hollow cubes for Z-scheme photocatalytic water splitting. Angew Chem Int Ed 129:2728–2732

15. Xing M, Qiu B, Du M et al (2017) Spatially separated CdS shells exposed with reduction surfaces for enhancing photocatalytic hydrogen evolution. Adv Funct Mater 27:1702624
16. Qian J, Liu P, Xiao Y et al (2009) TiO_2-Coated multilayered SnO_2 hollow microspheres for dye-sensitized solar cells. Adv Mater 21:3663–3667
17. Chen X, Ye J, Ouyang S et al (2011) Enhanced incident photon-to-electron conversion efficiency of tungsten trioxide photoanodes based on 3D-photonic crystal design. ACS Nano 5:4310–4318
18. Qi DY, Lu LJ, Xi ZH et al (2014) Enhanced photocatalytic performance of TiO_2 based on synergistic effect of Ti3+ self-doping and slow light effect. Appl Catal B Environ 160–161:621–628
19. Dinh CT, Yen H, Kleitz F et al (2014) Three-dimensional ordered assembly of thin-shell Au/TiO_2 hollow nanospheres for enhanced visible-light-driven photocatalysis. Angew Chem Int Ed 53:6618–6623

Chapter 10
Heterogeneous Photo-Fenton Technology

10.1 Introduction

In recent years, a lot of organic compounds have been used in industries with the rapid growth of modern science and technology, and they are inevitably released into environment, which causes great harm to human beings due to their toxicity, mutagenicity, and potential carcinogenicity [1–5]. Therefore, it is essential to remove these organic contaminants from wastewater and natural water.

Advanced oxidation processes (AOPs) need less energy than direct oxidation to degrade organic compounds [6–10]. Fenton technology is one of the most widely used methods as AOPs for the treatment of organic pollutants in water [11, 12]. Fenton oxidation process, a catalytic reaction of H_2O_2 with iron ions, mainly produces $\bullet OH$ radicals to oxidize organic compounds [13, 14], as shown in Eq. (10.1), which is considered as the core reaction in the process:

$$H_2O_2 + Fe^{2+} \rightarrow Fe^{3+} + OH^- + \bullet OH \tag{10.1}$$

The generated Fe^{3+} can be reduced by other H_2O_2 to reproduce Fe^{2+} and more radicals $HO_2\bullet$), which is called Fenton-like reaction and may account for the reason that a small amount of iron added can continue catalyzing H_2O_2 to produce $\bullet OH$ radicals, as shown in Eq. (10.2).

$$Fe^{3+} + H_2O_2 \rightarrow Fe^{2+} + HO_2\bullet + H^+ \tag{10.2}$$

There are other several reactions in Fenton oxidation process as well, as shown in Eqs. (10.3), (10.4), and (10.5):

$$Fe^{2+} + \bullet OH \rightarrow Fe^{3+} + OH^- \tag{10.3}$$

J. Zhang et al., *Photocatalysis*, Lecture Notes in Chemistry 100,
https://doi.org/10.1007/978-981-13-2113-9_10

$$Fe^{2+} + \bullet O_2H \rightarrow Fe^{3+} + HO_2^{-} \quad (10.4)$$

$$Fe^{3+} + \bullet O_2H \rightarrow Fe^{2+} + O_2 + H^{+} \quad (10.5)$$

Therefore, hydroxyl radicals are continuously produced and are able to degrade organic compounds uninterruptedly.

However, traditional Fenton reaction has some shortcomings that are focused on (1) low efficiency in utilizing H_2O_2, (2) narrow pH range (almost Fenton reaction is conducted at pH below 3.0), (3) great loss of iron ions and formation of solid sludge ($Fe(OH)_3$ is mainly included) so that the degradation rate subsequently decreases, and (4) difficulty in recycling catalysts [15–19].

Graphene is a two-dimensional lamellar structure with one-atom-thick, which is formed as hexagon rings by sp^2-hybridized carbon atoms. The graphene was first produced by a technique called micromechanical cleavage in 2004 [20] which subsequently raised the research climax. Subsequently, several other methods based on redox process are invented, including SiC epitaxial method and chemical vapor deposition method, which makes easier to prepare graphene. Due to its honeycomb network structure, graphene has a high surface area (~2630 $m^2\ g^{-1}$), high current density (10^8 A cm^{-2}), superior mechanical properties, high thermal conductivity (~2000–5000 W m K^{-1}) [21], excellent mobility of charge carriers (~100,000 $cm^2\ V^{-1}\ s^{-1}$) [22], optical transmittance [23], super hydrophobicity at nanometer scale, etc. These excellent properties make graphene as a unique material in wide applications including batteries [24–26], solar cells [27–29], sensors [30–32], catalysts, water treatment [33–41], and so on.

Recently, graphene has been strongly focused on its application in Fenton reaction to overcome the shortcomings of traditional Fenton technique. 2D π-π conjugation of graphene can reduce the recombination rate of electrons and holes. Graphene is a prominent electron acceptor to attract the excited photoelectrons from semiconductors, which are eventually segregated with the holes. The incorporation of semiconductor nanoparticles into graphene surface limits the restacking and agglomeration of graphene, which enlarges the surface area of the composites. Meanwhile, the functional groups and defect sites of graphene as building blocks provide the nucleation and growth sites for semiconductor nanoparticles, which lead to less aggregation of semiconductor nanoparticles. Therefore, the catalytic activity of the catalyst is also enhanced due to synergistic effects between graphene and semiconductor in Fenton reaction. What is more, the combination of two materials makes the nanoparticles tightly anchored on graphene which can effectively prevent the leaching of the catalyst and make recovery easier [42–44]. Figure 10.1 illustrates the mechanism of electron transfer of graphene/metal oxide composite.

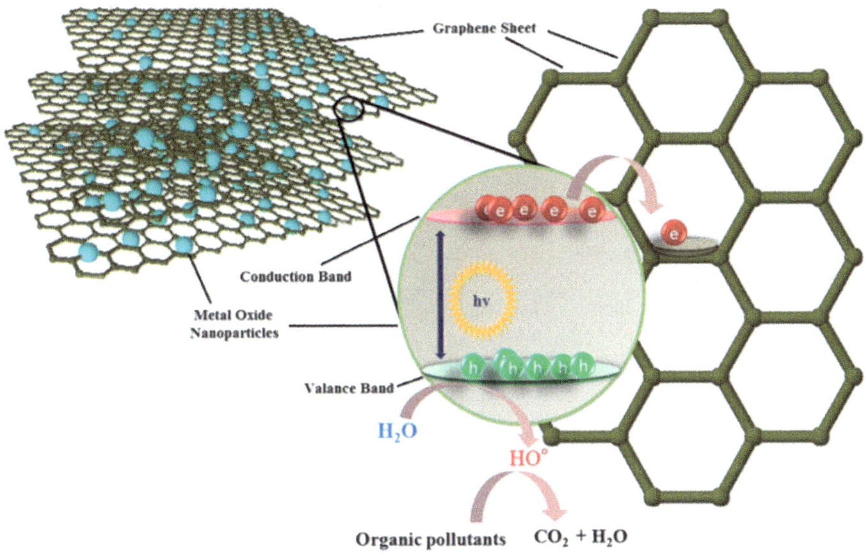

Fig. 10.1 Electron transfer from conduction band of metal oxide to graphene through percolation mechanism

10.2 Graphene/Iron (Hydr)oxide Composites Applied in Fenton Reaction

To date, a series of studies has been reported on the application of graphene/iron (hydr)oxide composites as photocatalysts to degrade organic pollutants in wastewater, especially dye pollution, which is attributed to their excellent conductivity, adsorptivity, stability, reusability, etc. Graphene has been used as an ideal support material to combine with various iron (hydr)oxides such as Fe_2O_3, Fe_3O_4, FeOOH, $ZnFe_2O_4$, and so on [45–48]. The synergistic effects of restricting electron–hole recombination and preventing the leaching of iron oxides strongly enhance the photocatalytic activity of the composite. Besides, the large plane structure of graphene makes aromatic pollution easier to be adsorbed and accelerates the process of catalysis. Thus, graphene/Fe_2O_3, graphene/Fe_3O_4, and other graphene/iron (hydr) oxides have been discussed and their photocatalytic activity has been evaluated in the following parts and tabulated in Table 10.1.

10.2.1 Graphene/Fe_2O_3 Composite as Photocatalyst in Fenton Reaction

α-Fe_2O_3 is the most widely used crystalline structure of iron oxide as it is abundant, inexpensive, and environmentally benign. α-Fe_2O_3 is an n-type semiconductor and

Table 10.1 The photocatalytic activity of various graphene/iron (hydr)oxide composites

Photocatalysts	Light source	Photocatalytic application	Photocatalytic activity enhancement	Reference
Fe_2O_3/NG	Visible light ($\lambda > 420$ nm) xenon lamp	Degradation of MB and glyphosate	1.5-fold and 2.3-fold of α-Fe_2O_3	[50]
Fe_2O_3/GO	Visible light ($\lambda > 420$ nm), 300 W Dy lamp	Degradation of Rhodamine B and 4-nitrophenol	–	[51]
α-Fe_2O_3/G	350 W xenon lamp	Degradation of RhB	2.3-fold of α-Fe_2O_3	[52]
α-Fe_2O_3/GO	UV light nm), 100 W high-pressure mercury lamp	Degradation of MB	2.9-fold of TiO_2 and 2.4-fold of α-Fe_2O_3	[53]
Fe_2O_3/GAs	Visible light (AM 1.5G), 300 W xenon lamp	Degradation of MO	–	[54]
Fe_3O_4/GO	–	Degradation of isatin	–	[61]
Fe_3O_4/RGO	Visible light (AM 1.5G), 300 W xenon lamp	Degradation of MO	More high and stable to Fe_3O_4	[62]
Fe_3O_4/RGO	Visible light (AM 1.5G), 300 W xenon lamp	Degradation of MB	–	[63]
Fe_3O_4/GO	Visible light (AM 1.5G), 500 W xenon lamp	Degradation of phenol	–	[64]
Fe_3O_4/AG	Sunlight (10.00 am ~ 2.00 pm)	Degradation of phenol, 2-NP, and 2-CP	Remarkable to the other reported nanomaterials	[65]
Fe_3O_4/HG	500 W high-voltage mercury lamp	Degradation of MO		[66]
Fe^0–Fe_3O_4–RGO	–	Degradation of MB	–	[67]
Fe_3O_4/GO	–	Degradation of AO7	–	[68]
$ZnFe_2O_4$/G	Visible light ($\lambda > 420$ nm), 300 W xenon lamp	Degradation of MB	20-fold of spinel-based photocatalysts, 4-fold of TiO_2-based photocatalysts, and 4-fold of other photocatalysts	[72]

(continued)

Table 10.1 (continued)

Photocatalysts	Light source	Photocatalytic application	Photocatalytic activity enhancement	Reference
$ZnFe_2O_4$/G	Visible light ($\lambda > 420$ nm), 500 W xenon lamp	Degradation of MB	–	[73]
$ZnFe_2O_4$/G	Visible light ($\lambda > 420$ nm), 500 W xenon lamp	Degradation of RhB, MO, and MB	–	[74]
α-FeOOH/RGO	Solar light	Degradation of phenol	–	[75]
α-FeOOH/GCA	UV light (365 nm), 125 W high-pressure mercury lamp	Degradation of MB, RdB, OII, phenol, and BPA	–	[76]

has a low bandgap of 2.2 eV so that it can absorb visible light within 560 nm, indicating it can utilize most of the visible light to oxidize organic contaminants. However, the shortcoming of α-Fe_2O_3 lies in the high electron–hole recombination rate and slow conversion of Fe (II) and Fe (III), so the catalytic activity of α-Fe_2O_3 is much poor compared to that of γ-Fe_2O_3, Fe_3O_4, and some other iron catalyst [49]. Several methods including fabricating composite with carbon material and forming yolk–shell structure with CdS realized the effective suppression of photo-generated electron and hole of α-Fe_2O_3 and make it more active in Fenton reaction.

Graphene as a single-layer carbon material has been frequently explored to prepare composites with α-Fe_2O_3 as α-Fe_2O_3/graphene composites. The large contact interface and strong interaction between graphene and α-Fe_2O_3 promote the electron transfer from α-Fe_2O_3 to graphene, which results in an enhanced photocatalytic activity. Liu et al. [50] have prepared the composite of α-Fe_2O_3 anchored on the graphene oxide (GO) nanosheet (α-Fe_2O_3/GO) (Fig. 10.2) and found that the photocatalytic activity of the composite has been enhanced, which led to approximately 2.9-fold that of classical Degussa P25 TiO_2 and 2.4-fold that of α-Fe_2O_3 for the degradation of methylene blue in Photo-Fenton reaction. Guo et al. [51] have developed a method to synthesize Fe_2O_3/GO composite at low temperature (60 °C) and found that the degradation rate of Rhodamine B and 4-nitrophenol was efficiently improved and the catalyst was potential for its good stability, little iron leaching, simple separation, stable catalytic activity, and wide pH range.

What is further reported is graphene content of the composite, which intensively influences its photocatalytic activity and other performance. Generally speaking, there is an optimal value for the composite quantity of graphene. The photocatalytic activity could be improved using graphene within a certain quantity, but it will decrease when graphene content is beyond the threshold value through enhancing absorption and scattering of photons by excess carbon content present in the composite. Han et al. [52] have explored the photocatalytic activities of the

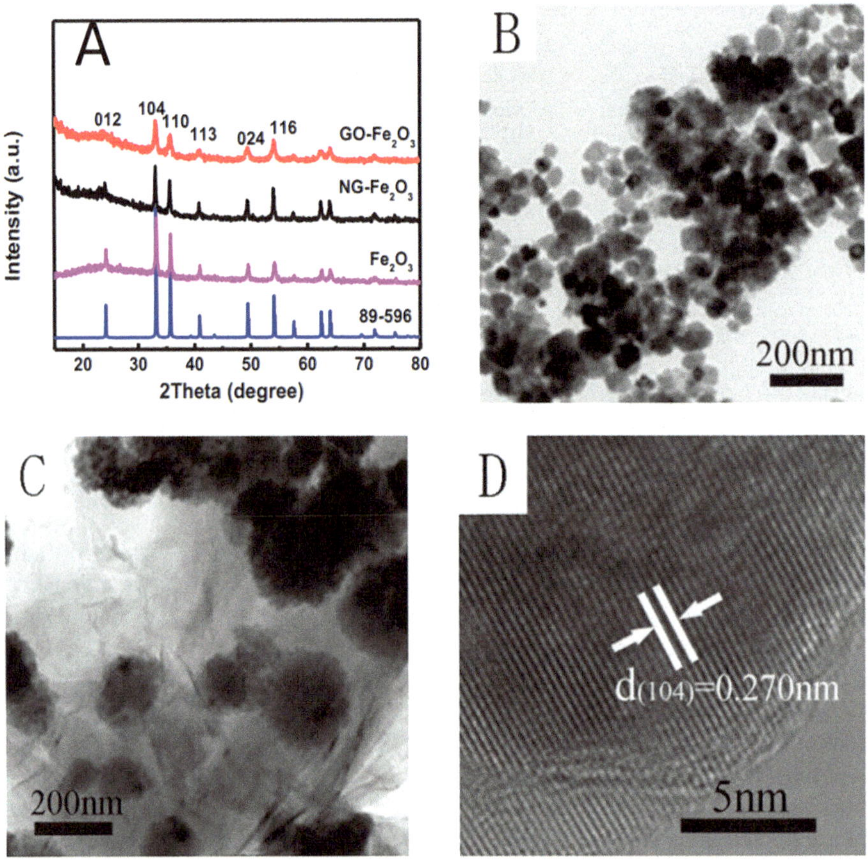

Fig. 10.2 XRD patterns (**a**), TEM image of GO-Fe_2O_3 (**b**), NG-Fe_2O_3 (**c**) and HRTEM image of NG-Fe_2O_3 (**d**). Reprinted with permission from ref. [50]. Copyright 2017, Elsevier

α-Fe_2O_3/GO composites with a serious gradient ratio of GO to α-Fe_2O_3 (0.1%, 1.0%, 2.0%, 5.0%, and 8.0%, respectively), comparing with the pure α-Fe_2O_3 nanoplates. As Fig. 10.3a shows, the absorption peak at 550 nm decreased in intensity as the time prolonged. And compared to pure α-Fe_2O_3, the value of c/c_0 decreased faster within a certain period of time when the ratio of GO increased and the composite with 5.0 wt.% of GO had the best catalytic performance shown in Fig. 10.3b and 3c, respectively. As a result, the degradation rate constant of the composite with the optimal ratio of 5 wt% graphene was almost four times faster than the pure α-Fe_2O_3 nanoplates and 98% of Rhodamine (RhB) was decomposed with 20 min of irradiation (Fig. 10.3d).

Moreover, the structure and the morphology of the graphene also affect the photocatalytic activity of the composites. A composite of α-Fe_2O_3 and graphene with N-doping has exhibited higher photocatalytic efficiency. Liu et al. [53] have reported a pyrrolic N-doped graphene oxide/Fe_2O_3 mesocrystal (NG-Fe_2O_3)

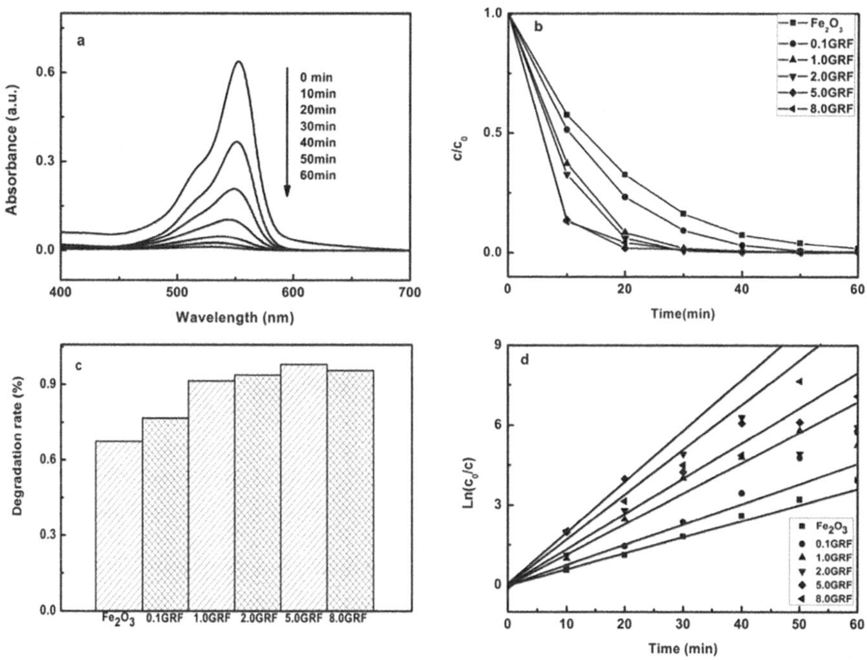

Fig. 10.3 (**a**) Time-dependent UV-vis absorption spectra in the presence of α-Fe_2O_3 under Xe light irradiation. (**b**) Photodegradation of RhB by α-Fe_2O_3 and α-Fe_2O_3/graphene composites under Xe light irradiation 60 min. (**c**) Comparison of photocatalytical performance of α-Fe_2O_3 and α-Fe_2O_3/graphene composites in 20 min. (**d**) Kinetic curves of the degradation of RhB by α-Fe_2O_3 and α-Fe_2O_3/graphene composites. Reprinted with permission from ref. [52]. Copyright 2014, Wiley

nanocomposite which showed 1.5 times higher on degrading methyl blue and 2.3 times higher on converting glyphosate. There are two reasons to explain: (1) the content of the oxygen-containing groups on GO had been adjusted and the morphology of NG-Fe_2O_3 has been changed a larger BET surface area compared to that of bare Fe_2O_3 and (2) the intimating contact between Fe_2O_3 and NG accelerated the electron transfer from Fe_2O_3 to NG so that the holes in Fe_2O_3 were increased to accept electrons from H_2O_2 to generate hydroxyl radicals in Photo-Fenton reaction. Thus, the photocatalytic efficiency was improved as observed.

Recently, Fe_2O_3/3D graphene aerogels (GAs) have been widely studied, but most of researches have been done on lithium ion batteries due to the difficulty in dispersing Fe_2O_3 particles on 3D-graphene, which limits its application in Photo-Fenton reaction. Qiu et al. [54] have solved the problem of dispersion of Fe_2O_3 particles on GAs by a modified Stöber-like method. Such 3D network structure inhibited Fe(II) loss and stabilized the conversion of Fe(III)/Fe(II) in Photo-Fenton reaction. Besides, the composite was full of elasticity and easy to be recycled shown in Fig. 10.4, and the loss of Fe(II) of Fe_2O_3/GR ordinary composite remained high in acidic solution, which led to the deactivation of the catalyst and degradation rate of organic contaminants. Thus, compared to pure Fe_2O_3 and Fe_2O_3/GR with the similar

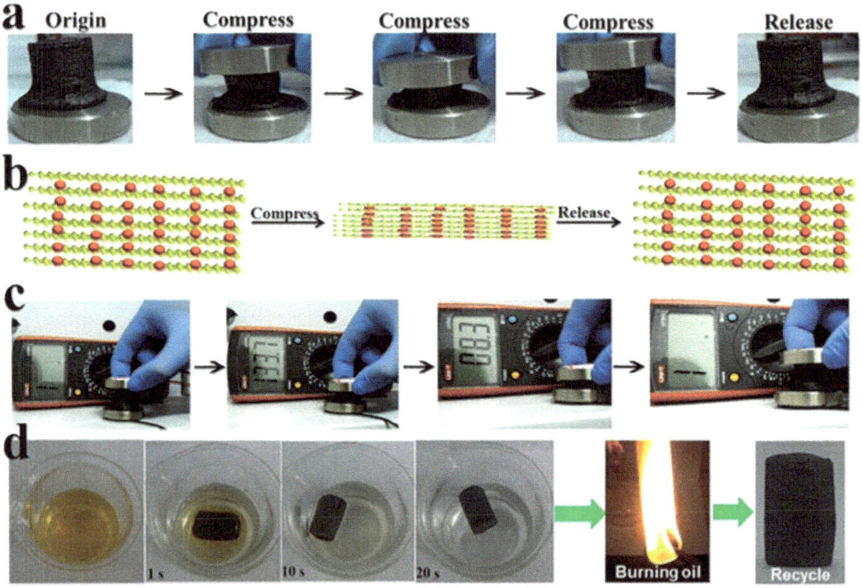

Fig. 10.4 (**a**) Compression test of Fe_2O_3/GAs. (**b**) Schematic illustration of the structure of Fe_2O_3/GAs during the compression test. (**c**) The changing impedance of Fe_2O_3/GAs during the compression process. (**d**) Sequential images of Fe_2O_3/GAs absorbing pump oil (dyed with Sudan III) on a water surface and the recycling of Fe_2O_3/GAs through burning off the oil. Reprinted with permission from ref. [54]. Copyright 2015, Royal Society of Chemistry

particle size and concentration, Fe_2O_3/GAs exhibited higher Photo-Fenton reaction activity after the several cycles.

10.2.2 Graphene/Fe_3O_4 Composite as Photocatalyst in Fenton Reaction

Fe_3O_4, a magnetic material with a structure of inverse spinal, is regarded as the most promising catalyst in Fenton-like reaction due to its decent magnetic, electric, and catalytic properties, biocompatibility, and low toxicity [55, 56] The octahedral structure contains both Fe^{2+} and Fe^{3+} and the electrons can move fast between them, allowing the Fe species to be selectively reduced or oxidized and keeping the structure invariant at the same time. In addition, the narrow bandgap (0.1 eV) of Fe_3O_4 is of great importance to electron carrier and magnetic properties of Fe_3O_4 cause it to be easily dispersed by an external magnetic field, both of which can enhance photocatalytic activity. However, the nanoscaled Fe_3O_4 particles are prone to aggregate to become larger particles that will lose its initial huge surface area and dispersibility in the aqueous solution and finally diminish the photocatalytic activity toward organic pollutants. Meanwhile, the slow conversion rate of Fe(II) and Fe(III) limits the reaction rate of Fenton process which is the control step and key procedure to enhance the catalytic capacity of Fe_3O_4 NPs [57–60].

In order to overcome these drawbacks, methods like immobilizing Fe_3O_4 nanoparticles onto support materials or encapsulating them within thin protective layer to prevent their aggregation and electron acceptors are utilized to accelerate the electron transfer. Naturally, carbon materials, especially graphene, attract the focus of chemical researchers again, which not only can be supporters to anchor nanoparticles but also acceptors to receive electrons. Several researches have reported that the in situ growth of Fe_3O_4 NPs onto graphene can not only effectively inhibit the aggregation of magnetite but also enable them to contact with each other intimately, and such structure favored the transmission of photoactive electrons from Fe_3O_4 NPs to graphene. The magnetite nanoparticles are dispersed on the GO (or rGO) sheets and immobilized by ferric or ferrous ions bonded with oxygen-containing functional group as crystal nucleuses, which will be the active sites for catalysis. Herein, the surface area of magnetite is enlarged and the surface energy-derived agglomeration of nanosized particles is prevented by graphene sheets. In return, Fe_3O_4 NPs with certain size can avoid accumulation of adjacent graphene sheets. Zhou et al. [61] have successfully synthesized GO-Fe_3O_4 composite through in situ depositing cubic-phase Fe_3O_4 on the surface of GO and proved the existence of the C–O–Fe coordination bond by FTIR. The composite showed excellent performance on photocatalytic degrading organic contaminant isatin. In another study, Qiu et al. [62] reported a simple Stöber-like method without additional reductants and organic surfactants on the synthesis of rGO–Fe_3O_4 nanocomposite, which was environmental friendly and suitable for mass production. The Fe_3O_4 NPs were ultra-dispersed on the graphene sheets during the in situ growth process and the particle sizes were well controlled at an extremely small value (3–8 nm), which was clearly seen in Fig. 10.5. The results of Photo-Fenton experiments to degrade methyl orange, methylene blue, and Rhodamine B were satisfactory due to the high surface area and fast electron transfer.

The graphene content in the magnetite-based composites also plays one of the decisive factors for the performance of photocatalyst. There is no doubt that the graphene content of the composite has an optimal value for the sake of the highest photocatalytic activity similar to combination with Fe_2O_3 above. Zubir et al. [63] have found that the beneficial intercalation of GO within Fe_3O_4 nanoparticles was 10 wt% after a series of experiments using composites with different weight ratio of GO to degrade organic compounds, which showed 20% higher degradation rate of Acid Orange 7 than that of bare Fe_3O_4 nanoparticles, as well (Fig. 10.6a). This can be explained that at high GO loading, stacking of the graphene sheets may happen through the π-π interactions which correspond to the van der Waals and hydrophobic fields around the carbon basal plane of GO sheets. Therefore, the aggregation of Fe_3O_4 NPs on the exterior surface of GO stacking might hamper the effective diffusion and contact between the reactants toward the active sites and decrease the ample formation of hydroxyl radicals to decompose AO7 during the reaction. The pH range was extended to nearly neutral condition and the cyclicity of the composite was perfect as shown in Fig. 10.6b and 6c. Likewise, Yu et al. [64]

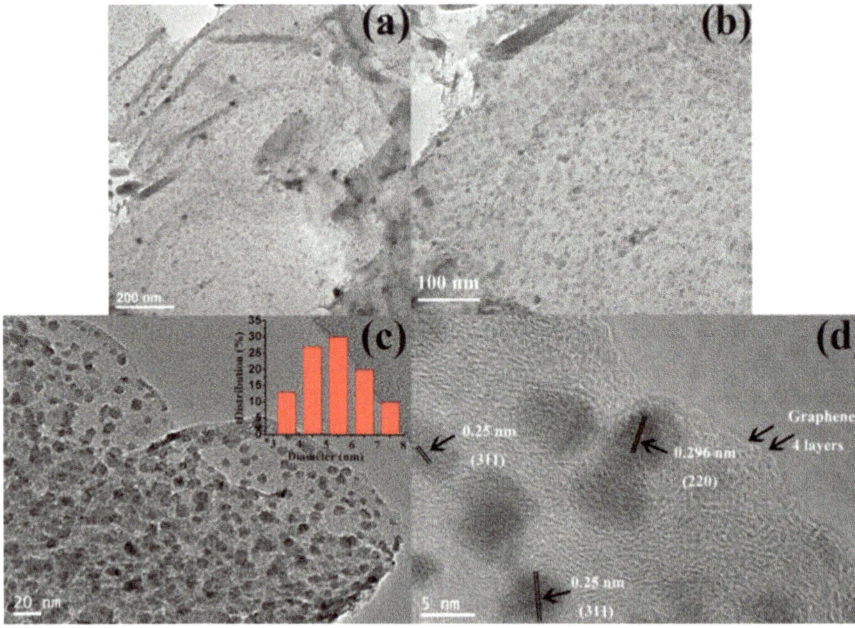

Fig. 10.5 TEM images (**a**, **b**) and HRTEM images (**c**, **d**) of Fe_3O_4/RGO composites. Inset of (**c**) is the corresponding particle size distribution of the loaded Fe_3O_4 nanoparticles derived from 100 of Fe_3O_4 nanoparticles in (**c**). Reprinted with permission from ref. [62]. Copyright 2016, Elsevier

comparatively studied the influence of the weight ratio of GO in Fe_3O_4-based composite ranging from 0 to 15 wt% on the photocatalytic activity. The result showed that the degradation rate reached the highest value when the GO content is 5 wt% in Fe_3O_4/GO composite regardless of increasing or decreasing GO content because the active sites may be covered and the contact with H_2O_2 may be hindered by superfluous GO.

In several reports, the composite of Fe_3O_4 nanoparticles combining with modified graphene shows the more efficient photocatalysis than ordinary Fe_3O_4/GO catalyst. The introduction of other functional groups or elements into graphene may facilitate the higher catalytic activity and conductivity of graphene and subsequently reinforce the capability of prohibiting the electron–hole recombination rate and adsorbing organic contaminants onto the sheets with compositing with Fe_3O_4 nanoparticles. Boruah et al. [65] synthesized AG/Fe_3O_4 composite that graphene was decorated with ammonia. The new catalyst exhibited efficient photocatalytic activity degradation of phenol, 2-nitrophenol (2-NP), and 2-chlorophenol (2-CP) and further high removal of three organic compounds under sunlight irradiation, which was attributed to the synergistic effect between AG and Fe_3O_4 NPs by preventing the recombination of electron–hole pair to enhance the catalytic performance. The synthetic route and the mechanism of degradation were presented in Fig. 10.7. Graphene could not form a good composite with Fe_3O_4 due to its hydrophobicity, so Wang et al. [66] prepared a type of hydrophilic graphene (HG) by GO reacting with

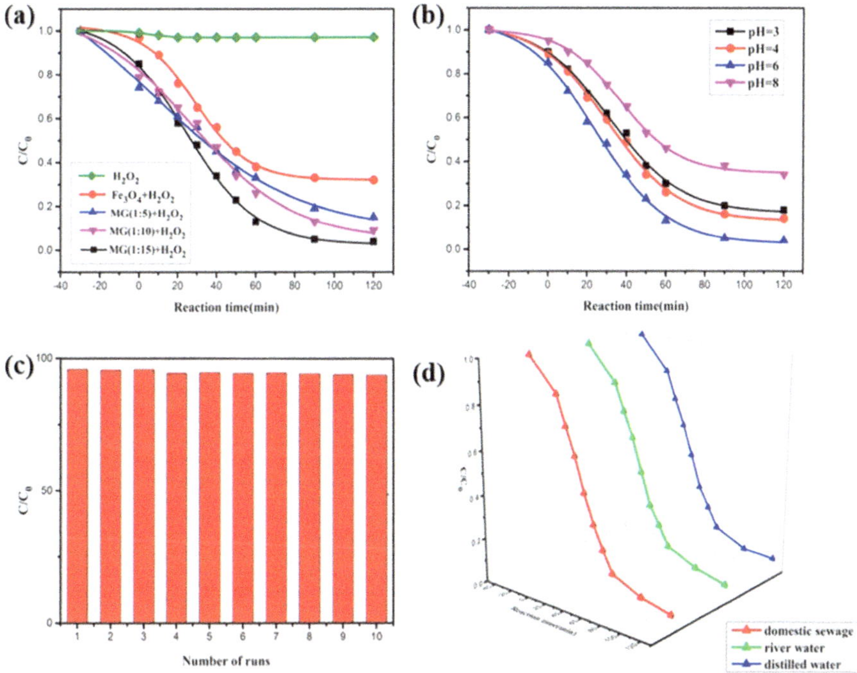

Fig. 10.6 (a) Effect of RGO content in RGO/Fe_3O_4 catalyst ($[MB]_0$ = 20 mg L^{-1}; $[H_2O_2]_0$ = 10 mmol L^{-1}; [catalyst] = 0.25 g L^{-1}; pH = 6; room temperature). (b) Degradation efficiency in different pH($[MB]_0$ = 20 mg L^{-1}; $[H_2O_2]_0$ = 10 mmol L^{-1}; [catalyst] = 0.25 g L^{-1}; room temperature). (c) Degradation rate for each run with RGO/Fe_3O_4 catalyst ($[MB]_0$ = 20 mg L^{-1}; $[H_2O_2]_0$ = 10 mmol L^{-1}; [catalyst] = 0.25 g L^{-1}; pH = 6; room temperature). (d) Degradation rate in actual water sample ($[MB]_0$ = 20 mg L^{-1}; $[H_2O_2]_0$ = 10 mmol L^{-1}; [catalyst] = 0.25 g L^{-1}; pH = 6; room temperature). Reprinted with permission from ref. [63]. Copyright 2017, Elsevier

phenylhydrazine-4-sulfonic acid to synthesize HG/Fe_3O_4 composite. The result showed that Fe_3O_4 nanoparticles were uniformly and tightly clinched onto the HG sheets, and the composites demonstrated paramagnetic characteristic, better stability in water, and higher Photo-Fenton activity.

After years of research, a variety of preparation methods of Fe_3O_4 and graphene-based catalyst have been invented in order to improve the photocatalytic efficiency in Fenton reaction, cost saving, and environmental friendly. Santhosh et al. [67] prepared G–Fe_3O_4 composite through a one-step solvothermal route, which was more convenient and less contaminative compared to two (or more)-step routes. Fe_3O_4 was proved to anchor firmly and well dispersedly on the graphene sheets in the resulting G–Fe_3O_4 composite which showed excellent performance on adsorbing heavy metal like lead ion and degrading methylene blue due to the inhibition of electron–hole recombination and more active sites for degradation. Jiang et al. [68] studied the fabrication of rGO–Fe_3O_4 nanocomposite through a creative procedure that utilized a brown alga (*Sargassum thunbergii*) as the solely reducing agent,

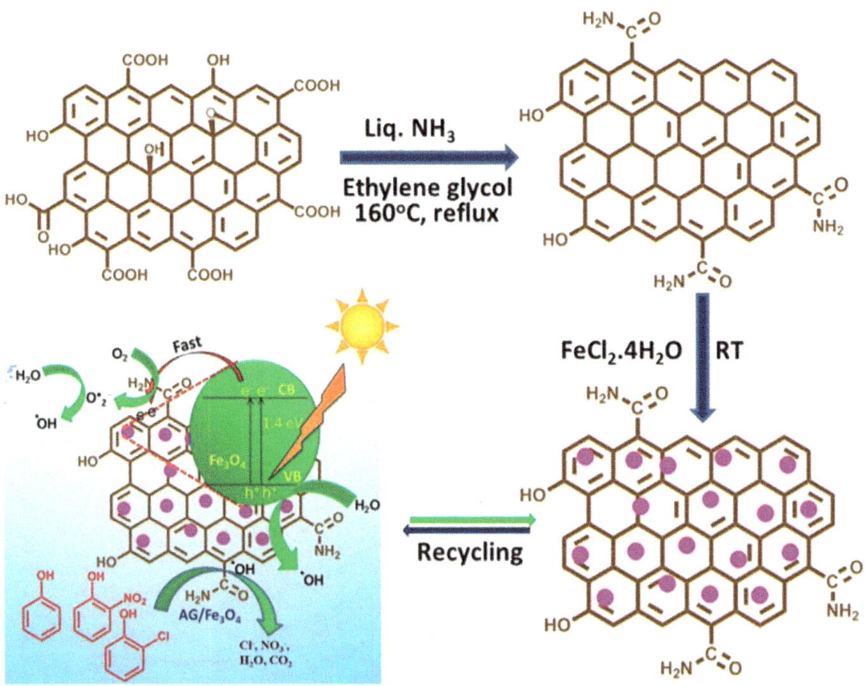

Fig. 10.7 Schematic representation for the synthesis of AG/Fe_3O_4 nanocomposite toward photocatalytic degradation of phenol, 2-NP, and 2-CP under sunlight irradiation. Reprinted with permission from ref. [65]. Copyright 2017, Elsevier

which replaced the toxic acids and amines to avoid polluting the environment. Besides, *Sargassum thunbergii* is so abundant across the coast of China that the photocatalyst can be mass prepared with low cost. This method can be regarded as the exploration on green synthetic technology and has potential of further research. The synthesized material displayed excellent catalytic performance, showing 96% degradation rate under mild conditions. Except the methods mentioned above, there are several other ideas such as hydrothermal method and co-participation method, etc., which facilitate the wide application of rGO–Fe_3O_4 as photocatalyst.

10.2.3 Graphene/Other Iron (Hydr)oxide Composite as Photocatalyst in Fenton Reaction

In addition to the above composites, other iron (hydr)oxides have also been incorporated with graphene to prepare highly effective photocatalysts for Photo-Fenton reaction. For instance, $ZnFe_2O_4$ [69], $MnFe_2O_4$ [70], and $CoFe_2O_4$ [71] with spinel structure have attracted much attention for their visible light-driven photoactivity.

Besides, α-FeOOH and β-FeOOH [48] are also explored to apply as catalyst for water purification.

$ZnFe_2O_4$ has a narrow bandgap of 1.9 eV so that it can absorb visible light up to 653 nm, which tremendously enhances the utilization of sunlight. $ZnFe_2O_4$ has great potential to be used as a high-performance photocatalyst to deal with water organic pollution owing to its high catalytic capability, low cost, and simple magnetic separation. Besides, graphene is utilized as support material and electron acceptor for $ZnFe_2O_4$. The nanosized $ZnFe_2O_4$ particles that grew on graphene have high surface area and low electron–hole recombination due to its suitable short diameter. The photo-generated electrons are rapidly transferred to graphene so that the holes are preserved to react with H_2O_2 and H_2O to generate hydroxyl radicals for decomposition of dye pollution. Furthermore, both the aggregation of $ZnFe_2O_4$ and the stack of graphene sheets were prevented. Yang et al. [72] reported a novel strategy to synthesize G-$ZnFe_2O_4$ composite through interface engineering. They used graphene as barrier to control the growth of $ZnFe_2O_4$, channels to transport photo-generated electrons, and vessels to inhibit the recombination of electrons and holes. The results showed that the G-$ZnFe_2O_4$ catalyst had an ultrafast degradation rate of methyl blue, which was 20 times higher than the previous reported spinel-based photocatalysts, 4 times higher than that of the TiO_2-based photocatalysts, and 4 times higher than those of the other photocatalysts. Fu et al. [73] prepared a magnetically separable $ZnFe_2O_4$–graphene nanocomposite photocatalyst through a facile one-step hydrothermal method (Fig. 10.8). The degradation rate of methyl blue was 88% in just 5 min and reached 99% after 90 min visible light irradiation but also concluded that compared with pure $ZnFe_2O_4$ catalyst, the as-prepared photocatalyst had a dual function of photocatalytic decomposing MB and photo-generating hydroxyl radicals. Lu and cooperators [74] obtained a $ZnFe_2O_4$–G composite with $ZnFe_2O_4$ nanocrystals simultaneously anchored on rGO by a facile one-pot solvothermal method. Similarly, the $ZnFe_2O_4$–G composite presented powerful photocatalytic activity for degradation of Rhodamine B, methyl orange, and methylene blue by Photo-Fenton reaction and could be easily recycled by an external magnetic field due to its excellent magnetism, for which the photocatalyst had the potential to be applied in water treatment.

Fig. 10.8 Images of $ZnFe_2O_4$–G(0.2) suspension with (a) and without (b) a magnetic field. Reprinted with permission from ref. [73]. Copyright 2011, American Chemical Society

In recent studies, α-FeOOH has been found to degrade some recalcitrant organic compounds at neutral to alkaline pH values under UV irradiation. It first absorbs UV light to produce electrons and holes and subsequently reacts with H_2O and H_2O_2 to produce •OH radicals. The hydroxyl radicals can also be generated by breaking up the FeO–OH bond through UV irradiation. Importantly, the incorporation of graphene and α-FeOOH enables the composite to have visible light response and absorption characteristics, and the photo-generated electrons are also transported to graphene sheets to reduce their recombination and enhance the photocatalytic activity. Wang et al. [75] prepared GO/α-FeOOH composite by reducing Fe (II) onto GO via in situ self-assembly process. They demonstrated that the as-prepared catalyst had a high activity on oxidizing phenol under visible light irradiation in a wide pH range. Liu et al. [76] synthesized α-FeOOH@GCA by incorporating GO-CNTs and α-FeOOH nanoparticles using a facile in situ hydrolysis route. Compared with pristine α-FeOOH, the composite was proved to show more effective in H_2O_2 activity and more excellent performance on degrading OII.

10.3 Conclusions and Outlook

In summary, various iron-based oxides/graphene composites have been explored to obtain more effective, environmental friendly, cheaper, and more toxic catalyst in order to solve the problem of increasingly dye pollution in water. The synergistic effect of graphene and iron (hydr)oxides semiconductors enhances the photocatalytic activity of iron (hydr)oxides to degrade a large number of water dye pollutants such as methyl blue, phenol, and so forth. The combination of the size-dependent property of nanomaterials and the excellent properties of graphene endow the composites exhibiting much more functionalities, such as large surface area, wide pH reaction range, wide sunlight response and absorption, high adsorption capacity, high speed of electron transportation, low recombination of electrons and holes, and so on. The iron (hydr)oxides have various bandgap ranging from 0.1 eV (Fe_3O_4) to 2.8 eV (FeOOH) and the potential of the minimum conductive band is more negative than the water reduction potential (H^+/H_2) or the potential of the maximum valence band is more positive than the water oxidation potential (H_2O/O_2). A variety of synthetic methods have been studied to produce iron (hydr)oxides-based composites. However, there are still some problems needed to be overcome for large-scale application. For instance, it is difficult to obtain high purified graphene with precisely controlling the defects and defect sites. The recombination of electrons and holes and restacking of graphene to graphite are not completely solved. It still needs to be further researched to find new methods to overcome the problems mentioned above.

The exploration of new catalysts and treatment technique of the organic pollutants in wastewater is an exciting and difficult work for scientific researchers. Considering the recycling of catalysts, magnetic materials may attract more attention due to that they not only possess the catalytic performance but also can be magnetic separated by an external magnetic field, which makes them better for recycling and reduces the secondary pollution to the environment. Besides, graphene-based materials are promising and potential for water purification and also considered as future materials, but there are still lots of uncertainties to completely understand its properties and phenomenon. Water quality is quite important for human health. Therefore, Fe_2O_3, Fe_3O_4, and other iron oxides coupled with graphene are discussed and reviewed in this article to summarize the former research achievements in related fields and also to understand the properties and performance of these composites.

References

1. Speth DR, In't Zandt MH, Guerrero-Cruz S et al (2016) Genome-based microbial ecology of anammox granules in a full-scale wastewater treatment system. Nat Commun 7:11172
2. Gago-Ferrero P, Schymanski EL, Bletsou AA et al (2015) Extended suspect and non-target strategies to characterize emerging polar organic contaminants in raw wastewater with LC-HRMS/MS. Environ Sci Technol 49(20):12333–12341
3. Ojha DP, Joshi MK, Kim HJ (2017) Photo-Fenton degradation of organic pollutants using a zinc oxide decorated iron oxide/reduced graphene oxide nanocomposite. Ceram Int 43 (1):1290–1297
4. Hoffmann MR, Martin ST, Choi W et al (1995) Environmental applications of semiconductor photocatalysis. Chem Rev 95(1):69–96
5. Legrini O, Oliveros E, Braun AM (1993) Photochemical processes for water treatment. Chem Rev 93(2):671–698
6. Wu K, Xie Y, Zhao J et al (1999) Photo-Fenton degradation of a dye under visible light irradiation. J Mol Catal A-Chem 144(1):77–84
7. Neyens E, Baeyens J (2003) A review of classic Fenton's peroxidation as an advanced oxidation technique. J Hazard Mater 98(1):33–50
8. Tian S, Tu Y, Chen D et al (2011) Degradation of Acid Orange II at neutral pH using Fe2 (MoO4)3 as a heterogeneous Fenton-like catalyst. Chem Eng J 169(1):31–37
9. Feng J, Hu X, Yue P (2004) Discoloration and mineralization of Orange II using different heterogeneous catalysts containing Fe: a comparative study. Environ Sci Technol 38 (21):5773–5778
10. Arnold SM, Hickey WJ, Harris RF (1995) Degradation of atrazine by Fenton's reagent: condition optimization and product quantification. Environ Sci Technol 29(8):2083–2089
11. Herney-Ramirez J, Vicente MA, Madeira LM (2010) Heterogeneous photo-Fenton oxidation with pillared clay-based catalysts for wastewater treatment: a review. Appl Catal B-Environ 98 (1):10–26
12. Pouran SR, Aziz ARA, Daud WMAW (2015) Review on the main advances in photo-Fenton oxidation system for recalcitrant wastewaters. J Ind Eng Chem 21:53–69
13. Babuponnusami A, Muthukumar K (2014) A review on Fenton and improvements to the Fenton process for wastewater treatment. J Environ Chem Eng 2(1):557–572
14. Dhakshinamoorthy A, Navalon S, Alvaro M et al (2012) Metal nanoparticles as heterogeneous Fenton catalysts. ChemSusChem 5(1):46–64
15. Li D, Yuranova T, Albers P et al (2004) Accelerated photobleaching of Orange II on novel (H5FeW12O40•10H2O)/silica structured fabrics. Water Res 38(16):3541–3550
16. Sabhi S, Kiwi J (2001) Degradation of 2, 4-dichlorophenol by immobilized iron catalysts. Water Res 35(8):1994–2002
17. Hu X, Liu B, Deng Y et al (2011) Adsorption and heterogeneous Fenton degradation of 17α-methyltestosterone on nano Fe3O4/MWCNTs in aqueous solution. Appl Catal B-Environ 107(3):274–283

18. Timofeeva MN, Hasan Z, Orlov AY et al (2011) Fe-containing nickel phosphate molecular sieves as heterogeneous catalysts for phenol oxidation and hydroxylation with H2O2. Appl Catal B-Environ 107(1):197–204
19. Song Z, Wang N, Zhu L et al (2012) Efficient oxidative degradation of triclosan by using an enhanced Fenton-like process. Chem Eng J 198:379–387
20. Novoselov KS, Geim AK, Morozov SV et al (2004) Electric field effect in atomically thin carbon films. Science 306(5696):666–669
21. Nika DL, Pokatilov EP, Askerov AS et al (2009) Phonon thermal conduction in graphene: role of Umklapp and edge roughness scattering. Phys Rev B 79(15):155413
22. Mayorov AS, Gorbachev RV, Morozov SV et al (2011) Micrometer-scale ballistic transport in encapsulated graphene at room temperature. Nano Lett 11(6):2396–2399
23. Li X, Zhu Y, Cai W et al (2009) Transfer of large-area graphene films for high-performance transparent conductive electrodes. Nano Lett 9(12):4359–4363
24. Wang H, Cui LF, Yang Y et al (2010) Mn3O4-graphene hybrid as a high-capacity anode material for lithium ion batteries. J Am Chem Soc 132(40):13978–13980
25. Yoo EJ, Kim J, Hosono E et al (2008) Large reversible Li storage of graphene nanosheet families for use in rechargeable lithium ion batteries. Nano Lett 8(8):2277–2282
26. Liang M, Zhi L (2009) Graphene-based electrode materials for rechargeable lithium batteries. J Mater Chem 19(33):5871–5878
27. Wang X, Zhi L, Müllen K (2008) Transparent, conductive graphene electrodes for dye-sensitized solar cells. Nano Lett 8(1):323–327
28. Yin Z, Wu S, Zhou X et al (2010) Electrochemical deposition of ZnO nanorods on transparent reduced graphene oxide electrodes for hybrid solar cells. Small 6(2):307–312
29. Wu J, Becerril HA, Bao Z et al (2008) Organic solar cells with solution-processed graphene transparent electrodes. Appl Phys Lett 92(26):237
30. Schedin F, Geim AK, Morozov SV et al (2007) Detection of individual gas molecules adsorbed on graphene. Nat Mater 6(9):652–655
31. Shao Y, Wang J, Wu H et al (2010) Graphene based electrochemical sensors and biosensors: a review. Electroanalysis 22(10):1027–1036
32. Shan C, Yang H, Song J et al (2009) Direct electrochemistry of glucose oxidase and biosensing for glucose based on graphene. Anal Chem 81(6):2378–2382
33. An J, Zhu L, Wang N et al (2013) Photo-Fenton like degradation of tetrabromobisphenol A with graphene BiFeO3 composite as a catalyst. Chem Eng J 219:225–237
34. Zhou Q, Xing A, Li J et al (2016) Synergistic enhancement in photoelectrocatalytic degradation of bisphenol A by CeO2 and reduced graphene oxide co-modified TiO2 nanotube arrays in combination with Fenton oxidation. Electrochim Acta 209:379–388
35. Qiu B, Deng Y, Du M et al (2016) Ultradispersed cobalt ferrite nanoparticles assembled in graphene aerogel for continuous photo-Fenton reaction and enhanced lithium storage performance. Sci Rep-UK 6:29099
36. Babu SG, Vinoth R, Neppolian B et al (2015) Diffused sunlight driven highly synergistic pathway for complete mineralization of organic contaminants using reduced graphene oxide supported photocatalyst. J Hazard Mater 291:83–92
37. Wan Z, Hu J, Wang J (2016) Removal of sulfamethazine antibiotics using Ce Fe-graphene nanocomposite as catalyst by Fenton-like process. J Environ Manag 182:284–291
38. Wang Y, Yu Y, Deng C et al (2015) Synthesis of mesoporous MCM-41 supported reduced graphene oxide-Fe catalyst for heterogeneous Fenton degradation of phenol. RSC Adv 5 (126):103989–103998
39. Yun S, Lee YC, Park HS (2016) Phase-controlled iron oxide nanobox deposited on hierarchically structured graphene networks for lithium ion storage and photocatalysis. Sci Rep 6:19959
40. Zhou Y, Xiao B, Liu SQ et al (2016) Photo-Fenton degradation of ammonia via a manganese–iron double-active component catalyst of graphene–manganese ferrite under visible light. Chem Eng J 283:266–275
41. He H, He Z, Shen Q (2017) Photocatalysis of novel reduced graphene oxide-CoSe nanocomposites with efficient interface-induced effect. Compo Interf 24(1):85–97

42. Lee C, Wei X, Kysar JW et al (2008) Measurement of the elastic properties and intrinsic strength of monolayer graphene. Science 321(5887):385–388
43. Zheng Y, Wei N, Fan Z et al (2011) Mechanical properties of grafold: a demonstration of strengthened graphene. Nanotechnology 22(40):405701
44. Geim AK, MacDonald AH (2007) Graphene: exploring carbon flatland. Phys Today 60 (8):35–41
45. Liu Q, Guo Y, Chen Z et al (2016) Constructing a novel ternary Fe (III)/graphene/gC3N4 composite photocatalyst with enhanced visible-light driven photocatalytic activity via interfacial charge transfer effect. Appl Catal B-Environ 183:231–241
46. Zubir NA, Motuzas J, Yacou C et al (2016) Graphene oxide with zinc partially substituted magnetite (GO–Fe1−xZnxOy) for the UV-assisted heterogeneous Fenton-like reaction. RSC Adv 6(50):44749–44757
47. Lee YC, Chang SJ, Choi MH et al (2013) Self-assembled graphene oxide with organo-building blocks of Fe-aminoclay for heterogeneous Fenton-like reaction at near-neutral pH: a batch experiment. Appl Catal B-Environ 142:494–503
48. Xiao F, Li W, Fang L et al (2016) Synthesis of akageneite (beta-FeOOH)/reduced graphene oxide nanocomposites for oxidative decomposition of 2-chlorophenol by Fenton-like reaction. J Hazard Mater 308:11–20
49. Mishra M, Chun DM (2015) α-Fe2O3 as a photocatalytic material: a review. Appl Catal A-Gen 498:126–141
50. Liu B, Tian L, Wang R et al (2017) Pyrrolic-N-doped graphene oxide/Fe2O3 mesocrystal nanocomposite: efficient charge transfer and enhanced photo-Fenton catalytic activity. Appl Surf Sci 2017 412:207–213
51. Guo S, Zhang G, Guo Y et al (2013) Graphene oxide-Fe2O3 hybrid material as highly efficient heterogeneous catalyst for degradation of organic contaminants. Carbon 60:437–444
52. Han S, Hu L, Liang Z et al (2014) One-step hydrothermal synthesis of 2D hexagonal nanoplates of α-Fe2O3/graphene composites with enhanced photocatalytic activity. Adv Funct Mater 24 (36):5719–5727
53. Liu Y, Jin W, Zhao Y et al (2017) Enhanced catalytic degradation of methylene blue by α-Fe2O3/graphene oxide via heterogeneous photo-Fenton reactions. Appl Catal B-Environ 206:642–652
54. Qiu B, Xing M, Zhang J (2015) Stöber-like method to synthesize ultralight, porous, stretchable Fe2O3/graphene aerogels for excellent performance in photo-Fenton reaction and electrochemical capacitors. J Mater Chem A 3(24):12820–12827
55. Santhosh C, Kollu P, Doshi S et al (2014) Adsorption, photodegradation and antibacterial study of graphene-Fe3O4 nanocomposite for multipurpose water purification application. RSC Adv 4 (54):28300–28308
56. Shen J, Li Y, Zhu Y et al (2016) Aerosol synthesis of Graphene-Fe3O4 hollow hybrid microspheres for heterogeneous Fenton and electro-Fenton reaction. J Environ Chem Eng 4 (2):2469–2476
57. Hua Y, Wang S, Xiao J et al (2017) Preparation and characterization of Fe3O4/gallic acid/ graphene oxide magnetic nanocomposites as highly efficient Fenton catalysts. RSC Adv 7 (46):28979–28986
58. Zubir NA, Yacou C, Zhang X et al (2014) Optimisation of graphene oxide–iron oxide nanocomposite in heterogeneous Fenton-like oxidation of Acid Orange 7. J Environ Chem Eng 2(3):1881–1888
59. Chang YH, Yao YF, Luo H et al (2014) Magnetic Fe3O4-GO nanocomposites as highly efficient Fenton-like catalyst for the degradation of dyes. Int J Nanomanuf 10(1–2):132–141
60. Liu W, Qian J, Wang K et al (2013) Magnetically separable Fe3O4 nanoparticles-decorated reduced graphene oxide nanocomposite for catalytic wet hydrogen peroxide oxidation. J Inorg Organomet P 23(4):907–916
61. Zhou Z, Su M, Shih K (2017) Highly efficient and recyclable graphene oxide-magnetite composites for isatin mineralization. J Alloy Compd 725:302–309

62. Qiu B, Li Q, Shen B et al (2016) Stöber-like method to synthesize ultradispersed Fe3O4 nanoparticles on graphene with excellent Photo-Fenton reaction and high-performance lithium storage. Appl Catal B-Environ 183:216–223
63. Jiang X, Li L, Cui Y et al (2017) New branch on old tree: green-synthesized RGO/Fe3O4 composite as a photo-Fenton catalyst for rapid decomposition of methylene blue. Ceram Int 43 (16):14361–14368
64. Yu L, Chen J, Liang Z et al (2016) Degradation of phenol using Fe3O4-GO nanocomposite as a heterogeneous photo-Fenton catalyst. Sep Purif Technol 171:80–87
65. Boruah PK, Sharma B, Karbhal I et al (2017) Ammonia-modified graphene sheets decorated with magnetic Fe3O4 nanoparticles for the photocatalytic and photo-Fenton degradation of phenolic compounds under sunlight irradiation. J Hazard Mater 325:90–100
66. Wang P, Wang L, Sun Q et al (2016) Preparation and performance of Fe3O4@hydrophilic graphene composites with excellent Photo-Fenton activity for photocatalysis. Mater Lett 183:61–64
67. Yang B, Tian Z, Zhang L et al (2015) Enhanced heterogeneous Fenton degradation of Methylene Blue by nanoscale zero valent iron (nZVI) assembled on magnetic Fe3O4/reduced graphene oxide. J Water Process Eng 5:101–111
68. Zubir NA, Yacou C, Motuzas J et al (2014) Structural and functional investigation of graphene oxide-Fe3O4 nanocomposites for the heterogeneous Fenton-like reaction. Sci Rep 4:4594
69. He H, Huang J, Lu J (2016) Photo-and Photo-Fenton-like catalytic degradations of Malachite Green in a water using magnetically separable ZnFe2O4-reduced graphene oxide hybrid nanostructures. J Sci Res Rep 10(2):1–11
70. Yao Y, Cai Y, Lu F et al (2014) Magnetic recoverable MnFe2O4 and MnFe2O4-graphene hybrid as heterogeneous catalysts of peroxymonosulfate activation for efficient degradation of aqueous organic pollutants. J Hazard Mater 270:61–70
71. Yao Y, Yang Z, Zhang D et al (2012) Magnetic CoFe2O4-graphene hybrids: facile synthesis, characterization, and catalytic properties. Ind Eng Chem Res 51(17):6044–6051
72. Yang D, Feng J, Jiang L et al (2015) Photocatalyst interface engineering: spatially confined growth of ZnFe2O4 within graphene networks as excellent visible-light-driven photocatalysts. Adv Funct Mater 25(45):7080–7087
73. Fu Y, Wang X (2011) Magnetically separable ZnFe2O4-graphene catalyst and its high photocatalytic performance under visible light irradiation. Ind Eng Chem Res 50 (12):7210–7218
74. Lu D, Zhang Y, Lin S et al (2013) Synthesis of magnetic ZnFe2O4/graphene composite and its application in photocatalytic degradation of dyes. J Alloy Compd 579:336–342
75. Wang Y, Fang J, Crittenden JC et al (2017) Novel RGO/α-FeOOH supported catalyst for Fenton oxidation of phenol at a wide pH range using solar-light-driven irradiation. J Hazard Mater 329:321–329
76. Liu Y, Liu X, Zhao Y et al (2017) Aligned α-FeOOH nanorods anchored on a graphene oxide-carbon nanotubes aerogel can serve as an effective Fenton-like oxidation catalyst. Appl Catal B-Environ 213:74–86

Chapter 11
Photo-Fenton Reaction

Fenton reaction was named after Henry J. Fenton, who first reported the activation of H_2O_2 by Fe^{2+} for the oxidization of tartaric acid [1]. The reaction system has now expanded to heterogeneous system based on metal or metal oxide and versatile peroxides including H_2O_2, persulfate, monopersulfate, etc. It has become the most useful method for the degradation of organic pollutants in aqueous solution.

The classical mechanism of Fenton process is a simple redox reaction, where Fe^{2+} is oxidized to Fe^{3+} and H_2O_2 is reduced to hydroxide ion (OH^-), simultaneously forming the hydroxyl radical ($HO^•$) with strong oxidation potential ($E^0 = 2.73$ V vs NHE, Eq. 11.1). In the conventional Fenton reaction, the ferric ion produced in Eq. 11.1 can be reduced back to ferrous ion by H_2O_2 (Eq. 11.2). The Fe^{3+} reduction (0.02 M^{-1} s^{-1}, Eq. 11.2) is much slower than the Fe^{2+} oxidation (40–80 M^{-1} s^{-1}, Eq. 11.1), which causes the slowing down of the Fenton reaction and requires the great amount of Fe^{2+} in the initial step to achieve the complete mineralization of organics [2]. Hydroxyl radical can be scavenged by H_2O_2 (Eq. 11.3) and Fe^{2+} (Eq. 11.4).

$$Fe^{2+} + H_2O_2 \rightarrow Fe^{3+} + HO^{\bullet} + OH^- \quad (11.1)$$

$$Fe^{3+} + H_2O_2 + H_2O \rightarrow Fe^{2+} + H_3O^+ + HO_2{}^{\bullet -} \quad (11.2)$$

$$HO^{\bullet} + H_2O_2 \rightarrow HO_2{}^{\bullet -} + H_2O \quad (11.3)$$

$$HO^{\bullet} + Fe^{2+} \rightarrow Fe^{3+} + OH^- \quad (11.4)$$

For the degradation of organic molecules, the optimum pH for the Fenton reaction is typically in the range of 2–3 and the optimum mass ratio of iron to H_2O_2 is ca. 1.5. Due to the slow regeneration of Fe^{2+}, a great amount of iron ion needs to be used, thus resulting in the large-scale sludge generation from iron and poor reusability of the catalyst. It was later realized that Fenton reactions were markedly accelerated by light (photo-Fenton reactions) concerning the reaction speed and the mineralization degree of organics. This chapter discusses the complex mechanism of photo-Fenton

J. Zhang et al., *Photocatalysis*, Lecture Notes in Chemistry 100,
https://doi.org/10.1007/978-981-13-2113-9_11

reactions and the important factors influencing the efficiency of these processes in applications to wastewater treatment. An effort is made here to inform the reader on the current state of the art, the remaining challenges for the development of these processes.

11.1 Homogeneous Photo-Fenton Reaction

Polychromic UV light was first utilized to assist Fenton process ($FeCl_2/H_2O_2$), where the photolysis of H_2O_2 in the absence of Fe_2Cl was used as a control [3]. Decomposition of 4-nitrophenol was observed in both cases, and the better result was obtained from UV light-assisted Fenton system, which was ascribed to the photo-triggered reduction of Fe^{3+} to Fe^{2+}. The concept of "photo-Fenton reactions" was formally proposed in 1992 [4], where the yield of $HO^{\bullet}$ formed per Fe^{2+} oxidized by H_2O_2 was determined as a function of pH. In this case, Fe^{2+} was produced from the photoreduction of Fe^{3+} ($\lambda = 436$ nm) in order to avoid artifacts that can result from locally high concentrations of the reagents that are inherent in the initial stages of the mixing process. The photo-generated Fe^{2+} and its oxalate, citrate, and phosphate complexes react with H_2O_2 efficiently, producing $HO^{\bullet}$ in aqueous solutions with pH ranging from 3 to 8.

The study on the oxidation of reactive black 5 (RB5) in aqueous solution demonstrated that both Fenton (H_2O_2/Fe^{2+}) and photo-Fenton (H_2O_2/Fe^{2+}/UV) processes can effectively decolorize RB5 with a little difference between them [5]. In contrast, there is a significant increment for TOC removal through photo-Fenton process (46.4%) compared to Fenton process (21.6%). This fact indicates that although UV low-pressure mercury lamp has little effect on dye decolorization, it is particularly important in dye mineralization.

The cost associated with the use of artificial irradiation sources has hindered the industrial application of the photo-Fenton process using Fe^{2+}, H_2O_2, and UV irradiation as a source of $HO^{\bullet}$ for the oxidation of organics. Solar radiation for the photodegradation of raw gasoline in water has been studied to reduce the cost. The photo-Fenton process was also applied to a real effluent, i.e., oil field-produced water, and the experimental results demonstrate the feasibility of employing solar irradiation to degrade this complex saturated-hydrocarbon-containing system [6].

Besides organics, the inactivation of the coliphage MS2 (a human virus indicator) by solar photo-Fenton was evaluated at near-neutral pH in carbonate buffer solution matrix. The effects of reactant concentration (H_2O_2, Fe^{2+}, Fe^{3+}) and solar irradiance on the photo-Fenton process were studied. Specifically, the solar exposure/Fe^{3+} treatment showed a strong dependence on the iron concentration and solar irradiance intensity. The MS2 inactivation observed with the photo-Fenton process (solar exposure/$H_2O_2/Fe^{2+/3+}$) carried out with Fe^{3+} was faster than with Fe^{2+} (detection limit achieved at 20 min and 50 min, respectively). Moreover, virus inactivation by photo-Fenton under different solar irradiance values (15, 30, and 45 W m^{-2}), H_2O_2 and Fe^{3+} concentrations (0.1, 0.5, and 1 mg L^{-1}), and different pH values (6, 7, and 8) was evaluated. In order to validate the efficiency of the photo-Fenton process at near-neutral pH on virus inactivation, the photocatalytic treatment was carried out in natural water (Lake Geneva, Switzerland) and with a human virus (echovirus). Finally, a conceptual mechanism interpretation was proposed regarding how solar photo-Fenton acts on viruses in water, involving the key species Fe^{2+}, Fe^{3+}, and H_2O_2, solar irradiance, organic matter, and their possible reactions and interactions, where the probable virus adsorption onto iron particles has a significant effect on the inactivation efficiency of MS2 coliphage by photo-Fenton treatment at near-neutral pH [7].

11.2 Heterogeneous Photo-Fenton Reaction

Many homogeneous catalysts have been reported as efficient toward the Fenton reaction. However, the main drawbacks of homogeneously catalyzed Fenton reactions include the need for large amounts of transition metal salts, the demand for strong acidic system (pH = 3–4), the subsequent management of massive sludge, as well as the acidification of the effluents before the reaction, and the neutralization of treated solutions before disposal. Heterogeneous catalysts offer the advantage of allowing easier recycling from the effluent.

11.2.1 Membrane and Fabrics

Nafion film contains $-SO_3Na$ group with Na ions replaceable by Fe^{3+} through a simple ion exchange reaction. For example, Fe^{3+}/Nafion catalyst was used for the photo-Fenton degradation of Orange II in the presence of H_2O_2/visible light [8]. However, it should be stressed that the Nafion film-based catalyst is too expensive to be used for a practical application even though the catalyst can be separated easily from solution. Yuranova et al. immobilized Fe^{3+} ions on carbon fabric for the discoloration of Orange II under visible light irradiation [9]. It was observed that surface iron oxide and iron hydroxide are the catalytically active sites on carbon fabric during the reaction. Bozzi et al. reported the immobilization of Fe^{3+} on silica fabrics to degrade oxalates under visible light irradiation [10]. They observed the existence of Fe^{3+}/Fe^{2+} mixed complex with oxalic acid on the surface of the catalyst with time and reported the formation and disappearance of Fe–carboxylates by infrared attenuated total reflection (IRATR) Fourier transform spectroscopy and concomitant recycling of the resulting Fe^{3+} back to the catalyst surface.

11.2.2 Resin

Ma et al. reported the immobilization of Fe^{3+} on cationic exchange resin (Amberlite IRA200) as a photocatalyst for the degradation of organic pollutants in aqueous H_2O_2, and XPS studies revealed the presence of mixed Fe^{3+}/Fe^{2+} ionic states before and after the reaction [11]. Li et al. explored the supported $(Fe(bpy)_3)^{2+}$ complex on cationic exchange resin as a photocatalyst for the degradation of organic pollutants by activating with molecular oxygen under visible light irradiation and suggested the involvement of active superoxide species ($Fe(bpy)_3{}^{3+}O_2$) by EPR studies [12].

11.2.3 Clay Base

An efficient Fe_2O_3-pillared rectorite (Fe-R) clay was successfully developed as a heterogeneous catalyst for photo-Fenton degradation of organic contaminants. X-ray diffraction analysis and high-resolution transmission electron microscope analysis clearly showed the existence of the Fe_2O_3 nanoparticles in the Fe-R catalyst. The catalytic activity of the Fe-R catalyst was evaluated by the discoloration and chemical oxygen demand (COD) removal of an azo dye Rhodamine B (RhB, 100 mg/L) and a typical persistent organic pollutant 4-nitrophenol (4-NP, 50 mg/L) in the presence of hydrogen peroxide (H_2O_2) under visible light irradiation ($\lambda > 420$ nm). It was found that the discoloration rate of the two contaminants was over 99.3%, and the COD removal rate of the two contaminants was over 87.0%. The Fe-R catalyst showed strong adsorption for the RhB in the aqueous solution. Moreover, the Fe-R catalyst still showed good stability for the degradation of RhB after five recycles [13].

Discoloration and mineralization of Reactive Red HE-3B were studied by using a laponite clay-based Fe nanocomposite (Fe-Lap-RD) as a heterogeneous catalyst in the presence of H_2O_2 and UV light [14]. Fe-Lap-RD mainly consists of Fe_2O_3 (maghemite) and $Fe_2Si_4O_{10}(OH)_2$ (iron silicate hydroxide) which have tetragonal and monoclinic structures, respectively, and has a high specific surface area (472 m^2/g) as well as a high total pore volume (0.547 cm^3/g). It was observed that discoloration of HE-3B undergoes a much faster kinetics than mineralization of HE-3B. It was also found that initial HE-3B concentration, H_2O_2 concentration, UV light wavelength and power, and Fe-Lap-RD catalyst loading are the four main factors that can significantly influence the mineralization of HE-3B. At optimal conditions, complete discoloration of 100 mg/L HE-3B can be achieved in 30 min and the total organic carbon removal ratio can be attained 76% in 120 min, illustrating that Fe-Lap-RD has a high photocatalytic activity in the photo-assisted UV light (254 nm) and H_2O_2.

Iron–montmorillonite (Fe–Mt) with delaminated structures was synthesized via the introduction of iron oxides into Na–montmorillonite [15]. Fe–Mt showed significant increases in the available iron content, surface area, and pore volume, along with a slight increase in the basal spacing from $d_{001} = 1.26$ (Na–Mt) to 1.53 nm (Fe–Mt). The Fenton process was efficient for phenol removal using Fe–Mt as a catalyst under visible light irradiation, and the process had two-stage pseudo-first-order kinetics. The overall reaction had a higher degradation rate even when it was only irradiated with visible light for the first 40 min period. Further investigation confirmed that the irradiation increased the presence of certain intermediates. Among them, 1, 4-benzoquinone, hydroquinone, and catechol all enhanced the Fenton reaction rates. Either catechol or hydroquinone was added to the Fenton system with Fe–Mt/H_2O_2 with or without visible light irradiation, and they both accelerated phenol degradation because catechol and hydroquinone are capable of reductively and effectively transforming Fe^{3+} into Fe^{2+}. The concentrations of dissolved total Fe increased with the increase in the oxalic acid concentration, which can strongly chelate Fe^{3+}. Hence, iron was released from Fe–Mt, and reductive transformation played an important role in promoting the Fenton reaction process for phenol removal.

11.2.4 Porous Carrier

Bossmann et al. investigated the heterogeneous photoenhanced Fenton oxidation of PVA and found that contrary to the homogeneous reaction mechanism [16], the degradation of PVA using the system zeolite Y/Fe^{3+}/H_2O_2 generates low molecular weight reaction products because DOC removal remains incomplete after 120 min reaction. In addition, they confirmed that Fe^{3+} does not form complexes with PVA and its oxidation products. Most likely, the Fe^{3+} remains bound inside the zeolite Y framework. However, it should be noted that a heterogeneous catalyst for photo-Fenton reaction with a high efficiency as well as an acceptable cost has not been fully established. Hence, there is no doubt that to develop such a catalyst has not only academic significance but also industrial applications.

Photo-Fenton-like processes using two types of Fe-zeolites (Fe-ZSM5 and Fe-Beta) as heterogeneous catalysts were carried out in order to treat contaminated effluents with organic compounds at neutral pH [17]. It was proved that light (solar and artificial) improves significantly the DOC removal in this kind of processes. A possible contribution by homogeneous photo-Fenton reaction catalyzed by the iron leached during the reaction was insignificant. This study also proves that the catalytic activity of Fe-zeolites is improved by photo-Fenton-like processes using solar light in a pilot plant equipped with compound parabolic collectors.

A novel α-FeOOH/mesoporous carbon (α-FeOOH/MesoC) composite prepared by in situ crystallization of adsorbed ferric ions within carboxyl functionalized mesoporous carbon was developed as a novel visible light-assisted heterogeneous Fenton-like catalyst [18]. The visible light active α-FeOOH nanocrystals were encapsulated in the mesoporous frameworks accompanying with surface attached large α-FeOOH microcrystals via C–O–Fe bonding. Assisting with visible light irradiation on α-FeOOH/MesoC, the mineralization efficiency increased owing to

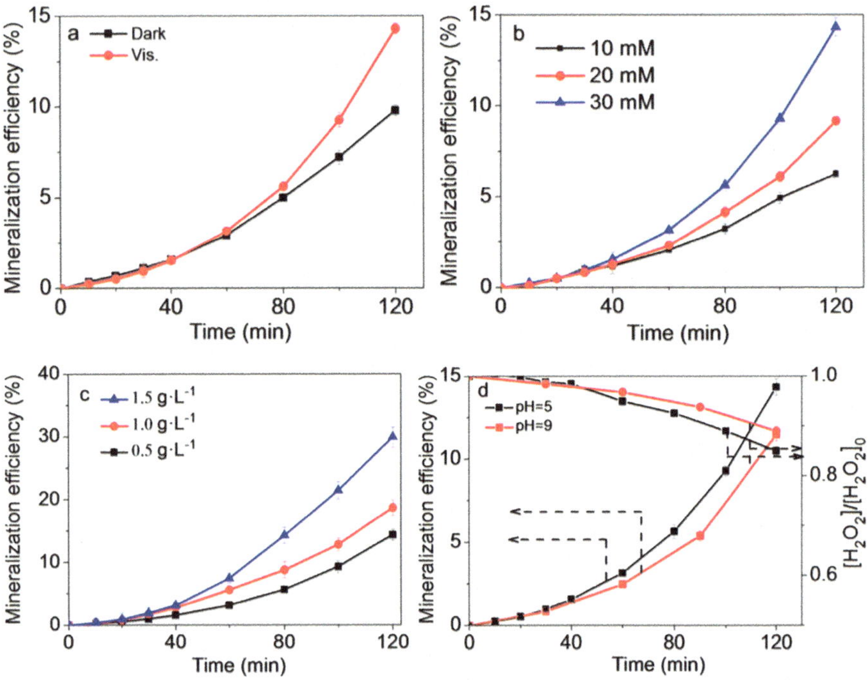

Fig. 11.1 Mineralization efficiency of phenol on the α-FeOOH/MesoC in dark and under visible light irradiation, respectively (**a**); Effects of H_2O_2 dosage, catalyst concentration, and pH value on phenol mineralization and the H_2O_2 consumption ratio in the α-FeOOH/MesoC suspension (**b**–**d**). Experimental conditions: 100 mg/L phenol, 0.5 g/L catalyst, initial pH of 5, 30 mM H_2O_2, and 45 °C. Reprinted with the permission from ref. [18]. Copyright 2011 American Chemical Society

the photocatalytic promoted catalyzing H_2O_2 beyond the photothermal effect (Fig. 11.1). The synergistic effect between α-FeOOH and MesoC in α-FeOOH/MesoC composite improved the mineralization efficiency than the mixture catalyst of α-FeOOH and MesoC. The iron leaching is greatly suppressed on the α-FeOOH/MesoC composite. Interestingly, the reused α-FeOOH/MesoC composites showed much higher phenol oxidation and mineralization efficiencies than the fresh catalyst and homogeneous Fenton system ($FeSO_4/H_2O_2$). The XPS, XRD, FTIR, and textural property results reveal that the great enhancement comes from the interfacial emerged oxygen containing groups between α-FeOOH and MesoC after the first heterogeneous Fenton-like reaction. Visible light-induced photocatalysis-assisted heterogeneous Fenton-like process in the α-FeOOH/MesoC composite system improved the $HO^•$ production efficiency and Fe(III)/Fe(II) cycle and further activated the interfacial catalytic sites, which finally realized an extraordinary higher degradation and mineralization efficiency.

The evaluation of heterogeneous Fenton degradation on dye pollutant, Acid Blue 29 (AB29), has been investigated [19]. The solid catalyst prepared by both sol–gel and incipient wetness impregnation methods was developed by occlusion of Fe^{3+}

ions on synthesized mesoporous silica from sodium silicate. The prepared catalysts were characterized for their textural and surface morphology. High concentration of soluble metal precursor with 8.0 wt% Fe^{3+} can be easily deposited on silica. The results showed that the Fe–SiO_2 catalyst demonstrated good performance in the degradation of 50 ppm Acid Blue 29 (AB29) which was nearly completed in 100 min under visible light irradiation with optimum operating conditions at 0.4 g Fe–SiO_2/L, pH 3.0, and 10 mM H_2O_2. The catalyst is reusable over four consecutive cycles and minimal leaching of iron ions (<0.5 ppm) was observed.

A sewage sludge-derived porous carbon (SC), which was prepared by physicochemical activation and carbonization (600 °C), was applied for the adsorption and degradation of 1-diazo-2-naphthol-4-sulfonic acid (1, 2, 4-acid) in the presence of H_2O_2 and the performance was compared to that of pure Fe_3O_4 magnetic nanoparticles (MNPs). The prepared SC showed mesoporous structure with magnetic property, which made it favorable for solid–liquid separation application. Further experiments revealed that SC had a higher adsorption capacity and degradation efficiency of 1, 2, 4-acid than bare Fe_3O_4. The Langmuir and Freundlich model fitted the isotherm data and illustrated that the equilibrium adsorption amount of 1, 2, 4-acid onto SC (95.1 mg g^{-1}) was quadruple as large as that on Fe_3O_4 (26.4 mg g^{-1}). The subsequent degradation experiments were conducted at pH = 5.0 in the presence of 15 mM H_2O_2 with regard to 1, 2, 4-acid degradation efficiency and metal ions leach. The 120 min treatment in SC/H_2O_2 system achieved 94% of 1, 2, 4-acid (from 150 mg L^{-1} after adsorption equilibrium to 9 mg L^{-1}) and 48.1% TOC reduction, far higher than the efficiency of 46% and 24.3% by using Fe_3O_4 MNPs. Further analysis evidenced the cocatalytic effect of iron, carbon, silicon, and aluminum, which existed in large quantities in sludge-derived SC. The carbonaceous phase along with silica contributes to an increase in the dispersion of catalytic centers and an adsorbent to concentrate organic pollutant, whereas the iron oxide as well as alumina provides the catalytic centers for a Haber–Weiss initiated reactions [20].

11.2.5 Graphene

A facile Stöber-like method was used to prepare the ultra-dispersed Fe_3O_4 nanoparticles (3–8 nm) on the reduced graphene oxide (RGO) sheet by using (Fe (acac)$_3$) as the iron precursor. This strategy provides a facile and environmentally friendly method for the large-scale synthesis of Fe_3O_4/RGO without any additional reductants and organic surfactants. The prepared hybrid materials were used as the photo-Fenton catalyst, which displayed a high and stable performance for the recyclable degradation of methyl orange pollutant, owing to the high conversion efficiency of Fe^{3+}/Fe^{2+} and the magnetic property of Fe_3O_4 [21]. Three-dimensional (3D) graphene aerogel (3D-GA)-supported Fe_2O_3 nanocrystals were prepared through a Stöber-like method [22]. Fe_2O_3/GAs have a 3D network structure with a high surface area of 316 m^2 g^{-1} and physicochemical stability. 3D-GAs can inhibit

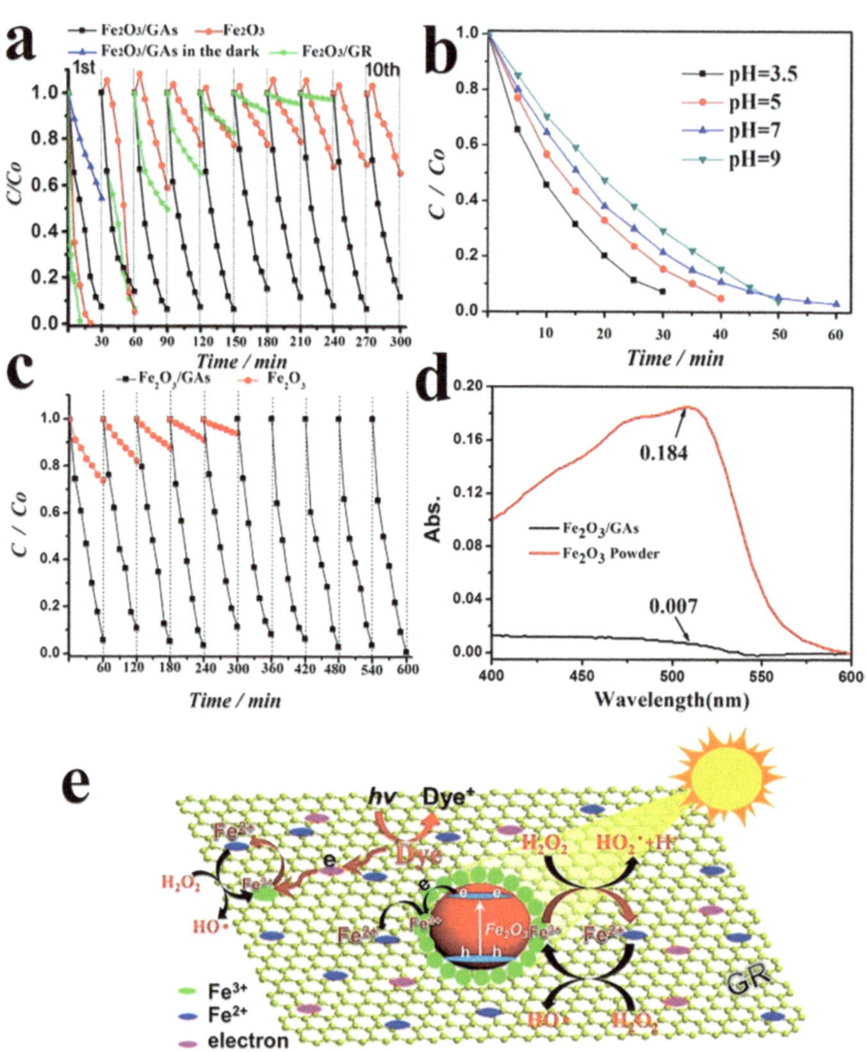

Fig. 11.2 (**a**) Cycle test for the solar-driven degradation of methyl orange (75 mL MO, 10 mg L 1) under different reaction systems (1.2 mL H_2O_2 (30 wt%); the initial pH was 3.5). (**b**) Solar-driven degradation of MO on Fe_2O_3 /GAs (1.2 mL H_2O_2 (30 wt%); the initial pH was adjusted from 3.5 to 9.0 by the addition of 0.1 M HCl). (**c**) Cycle test for the solar-driven degradation of MO on different catalysts with an initial pH of 7.0. (**d**) Absorption spectra of MO filter liquor, photo-degraded by Fe_2O_3 powders, and Fe_2O_3 /GAs in the presence of 1,10-phenanthroline monohydrate (the initial pH was 3.5). (**e**) Photo-Fenton reaction model of Fe_2O_3 /GAs in a system containing dye pollutants and H_2O_2. Reproduced from Ref. [22] by permission of The Royal Society of Chemistry

the loss of Fe^{2+} and stabilize the conversion of Fe^{3+}/Fe^{2+} in the photo-Fenton reaction. Compared with Fe_2O_3 and Fe_2O_3/2D-graphene (Fe_2O_3/GR), Fe_2O_3/GAs exhibit an ultrastable, solar-driven Fenton activity over a wide pH range of 3.5–9.0 (Fig. 11.2).

11.2.6 Iron Oxide–Semiconductor

To gain better enrichment ability for pollutant as well as avoiding aggregation, inert materials such as carbon and SiO_2 are introduced to fabricate loading or coating type catalyst. Among them, yolk–shell (Y−S)-structured Fe_3O_4@void@shell composites composed of a movable Fe_3O_4 core and a layer of permeable shell have recently gained increasing attention in the fields of catalysis, biology, and energy storage, where the interior core is efficiently protected from agglomeration and accessible for small molecules. In the context of catalysis, the cavity between the core and shell with a flexibly tunable volume provides a nanoreactor for a variety of reactions, in which the confined reactants may result in improved reaction rate or altered synthetic route. On the other hand, the limited working pH range around 3 commonly encountered by most Fenton agents has significantly obstructed the wide application. Although the heterogeneous Fenton agent generally can be applied in less acidic conditions, the regeneration efficiency of Fe^{2+} from Fe^{3+} and the oxidation potential of hydroxyl radicals have actually both been decreased. The combination of semiconductor and Fe-containing compound has proven to be effective on reducing Fe^{3+} to Fe^{2+} by the photo-generated electron from semiconductor [23]. Considering the priority of the Y−S-structured composite as a nanoreactor, it is desirable to apply it in the ph-F process, which, however, has been retarded by the tedious preparation process including layer-by-layer coating and removal of sacrificing layer, as well as the underdeveloped techniques on the component-tailoring. A simple and general synthesis technique for Y−S-structured composite is extremely desired to put forward its application in the ph-F treatment of sewage water.

Y−S-structured Fe_3O_4@void@CdS nanoparticles (NPs) were synthesized through a one-pot coating–etching process with Fe_3O_4@ SiO_2 as the core [24], where the coating of an outer CdS shell from a chemical bath deposition (CBD) process is simultaneously accompanied by the gradual etching of an inner SiO_2 shell. The as-prepared Fe_3O_4@void@CdS NPs (ca. 200 nm) possess good monodispersity and a uniform CdS shell of ca.15 nm (Fig. 11.3). This composite exhibits excellent photo-Fenton activity toward the degradation of methylene blue (MB) in a wide pH working range of 4.5–11 under the visible light irradiation (Fig. 11.4). A series of control experiments demonstrate the unique Y−S structure contributes to the enhanced activity, where the separation of hole−electron pair from CdS and the reduction of Fe^{2+} from Fe^{3+} are mutually promoted. The similar efficiency can also be achieved when the shell component changes to TiO_2 or CeO_2 [25], demonstrating a general strategy for the design of robust photo-Fenton agent.

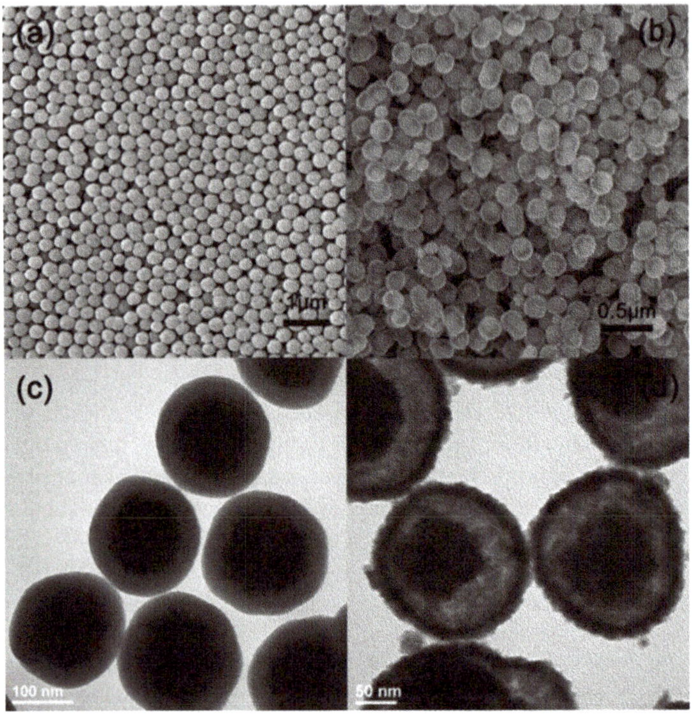

Fig. 11.3 FESEM images of (**a**) Fe_3O_4@SiO_2 and (**b**) Fe_3O_4@void@CdS NPs; TEM images of (**c**) Fe_3O_4@SiO_2 and (**d**) Fe_3O_4@void@CdS NPs. Reprinted with the permission from ref. [24]. Copyright 2011 American Chemical Society

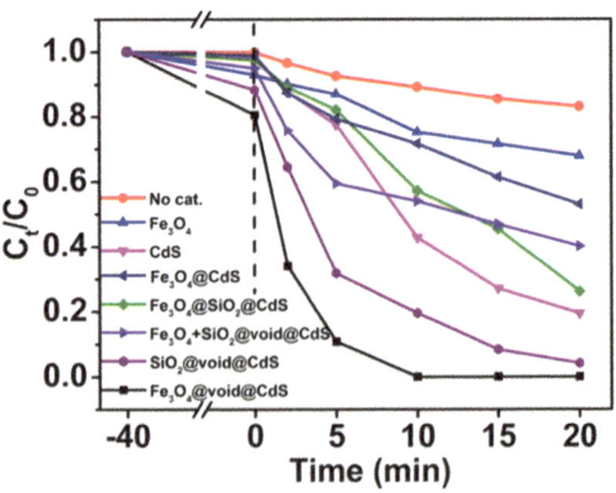

Fig. 11.4 Ph-F degradation of MB by different catalysts (no cat., Fe_3O_4, CdS, Fe_3O_4@CdS, Fe_3O_4@SiO_2@CdS, physical mixture of Fe_3O_4 and SiO_2@void@CdS, SiO_2@void@CdS, and Fe3O4@void@ CdS. Cat., 0.2 g L^{-1}; MB, 10 mg L^{-1}; H_2O_2, 300 mM; Vol., 50 mL). Reprinted with the permission from ref. [24]. Copyright 2011 American Chemical Society

11.2.7 Zerovalent Iron

Zerovalent iron (ZVI) powder is an environmental friendly reducing agent which has been widely studied for environmental remediation in recent years. ZVI is readily oxidized to Fe^{2+} ion in the presence of oxidants. In the aqueous systems, this phenomenon leads to the dissolution of solid which is the primary cause of metal corrosion. ZVI was used to degrade the azo dye in the aqueous medium by Devi et al. [26]. The influence of various reaction parameters like effect of iron dosage, concentration of H_2O_2/ammonium per sulfate (APS), initial dye concentration, and effect of pH and the influence of radical scavenger are studied, and optimum conditions are reported. The degradation rate decreased at higher iron dosages and also at higher oxidant concentrations due to the surface precipitation which deactivates the iron surface. The rate constant for the processes Fe^0/UV and Fe^0/APS/UV is twice compared to their respective Fe^0/dark and Fe^0/APS/dark processes. The rate constant for Fe^0/H_2O_2/UV process is four times higher than Fe^0/H_2O_2/dark process. The increase in the efficiency of Fe^0/UV process is attributed to the cleavage of stable iron complexes which produces Fe^{2+} ions that participates in cyclic Fenton mechanism for the generation of hydroxyl radicals. The increase in the efficiency of Fe^0/APS/UV or H_2O_2 compared to dark process is due to continuous generation of hydroxyl radicals and also due to the frequent photo reduction of Fe^{3+} to Fe^2. Though H_2O_2 is a better oxidant than APS in all respects, it is more susceptible to deactivation by $HO^•$ scavengers. The decrease in the rate constant in the presence of $HO^•$ scavenger is more for H_2O_2 than APS. Iron powder retains its recycling efficiency better in the presence of H_2O_2 than APS. The decrease in the degradation rate in the presence of APS as an oxidant is due to the fact that generation of free radicals on iron surface is slower compared to H_2O_2. Also, the excess acidity provided by APS retards the degradation rate as excess H^+ ions acts as hydroxyl radical scavenger. The degradation of methyl orange (MO) using Fe^0 as an acid-driven process shows higher efficiency at pH 3. The efficiency of various processes for the decolorization of MO dye is of the following order: Fe^0/H_2O_2/UV > Fe^0/H_2O_2/dark > Fe^0/APS/UV > Fe^0/UV > Fe^0/APS/dark > H_2O_2/UV ≈ Fe^0/dark > APS/UV. Dye resisted to degradation in the presence of oxidizing agent in dark. The degradation process was followed by UV–vis and GC–MS spectroscopic techniques. The complete degradation of the dye was achieved in the presence of oxidizing agent when the system was amended with iron powder under UV light illumination. The concentration of Fe^{2+} leached at the end of the optimized degradation experiment is found to be 2.78×10^{-3} M. With optimization, the degradation using Fe^0 can be effective way to treat azo dyes in aqueous solution.

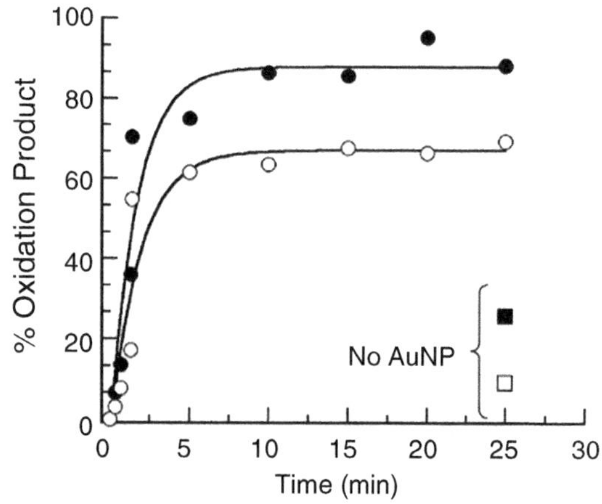

Fig. 11.5 Acetophenone (black dot) (λmonitor = 245 nm) and benzaldehyde (blank dot) (λmonitor = 250 nm) conversion as a function of LED photoexcitation time. Note the reduced conversion to acetophenone (black square) and benzaldehyde (blank square) in the absence of AuNP. Reprinted with the permission from ref. [29]. Copyright 2011 American Chemical Society

11.2.8 Noble Metal

Degradation of phenol with a small excess of H_2O_2 can be achieved by using gold nanoparticles supported on Fenton-treated diamond nanoparticles (Au/OH-npD) as a highly selective (at least 79%) and efficient (turnover number: 321,000) catalyst for the Fenton reaction at room temperature [27]. The catalytic activity of Au/OH-npD for the Fenton degradation of phenol with hydrogen peroxide can be increased over one order of magnitude by irradiation at 532 nm [28]. Moreover, there is a linear relationship between the initial reaction rate and the incident photon flux. This photo-enhancement allows promoting Fenton activity at pH 8 in which the catalytic activity of Au/OH-npD is negligible. The same photo enhancement activity for the Fenton degradation of phenol was observed for other supported gold catalysts including those that do not exhibit microsecond transients in the nanosecond laser flash photolysis (Au/TiO_2 and Au/SiO_2) due to their lifetime shorter than microseconds. It is proposed that the photo enhancement should be a general phenomenon in gold catalysis for those reaction mechanisms involving positive and/or negative gold species.

Surface plasmon excitation of aqueous colloidal gold nanoparticles with visible light in the presence of H_2O_2 led to rapid and selective oxidation of *sec*-phenethyl and benzyl alcohols to acetophenone and benzaldehyde (Fig. 11.5), respectively. Laser drop, light emitting diode, and microwave irradiation have been used as energy sources. Interestingly, *sec*-phenethyl alcohol conversion was calculated to be 95% in 20 min when monochromatic 530 nm LEDs were used, being as good or better yield than the corresponding laser and microwave techniques. These results demonstrate the versatility of this inexpensive arrangement. Further attention was placed on the possible mechanism for Au nanoparticle plasmon-mediated alcohol oxidations in the presence of H_2O_2 [29].

11.3 Thermal-Assisted Photo-Fenton Reaction

The effect of the reaction temperature on the degradation rate of formic acid, using the Fenton and photo-Fenton processes, was investigated [30]. First, for both reactions, a stirred tank laboratory photoreactor irradiated from the bottom was used to evaluate the kinetic parameters between 298 and 328 K. Afterward, the proposed kinetic model was used to predict the conversion of the organic pollutant in a flat-plate solar photoreactor. The previously reported radiation field and mass balances have been used to compute the formic acid and hydrogen peroxide concentrations as a function of time in the solar reactor. Theoretical and experimental results show that UV solar radiation improves the effectiveness of the Fenton process. At lower temperatures, the pollutant conversion enhancement is significant, but this effect is less important at higher temperatures. For instance, experimental conversion enhancements after 20 min are 186.0, 74.0, and 7.4% for 298, 313, and 328 K, respectively.

11.4 The Detrimental Factors

Although $HO^{\bullet}$ is a powerful species that can nonselectively react with most organics, the Fenton reaction could not reach deep mineralization of target pollutants (usually less than 50% CO_2 yield). The final products formed are usually low-molecular-weight organic acids (LMWOAs), such as oxalic, malonic, and acetic acids. Consistent with this is that the Fenton's reagent is powerful at initial stage and gradually loses its capability for organic destruction with reaction time. It is commonly accepted that such deactivation is due to the interaction of Fe^{3+} with LMWOAs (the degradation intermediates), which is unfavorable to either Fe^{3+}/Fe^{2+} recycling or to $HO^{\bullet}$ generation, although the exact mechanism is still obscure. Effective circumvention of the blockage from these LMWOA intermediates by an appropriate pathway may improve the iron cycle and achieve continuous production of $HO^{\bullet}$ for target pollutant mineralization.

The influence of low-molecular-weight organic acids (LMWOAs), such as malonic acid, ethylenediaminetetraacetic acid, and oxalic acid, on the Fenton degradation of organic pollutants was examined under visible irradiation ($\lambda > 450$ nm) [31]. The Fenton degradation of malachite green in the dark was completely blocked in the presence of LMWOAs. It was found that either visible light irradiation or the addition of hydroquinone could initiate the dye degradation, but the mineralization yield was almost zero. An important result was that the dye mineralization in the presence of LMWOAs could be achieved when both visible irradiation and hydroquinone were introduced (Fig. 11.6). Similar results were obtained with colorless pollutants, such as benzyltrimethylammonium chloride and 2, 4, 5-trichlorophenol. The coupling visible irradiation and hydroquinone could be a strong and universal

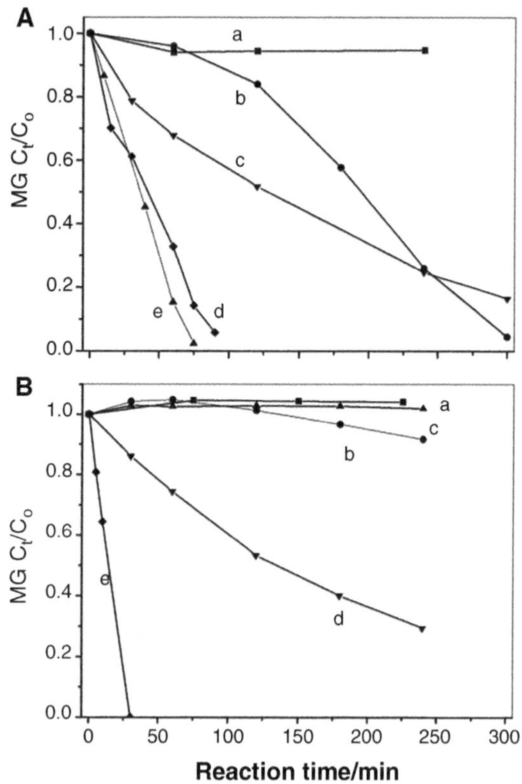

Fig. 11.6 Fenton degradation of MG in the presence of (**A**) malonic acid and (**B**) EDTA: (a) in the dark, (b) under visible irradiation, (c) in the presence of hydroquinone in the dark, (d) in the presence of hydroquinone under visible irradiation, (e) under UV irradiation. Initial concentrations for (**A**): 1×10^{-5} M Fe^{3+}; 1×10^{-2} M H_2O_2; 5×10^{-5} M malonic acid; 2×10^{-5} M hydroquinone; 2×10^{-5} M MG. Initial concentrations for (**B**): 2×10^{-4} M Fe^{3+}; 2×10^{-3} M H_2O_2; 2×10^{-5} M hydroquinone; 2×10^{-5} M MG; 5×10^{-4} M EDTA. Reprinted with the permission from ref. [31]. Copyright 2011 American Chemical Society

driving force in the Fenton reaction for the complete degradation and mineralization of organic pollutants, even in the presence of LMWOAs.

Orange II, the 4-(2-hydroxy-1-naphthylazo)benzenesulfonic acid Na salt was taken as a model for the oxidation of organic compounds in photo-assisted Fenton process in the presence of Cl^- in solution [32]. The $HO^\bullet$ radicals seem to originate due to the photolysis of $Fe(OH)^{2+}$ complexes in solution, whereas the $Cl_2^{-\bullet}$ radical was a product of the photolysis of the $FeCl_2^+$ complexes. The rate constants for the reaction of $HO^\bullet$ and $Cl_2^{-\bullet}$ radicals with Orange II were determined by laser kinetic spectroscopy. A significant decrease in the decoloration rate was observed for Orange II upon addition of Cl^- (10 mM), but further addition of Cl^- only marginally affected the latter reaction rate. Chlorinated hydrocarbons were observed as the products of Orange II oxidation in photo-assisted and dark Fenton processes in the presence of Cl^- anions. Light irradiation decreased the amount of chlorinated organic products in solution as compared to dark Fenton processes. The results obtained are in agreement with the radical reaction theory for the particular case of the Fenton reaction. Kinetic modeling of the Orange II decoloration was carried out taking into consideration the rate constants found for the reactions with $HO^\bullet$ and $Cl_2^{-\bullet}$, respectively.

References

1. Fenton HJH (1894) Oxidation of tartaric acid in presence of iron. J Chem Soc 65:899–910
2. Walling C, Goosen A (1973) Mechanism of the ferric ion catalyzed decomposition of hydrogen peroxide. Effect of organic substrates. J Am Chem Soc 95:2987–2991
3. Lipczynska-Kochany E (1991) Novel method for a photocatalytic degradation of 4-nitrophenol in homogeneous aqeuous solution. Environ Technol 12:87–92
4. Zepp RG, Faust BC, Hoigne J (1992) Hydroxyl radical formation in aqueous reactions (pH 3-8) of iron(II) with hydrogen peroxide: the photo-Fenton reaction. Environ Sci Technol 26:313–319
5. Lucas MS, Peres JA (2006) Decolorization of the azo dye Reactive Black 5 by Fenton and photo-Fenton oxidation. Dyes Pigments 71:236–244
6. Moraes JEF, Silva DN, Quina FH et al (2004) Utilization of solar energy in the photodegradation of gasoline in water and of oil-field-produced water. Environ Sci Technol 38:3746–3751
7. Ortega-Gómez E, Ballesteros Martín MM, Carratalà A et al (2015) Principal parameters affecting virus inactivation by the solar photo-Fenton process at neutral pH and μM concentrations of H_2O_2 and $Fe^{2+/3+}$. Appl Catal B Environ 174–175:395–402
8. Fernandez J, Bandara J, Kiwi J et al (1998) Efficient photo-assisted Fenton catalysis mediated by Fe ions on Nafion membranes active in the abatement of non-biodegradable azo-dye. Chem Commun 14:1493–1494
9. Yuranova T, Enea O, Mielczarski E et al (2004) Fenton immobilized photo-assisted catalysis through a Fe/C structured fabric. Appl Catal B Environ 49:39–50
10. Bozzi A, Yuranova T, Mielczarski J et al (2002) Abatement of oxalates catalyzed by Fe-silica structured surfaces *via* cyclic carboxylate intermediates in photo-Fenton reactions. Chem Commun 0:2202–2203
11. Ma W, Huang Y, Li J et al (2003) An efficient approach for the photodegradation of organic pollutants by immobilized iron ions at neutral pHs. Chem Commun 0:1582–1583
12. Li J, Ma W, Huang Y et al (2004) Oxidative degradation of organic pollutants utilizing molecular oxygen and visible light over a supported catalyst of $Fe(bpy)_3^{2+}$ in water. Appl Catal B Environ 48:17–24
13. Zhang G, Gao Y, Zhang Y et al (2010) Fe_2O_3-pillared rectorite as an efficient and stable Fenton-like heterogeneous catalyst for photodegradation of organic contaminants. Environ Sci Technol 44:6384–6389
14. Feng J, Hu X, Yue PL et al (2003) Discoloration and mineralization of Reactive Red HE-3B by heterogeneous photo-Fenton reaction. Water Res 37:3776–3784
15. Wei X, Wu H, He G et al (2017) Efficient degradation of phenol using iron-montmorillonite as a Fenton catalyst: importance of visible light irradiation and intermediates. J Hazard Mater 321:408–416
16. Bossmann SH, Oliveros E, Gob S et al (2001) Degradation of polyvinyl alcohol (PVA) by homogeneous and heterogeneous photo catalysis applied to the photochemically enhanced Fenton reactions. Water Sci Technol 44:257–262
17. Gonzalez-Olmos R, Martin MJ, Georgi A et al (2012) Fe-zeolites as heterogeneous catalysts in solar Fenton-like reactions at neutral pH. Appl Catal B Environ 125:51–58
18. Qian X, Ren M, Zhu Y et al (2017) Visible light assisted heterogeneous Fenton-like degradation of organic pollutant via α-FeOOH/mesoporous carbon composites. Environ Sci Technol 51:3993–4000
19. Song AN, Hameed BH (2013) Degradation of Acid Blue 29 in visible light radiation using iron modified mesoporous silica as heterogeneous photo-Fenton catalyst. Appl Catal A: Gel 450:96–105
20. Gu L, Zhu N, Guo H et al (2013) Adsorption and Fenton-like degradation of naphthalene dye intermediate on sewage sludge derived porous carbon. J Hazard Mater 246–247:145–153

21. Qiu BC, Li QY, Shen B et al (2016) Stöber-like method to synthesize ultradispersed Fe3O4 nanoparticles on graphene with excellent Photo-Fenton reaction and high-performance lithium storage. Appl Catal B Environ 125:216–223
22. Qiu BC, Xing MY, Zhang JL (2015) Stöber-like method to synthesize ultralight, porous, stretchable Fe_2O_3/graphene aerogels for excellent performance in photo-Fenton reaction and electrochemical capacitors. J Mater Chem A 3:12820–12827
23. Yang X, Chen W, Huang J et al (2015) Rapid degradation of methylene blue in a novel heterogeneous Fe_3O_4@rGO@TiO_2-catalyzed photo-Fenton system. Sci Rep 5:10632
24. Shi W, Du D, Shen B et al (2016) Synthesis of yolk–shell structured Fe_3O_4@void@CdS nanoparticles: a general and effective structure design for photo-Fenton reaction. ACS Appl Mater Interf 8:20831–20838
25. Du D, Shi W, Wang LZ et al (2017) Fenton-like catalyst for the extremely efficient elimination of tetracycline. Appl Catal B Environ 200:484–492
26. Devi LG, Kumar SG, Reddy KM et al (2009) Photo degradation of methyl orange an azo dye by advanced Fenton process using zero valent metallic iron: influence of various reaction parameters and its degradation mechanism. J Hazard Mater 164:459–467
27. Navalon S, Matin R, Alvaro M et al (2010) Gold on diamond nanoparticles as a highly efficient Fenton catalyst. Angew Chem Inter Ed 49:8403–8407
28. Navalon S, De Miguel M, Martin R et al (2011) Enhancement of the catalytic activity of supported gold nanoparticles for the Fenton reaction by light. J Am Chem Soc 133:2218–2226
29. Hallett-Tapley GL, Jazmín Silvero M, González-Béjar M et al (2011) Plasmon-mediated catalytic oxidation of sec-phenethyl and benzyl alcohols. J Phys Chem C 115:10784–10790
30. Farias J, Rossetti GH, Albizzati ED et al (2007) Solar degradation of formic acid: temperature effects on the photo-Fenton reaction. Ind Eng Chem Res 46:7580–7586
31. Ma J, Ma W, Song W et al (2006) Fenton degradation of organic pollutants in the presence of low-molecular-weight organic acids: cooperative effect of quinone and visible light. Environ Sci Technol 40:618–624
32. Kiwi J, Lopes A, Nadtochenko V (2000) Mechanism and kinetics of the OH^-Radical intervention during Fenton oxidation in the presence of a significant amount of radical scavenger (Cl^-). Environ Sci Technol 34:2162–2168

Chapter 12
Roles and Properties of Cocatalysts in Semiconductor-Based Materials for Efficient CO_2 Photoreduction

12.1 Introduction

Since the industrial revolution, the rapid development of the economy resulting in the emission amount of CO_2 to the atmosphere increased year by year. As a greenhouse gas, CO_2 could adsorb the solar energy and raise the temperature itself. Consequently, the excessive CO_2 emission would cause global warming and lead to serious environmental problems such as ice mountain melting, sea level rise, ecological balance break, etc., which have gain great attention from the whole world [1]. Recently, the Intergovernmental Panel on Climate Change (IPCC) reports that currently global warming has raised the average temperature of 1.5 °C compared with the preindustrial level [2]; no wonder, it is urgent for us to find possible solutions to cope with the CO_2 excessive emission problem. Therefore, the efficient capture and storage of CO_2 to further transform it into desirable chemicals with renewable energy input could address the above problems ideally.

Since Fujishima and Honda observed the H_2 production by TiO_2 electrode under light irradiation [3], semiconductor-based photocatalysts and its applications in environment management and energy transformation have gained much attention [4]. Some reported semiconductors like TiO_2, C_3N_4, CdS, etc.. possessed suitable bandgaps and reduction potentials which can trigger the CO_2 photoreduction reaction (CO2PR) with H_2O. However, as a very stable molecule, CO_2 photoreduction with H_2O is difficult to occur, owing to this process involved with multiple steps like breaking the C=O bonds and forming the C-H bonds, which both are endothermic reactions and also required abundant electrons and corresponding H protons. In addition, as a major competing reaction, H_2O reduction to H_2 is more easily to occur in both thermodynamics and kinetics [5–7]. In this way, bare semiconductors without modification often show low CO_2 photoreduction activity and selectivity.

In the use of semiconductor as the photocatalyst to trigger CO_2 conversion with H_2O, the activity is mainly affected by three key factors: (i) light adsorption and the excitation of semiconductors, (ii) photo-generated electron-hole pairs' separation

J. Zhang et al., *Photocatalysis*, Lecture Notes in Chemistry 100,
https://doi.org/10.1007/978-981-13-2113-9_12

and transfer, and (iii) surface catalytic reactions. Based on these principles, semiconductors modified with suitable cocatalysts can greatly boost the CO2PR activity and selectivity [4, 8]. In Table 12.1 we summarized the semiconductor-based photocatalysts with various cocatalysts loaded in CO2PR based on the researches in recent years. The content of Table 12.1 including the optimal products (such as CH_4, CO, CH_3OH, HCOOH, and H_2) yields brief introduction of photocatalysts' preparation, CO2PR evaluation details, etc. Among these cocatalysts, for example, noble metal NPs like Pt [8–13], Au [14], Pd [15–18], Ag [19–21], etc., lower Fermi level is often considered as efficient cocatalysts for electron trapping and active site with suitable binding energy to some intermediates [5]; combined semiconductors with suitable band structures together can form p–n junction or Z-scheme and thus favored the light adsorption and charge separation [22–25]; the semiconductor–MOF or semiconductor–graphene composites often show enhanced CO_2 adsorption performance and effective CO_2 activation properties, etc.

In this chapter, we plan to present a short review to discuss about the cocatalysts including metal NPs, metal alloy NPs, graphene, carbon nanodots, MOFs, semiconductors, etc., synthesis methods, and basic roles in CO_2 photoreduction based on recent research progress. The advanced techniques such as time-resolved DRIFTS, time-resolved PL decay, EPR, and DFT calculations applied to clarify the charge transfer mechanism and surface catalytic reaction pathways are also discussed in detail.

12.2 Basic Principles of CO_2 Photoreduction

Generally, semiconductors can be excited by photon carriers with energy higher than its bandgap energy; after the excitation, the photo-generated electron-hole pairs would migrate to the surface for certain reduction/oxidation reactions or recombine together and release energy by means of heat. In order to catalyze the CO2PR reaction with H_2O, the photo-generated electron-hole pairs must possess suitable reduction and oxidation potential (Fig. 12.1). According to previous reports, different standard reduction potentials of CO_2 reduction with H protons to yield different products are shown in Eqs. 12.1, 12.2, 12.3, 12.4, and 12.5 [5, 26].

$$CO_2 + 2H^+ + 2e^- \rightarrow HCOOH \qquad E_{redox}{}^0 = -0.61V \tag{12.1}$$

$$CO_2 + 2H^+ + 2e^- \rightarrow CO \qquad E_{redox}{}^0 = -0.53V \tag{12.2}$$

$$CO_2 + 6H^+ + 6e^- \rightarrow CH_3OH + H_2O \qquad E_{redox}{}^0 = -0.38V \tag{12.3}$$

$$CO_2 + 8H^+ + 8e^- \rightarrow CH_4 + 4H_2O \qquad E_{redox}{}^0 = -0.24V \tag{12.4}$$

$$CO_2 + e^- \rightarrow CO_2{}^- \qquad E_{redox}{}^0 = -1.90V \tag{12.5}$$

Table 12.1 Semiconductor-based photocatalysts with cocatalysts in CO_2 photoreduction

	Photocatalysts	Cocatalysts	Optimal product yield (μmol/g·h exception unit stated otherwise)	Brief photocatalyst preparation method	CO_2 photoreduction evaluation details	Refs.
1	MgO–Pt–TiO_2	MgO, Pt	CH_4 (8.9); H_2 (8.4)	Impregnation method for MgO modification and photo-deposition for Pt load	20 mg catalyst placed on a Teflon holder, 0.2 MPa CO_2 and 4 mL H_2O, temperature was kept at 323 K, 100 W Xe lamp: 60 mW/cm^2 at UV range was used as the light source	[10]
2	Pt–TiO_2 single crystal	Pt	CH_4 (1361)	tilted-target sputtering (TTS) configuration	Catalyst film, CO_2 and H_2O vapor, 400 W Xe lamp was used as the light source, continuous flow	[9]
3	Pd/C_3N_4	Pd	CO (20.3); H_2 (9.7); CH_4 (0.28); C_2H_5OH (2.2)	Solution-phase method using different facet-selective capping agents and PVP as reductant and stabilizer	10 mg catalyst dispersed on a porous holder made of quartz sand, 0.15 MPa CO_2 and 3 mL H_2O, 300 W Xe lamp with 400 nm cutoff filter: 100 mW/cm^2 was used as the light source	[15]
4	Pd/TiO_2 sheet	Pd	CO (12.6); CH_4 (3.0)	Self-assembly of pre-synthesized Pd nanosheets onto the TiO_2 nanosheets	10 mg catalyst dispersed on a porous holder made of quartz sand, 0.15 MPa CO_2 and 3 mL H_2O, 300 W Xe lamp with 400 nm cutoff filter: 100 mW/cm^2 was used as the light source	[16]
5	Ag/Brookite nanocube	Ag	CH_4 (11.5); CO (128.8)	$AgNO_3$ was mixed with brookite and reduced by $NaBH_4$	0.15 g catalyst, 28 cm^2, CO_2, and H_2O vapor generate from $NaHCO_3$ + H_2SO_4; 300 W Xe lamp was used as the light source	[21]
6	Metal NPs/TiO_2 nanotube	Au, Ru, ZnPd	CH_4 (58.47, 26.37, 26.83, respectively)	Stepwise impregnation(colloid synthesis first) of metal NPs	Stainless steel reaction chamber with water droplets, solar simulator equipped with class A filters	[28]

(continued)

Table 12.1 (continued)

	Photocatalysts	Cocatalysts	Optimal product yield (μmol/g·h exception unit stated otherwise)	Brief photocatalyst preparation method	CO_2 photoreduction evaluation details	Refs.
7	Pt-CdS-Cu_2S	Pt	CO (3.02)	Site-selective growth of Pt on the CdS tips followed by light-induced cation exchange of CdS to Cu_2S	Pt-CdS, 0.1 M Na_2SO_3 as hole scavenger, 1 M Na_2CO_3 as CO_2 source, and 0.63 mM $CuCl_2$ as copper precursor; pH = 11; 450 W Xe lamp was used as the light source	[12]
8	Ru/$NaTaO_3$	Ru	CH_4 (51.8)	Photo-deposition of Ru on $NaTaO_3$	0.07 g sample was loaded in the reactor, 3 mL H_2O, 40 kPa CO_2, and 40 kPa H_2; 300 W Xe lamp was used as the light source	[29]
9	Ag/TiO_2	Ag	CH_4 (1.40)	Silver mirror method by using HCHO and $AgNO_3$	0.1 g catalyst-4.2cm^2, 0.4 mL H_2O, CO_2 and H_2O vapor; the system was placed in the dark for 10 h before the light irradiation; 300 W Xe lamp was used as the light source	[30]
10	V_o (oxygen vacancies)-rich Pt/Ga_2O_3	Pt	CO (105); CH_4 (4.75)	Photo-deposition of Pt on Ga_2O_3	200 mg catalyst dispersed in 100 mL H_2O, closed gas circulation–evacuation reactor, 1.01 bar CO_2; 300 W Xe lamp was used as the light source	[31]
11	Pd/TiO_2	Pd	CH_4 (118.54); CO (15.45); H_2 (13.23) average of 3 h	Na_2PdCl_4 was mixed with TiO_2 and reduced by glucose	0.1 g catalyst dispersed onto 0.1 g silica wool in a miniature visual autoclave,150w Hg lamp was used as the light source, 2.5 MPa H_2/CO_2 = 4:1	[17]
12	Pd_7Cu_1 alloy-TiO_2	Pd_7Cu_1 alloy	CH_4 (19.6)	Liquid phase synthesis using K_2PdCl_4 and $CuCl_2$ as precursor	5 mg catalyst mounted above a small quartz beaker placed in a quartz tube, 0.2 Mpa CO_2 and 1 mL H_2O, 300 W Xe lamp was used as the light source	[18]

13	CuPt–TiO_2	Cu-Pt alloy	CH_4 (11.3)	Varied Pt and Cu precursor mixed with TiO_2 then anneal at air and H_2 at 673 K	1 cm^2 0.4 mg catalyst film placed vertically in reaction chamber with 50uL H_2O and 1.2 atm CO_2, 150 W Xe lamp was used as the light source	[32]
14	Au–Cu/TiO_2 (p25)	Au–Cu alloy	CH_4 (2200); H_2 (286)	Stepwise deposition–precipitation for each metal and then reduced in H_2 flow	0.6 mg catalyst, 1.25 cm^2 film, 1.7 atm H_2O saturated CO_2	[14]
15	Au–Cu@$SrTiO_3$/TiO_2	Au–Cu alloy	CH_4 (421.2); CO (3770)	Microwave-assisted solvothermal method	5 mg catalyst-2.5 cm^2, CO_2 saturated N_2H_4 and CO_2; 300 W Xe lamp was used as the light source	[33]
16	ZnO@Co_3O_4	Co_3O_4	CH_4 (0.99)	Synthesize ZIF-8@ZIF-67 first, then followed by annealing treatment at N_2 and air, respectively	0.1 g sample deposited in reaction cell; 3 mL H_2O and 80 kPa CO_2 were introduced into the reactor; 300 W Xe lamp was used as the light source	[34]
17	ZnO-CuO nanowire	ZnO	CO (1.98 mmol/g·h)	Atomic layer deposition(ALD) of ZnO on CuO nanowire	Catalyst film, CO_2 passed through a water bubbler to generate CO_2 and H_2O mixture, 400 W Xe lamp: 300 mW/cm^2 was used as the light source, continuous flow reactor	[35]
18	g-C_3N_4/ZnO	ZnO	CH_3OH (0.6)	Mixing urea and zinc nitrate hexahydrate first then followed by one-step facile calcination method	100 mg sample was deposited onto the bottom of 200 mL reactor, 0.12 g $NaHCO_3$ and 0.25 mL-4 M HCl was introduced into the reactor as the CO_2 and H_2O source, 300 W Xe lamp was used as the light source	[36]
19	Hybrid CuO–$TiO_{2-x}N_x$ hollow nanocubes	Cu_2O	CH_4 (41.3)	Calcination of TiO_2@Cu_3N nanotubes as described in the paper	100 mg catalyst in a reaction chamber, CO_2/H_2O mixture, 300 W Xe lamp: 100 mW/cm^2 was used as the light source	[37]

(continued)

Table 12.1 (continued)

	Photocatalysts	Cocatalysts	Optimal product yield (μmol/g·h exception unit stated otherwise)	Brief photocatalyst preparation method	CO_2 photoreduction evaluation details	Refs.
20	Hierarchical Z-scheme CdS–WO_3	CdS	CH_4 (1.02)	Deposition precipitation method using $Cd(NO_3)_2$ as precursor and Na_2S as precipitant	100 mg catalyst added to 10 mL distilled water in a glass reactor, 1 atm CO_2, 300 W Xe lamp:150 mW cm^{-2} was used as the light source, ambient temperature	[38]
21	Z-scheme BiOI/g-C_3N_4	BiOI	CO (4.86); CH_4 (0.18); O_2 (2.78)	Deposition method using $Bi(NO_3)$ as precursor and KI as precipitant	0.10 g catalyst, CO_2 and H_2O vapor, 300 W Xe lamp with a UV-cutoff flitter ($\lambda > 400$ nm) was used as the light source	[39]
22	Pt-MnO_x/CeO_2	Pt, MnO_x	CH_4 (1.12)	One-step photo-deposition method	0.1 g catalyst-4.2 cm^2, 0.4 mL H_2O, CO_2 and H_2O vapor, the system was placed in dark to reach adsorption-desorption equilibrium before the light irradiation; 300 W Xe lamp was used as the light source	[40]
23	Pt/HCTSO	Pt, CoO_x	CH_4 (9.3); CO (0.3); H_2 (14.1)	"EISA" method for co-species incorporation and alcohol reduction for Pt loading	30 mg filmlike catalyst on 6 cm^2 diameter Petri dish, CO_2, and 1 mL H_2O; 300 W Xe lamp was used as the light source	[8]
24	Pt@CuO_x/TiO_2 (p25)	Pt@CuO_x	CH_4 (33); CO (8.3); H_2 (25)	Stepwise photo-deposition of Pt and CuO_x using a 300 W Hg lamp	0.02 g catalyst, 0.2 MPa CO_2, 4 mL H_2O, 200 W Xe lamp was used as the light source	[41]
25	Pt-RuO_2/ Zn_2GeO_4 nanobelt	Pt, RuO_2	CH_4 (25) first hour	Photoreduction of H_2PtCl_4 under a 300 W Xe lamp	0.1 g catalyst-4.2cm^2, 1 atm, 1 mL H_2O, CO_2 and H_2O vapor, 300 W Xe lamp was used as the light source	[42]

26	CdSe–Pt–TiO_2	CdSe, Pt, Pd	CH_3OH (90.22 ppm/g. h); $HCOOCH_3$ (225.4 ppm/g.h)	Wet impregnation method for Pd or Pt loading; adding commercial CdSe QDs to TiO_2 then anneal in N_2	Stainless steel reaction chamber with 100–140 mg catalyst; CO_2 passes through a water bubbler (25 °C) and then enters the reactor; 6 W UV-B lamp with 2 mW/cm^2 was used as the light source	[43]
27	Pt@CdS/inverse opal TiO_2	Pt@CdS	CH_4 (36.8)	Gas bubbling-assisted membrane reduction-precipitation (GBMR/P) method	20 mg catalyst spin coating in 6.5 cm diameter reactor, CO_2 and H_2O, 300 W Xe lamp: 100 mW/cm^2 was used as the light source	[44]
28	Au@CdS/inverse opal TiO_2	Au@CdS	CH_4 (41.6)	Gas bubbling-assisted membrane reduction-precipitation (GBMR/P) method	20 mg catalyst spin coating in 6.5 cm diameter reactor, CO_2 and H_2O, 300 W Xe lamp: 100 mW/cm^2 was used as the light source	[22]
29	Z-scheme Ag_3PO_4/g-C_3N_4	Ag, $AgPO_4$	CO (~45); CH_3OH (~10)	In situ deposition of Ag^+ using Na_2HPO_4 as precipitant	10 mg catalyst placed in a stainless steel reactor (volume~132 mL), CO_2, and 4 mL H_2O; 500 W Xe lamp was used as the light source	[25]
30	Z-scheme WO_3/Au/In_2S_3 nanowire arrays	In_2S_3, Au	CH_4 (0.42)	Plasma sputtering method used for loading Au to WO_3 and thermal growth for In_2S_3 to coat on Au surface	10 cm^2 sample placed on the bottom of a Pyrex glass reactor, ambient pressure CO_2, 0.4 mL H_2O; 300 W Xe lamp with a UV cutoff filter ($\lambda > 420$ nm) was used as the light source	[45]
31	Cu-loaded graphene oxide–TiO_2 composite	Cu(I) species	C_2H_5OH (144.7 at pH 11.0); CH_3OH (47.0 at pH 4.0)	Add the Cu species to the precursor of graphene oxide-TiO_2 followed with thermal treat	250 mg catalyst dispersed in 250 mL solutions of different pH in a tube photoreactor, UV–vis Heraeus TQ 150 medium-pressure mercury vapor lamp, 27 cm^3 min^{-1} he and 3 cm^3 min^{-1} CO_2, 25 °C	[46]

(continued)

Table 12.1 (continued)

	Photocatalysts	Cocatalysts	Optimal product yield (μmol/g·h exception unit stated otherwise)	Brief photocatalyst preparation method	CO_2 photoreduction evaluation details	Refs.
32	Pt/carbon-coated In_2O_3 nanobelt	Pt, C	CO (126.6); CH_4 (27.9)	Hydrothermal synthesis for carbon layer using glucose as precursor and photo-deposition method for depositing Pt from H_2PtCl_4	200 mg catalyst dispersed in a water solution (200 mL) with 10 vol% triethanolamine placed in a closed gas circulation–evacuation reactor, 1.01 bar CO_2; 300 W Xe lamp was used as the light source, 25 °C	[47]
33	C/H-g-C_3N_4	C	CO (22.6); CH_4 (12.5) in 9 h	One-step pyrolysis method using melamine and natural soybean oil as precursors	0.1 g catalyst in stainless steel reactor, CO_2 and H_2O mixture; 500 W Xe lamp was used as the light source, temperature was kept at 303 K, and pressure was maintained at 110 KPa	[48]
34	TiO_2/graphene sandwich structure	Graphene	CH_4 (8); C_2H_6 (16.8)	One-step hydrothermal method	0.1 g catalyst placed on a glass reactor with 4.2 cm^2 area, CO_2 and 1 mL H_2O; 300 W Xe lamp was used as the light source	[49]
35	G-$Ti_{0.91}O_2$ hollow sphere	Graphene	CH_4 (1.14); CO (8.91)	The LBL self-assembly approach using GO nanosheets and $Ti_{0.91}O_2$ nanosheets as inorganic shell building blocks	0.1 g catalyst placed on a glass reactor with 4.2 cm^2 area, CO_2 and 1 mL H_2O; 300 W Xe lamp was used as the light source	[50]
36	Co_3O_4 hexagonal platelets	$[Ru(bpy)_3]Cl_2$	CO (3523)	Co_3O_4 hexagonal platelets was obtained by calcinate β-$Co(OH)_2$ at 400 °C, commercial $[Ru(bpy)_3]Cl_2$	10 mg Co_3O_4 platelets and 10 mg $[Ru(bpy)_3]Cl_2.6H_2O$, 30 mL solvent (MeCN/TEOA/H_2O = 3:1:1 vol/vol); CO_2 1 bar, 300 W Xe lamp with 420 filter (293.61 mW/cm^2) was used as the light source	[51]

37	$CsPbBr_3$ perovskite quantum dot/graphene oxide	Graphene oxide	CO (58.7); CH_4 (28.6); H_2 (1.58)	Modified Hummers' method for GO preparation and then GO–DMF solution was used to synthesize $CsPbBr_3$ QD/GO composite	4 mg catalyst, 10 mL ethyl acetate, CO_2, 100 W Xe lamp with an AM 1.5G filter for simulating solar light: 150 mW/cm^2 was used as the light source	[52]
38	$Cu_3(BTC)_2$@TiO_2	$Cu_3(BTC)_2$ BTC is benzene-1,3,5-tricarboxylate	CH_4 (2.64 μmol/g-TiO_2. h)	Synthesize $Cu_3(BTC)_2$ first then coating TiO_2 shell using hydrothermal method.	100 mL quartz tube reactor; 300 mg $Cu_3(BTC)_2$@TiO_2 was placed on quartz boat containing 5 ml H_2O, 0.15 MPa CO_2; and 300 W Xe lamp was used as the light source	[53]
39	Reduced graphene oxide-CdS nanorod	Reduced graphene oxide	CH_4 (2.51)	Modified Hummers' method for GO preparation, microwave-solvothermal method for rGO-CdS nanorod synthesis	100 mg sample was deposited onto the bottom of 200 mL reactor, 0.12 g $NaHCO_3$ and 0.25 mL-4 M HCl was introduced into the reactor as the CO_2 and H_2O source, 300 W Xe lamp with UV-cutoff ($\lambda > 420$ nm) was used as the light source	[54]
40	CND/pCN	Carbon nanodots	CH_4 (29.23); CO (58.82)	Alkali-assisted ultrasonication method to prepare CNDs first and then mixing CNDs and pCN and followed by hydrothermal treatment	50 mg sample fixed into quartz tube, CO_2 passed through a water bubbler to generate CO_2 and H_2O mixture, 1 atm and ambient temperature; 500 W Xe lamp was used as the light source	[55]
41	Uio-66/C_3N_4 nanosheet	Uio-66	CO (9.9)	C_3N_4 nanosheet was obtained by liquid exfoliating bulk C_3N_4 in water and then the C_3N_4 nanosheet self-assemble with Uio-66 in water	The photocatalyst was dispersed in the reactor cell with 5 mL (MeCN/TEOA = 4:1) solution, after evacuation, 80 kPa CO2 was injected into the system, a 300 W Xe lamp with two filters ($400 < \lambda < 800$ nm) was used as the light source	[56]

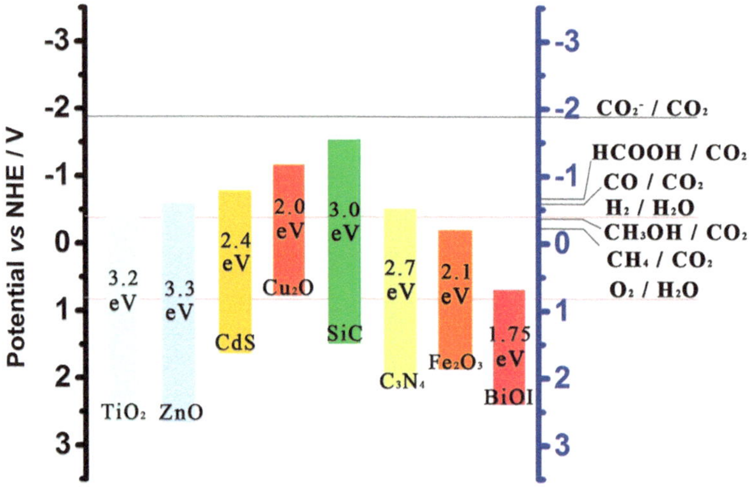

Fig. 12.1 Band structures of several common semiconductors and corresponding standard reduction potential involved with CO_2 and H_2O photocatalytic reduction at pH =7

In most semiconductor-based CO2PR reactions, H_2O is selected as the sole electron donor, in this case, the H_2O oxidation reaction to release H protons and O_2; hydrogen evolution reaction (HER) to generate H_2 has also two major steps: Eqs. 12.6 and 12.7:

$$2H_2O + 4h^+ \rightarrow 4H^+ + O_2 \qquad E_{redox}{}^0 = -0.82V \tag{12.6}$$

$$2H^+ + 2e^- \rightarrow H_2 \qquad E_{redox}{}^0 = -0.41V \tag{12.7}$$

Owing to the complexity of CO_2 reduction reaction that involved multi-electrons and protons' participation and various intermediates' adsorption/desorption processes, the deep understanding of the CO_2 reduction is still unclear both experimentally and theoretically. To date, researchers have proposed two classic CO_2 reduction pathways which are called formaldehyde pathway (also known as multiple proton-coupled electron transfer) and carbene pathway (also known as deoxygenation path), respectively [5, 27]. However, the carbene pathway is initially involved with one electron's CO_2 reduction to generate $CO_2{}^-$ (Eq. 12.5) which cannot be accomplished by many semiconductors owing to the limited reduction potential (Fig. 12.1) [5, 26]. Recently, Ji et al. [27] using TiO_2 (101) as the prototype proposed a new mechanism which involved CO_2 photoreduction at Ti site and oxygen vacancy site based on DFT calculation. Their result shows that the O vacancy (O_v) served as the active site to bind the intermediates like CH_2O or CH_3O and facilitate the CH_4 or CH_3OH generation; besides, the O_v also offers two electrons to protect the intermediates from reoxidation. Meanwhile, the intermediates adsorbed on Ti site could be oxidized by holes rapidly and result in low CO_2 photoreduction efficiency. Still, considering the alternative catalytic conditions among different photocatalysts, the study of the reaction mechanism of CO2PR remains challengeable.

12.3 Cocatalysts in Semiconductor-Based CO_2 Photoreduction

12.3.1 Preparations

12.3.1.1 Metal–Semiconductor Composites

There are numerous methods that are involved with the metal precursors' reduction and further deposit metal NPs on the semiconductors' surface, leading to the formation of metal–semiconductor composite. The common methods include photo-deposition method, alcohol reduction method, chemical reduction method (common reducing agents such as $NaBH_4$, ascorbic acid, glucose, trisodium citrate, hydrazine, etc.), deposition–precipitation (DP) method, atomic layer deposition (ALD) method, etc. (Table 12.1).

Xie et al. [10] synthesized Pt–TiO_2 composites with the use of three different methods: (1) photo-deposition of Pt NPs on titania with 300 W Xe lamp as light source and methanol as the sacrificial reagent, (2) impregnation of Pt precursor with titania and followed by calcination treatment in H_2 at 673 K, and (3) hydrazine reduction of H_2PtCl_6 an aqueous solution containing titania. The TEM results indicate that the photo-deposition method and hydrazine reduction method both result in smaller-sized Pt particles (mean size, 3.7 and 4.2 nm, respectively) and the impregnation method followed by H_2 calcination will result in bigger-sized Pt particles (6.8 nm). Wang et al. [9] adopted a unique tilted-target sputtering (TTS) method for the ultrafine Pt cluster deposition on 1D TiO_2 single-crystal film. The loading amount of Pt and cluster size (0.5–2 nm) were manipulated by adjusting the deposition time (5–60s). Song et al. [15] studied the shape-dependent Pd/C_3N_4 few-layer composites in CO2PR. During the synthesis process, the author used HCHO and $Na_2C_2O_4$ to promote the formation of Pd (111) facets, while the Br^- and I^- were introduced to stabilize the Pd (100) facet. As a result, Pd cube/C_3N_4 and Pd nanotetrahedron/C_3N_4 can be well obtained through a solution-phase solvothermal method. Despite their shape, these two Pd polyhedrons have comparable particle size (4–6 nm), which are the smallest Pd nanocrystals with specific facets obtained in aqueous phase by now.

Compared with single-unit metal–semiconductor composite, binary metal alloy NPs with diverse surface active sites and metal-support interfaces thus could possess more potential in photocatalysis. In the synthesis of alloy NPs involved with at least two metal precursors, Long et al. [18] synthesized a series of PdxCu1 fcc-phased NPs in situ growth on the TiO_2 nanosheet in the presence of ascorbic acid and PVP. Through varying the ratio of K_2PdCl_4 to $CuCl_2$, sphere-like NPs of CuPd1Cu1, Pd3Cu1, Pd5Cu1, Pd7Cu1, and Pd11Cu1 could be obtained. Neaţu et al. [14]

adopted the stepwise deposition–precipitation method to prepare Au–Cu alloy/TiO_2 photocatalyst, using 0.2 M NaOH to tune the pH value of TiO_2-$HAuCl_4$ and Au/TiO_2-$Cu(NO_3)_2$ aqueous slurry to 8.5 and annealing in air and H_2 atmosphere, respectively. The Au–Cu alloy NPs were characterized by HR-TEM which constitute Au–Cu lattice fringes; meanwhile, the redshift of Au SPR adsorption peak in UV-vis DRS spectra also suggests the Au and Cu formed alloy status. It should be noted that, using this synthesis method, the unalloyed Au and Cu NPs also could be detected.

12.3.1.2 Semiconductor Heterojunctions and Z-Scheme Composites

In order to develop the economic noble metal-free photocatalysts with high efficiency in CO2PR, some semiconductor junction composites have been developed; the common methods include hydrothermal/solvothermal method, impregnation method, self-template method, ALD method, etc. (Table 12.1).

Jin et al. [38] reported a hierarchical-structured Z-scheme CdS-WO_3 photocatalyst and applied it in CO2PR. The hierarchical hollow WO_3 spheres were formed by immersing $SrWO_4$ in HNO_3 at first, and then the precipitate was washed and calcined at 500 °C in air. The as-prepared WO_3 spheres were negatively charged at pH = 7; therefore, stepwise adding Cd^{2+} and S^- source slowly will generate heterostructure CdS–WO_3 composite. Wang et al. [34] developed a porous ZnO@Co_3O_4 composite by using ZIF-8 and ZIF-67 as precursor templates (Fig. 12.2). First, the ZIF-8@ZIF-67 core–shell structure was synthesized through a solvothermal process and then followed by a N_2-400 °C 2-h calcination and air-400 °C 2 h calcination treatment. The two-step calcination process was determined by the TG/DTA analysis, while one-step calcination under air atmosphere will lead to nonporous ZnO NPs.

In et al. [37] designed a novel CuO–$TiO_{2-x}N_x$ hybrid hollow nanocubes with the use of CuN_3 nanocubes as reactive templates (Fig. 12.3a). After slow hydrolysis of titanium-n-butoxide on the surface of CuN_3, the calcination treatment at 450 °C was carried out. During the calcination process, the CuN_3 reacts with the oxygen and form hollow CuO nanocube; meanwhile, the nitrogen diffuses outward and reacts with the crystalline TiO_2 to form $TiO_{2-x}N_x$. Park and coworkers [57] once proposed a novel Cu_xO–TiO_2 p–n heterojunction (Fig. 12.3b). The Cu/Cu_2O nanoparticles were first synthesized through a thermal decomposition method, and then the $TiCl_4$ was mixed with the Cu/Cu_2O NPs in argon; after calcination in air, the $TiCl_4$ crystallized to TiO_2, and the Cu/Cu_2O NPs were oxidized to Cu_xO with organic ligands removed at the same time; at last, the mesoporous Cu_xO–TiO_2 composites were obtained.

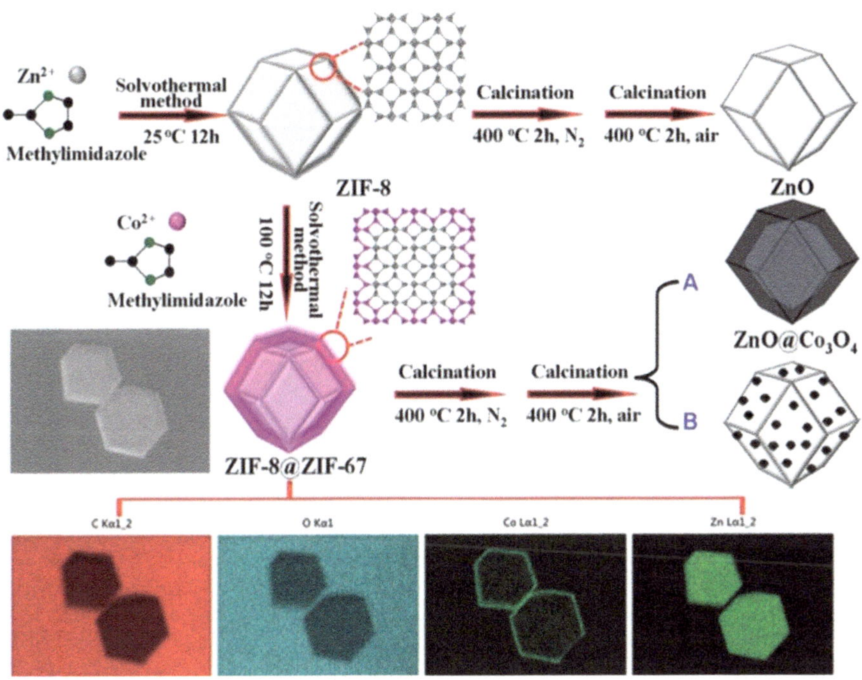

Fig. 12.2 Schematic illustration of the synthesis of polyhedral ZnO and ZnO@Co_3O_4 originated from ZIF8 and ZIF-8@ZIF-67, respectively. (Reprinted with permission from Ref. [34]. Copyright 2016, Royal Society of Chemistry)

12.3.1.3 Dual Cocatalysts

Considering the precise construct of the configuration of photocatalyst with synergistic dual cocatalysts will enhance the CO2PR efficiency greatly. Recently, more and more researches have focused on the dual cocatalysts deposition with advanced structures. Generally, compared with sole cocatalyst composites, the synthesis of dual cocatalysts is more complicated, which need involve with stepwise deposition of dual units. The spatial locations of dual cocatalysts should depend on the functions of these two species which are either separated or combined with each other.

The construction of spatial separated electron trapping agents and hole collectors could greatly promote the charge separation efficiency of the photocatalyst. Dong et al. [8] developed a 3D hierarchical structured TiO_2–SiO_2 with CoO_x and Pt growing inside and outside of the skeleton. Firstly, the $Co(AC)_2.4H_2O$ and Ti–Si sol were mixed together and underwent a synchronizing self-assemble process; after the 500 °C calcination, the hydrolyzed $Co(OH)_x$ transformed into CoO_x NPs embedded under the hierarchical TiO_2–SiO_2 skeleton homogenerously (denoted as the HCTSO). Subsequently, the Pt NPs were growing in situ on the outer surface of the HCTSO via alcohol reduction of H_2PtCl_6. In order to improve the CO_2

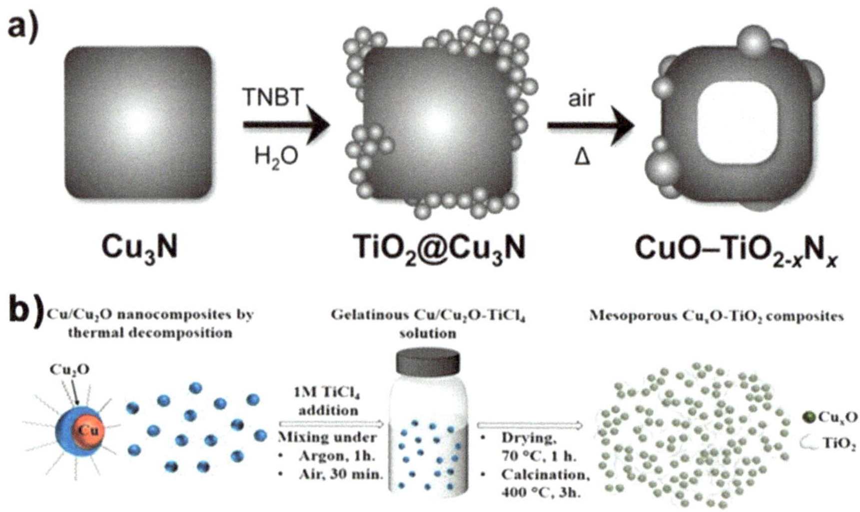

Fig. 12.3 (**a**) Scheme illustration of multistep template strategy to convert Cu_3N nanocube into $TiO_2@Cu_3N$ and hollow $CuO–TiO_{2-x}N_x$ nanocubes. Reprinted with permission from Ref. [37]. Copyright 2012 Jonh Wiley & Sons, Inc. (**b**) Scheme diagram of the step-by-step synthesis of mesoporous $Cu_xO–TiO_2$ composite. (Reprinted with permission from Ref. [57]. Copyright 2016, American Chemical Society)

adsorption and the selectivity toward CH_4, Pan et al. [47] developed a 5-nm-thick carbon layer coated on In_2O_3 nanobelt coupled with Pt NP+ loadings. The glucose was used as the carbon source and the carbon layer was formed at 600 °C under the Ar atmosphere. Afterward, the Pt NPs were deposited on the surface of carbon layer through a photo-deposition method. The loading amount of carbon layer and Pt are 8% and 2%, respectively. Another classic dual cocatalysts structure of $Pt@Cu_xO$ loaded on TiO_2 (p25) was proposed by Zhai and coworkers [41]. At first, $Pt–TiO_2$ was first prepared by photoreduction of H_2PtCl_6 in the TiO_2 suspension; afterward, the Cu species were deposited on the Pt surface under the illumination and using the $CuSO_4$ as the precursor. The Cu is easily oxidized into Cu^1 in air; therefore, the Cu existed as Cu_xO form. Moreover, prolonging the photo-deposition time of Cu species will increase the Cu_xO coverage on Pt NPs; the 5-h irradiation will form a complete $Pt@Cu_xO$ core–shell structure.

The all-solid-state Z-scheme photocatalysts could take advantage of more negative reduction potential electrons and more positive oxidation potential holes from different semiconductors counterparts, thus attract more and more attention in CO2PR. Generally speaking, the photosystem II (oxidation part PSII) and photosystem I (reduction part PSI) are connected by a conductor. Li et al. [45] developed an elegant all-solid Z-scheme $WO_3/Au/In_2S_3$ nanowire photocatalyst; the WO_3 nanowire was first grown on the tungsten foil under Ar flow with WO_3 powder as precursor, then Au NPs were deposited on WO_3 nanowire by plasma sputtering method, and the In_2S_3 shell coated on Au surface was finally obtained through a chemical vapor deposition method (In_2S_3 powder and Au/WO_3 were placed in quartz furnace separately, the temperature of the furnace will be increased to 800 °C with certain Ar flow, and the deposition time is 10 min).

12.3.1.4 Carbon-Based Cocatalysts

Besides the metal nanoparticles and semiconductors cocatalysts, there are intensive research works focusing on the carbon-based cocatalysts in CO2PR. Graphene, carbon nanotube, carbon nanodots, carbon nanosheet layer, metal organic frameworks (MOFs), metal–ligand complex, etc. are common carbon-based cocatalysts in recent years (Table 12.1).

Graphene, owing to its flexible mechanical strength, remarkable electric conductivity, high surface area, etc., has been applied to many fields. In photocatalysis, graphene can transfer the photo-generated electrons, hence improving the electron-hole separation efficiency and prolonging the lifetime of charge carriers (Fig. 12.5). Tu et al. [49] fabricated a sandwich structure TiO_2–graphene nanosheet over a one-step hydrothermal method. The graphene oxide (GO) was prepared according to Hummers' method which is a common method for many reported graphene–semiconductor composites. During the hydrothermal process, the GO, Ti precursor, and solvent (ethylenediamine abbreviated as En/H_2O) underwent an in situ simultaneous reduction–hydrolysis process, the GO was reduced by En, and the Ti precursor hydrolyzed to form TiO_2 NPs. Different weight ratios of TiO_2/graphene were obtained by varying the GO amount during the synthesis. Ong et al. [58] adopted the electrostatistic self-assembly strategy to prepare the reduced graphene oxide (rGO)/protonated C_3N_4 (pCN) composites. Owing to abundant CN motifs existing on the g-C_3N_4 surface, the surface protonation by HCl could be easily conducted. After the HCl treatment, the pCN was positively charged according to the Zeta potential test, which could spontaneously assemble on the negative-charged GO (prepared by Hummers' method). Finally, the GO was reduced to rGO by $NaBH_4$ to form 2D/2D rGO/pCN composite. Unlike 2D graphene–semiconductor composites, Zhang proposed that encapsulation by graphene-like carbon sheet could enhance the confinement effect of the core nanoparticles compared with its naked counterparts. Therefore, Fe@C NPs were fabricated for the use of MIL-101 as self-sacrificing template and precursor. During the synthesis, two-step calcination method was utilized; first, MIL-101(Fe) was collapsed and formed Fe_3C and Fe_3O_4 in Ar-500 °C; meanwhile, the Fe species could avoid sintering into large NPs and , subsequently, the temperature raised to 700 °C to obtain the Fe@C NPs. It should be noted that rational regulate the calcination temperature and retention time is the key to control the particle size and graphite carbon layer's thickness.

Metal organic frameworks (MOFs), as one class of porous nanocrystals, possess huge surface area, tunable surface functional groups, and alternative compositions which have been applied to multiple fields such as catalysis, gas capture and separation, drug delivery, molecule identification, etc. Due to strong CO_2 adsorption capability of UiO-66, cooperation with some narrow bandgap semiconductor could improve the CO2PR efficiency. Shi et al. [56] developed an electrostatic self-assembly strategy to fabricate the UiO-66/C_3N_4 composite. Firstly, the C_3N_4 nanosheet (CNNS) was prepared by liquid-state ultrasound exfoliation method; after the centrifuge to remove the large bulk C_3N_4, the CNNS was obtained. Secondly, the as-prepared UiO-66 and CNNS were mixed in water, because the CNNS is negatively charged in water with −35.91 mV Zeta potential and + 7.71 mV for UiO-66; that is the reason why electronic statistic self-assembly happened. Li et al. [53] developed a $Cu_3(BTC)_2$ (HKUST-1)@TiO_2 core–shell structured composite; the solid Cu precursor and the involvement of PVP are key to coat TiO_2 on the $Cu_3(BTC)_2$ nanocrystals uniformly. During the control experiments, using Cu $(OH)_2$ as Cu precursor and in the absence of PVP, the TiO_2 cannot be coated on the $Cu_3(BTC)_2$ uniformly; besides, using unsolid $Cu(NO_3)_2$ as precursor, the thermal stability of $Cu_3(BTC)_2$ is low; in this case, it will decompose at the 180 °C coating process and also cannot get desirable result.

Carbon nanodots, including carbon quantum dots, carbon dots, and graphene quantum dots, which are a new class of zero-dimensional (0D) carbon materials have attracted people's attention over the past few decades; the unique properties of carbon dots such as superior up-conversion and size-dependent photoluminescence, high stability, low cytotoxicity, earth abundance, etc., thus made it a plausible candidate in many fields. The synthesis of carbon nanodots can be roughly classified into two approaches: bottom-up approach and top-down approach. Kang et al. [59] reported a facile electrochemical approach to synthesize large-scale high-quality carbon dots. The authors used two graphite rods as the counter electrode and ultrapure water as the electrolyte; statistic potential with 15–60 V was applied to the two electrodes; after 120-h electrolysis, a dark-yellow solution was formed, and the water-soluble carbon dots were obtained after filter and centrifuge. Ong et al. [55] adopted glucose as the carbon source using alkali-assisted ultrasonication method to prepare carbon nanodots (CND). Briefly, glucose and NaOH solution was mixed together and sonicated for 2 h and resulted in a dark-brown solution; after neutralization and filter, a brown carbon dot solution was obtained. Owing to the natural properties of same negative polarity of C_3N_4 and carbon dots, the coupled CND/C_3N_4 in this paper was obtained by protonation C_3N_4 in HCl solution in advance; after that, the C_3N_4 is positively charged thus can attract the CND by electrostatic force.

12.4 Roles and Properties of Different Cocatalysts

12.4.1 Promote the Charge Separation and Transfer

It is well-known that the noble metal NPs such as Pt, Pd, Au, Ag, Ru, etc. loaded onto the semiconductors could trap the photo-generated electrons and promote the separation of charge carriers. The reason could be attributed to the Fermi level of metal NPs which lies energetically below the conduction band level of its semiconductor counterpart; besides, the Schottky barrier formed at the metal–semiconductor interface thus could prevent the electrons from flowing backward. In this way, the surface sites of metal NPs become the active sites for the CO2PR reduction, and the performance of the metal–semiconductor composite is highly depending on the electron trapping ability of the supported metal NPs.

Xie et al. [10] compared the activity of five noble metals (Ag, Au, Rh, Pt, Pd) supported on TiO_2 in CO2PR. The yield of CH_4 and the rate of total electrons' consumption in the CO2PR increase with the order of TiO_2 < Ag-TiO_2 < Rh-TiO_2 < Au-TiO_2 < Pd-TiO_2 < Pt–TiO_2, which equals with the same trend of the work function of these noble metals. This result reflects the fact that the electron trapping ability contributes to the reductive performance of supported metal catalyst and Pt–TiO_2 in this evaluation system is superior to the others. Since Pt is a very efficient cocatalyst in photocatalysis, the rational designation of Pt NPs with suitable shape (expose certain facets) and particle size (both geometric and electronic) is important. As we mentioned before, Wang et al. [9] synthesized a series of different-sized Pt NPs ranging from 0.5 to 1.5 nm loaded on the 1D TiO_2 single crystals through a TTS method, and the 1 nm Pt NPs show the highest CH_4 yield. The author claims that the ultrasmall Pt NPs (less than 1 nm) could prevent the electrons transferring from TiO_2 because of its higher energy band compared with the CB of TiO_2; on the contrary, bigger Pt NPs act as electron-hole recombination center which is also detrimental in the photocatalysis. Furthermore, the author adopted the femtosecond time-resolved TA spectroscopy to elucidate the charge transfer dynamics. After liner fitting, Pt–TiO_2 shows a greater slope compared with its TiO_2 counterpart which directly reflects the Pt NPs suppress the charge recombination process.

In order to replace the expensive noble metal cocatalysts into some earth-abundant materials, the development of noble metal-free cocatalysts with comparable performance is essential. Among them, carbon dots and graphene also play key role in promoting the charge separation in photocatalysis. Ong et al. [55] report carbon nanodots (CND) supported on protonated C_3N_4 composite. The obtained CND/pCN shows the CND with 4.4 nm diameter dispersed well on the pCN surface, and the CND did not affect the adsorption edge of C_3N_4 but act as conductive electron channel for charge separation (Fig. 12.4a, b). The author adopted steady-state PL spectroscopy and time-resolved transient PL decay to verify the charge separation kinetics. The pure p-C_3N_4 shows an intensive and broad PL emission peak which means a great extent of electron-hole recombination; the CND/pCN hybrids on the other hand show obvious decrease of peak intensity which suggests the recombination of charge carriers were suppressed (Fig. 12.4c). The emission lifetime of CND/pCN reduced compared with pCN which means the rapid interfacial electrons inject from pCN to CNDs and participate in the CO2PR reaction (Fig. 12.4d). Besides the electrons' trapping ability, CNDs also can serve as the photosensitizer. Yu et al. [60] reported CDs/TiO_2 composite with enhanced visible light hydrogen production rate. The author claimed that π-conjugated CDs sensitize TiO_2 by forming C-O-Ti bond and donate electrons under visible light irradiation.

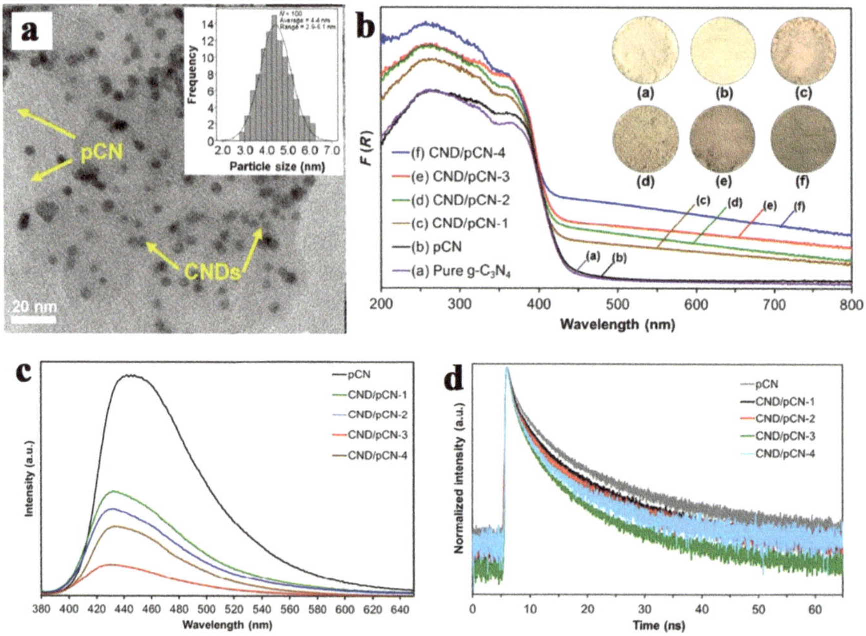

Fig. 12.4 (**a**) TEM image of CND/pCN-3 sample, the inset image shows the particle size distribution of CNDs deposited on the pCN nanosheet. (**b**) UV–vis DRS spectra of different samples; the digital photographs showing the colors of different samples are inset. (**c**) PL spectra of pCN and CND/pCN samples. (**d**) Time-resolved transient PL decay curves of pCN and CND/pCN samples excited at 405 nm. (Reprinted from Ref. [55]. Copyright 2017, Wee-Jun Ong et al. licensee Springer)

When graphene was introduced as the cocatalyst, Yu et al. [54] developed a metal-free CdS/rGO composite with enhanced CH_4 generation rate (2.51 μmol/g·h) which is ten times higher than pure CdS rods and overperforms the Pt/CdS. The enhanced performance was attributed to the π–π conjugate interaction between CO_2 and graphene and thus improves the CO_2 adsorption amount and destabilizes CO_2; besides, the rGO promotes the electron transfer, and storage was confirmed by conducting transient photocurrent and impedance analysis (Fig. 12.5b). Recently, Xu et al. [52] reported a $CsPbBr_3$ perovskite graphene composite which could efficiently convert CO_2 into CH_4 with 99.3% selectivity. The author also adopted steady-state PL and time-resolved PL decay to probe the electron transfer dynamic. Distinct PL intensity quenching of $CsPbBr_3$ QD and the PL decay time of $CsPbBr_3$ QD/GO composite are shorter compared with $CsPbBr_3$ QDs, which both reflect the introduction of GO benefits to the electron transfer and suppress the electron-hole recombination (Fig. 12.5d).

Construction of the semiconductors' heterojunction is another strategy to improving the charge separation efficiency in CO2PR. Shi et al. [61] reported a visible light responsive $g\text{-}C_3N_4/NaNbO_3$ nanowire with higher CO2PR activity than either $g\text{-}C_3N_4$ or $NaNbO_3$. From Fig. 12.6 (a), the HR-TEM image shows the obvious

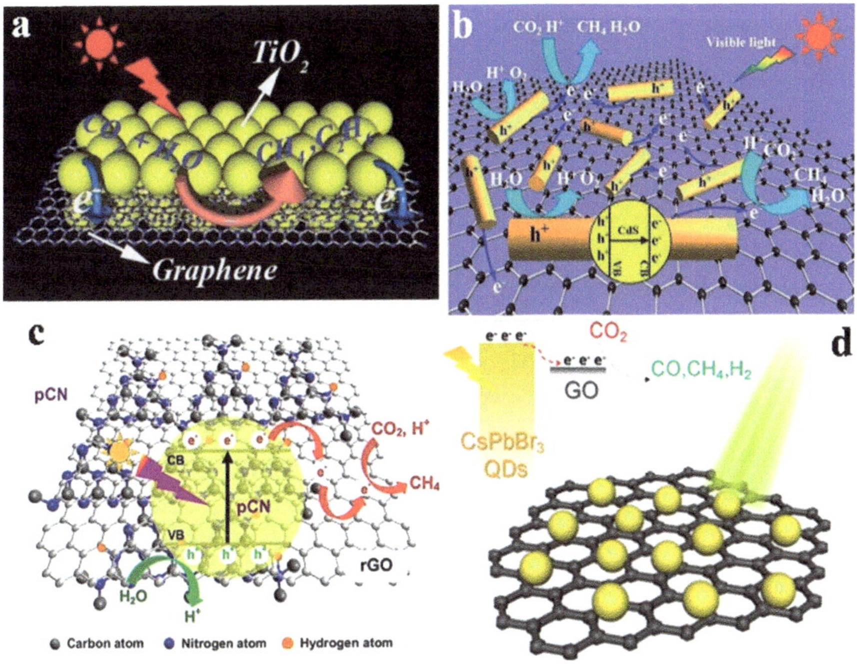

Fig. 12.5 (**a**) Schematic illustration of charge separation and transfer in G-TiO_2 system and photoreduction of CO_2 and H_2O. Reprinted with permission from Ref. [49]. Copyright 2013 Jonh Wiley & Sons, Inc. (**b**) Schematic illustration of charge separation and transfer in CdS-rGO composite. Reprinted with permission from Ref. [54]. Copyright 2014 Royal Society of Chemistry. (**c**) Schematic illustration of the charge transfer and separation in rGO/pCN nanocomposite for CO_2 photoreduction with H_2O to CH_4. Reproduced from Ref. [58] with permission of Elsevier. (**d**) Schematic diagram of CO_2 photoreduction over $CsPbBr_3$ QDs/rGO. (Reprinted with permission from Ref. [52]. Copyright 2017 American Chemical Society)

intimate interface, which implies the existence of heterojunction between C_3N_4 and $NaNbO_3$. The band structures of g-C_3N_4 and $NaNbO_3$ were determined by UV–vis DRS and VB-XPS in Fig. 12.6 (b, c). The wavelength (λ) of adsorption edge of g-C_3N_4 and $NaNbO_3$ is 365 nm and 450 nm, respectively; therefore, the E_g (bandgap energy = 1240/λ) was calculated to be 3.4 eV and 2.7 eV, respectively. Meanwhile, the VB XPS result shows that the E_{VB} (valence band energy) of g-C_3N_4 and $NaNbO_3$ was located at ~1.57 eV and 2.7 eV, respectively; therefore, the E_{CB} (conduction band energy) of g-C_3N_4 and $NaNbO_3$ was calculated to be −1.13 eV and − 0.77 eV based on the equation $E_{CB} = E_{VB} - E_g$. Since the CB level of C_3N_4 is more negative than $NaNbO_3$, the photo-excited electrons from C_3N_4 could migrate to $NaNbO_3$ and suppress the electron-hole carriers' recombination.

Although p–n heterojunctions greatly inhibit the recombination of photo-generated electrons and holes, however, after the photo-generated electrons migrate to the CB with lower reduction potential, the redox ability of the integral composite

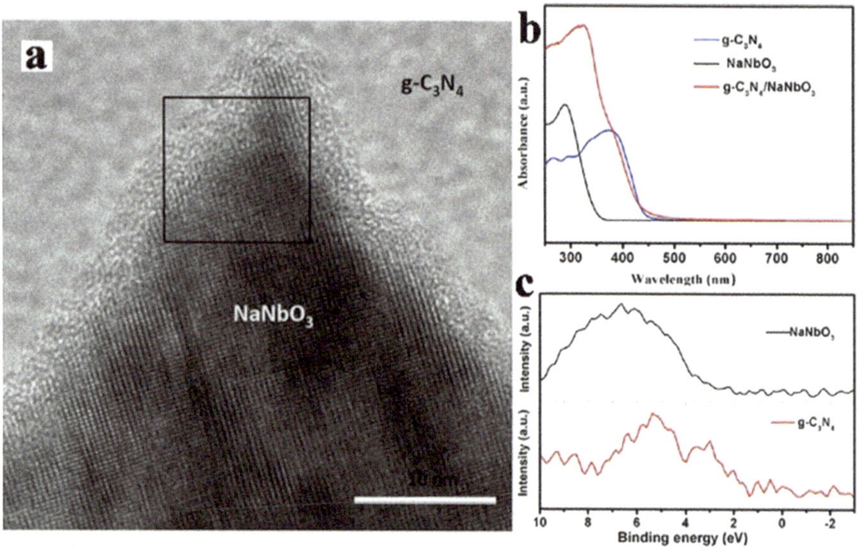

Fig. 12.6 (**a**) HR-TEM image of g-C_3N_4/$NaNbO_3$ heterojunction. (**b**) UV–vis DRS spectra of different samples. (**c**) VB-XPS spectra of g-C_3N_4 and $NaNbO_3$. (Reprinted with permission from Ref. [61]. Copyright 2014, American Chemical Society)

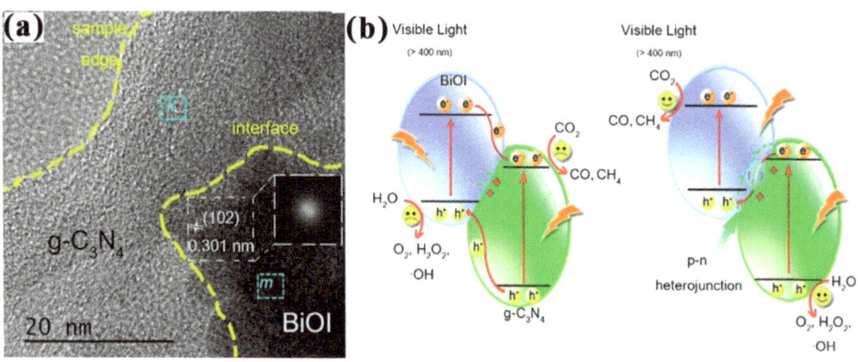

Fig. 12.7 (**a**) HR-TEM image of 7.4-BiOI/C_3N_4 composite. (**b**) Schematic illustrations of double-charge transfer mechanism (left) and Z-scheme charge transfer mechanism (right). (Reprinted with permission from Ref. [39]. Copyright 2016 American Chemical Society)

will be impaired [23]. Therefore, the all-solid-state Z-scheme system was developed to solve this problem. Generally, the all-solid-state Z-scheme photocatalytic system (PS) was divided into two components: PS-PS and three-component PS-conductor (C)-PS. Wang et al. [39] construct an indirect PS-PS Z-scheme BiOI/C_3N_4 composite with enhanced visible light CO2PR performance (Fig. 12.7a). In order to investigate the charge transfer modes, the author proposed two possible charge transfer routes: double-transfer mechanism and Z-scheme mechanism. The contrast experiment was adopted by using visible light as the light source; the result of no product yield proved that the electron transfer was not followed by the previous one. Furthermore, when the EDTA was added as the hole scavengers, the result shows an improved CO and H_2 yield but decreased O_2 yield which further reflects the enhanced electron-hole separation (Fig. 12.7b).

Compared with the PS–PS Z-scheme, the interface between two solids usually contains many defects, which may inhibit the charge transfer. Therefore, the PS-C-PS Z-scheme with a conductor insertion could reduce the electron transfer resistance and thus improve the CO2PR efficiency. He et al. [25] reported an Ag_3PO_4/C_3N_4 composite with enhanced CO2PR activity. Since the Ag_3PO_4 is not stable, therefore, Ag NPs were formed in situ within the composites under light irradiation and acted as the electron mediator. Taking into account of the CB level of Ag_3PO_4 (0.45 eV), if the composites followed the double-charge transfer mechanism, the introduction of Ag_3PO_4 cannot promote the CO2PR. So it is reasonable to believe that the charge transfer route followed Z-scheme mechanism. In this way, Ag accepts the photo-generated electrons from Ag_3PO_4 and recombines with holes from C_3N_4; subsequently, the photo-generated electrons with more negative reduction potential could be used into the CO2PR reaction. Wei et al. [44] developed a PS–C–PS Z-scheme structure photocatalyst that contains CdS (shell), Pt (core), and TiO_2 (support), which show enhanced CO_2 photoreduction activity and selectivity (36.8 μmol/g.h CH_4 yield and 98.1% CH_4 selectivity). The location of reduction sites and electron transfer route was confirmed by Ag photo-deposition method. The Ag NPs selectively deposited on the shell of CdS instead of TiO_2 surface which clearly demonstrates that the CdS acts as the reduction site and the electron transfer follows the $TiO_2 \rightarrow Pt \rightarrow CdS$ route. In order to present consolidate proof to Z-scheme charge transfer behavior, Li et al. [45] first adopt Kelvin probe force microscopy to detect surface potential change of In_2S_3-Au-WO_3. Compared with WO_3/In_2S_3, the SPV image (reflect the concentration of photo-generated holes) of $WO_3/Au/In_2S_3$ shows significant change from 10 mV to 30 mV; this difference vividly reflects the efficient charge separation and the role of Au as the electron mediator.

Another way to improve the charge separation efficiency in photocatalysis is the construction of double cocatalysts (usually refers to the electron trapping agent and hole collect agent) with spatial separated configurations. Domen and coworkers [62] first developed Ta_3N_5 photocatalyst hollow shell with Pt and CoO_x deposited inside and outside of the shell, respectively. Followed by this pioneered work, similar strategies have been proposed such as the thin heterojunction Pt–TiO_2@In_2O_3@MnO_x hollow shell structure, porous TiO_2 tube, or hollow C_3N_4 shell with spatial separated Pt and CoO_x NPs [63–65], etc. Recently, our group developed a new strategy to construct spatial configuration by introducing Pt NPs and CoO_x NPs outside and inside of the skeleton of hierarchical TiO_2-SiO_2 (HTSO) [8], abbreviated as Pt/HCTSO. The HR-TEM image clearly indicates that the Pt NPs and CoO_x NPs separated by the HTSO skeleton (Fig. 12.8a); on the other hand, EDS-mapping image shows the Pt and Co species are well-dispersed throughout the framework of HTSO and no aggregation happened (Fig. 12.8b). The CO_2 photoreduction evaluation result revealed that the 0.8% Pt/HCTSO (0.8%) shows enhanced

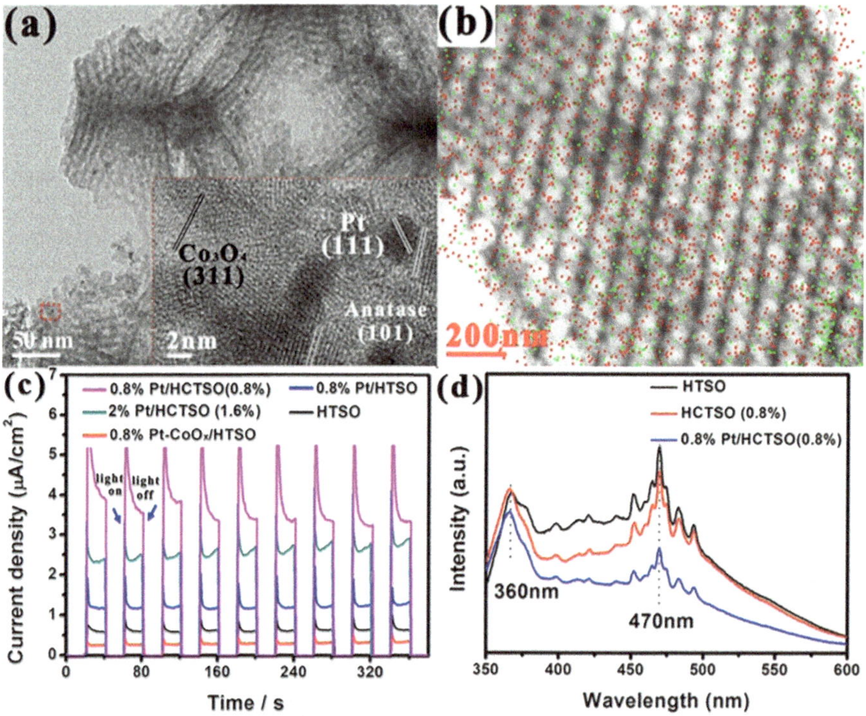

Fig. 12.8 (**a**) TEM and selective HR-TEM images of Pt/HRTSO. (**b**) Elemental mapping image of Pt/HRTSO, the red dots denote as Pt element and the green dots denote as Co element. (**c**) Transient photocurrent spectra of different samples (300 W Xe lamp with AM 1.5 filter was used as the light source and 0.5 M Na_2SO_4 solution is used as the electrolyte). (**d**) Room temperature PL spectra of different samples (excitation wavelength at 315 nm). (Reprinted with permission from Ref. [8]. Copyright 2016, Royal Society of Chemistry)

CH_4 yield and selectivity for CH_4 which are 1.9 and 4.4 times higher than 0.8% Pt/HTSO. To highlight the spatial locations of Pt and CoO_x in CO2PR, the Pt-CoO_x/HTSO was prepared by randomly loading Pt and CoO_x on the surface of HTSO; the CO2PR result shows even lower CO_2 reduction activity. The enhanced transient photocurrent response and decreased intensity of PL emission peaks (360 and 470 nm) all confirmed the spatial separated double cocatalysts promote the charge separation effectively (Fig. 12.8c, d). On the contrary, the random loading of Pt and CoO_x on the surface of HTSO results in many electron-hole recombination centers, which is detrimental to charge separation and thus shows poor performance in CO2PR.

12.4.2 Improve CO_2 Adsorption and Activation

The CO_2 adsorption and activation on the surface of photocatalyst are two important steps; however, without modification, common semiconductor-based photocatalysts often show low CO_2 uptake. Therefore, combining some unique cocatalysts with higher CO_2 adsorption amount should be a proper way to improve CO2PR efficiency. Xie et al. [66] in the use of MgO, a basic metal oxide, as the cocatalyst deposited on the TiO_2 surface, with the addition of Pt NPs, the Pt-MgO/TiO_2 composite shows an enhanced activity for CH_4 production. During the experiments, a linear relationship between different CO_2 chemisorption by different basic metal oxide-modified Pt–TiO_2 and CH_4 yields clearly demonstrates the important role of CO_2 adsorption; the MgO modification shows the highest CO_2 chemisorption compared with other basic metal oxides. Besides, the optimal MgO content is measured to be 1%; excess MgO adding will cause a thicker MgO layer and cover the Pt sites which are detrimental for CO_2 photoreduction. Li et al. [53] adopt MOF ($Cu_3(BTC)_2$) as the CO_2 adsorption cocatalyst and coat porous TiO_2 shell on the MOF crystals' surface. This unique design hybrid shows enhanced CH_4 yield and selectivity compared with bare TiO_2 counterpart. The CO_2 adsorption results between bare MOF and MOF@TiO_2 suggest the CO_2 molecules can easily pass the TiO_2 shell. In order to investigate the charge transfer and working mechanism, the author adopts TA analysis and first-principle simulation. The result indicates the photo-excited electrons can transfer to the MOF core; subsequently, the CO_2 molecules adsorbed in MOF can be activated and convert into CH_4 effectively. Similarly, Shi et al. [56] reported a C_3N_4/UiO-66 composite, in this work zirconium-based MOF: UiO-66 acts as both CO_2 absorber and a semiconductor-like material to promote the electron-hole separation. ESR was used to verify the electron transfer route. Specifically, signal of $g = 2.009$ is attribute to O_2^- which was found in C_3N_4/UiO-66 under visible light irradiation but absent in pristine UiO-66. This indicates that the C_3N_4 was performed as a photosensitizer; the photo-generated electrons transferred to UiO-66 and thus suppress the electron-hole recombination and enhance the CO2PR performance. Pan et al. [47] reported a carbon-coated In_2O_3 photocatalyst with the use of glucose as the carbon source; the 5 nm carbon layer could enhance the CO_2 chemisorption and suppress the hydrogen generation (Fig. 12.9). Compared with the pure In_2O_3 nanobelt, C-In_2O_3 shows enhanced CO_2 adsorption capacity compared with pristine In_2O_3 nanobelt, and the maximum CO_2 adsorption was reached with the use of 0.8 g glucose (Fig. 12.9b). The selectivity of CH_4 was studied by the thermodynamic and kinetic behavior of H proton transfer route in the assistance of DFT calculation. The result indicates that the H proton transfer to adsorbed CO_2 in Pt_2/C-In_2O_3 is easier than H_2 formation (endothermic); on the contrary, H proton reduction to H_2 on Pt_2/P-In_2O_3 is exothermic, which is easier than Pt_2/C-In_2O_3. This result well-explained the high CH_4 yield and CO2PR selectivity of Pt/C-In_2O_3 compared with Pt/P-In_2O_3.

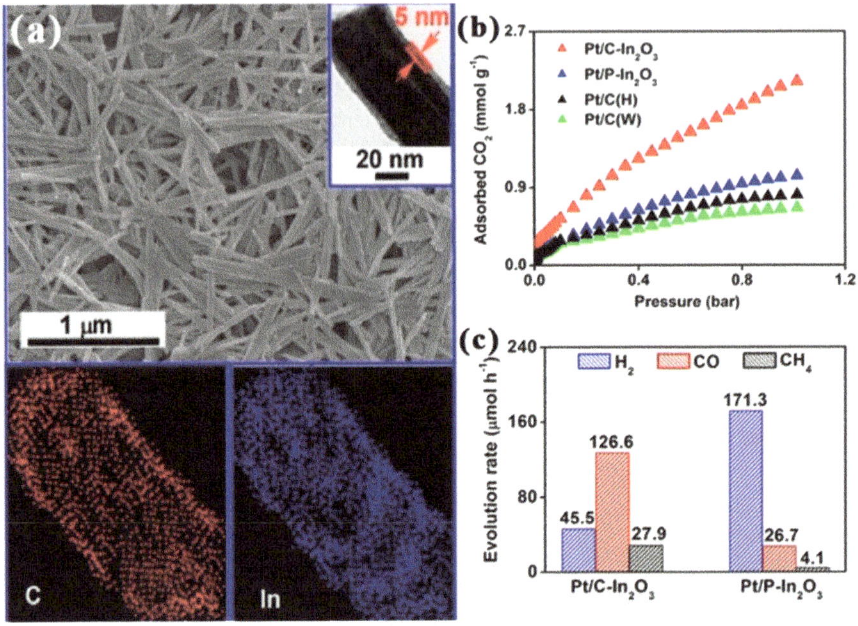

Fig. 12.9 (**a**) SEM and TEM (inset) images and EDX elemental mapping images of C-In_2O_3. (**b**) CO_2 adsorption capacities of In_2O_3-based samples. (**c**) H_2, CO, and CH_4 evolution rates from CO_2 photoreduction on Pt/C-In_2O_3 and Pt/P-In_2O_3. (Reprinted with permission from Ref. [47]. Copyright 2017, American Chemical Society)

12.4.3 Surface Active Sites in CO_2 Photoreduction

For better understanding the roles of cocatalysts in CO2PR, the identification of the active sites in cocatalysts and the study of reaction mechanism are very important. Only in this way can we develop photocatalysts with both high performance and selectivity in CO2PR. In order to investigate the active facet of Pd in CO_2 photoreduction, Bai et al. [15] synthesized Pd cube NPs (exposed mainly (100) facets) and Pd tetrahedron NPs (exposed mainly (111) facets) and deposit them on C_3N_4 layer separately. The size of Pd with different shapes is all around 4–6 nm; however, the activity was quite different, and the selectivity toward CO2PR of Pd nanotetrahedrons/C_3N_4 is obviously higher than Pd nanocubes/C_3N_4. Deep understanding of the shape-dependent selectivity of Pd was investigated by first-principle theory. Firstly, the adsorption energy for CO_2 and H_2O on Pd (111) is 0.23 eV and 0.37 eV and for Pd (100) is 0.064 and 0.554 eV, which indicates the CO_2 and H_2O tend to adsorb on Pd (111) and Pd (100), respectively. Secondly, when accepting two electrons, the Pd (111) shows a lower CO_2 activation energy barrier compared with Pd (100). This result reflects that the Pd (111) is the active site for CO_2 reduction and Pd (100) is more active for H_2O reduction. Generally, the active sites of supported metal catalysts rely on two factors: surface geometric structure and electronic structure, therefore, lattice engineering through alloy different metals could tuning the above two factors and further improve the activity and selectivity of the catalyst. Pd_7Cu_1 supported on TiO_2 nanoplates with isolation Cu in Pd lattice for CO2PR was reported by Long and coworkers [18]. In this research, when the Cu loading amount is below 12.5%, the XAFS results show that the oxidation of Cu (absent of Cu-O) could be inhibited effectively and Cu atoms were isolated in the Pd lattice (absent of Cu–Cu bonds). During the CO2PR evaluation, the Pd_7Cu_1/TiO_2 sample shows the optimal CO_2 reduction activity and CH_4 selectivity. In situ DRIFTS experiments show the enhanced signals of HCO_3^-, $CO_3^=$, and CO_2^- species over samples with isolation of Cu atoms; in addition, the first-principle theory also indicate the Pd–Cu pairs could enhance the CO_2 adsorption. Both experimental and theoretical results suggest the Pd-Cu pairs favor the CO_2 adsorption. The different d band centers of Cu in Pd_7Cu_1 and Pd_1Cu_1 revealed the Pd-surrounded environment could tune the electronic structure of Cu and improve the catalytic activity of Cu. Au–Cu alloy NPs supported on TiO_2 (p25) reported previously also show enhanced performance in CO2PR compared with Au/TiO_2 or Cu/TiO_2 (Fig. 12.10) [14].

To gain deep understanding of the reaction mechanism and intermediates along the CO2PR, time-resolved in situ FTIR was applied. During the reaction, the generation of Cu-CO band (2126 cm^{-1}) indicates the Cu favors the CO_2 reduction instead of H_2O reduction; also the $CO_2^{\cdot -}$ (1589 cm^{-1}) shows a continuously decreasing trend during the irradiation, which is assumed as a reactive species generated from the surface Ti^{3+} (Fig. 12.10c, d). Further studies were carried out by using two light sources (visible light and UV light) to trigger CO2PR. Under the visible light irradiation, CH_4 and H_2 were the main products over the optimal Au-Cu/TiO_2, indicates the hot electrons generated from the surface plasma resonance of Au NPs and reacts with the activated CO_2 to generate CH_4. However, when using UV light as the light source, the Au–Cu alloy NPs act as the electron sink and promote the charge separation, which generate higher amount of H_2.

Low-coordinated sites (i.e., edge or corner sites) in metal NP-supported catalyst are often treated as active sites. Generally, these low-coordinated sites possess unique properties like strong binding energy toward certain reaction intermediates and low free energy which to some rate determines steps. Combining the experimental results with the DFT calculation, Mistry et al. [67] proposed that, in CO_2 electroreduction reaction, Au NPs show the size-dependent activity; the smaller-sized Au with more low-coordinated sites is more active in HER than CRR. Gao et al. [68] studied the Pd NPs with variable size in CO_2 electroreduction, and the result shows that low-coordinated sites of Pd are more suitable for COOH* generation but HER is insensitive to different surface sites. Zhu et al. [16] synthesized Pd nanosheet with similar thickness but different size (TiO_2-Pd NSs-s, small; TiO_2-Pd NSs-m, middle; TiO_2-Pd NSs-l, large) and proposed the edge sites of Pd nanosheet are the active site for CO2PR. Keeping the Pd loading amount as constant, with decrease of the size of Pd nanosheet, results in increased Pd edge density, and the CO and CH_4 yield increased as well. So the edge sites of Pd may act as the active site in CO2PR; to further confirm this edge-dependent activity, the Pd nanorings with even smaller size and higher density of edge sites were prepared and deposited on TiO_2

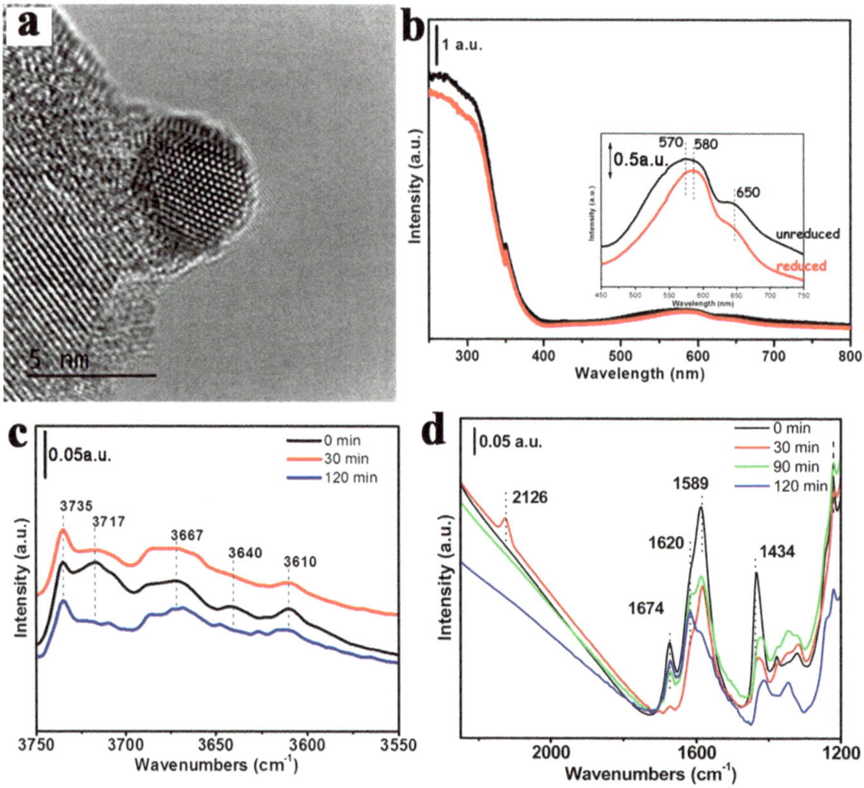

Fig. 12.10 (**a**) HR-TEM image of Au–Cu alloy loaded on TiO_2; the lattice spacing distance is 0.222 nm, which is different from Au or Cu. (**b**) UV–vis DRS spectra of Au–Cu/TiO_2 (Au/Cu = 1:2) sample before and after reduction in 400 °C H_2 atmosphere calcination. (**c** and **d**) Time-resolved in situ FTIR spectra of intermediates that generate from adsorption and evolution during irradiation of CO_2 and H_2O bounded on Au–Cu/TiO_2 (Au/Cu = 1:2). (Reprinted with permission from Ref. [14]. Copyright 2014, American Chemical Society)

(denoted as TiO_2-Pd NRs-s). The obtained TiO_2-Pd NRs-s show a lower TiO_2-Pd interface to Pd volume but higher edge to volume ratio compared with TiO_2-Pd NSs-s. Consequently, the TiO_2-Pd NRs-s show a higher CO2PR activity but lower HER yield (lower electron transfer ability). As a result, the density of edge sites of Pd is highly related with the CO2PR performance which is reasonable to assume as the active sites. To verify the roles of metal active sites in CO2PR more specifically, with the assistance of DFT calculation, Gao et al. [69] report a step-by-step CO_2 photoreduction over single-atom Pt or Pd supported on g-C_3N_4. The calculated relative binding energy between Pd and Pt within C_3N_4 sixfold cavity proved the existence of charge transfer and strong interaction between the metal and support. Two possible product pathways, HCOOH and CH3OH, are studied for Pd/C_3N_4. The calculated desorption energy barrier for the key intermediate HCOOH* on Pd/C3N4 is 0.46 eV, which is much lower than the formation of HCHO*, suggesting that HCOOH is the more preferred product than CH_3OH. For Pt/C_3N_4, the strong interaction between Pt and HCOOH* (1.06 eV) and the favorable CH_2* and H_2O* generation instead of CH_2OH* hydrogenation made it the suitable candidate for CH_4 production.

12.5 Summary and Perspective

So far, the synthesis of photocatalyst with cocatalysts incorporation and the unique properties of various cocatalysts in CO2PR have been carefully summarized. The roles of these cocatalysts such as promote the charge separation efficiency, improve the adsorption of the CO_2 amount, expand the light harvesting range, provide active sites for the activation of CO_2 or other intermediates, etc. also have been briefly discussed. Besides, the important roles of spatial configurations of the photocatalyst composite and the deposition amount of the cocatalysts are also illustrated carefully: inappropriate incorporation of cocatalysts would lead to negative effect of the photocatalyst's performance; on the contrary, rational structure design such as the Z-scheme model or cocatalysts with spatial separated configurations could enhance the performance of the photocatalyst. It should be noted that we mainly focus on the solid-state cocatalysts in this chapter; beside this, the molecular-state cocatalysts like metal complex and dyes also could act as the cocatalyst in the CO2PR; however, this type of photocatalytic system is often conducted in the liquid phase and in the presence of hole scavenger, which is quite different from the solid-phase cocatalysts, so these types of cocatalysts are not discussed in here.

Although numerous efforts have been done in the selection of suitable cocatalysts and the development of fine structures of photocatalyst in CO2PR, many problems still existed and need to be answered and improved:

1. The CO2PR evaluation method is alternative among different research groups; therefore, the product yields comparison of different photocatalysts which is problematic; other evaluation methods such quantum yield efficiency and turnover number (TON) are highly encouraged in the following studies.
2. The origination of the products should be verified carefully; the organic impurities or carbon-involved species also could be converted into the products and cause the illusion result; therefore, control experiment of CO_2 photocatalytic reduction reaction should be conducted with the use of isotope-labeled $^{13}CO_2$ as the reactant for comparison.
3. The reaction pathways and mechanism in CO2PR are still ambiguous; deep understanding of the CO2PR could bring inspiration to the researchers to design highly efficient and selective catalysts; in this case, the DFT calculation along with the in situ characterizations is highly advocated.
4. The stability of the cocatalysts in the long-term CO2PR reaction is another concern; many photocatalysts suffer from low stability due to the carbon-involved species accumulation and deactivate gradually; therefore, the

development of highly efficient and stable photocatalyst and the study of the reason of catalysts' deactivation are important.

References

1. Marszewski M, Cao S, Yu J, Jaroniec M (2015) Semiconductor-based photocatalytic CO_2 conversion. Mater Horiz 2(3):261–278
2. Walsh B, Ciais P, Janssens IA, JP ˜u, Riahi K, Rydzak F, DPv V, Obersteiner M (2017) Pathways for balancing CO_2 emissions and sinks. Nat Commun 8:14856–14868
3. Fujishima A, Honda K (1972) Electrochemical photolysis of water at a semiconductor electrode. Nature 238:37–38
4. Ma Y, Wang X, Jia Y, Chen X, Han H, Li C (2014) Titanium dioxide-based nanomaterials for photocatalytic fuel generations. Chem Rev 114(19):9987–10043
5. Habisreutinger SN, Schmidt-Mende L, Stolarczyk JK (2013) Photocatalytic reduction of CO_2 on TiO_2 and other semiconductors. Angew Chem Int Ed 52(29):7372–7408
6. Dong C, Xing M, Zhang J (2016) Economic hydrophobicity triggering of CO_2 photoreduction for selective CH_4 generation on noble-metal-free TiO_2-SiO_2. J Phys Chem Lett 7:2962–2966
7. Li K, Peng B, Peng T (2016) Recent advances in heterogeneous photocatalytic CO_2 conversion to solar fuels. ACS Catal 6:7485–7527
8. Dong C, Xing M, Zhang J (2016) Double-cocatalysts promote charge separation efficiency in CO_2 photoreduction: spatial location matters. Mater Horiz 3(6):608–612
9. Wang W, An WJ, Ramalingam B, Mukherjee S, Niedzwiedzki DM, Gangopadhyay S, Biswas P (2012) Size and structure matter: enhanced CO_2 photoreduction efficiency by size-resolved ultrafine Pt nanoparticles on TiO_2 single crystals. J Am Chem Soc 134(27):11276–11281
10. Xie S, Wang Y, Zhang Q, Deng W, Wang Y (2014) MgO- and Pt-promoted TiO_2 as an efficient photocatalyst for the preferential reduction of carbon dioxide in the presence of water. ACS Catal 4(10):3644–3653
11. Feng X, Sloppy JD, LaTempa TJ et al (2011) Synthesis and deposition of ultrafine Pt nanoparticles within high aspect ratio TiO_2 nanotube arrays: application to the photocatalytic reduction of carbon dioxide. J Mater Chem 21:13429–13433
12. Manzi A, Simon T, Sonnleitner C, Doblinger M, Wyrwich R, Stern O, Stolarczyk JK, Feldmann J (2015) Light-induced cation exchange for copper sulfide based CO_2 reduction. J Am Chem Soc 137(44):14007–14010
13. Mao J, Ye L, Li K et al (2014) Pt-loading reverses the photocatalytic activity order of anatase TiO_2 {001} and {010} facets for photoreduction of CO_2 to CH_4. Appl Catal B-Environ 144:855–862
14. Neatu S, Macia-Agullo JA, Concepcion P, Garcia H (2014) Gold-copper nanoalloys supported on TiO_2 as photocatalysts for CO_2 reduction by water. J Am Chem Soc 136(45):15969–15976
15. Bai S, Wang X, Hu C, Xie M, Jiang J, Xiong Y (2014) Two-dimensional g-C_3N_4: an ideal platform for examining facet selectivity of metal co-catalysts in photocatalysis. Chem Commun (Camb) 50(46):6094–6097
16. Zhu Y, Xu Z, Jiang W et al (2017) Engineering on the edge of Pd nanosheet cocatalysts for enhanced photocatalytic reduction of CO_2 to fuels. J Mater Chem A 5(6):2619–2628
17. Li N, Liu M, Yang B et al (2017) Enhanced photocatalytic performance toward CO_2 hydrogenation over nanosized TiO_2-loaded Pd under UV irradiation. J Phys Chem C 121(5):2923–2932
18. Long R, Li Y, Liu Y et al (2017) Isolation of Cu atoms in Pd lattice: forming highly selective sitesfor photocatalytic conversion of CO_2 to CH_4. J Am Chem Soc 139:4486–4492
19. Kong D, Tan JZY, Yang F, Zeng J, Zhang X (2013) Electrodeposited Ag nanoparticles on TiO_2 nanorods for enhanced UV visible light photoreduction CO_2 to CH_4. Appl Surf Sci 277:105–110

20. Kuriki R, Matsunaga H, Nakashima T, Wada K, Yamakata A, Ishitani O, Maeda K (2016) Nature-inspired, highly durable CO_2 reduction system consisting of a binuclear Ruthenium (II) complex and an organic semiconductor using visible light. J Am Chem Soc 138 (15):5159–5170
21. Li K, Peng T, Ying Z, Song S, Zhang J (2016) Ag-loading on brookite TiO_2 quasi nanocubes with exposed {210} and {001} facets: activity and selectivity of CO_2 photoreduction to CO/CH_4. Appl Catal B-Environ 180:130–138
22. Wei Y, Jiao J, Zhao Z, Liu J, Li J, Jiang G, Wang Y, Duan A (2015) Fabrication of inverse opal TiO_2-supported Au@CdS core-shell nanoparticles for efficient photocatalytic CO_2 conversion. Appl Catal B-Environ 179:422–432
23. Zhou P, Yu J, Jaroniec M (2014) All-solid-state Z-scheme photocatalytic systems. Adv Mater 26(29):4920–4935
24. Ong WJ, Putri LK, Tan LL, Chai SP, Yong ST (2016) Heterostructured $AgX/g\text{-}C_3N_4$ (X=Cl and Br) nanocomposites via a sonication-assisted deposition-precipitation approach: emerging role of halide ions in the synergistic photocatalytic reduction of carbon dioxide. Appl Catal B-Environ 180:530–543
25. He Y, Zhang L, Teng B, Fan M (2015) New application of Z-scheme $Ag_3PO_4/g\text{-}C_3N_4$ composite in converting CO_2 to fuel. Envirom Sci Technol 49(1):649–656
26. Chang X, Wang T, Gong J (2016) CO_2 photo-reduction: insights into CO_2 activation and reaction on surfaces of photocatalysts. Energy Environ Sci 9:2177–2196
27. Ji Y, Luo Y (2016) New mechanism for photocatalytic reduction of CO_2 on the anatase TiO_2 (101) surface: the essential role of oxygen vacancy. J Am Chem Soc 138(49):15896–15902
28. Kar P, Farsinezhad S, Mahdi N, Zhang Y, Obuekwe U, Sharma H, Shen J, Semagina N, Shankar K (2016) Enhanced CH_4 yield by photocatalytic CO_2 reduction using TiO_2 nanotube arrays grafted with Au, Ru, and ZnPd nanoparticles. Nano Res 9(11):3478–3493
29. Li M, Li P, Chang K, Wang T, Liu L, Kang Q, Ouyang S, Ye J (2015) Highly efficient and stable photocatalytic reduction of CO_2 to CH_4 over Ru loaded $NaTaO_3$. Chem Commun (Camb) 51(36):7645–7648
30. Yu B, Zhou Y, Li P, Tu W, Li P, Tang L, Ye J, Zou Z (2016) Photocatalytic reduction of CO_2 over Ag/TiO_2 nanocomposites prepared with a simple and rapid silver mirror method. Nanoscale 8(23):11870–11874
31. Pan Y, Sun Z, Cong H, Men Y, Xin S, Song J, Yu S (2016) Photocatalytic CO_2 reduction highly enhanced by oxygen vacancies on Pt-nanoparticle-dispersed gallium oxide. Nano Res 9 (6):1689–1700
32. Lee S, Jeong S, Kim WD, Lee S, Lee K, Bae WK, Moon JH, Lee S, Lee DC (2016) Low-coordinated surface atoms of CuPt alloy cocatalysts on TiO_2 for enhanced photocatalytic conversion of CO_2. Nanoscale 8(19):10043–10048
33. Kang Q, Wang T, Li P, Liu L, Chang K, Li M, Ye J (2015) Photocatalytic reduction of carbon dioxide by hydrous hydrazine over Au-Cu alloy nanoparticles supported on $SrTiO_3/TiO_2$ coaxial nanotube arrays. Angew Chem Int Ed 54(3):841–845
34. Wang T, Shi L, Tang J, Malgras V, Asahina S, Liu G, Zhang H, Meng X, Chang K, He J, Terasaki O, Yamauchi Y, Ye J (2016) A Co_3O_4-embedded porous ZnO rhombic dodecahedron prepared using zeolitic imidazolate frameworks as precursors for CO_2 photoreduction. Nanoscale 8(12):6712–6720
35. Wang WN, Wu F, Myung Y et al (2015) Surface engineered CuO nanowires with ZnO islands for CO_2 photoreduction. ACS Appl Mater Interfaces 7(10):5685–5692
36. Yu W, Xu D, Peng T (2015) Enhanced photocatalytic activity of $g\text{-}C_3N_4$ for selective CO_2 reduction to CH_3OH via facile coupling of ZnO: a direct Z-scheme mechanism. J Mater Chem A 3:19936–19947
37. In S-I, Dimitri D, Vaughn II, Schaak RE (2012) Hybrid CuO-TiO_2-xNx hollow nanocubes for photocatalytic conversion of CO_2 into methane under solar irradiation. Angew Chem Int Ed 124:3981–3984

38. Jin J, Yu J, Guo D, Cui C, Ho W (2015) A hierarchical Z-scheme CdS-WO_3 photocatalyst with enhanced CO_2 reduction activity. Small 11(39):5262–5271
39. Wang J, Yao H, Fan Z, Zhang L, Wang J, Zang S, Li Z (2016) Indirect Z-scheme BiOI/g-C_3N_4 photocatalysts with enhanced photoreduction CO_2 activity under visible light irradiation. ACS Appl Mater Interfaces 8(6):3765–3775
40. Li P, Zhou Y, Zhao Z, Xu Q, Wang X, Xiao M, Zou Z (2015) Hexahedron prism-anchored octahedronal CeO_2: crystal facet-based homojunction promoting efficient solar fuel synthesis. J Am Chem Soc 137(30):9547–9550
41. Zhai Q, Xie S, Fan W, Zhang Q, Wang Y, Deng W, Wang Y (2013) Photocatalytic conversion of carbon dioxide with water into methane: platinum and copper(I) oxide co-catalysts with a core-shell structure. Angew Chem Int Ed 125(22):5888–5891
42. Liu Q, Zhou Y, Kou J et al (2010) High-yield synthesis of ultralong and ultrathin Zn_2GeO_4 nanoribbons toward improved photocatalytic reduction of CO_2 into renewable hydrocarbon fuel. J Am Chem Soc 132(41):14385–14387
43. Sarkar A, Gracia-Espino E, Wågberg T, Shchukarev A, Mohl M, Rautio AR, Pitkänen O, Sharifi T, Kordas K, Mikkola JP (2016) Photocatalytic reduction of CO_2 with H_2O over modified TiO_2 nanofibers: understanding the reduction pathway. Nano Res 9(7):1956–1968
44. Wei Y, Jiao J, Zhao Z, Zhong W, Li J, Liu J, Jiang G, Duan A (2015) 3D ordered macroporous TiO_2-supported Pt@CdS core-shell nanoparticles: design, synthesis and efficient photocatalytic conversion of CO_2 with water to methane. J Mater Chem A 3(20):11074–11085
45. Li H, Gao Y, Zhou Y et al (2016) Construction and nanoscale detection of interfacial charge transfer of elegant Z-scheme WO_3/Au/In_2S_3 nanowire arrays. Nano Lett 16(9):5547–5552
46. Pastrana-Martínez LM, Silva AMT, Fonseca NNC, Vaz JR, Figueiredo JL, Faria JL (2016) Photocatalytic reduction of CO_2 with water into methanol and ethanol using graphene derivative-TiO_2 composites: effect of pH and copper(I) oxide. Top Catal 59(15–16):1279–1291
47. Pan Y, You Y, Xin S, Li Y, Fu G, Cui Z, Men Y, Cao F, Yu S, Goodenough JB (2017) Photocatalytic CO_2 reduction by carbon-coated indium-oxide nanobelts. J Am Chem Soc 139 (11):4123–4129
48. Wang Y, Bai X, Qin H, Wang F, Li Y, Li X, Kang S, Zuo Y, Cui L (2016) Facile one-step synthesis of hybrid graphitic carbon nitride and carbon composites as high-performance catalysts for CO_2 photocatalytic conversion. ACS Appl Mater Interfaces 8(27):17212–17219
49. Tu W, Zhou Y, Liu Q, Yan S, Bao S, Wang X, Xiao M, Zou Z (2013) An in situ simultaneous reduction-hydrolysis technique for fabrication of TiO_2-graphene 2D sandwich-like hybrid nanosheets: graphene-promoted selectivity of photocatalytic-driven hydrogenation and coupling of CO_2 into methane and ethane. Adv Funct Mater 23(14):1743–1749
50. Tu W, Zhou Y, Liu Q, Tian Z, Gao J, Chen X, Zhang H, Liu J, Zou Z (2012) Robust hollow spheres consisting of alternating titania nanosheets and graphene nanosheets with high photocatalytic activity for CO_2 conversion into renewable fuels. Adv Funct Mater 22 (6):1215–1221
51. Gao C, Meng Q, Zhao K et al (2016) Co_3O_4 hexagonal platelets with controllable facets enabling highly efficient visible-light photocatalytic reduction of CO_2. Adv Mater 28 (30):6485–6490
52. Xu Y, Yang M, Chen B, Wang X, Chen H, Kuang D, Su C (2017) A $CsPbBr_3$ perovskite quantum dot/graphene oxide composite for photocatalytic CO_2 reduction. J Am Chem Soc 139 (16):5660–5663
53. Li R, Hu J, Deng M et al (2014) Integration of an inorganic semiconductor with a metal–organic framework: a platform for enhanced gaseous photocatalytic reactions. Adv Mater 26 (28):4783–4788
54. Yu J, Jin J, Cheng B, Jaroniec M (2014) A noble metal-free reduced graphene oxide–CdS nanorod composite for the enhanced visible-light photocatalytic reduction of CO_2 to solar fuel. J Mater Chem A 2(10):3407
55. Ong WJ, Putri LK, Tan YC, Tan LL, Li N, Ng YH, Wen X, Chai SP (2017) Unravelling charge carrier dynamics in protonated g-C_3N_4 interfaced with carbon nanodots as co-catalysts toward

enhanced photocatalytic CO_2 reduction: a combined experimental and first-principles DFT study. Nano Res 10(5):1673–1696
56. Shi L, Wang T, Zhang H, Chang K, Ye J (2015) Electrostatic self-assembly of nanosized carbon nitride nanosheet onto a zirconium metal-organic framework for enhanced photocatalytic CO_2 reduction. Adv Funct Mater 25(33):5360–5367
57. Park SM, Razzaq A, Park YH, Sorcar S, Park Y, Grimes CA, In SI (2016) Hybrid Cu_xO-TiO_2 heterostructured composites for photocatalytic CO_2 reduction into methane using solar irradiation: sunlight into fuel. ACS Omega 1(5):868–875
58. Ong WJ, Tan LL, Chai SP, Yong ST, Mohamed AR (2015) Surface charge modification via protonation of graphitic carbon nitride (g-C_3N_4) for electrostatic self-assembly construction of 2D/2D reduced graphene oxide (rGO)/g-C_3N_4 nanostructures toward enhanced photocatalytic reduction of carbon dioxide to methane. Nano Energy 13:757–770
59. Ming H, Ma Z, Liu Y, Pan K, Yu H, Wang F, Kang Z (2012) Large scale electrochemical synthesis of high quality carbon nanodots and their photocatalytic property. Dalton Trans 41 (31):9526–9531
60. Yu H, Zhao Y, Zhou C, Shang L, Peng Y, Cao Y, Wu L, Tung C, Zhang T (2014) Carbon quantum dots/TiO_2 composites for efficient photocatalytic hydrogen evolution. J Mater Chem A 2(10):3344
61. Shi H, Chen G, Zhang C, Zou Z (2014) Polymeric g-C_3N_4 coupled with $NaNbO_3$ nanowires toward enhanced photocatalytic reduction of CO_2 into renewable fuel. ACS Catal 4 (10):3637–3643
62. Wang D, Hisatomi T, Takata T, Pan C, Katayama M, Kubota J, Domen K (2013) Core/Shell photocatalyst with spatially separated co-catalysts for efficient reduction and oxidation of water. Angew Chem Int Ed 52(43):11252–11256
63. Zheng D, Cao X, Wang X (2016) Precise formation of a hollow carbon nitride structure with a janus surface to promote water splitting by photoredox catalysis. Angew Chem Int Ed 55 (38):11512–11516
64. Li A, Chang X, Huang Z et al (2016) Thin heterojunctions and spatially separated cocatalysts to simultaneously reduce bulk and surface recombination in photocatalysts. Angew Chem Int Ed 55(44):13734–13738
65. Zhang J, Yu Z, Gao Z, Ge H, Zhao S, Chen C, Chen S, Tong X, Wang M, Zheng Z, Qin Y (2017) Porous TiO_2 nanotubes with spatially separated platinum and CoO_x cocatalysts produced by atomic layer deposition for photocatalytic hydrogen production. Angew Chem Int Ed Eng 56(3):816–820
66. Xie S, Wang Y, Zhang Q, Fan W, Deng W, Wang Y (2013) Photocatalytic reduction of CO_2 with H_2O: significant enhancement of the activity of Pt-TiO_2 in CH_4 formation by addition of MgO. Chem Commun (Camb) 49(24):2451–2453
67. Mistry H, Varela AS, Kühl S, Strasser P, Cuenya BR (2016) Nanostructured electrocatalysts with tunable activity and selectivity. Nat Rev Mater 1(4):16009
68. Gao D, Zhou H, Wang J, Miao S, Yang F, Wang G, Wang J, Bao X (2015) Size-dependent electrocatalytic reduction of CO_2 over Pd nanoparticles. J Am Chem Soc 137(13):4288–4291
69. Gao G, Jiao Y, Waclawik ER, Du A (2016) Single atom (Pd/Pt) supported on graphitic carbon nitride as an efficient photocatalyst for visible-light reduction of carbon dioxide. J Am Chem Soc 138(19):6292–6297

Chapter 13
Syntheses and Applications of Silver Halide-Based Photocatalysts

13.1 Introduction

In the past decades, photocatalysis has received increasing attention for its potential to solve worldwide energy crisis (water splitting [1–4] and solar cell [5–7]), environmental pollution (water and air purification [8–12], and pathogen inactivation [13–15]), and greenhouse effect (CO_2 conversion [16–19]). Among various photocatalytic materials, TiO_2 has been the most widely researched for its excellent properties, such as high photocatalytic activity, low cost, high stability, and nontoxicity [20–24]. However, its two defects, no response to visible light and low photo-quantum efficiency, seriously limit its practical application. Hence, to full use of solar energy, the development of visible light-driven photocatalytic materials has become the most significant topic in the photocatalytic field.

Silver halide (AgX), a kind of well-known photographic material, was developed in 1839 with the daguerreotype [25, 26]. The photographic process in AgX is as follows: after absorbing a photon, silver halide can produce an electron and a hole, and subsequently the electron combines with an interstitial silver ion to form an Ag^0 atom [27, 28]. Upon repeated absorption of photons, a cluster of silver atoms (latent image) will be formed ultimately [29, 30]. Due to the instability under light, silver halides are seldom used as photocatalysts previously.

In 1996, the photocatalytic activity of AgX was firstly reported by Calzaferri et al. [31]. During the photocatalytic reaction, the photo-generated electron–hole pairs will react with sacrificial agent and water to evolve O_2 or H_2. In this period, the Ag nanoparticles formed on the surface of AgX were seen as electron trappers, which can capture photo-generated electrons by Schottky barrier between Ag and AgX. Therefore, the formation of Ag nanoparticles can not only enhance the photocatalytic activity of AgX by separating carriers but also improve the stability of AgX by decreasing the amount of electrons in AgX [32]. In 2008, Huang et al. [33] found that the Ag nanoparticles on AgX can also enhance the absorption of visible light by surface plasma resonance (SPR) effect, which triggered an upsurge

J. Zhang et al., *Photocatalysis*, Lecture Notes in Chemistry 100,
https://doi.org/10.1007/978-981-13-2113-9_13

Fig. 13.1 (**A**) Crystal, structure, and lattice parameter of AgCl, AgBr, and AgI. (**B**) The models of face-centered cubic crystal

(A)

Name	Crystal	Sturcture	Lattice paramater a (Å)
AgCl	fcc	rock-salt	5.5491
AgBr	fcc	rock-salt	5.7745
AgI(α)	fcc	rock-salt	6.4950

(B)

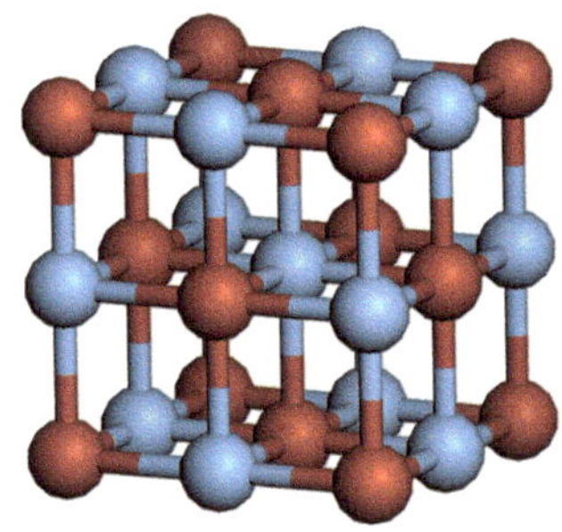

of researching Ag/AgX plasmonic photocatalysts. So far, a large number of researches involved in AgX have been carried out, including the morphology control, establishment of heterojunction or Z-scheme structure, and combining with recoverable material.

In this chapter, we will firstly introduce the properties and synthesis strategies of AgX materials and the preparation and characteristic of AgX with different morphologies. After that, the AgX-based heterojunction and Z-scheme structures will be classified by different composites and band structure, and each type will be presented detailedly. Finally, the present researches of recoverable AgX materials will be introduced.

13.2 Properties of AgX

AgCl, AgBr, and AgI(α) are face-centered cubic (fcc) crystal and rock salt (NaCl) structure, as shown in Fig. 13.1a, and their lattice parameters are 5.5491, 5.7745, and 6.4950 Å, respectively [34]. In detail, as shown in Fig. 13.1b, the Ag and Br atoms are all sixfold coordinated by Br and Ag atoms in the crystal. The solubility of AgCl, AgBr, and AgI(α) in water is extremely low. The Ksp for AgCl, AgBr, and AgI(α) are 1.77×10^{-10}, 5.35×10^{-13}, and 8.52×10^{-17}, respectively. The optical properties of AgCl, AgBr, and AgI(α) are dependent on their band gap. According to the literatures, the indirect band gaps of AgCl, AgBr, and AgI(α) are 3.30 [35], 2.65 [36], and 2.80 eV [37], respectively (Table 13.1). It should be noticed that the band gap of AgCl is over 2.95 eV, indicating that it can only absorb UV light. However, the Ag^+ on the surface of AgCl is easily reduced to Ag^0, which not only responds to visible light by SPR effect but also enhances the stability of AgCl via capturing photo-generated electrons.

Table 13.1 CB level, VB level, and band gap of AgCl, AgBr, and AgI(α)

Name	CB level (eV)	VB level (eV)	Band gap (eV)
AgCl	−1.20	2.10	3.30
AgBr	−0.50	2.15	2.65
AgI(α)	−0.42	2.38	2.80

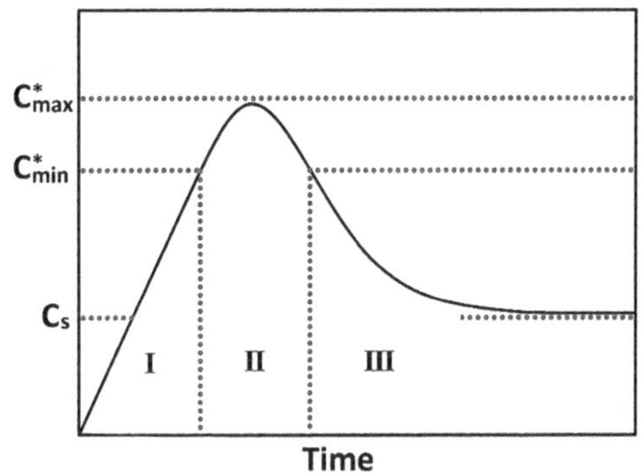

Fig. 13.2 Illustration of the nucleation and growth of AgX crystal during precipitation process, in which C_s is the solubility while C^*_{min} and C^*_{max} are the minimum and maximum concentration of nucleation, respectively

13.3 Synthesis Strategies

In past 20 years, many methods have been developed to prepare AgX photocatalytic materials. Most of these methods are mainly based on the extremely low solubility of AgX. Generally, these methods can be classified into three synthesis strategies: liquid–solid precipitation, in situ oxidation transformation, and ion exchange.

13.3.1 Liquid–Solid Precipitation

As we all know, in an aqueous solution, the activities of Ag ions $[Ag^+]$ and X ions $[X^-]$ are related to the solubility product K_{sp} ($[Ag^+]\ [X^+] = K_{sp}$). The relation between ion concentration and precipitation process of AgX is described in Fig. 13.2, in which C_s, C^*_{min}, and C^*_{max} represented the concentration corresponding to solubility, the critical supersaturation (the minimum concentration of nucleation), and the limiting supersaturation (the maximum concentration of nucleation), respectively. At the initial stage of adding Ag^+ and X^- ions (Stage I), no AgX nucleus can be formed because the concentration is below C^*_{min}. When the concentration exceeds C^*_{min} (Stage II), with the continuous addition of Ag^+ and X^-, the spontaneous nucleation of AgX takes place rapidly and the concentration begins to decrease. When the concentration is below C^*_{min} but higher than C_s (Stage III), the nucleation of AgX ceases and the AgX crystal nuclei grow gradually with the continuous addition of Ag^+ and X^- ions. In order to control the uniformity of crystal size, it is important to keep the concentration lower than C^*_{min} after the initial nucleation finished to prevent the occurrence of further nucleation.

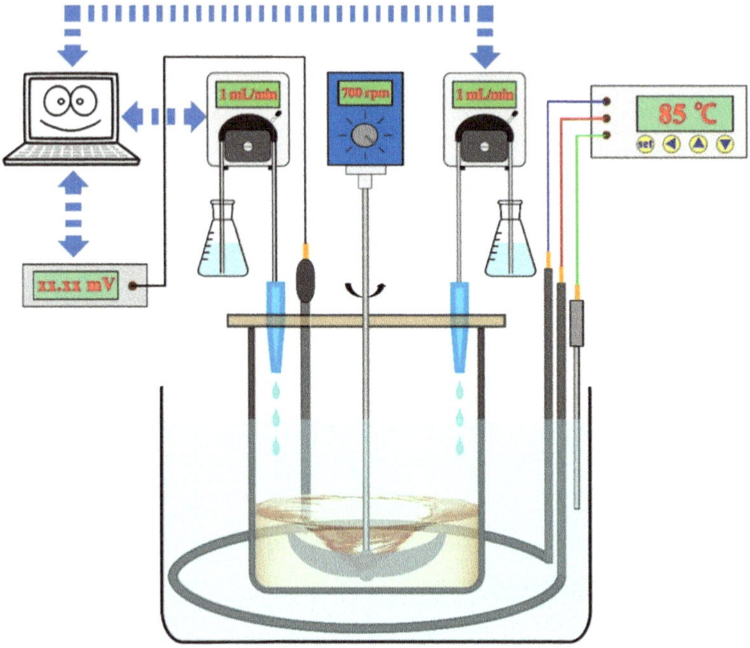

Fig. 13.3 Illustration of the apparatus for double-jet precipitation

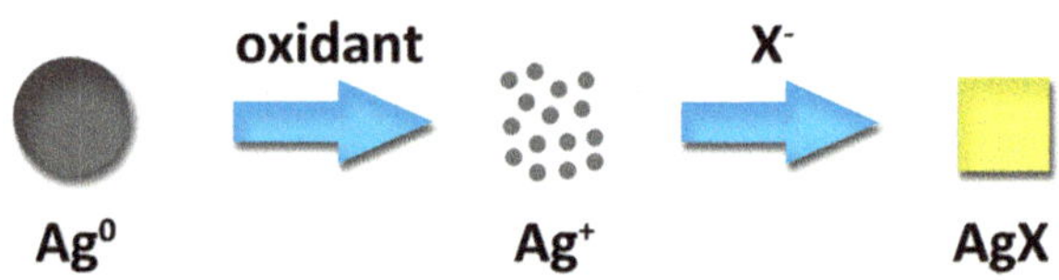

Fig. 13.4 Illustration of in situ oxidation transformation process

Figure 13.3 illustrates the computer-controlled double-jet precipitation apparatus for preparing AgBr crystals, by which the concentration of Ag^+ and X^- during the whole precipitation process can be precisely controlled [38–40]. As displayed in Fig. 13.3, the concentrations of Ag^+ and X^- ions are detected by ion selective electrodes and the injecting rates of Ag^+ and X^- ions are adjusted on the basis of the detecting results. Beneficial from this apparatus, the concentration can be controlled to below the C^*_{min} after nucleation process to control the uniformity of crystal size. By the double-jet precipitation apparatus, AgCl and AgBr crystals with different morphologies have been successfully synthesized [36, 41–44].

13.3.2 In Situ Oxidation Transformation

Different to the liquid–solid precipitation, the Ag source used in oxidation transformation is usually metallic silver (Ag^0) [45]. As shown in Fig. 13.4, the oxidant is

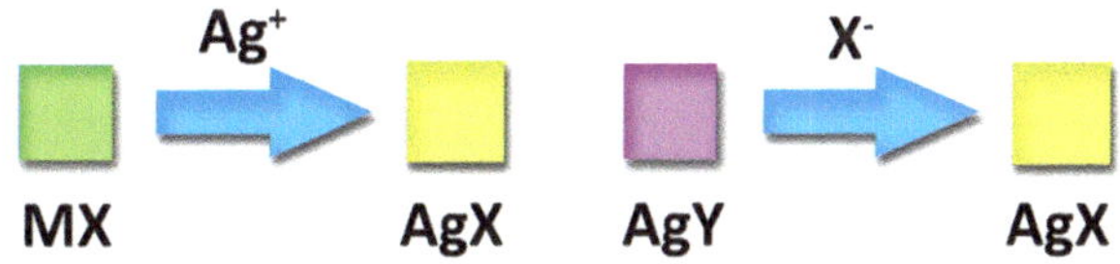

Fig. 13.5 Illustration of ion exchange process

added into the reaction system to oxide Ag^0 and release Ag^+. The released Ag^+ ions further react with X^- ions immediately to form AgX particles in situ. The advantage of this synthesis strategy is easily combining AgX with other semiconductors or substrates, because it is much easier to form Ag^0 on the surface of other materials than to directly form AgX.

13.3.3 Ion Exchange

According to the kind of exchanged ion, this synthetic strategy can be divided into cation exchange and anion exchange. During cation exchange process, the halide salts (MX) are used as the X sources. Then, Ag^+ will replace M^+ in MX due to the extremely low solubility of AgX and from AgX. Similarly, in the case of anion exchange process, the silver salts (AgY) acts as Ag sources. And X^- will exchange with Y^- in AgY. By this strategy, the obtained AgX can keep the original morphology of its precursor MX or AgY (Fig. 13.5).

13.4 Synthesis and Application of AgX with Different Morphologies

Morphology is a significant factor that influences the photoelectrochemical and photocatalytic performances of photocatalytic materials. As for AgX photocatalysts, the reported morphologies mainly include 1D, 3D, facet exposed, and porous structures.

13.4.1 1D Structure

Compared with bulk materials, one dimensional (1D) structured semiconductor materials, including nanowires, nanorods, nanotubes, and nanobelts, usually exhibit better electronic, optoelectronic, and electromechanical properties. These excellent properties will directly cause the enhancement in photocatalytic performances. In the past decades, several kinds of AgX-based 1D materials have been reported. In these works, there exist two synthetic strategies: (a) oxidation–halogenation method and (b) wet chemical method with dissolution and recrystallization progress.

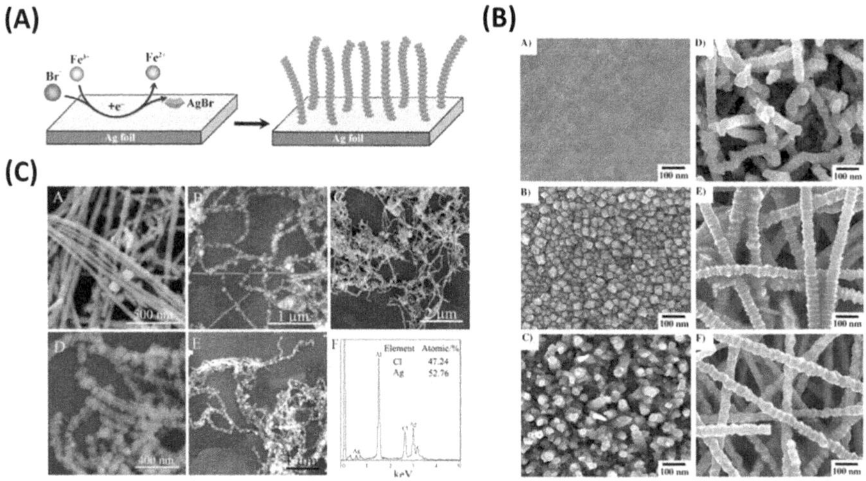

Fig. 13.6 (**a**) Schematic illustration of the heteroepitaxial growth process of AgBr nanowires on Ag substrates; (**b**) SEM images of AgBr nanowires during growth process at different reaction times (0, 0.5, 1, 2, 4, and 6 h) [46]. Reproduced from Ref. [46] by permission of John Wiley & Sons Ltd. (**c**) SEM images of the necklace-like Ag/AgCl nano-heterostructures with different AgCl amount (10%, 30%, 80%, 85%, and 100%) and EDS analysis of Ag/AgCl with 85% AgCl [50]. (Reproduced from Ref. [50] by permission of John Wiley & Sons Ltd)

1. *Oxidation–Halogenation Method*

Metallic Ag with different morphologies, foil, nanowire, or nanotube, can be utilized as the Ag source. In the oxidation–halogenation processes, an oxidation reaction (Ag^0 + oxidant → Ag^+ + reductant) and a halogenation reaction ($Ag^+ + X^-$ → AgX) will occur together.[46–54] Firstly, metallic Ag is oxidized into Ag^+ ion in the presence of oxidant such as Fe^{3+} or H_2O_2. Meanwhile, the dissolved X^- ion immediately combines with the released Ag^+ ion to form AgX. For example, Ye et al. largely synthesized uniform AgBr nanowires using commercial silver foils at room temperature [46]. It was revealed that the towerlike AgBr nanowire was produced by the following pathway: Firstly, the uniform AgBr nanocrystals with octahedron-like structures formed tidily on the surfaces of Ag foils. Then, with prolonging the reacting time, the formed AgBr nanocrystals attached to each other and self-assembled into towerlike nanostructure arrays under the direction of PVP (Fig. 13.6a, b). Using Ag nanowires as Ag source and $FeCl_3$ as oxidant, Huang et al. fabricated necklace-like Ag/AgCl nano-heterostructures [50]. By adjusting the addition of $FeCl_3$, the amount of AgCl in Ag/AgCl nanowires can be conveniently controlled, in Fig. 13.6c. They found that the ratio of Ag and AgCl plays an important role in determining the photocatalytic activity of Ag/AgCl nanowires. When the ratio of AgCl and Ag is 85:15, the Ag/AgCl nanowires exhibit the optimal photocatalytic activity for the decomposition of organic pollutants and water splitting to produce oxygen.

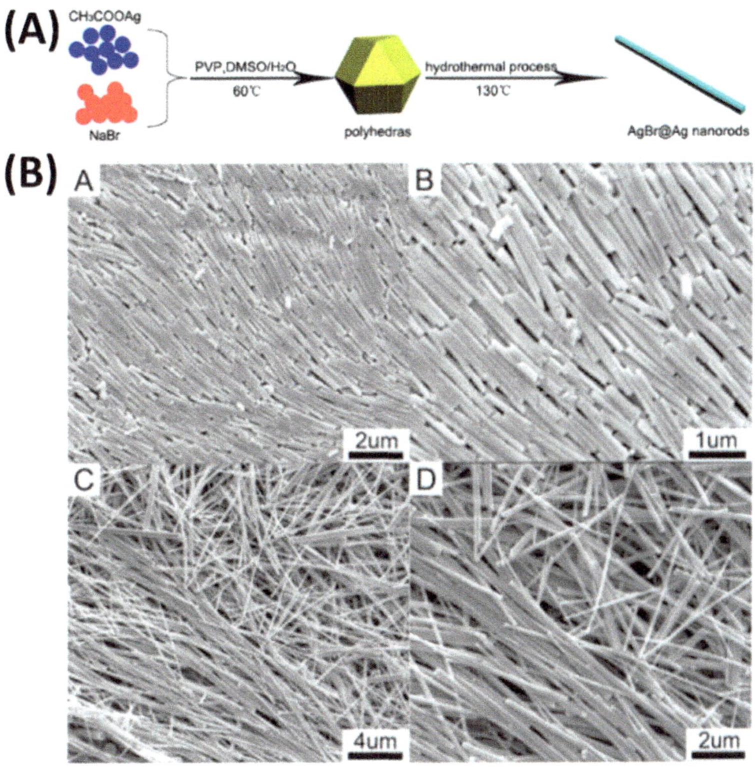

Fig. 13.7 (**a**) Schematic illustration for the formation of the AgBr@Ag nanorods; (**b**) SEM images of AgBr@Ag nanorods and nanowires at low and high magnification, respectively [55]. (Reprinted with the permission from Ref. [55]. Copyright 2013 American Chemical Society)

2. *Dissolution and Recrystallization Processes*

1D AgBr nanowires and nanorods can also be prepared by the two-step processes: dissolution and recrystallization [55]. As illustrated in Fig. 13.7a: Firstly, AgBr nanocrystallines were prepared by dropping CH_3COOAg water solution into NaBr mixture solution ($DMSO/H_2O$= 1:2). Then, the above suspension was directly added into autoclave for hydrothermal treatment. During the hydrothermal process, AgBr nanocrystallines act as the crystal seeds to form 1D structure by dissolution and recrystallization. In this process, both DMSO and PVP are indispensable for the formation of 1D structure. The former can not only enhance the dissolution of AgBr but also stabilize the AgBr {111} facets by interacting with the positively charged "Ag" with polarized functional group "–S=O." The latter can also absorb on the {111} facets of AgBr by "–C=O" to increase the exposure of {111} facets. From Fig. 13.7b, it can be seen that the AgBr nanorods can be changed to AgBr nanowires by slightly changing the experiment condition [55].

(a) Oxidation–halogenation method

Reactant	Oxidizing agent	Halide source	Parameter control	Morphology	References
Ag foil	$Fe(NO_3)_3$	NaBr	Reaction time	Towerlike AgBr nanowire 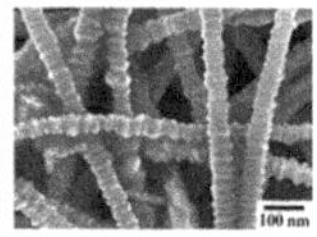	[46]
Ag nanowire	$FeCl_3$	$FeCl_3$	Reaction time	Ag/AgCl core–shell nanowire 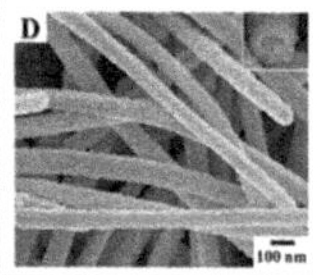	[48] [49]
Ag nanowire	$FeCl_3$	$FeCl_3$	Quantity of $FeCl_3$	Necklace-like Ag/AgCl nanowire 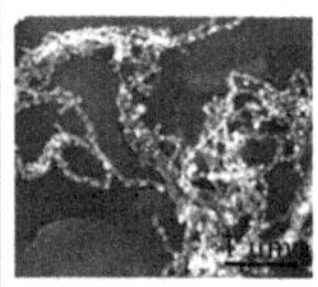	[50] [51] [52]
Ag nanowire	H_2O_2	HCl		Ag/AgCl nanowire 	[53]
Ag nanotube	$FeCl_3$	$FeCl_3$	Quantity of $FeCl_3$	Ag@AgCl nanotube 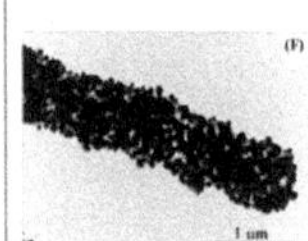	[54]

(b) Dissolution and recrystallization progress

Reactant	Reacting condition	Parameter control	Morphology	Reference
AgAc and NaBr	1. PVP, DMSO/H_2O 60 °C	1. Hydrothermal treatment time	AgBr@Ag nanowire 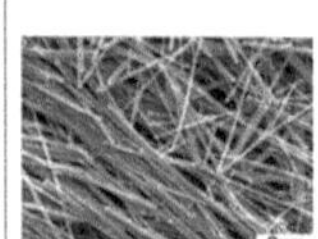	[55]
	2. Hydrothermal treatment 130 °C	2. Ratio of DMSO/H_2O		

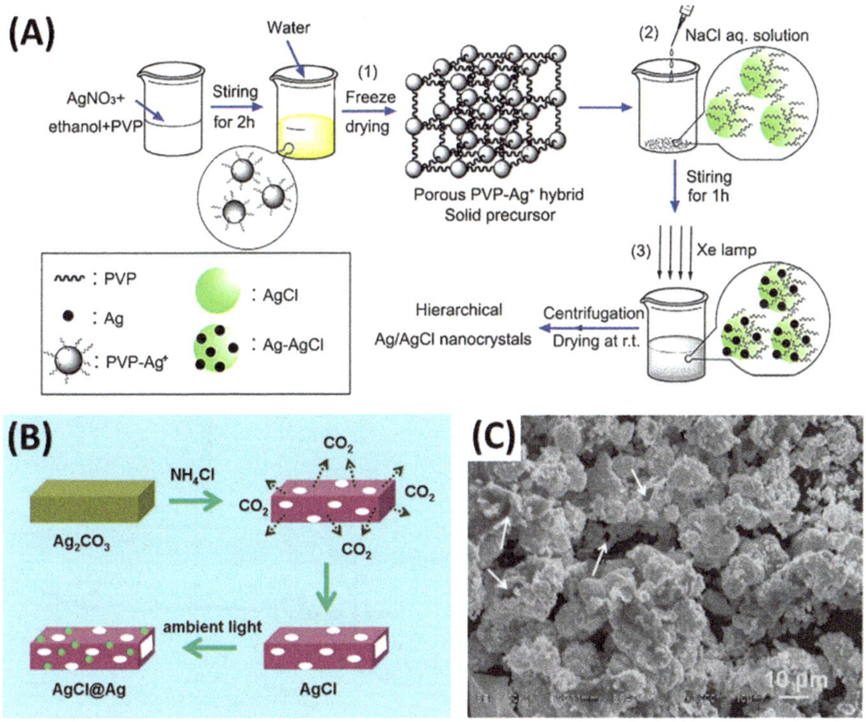

Fig. 13.8 Schematic diagram for the synthesis of (**a**) hierarchical Ag/AgCl nanocrystals via a freeze-drying route.[56] Reprinted from Ref. [56], Copyright 2014, with permission from Elsevier. (**b**) AgCl@Ag hollow architectures; (**c**). SEM images of AgCl@Ag hollow architectures [57]. (Reprinted from Ref. [57], Copyright 2013, with permission from Elsevier)

13.4.2 3D Structure

Different to the 1D structured AgX, 3D structured AgX have more complex framework. This framework can not only increase the surface area to provide more active sites but also promote the transfer of reactants from outside to inside. So far, the strategies for synthesizing AgX 3D structure can be classified into three types:

1. *Anion Exchange Method*

Anion exchange method is one of the effective routes to prepare 3D AgX-based photocatalysts. For this method, the formation mechanism of AgX is based on the principle that the solubility of AgX is lower than that of their precursor Ag salts. For example, Chen et al. firstly synthesized porous PVP–Ag^+ hybrid compounds by a freeze-dying route (Fig. 13.8a). Then, the obtained PVP–Ag^+ were transformed to hierarchical Ag/AgCl nanocrystals through a liquid–solid precipitation reaction with addition of NaCl solution. The photocatalytic performance of hierarchical Ag/AgCl nanocrystals was higher than that of P25 for the photo-degradation of organic dyes

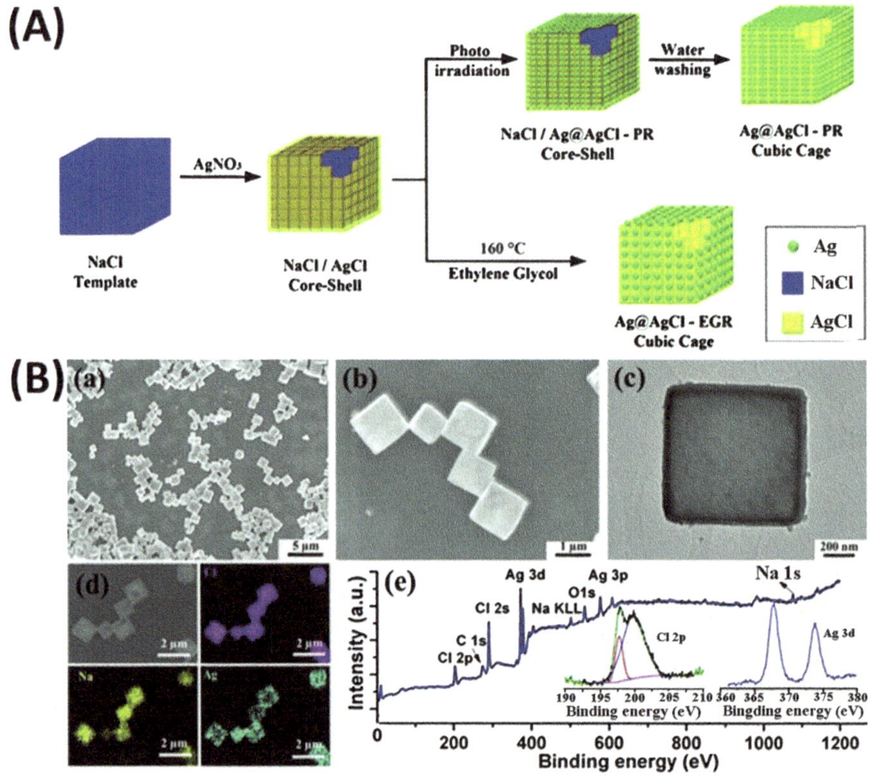

Fig. 13.9 (**a**) Schematic illustration of the water-soluble sacrificial salt–crystal–template (SCT) route for the formation of Ag@AgCl cubic cages. Two methods have been selected to generate Ag NPs: photoreduction (PR) and ethylene glycol-assisted reduction (EGR); (**b**). (*a, b*) Typical FESEM images of NaCl/AgCl core–shell cubes. (*c*) TEM image of an individual NaCl/AgCl core–shell cube [58]. (Reproduced from Ref. [58] by permission of John Wiley & Sons Ltd)

(RhB, MO, and MB) and alcohols (methanol and isopropanol) under visible light. [56] Similarly, Jiang et al. prepared AgCl@Ag hollow architectures by employing NH_4Cl as a reactive acidic etching agent to etch Ag_2CO_3 particles (Fig. 13.8b).[57] During the etching process, the surface of Ag_2CO_3 would be chlorinated and the poles would be created. Finally, the hierarchical porous AgCl@Ag hollow architectures were formed (Fig. 13.8c). The obtained hierarchical porous structure not only increases the adsorption of contaminants but also enhances the harvesting efficiency of light.

2. *Cation Exchange Method*

Similar to anion exchange method, the difference of solubility between AgX and MX (M means other metal element) is also used in cation exchange method. Different to the former method, the MX used in cation exchange method is usually with uniform morphology, such as cube and sphere. Consequently, the morphology of prepared 3D AgX is usually regular [58–62]. For instance, Chen et al. [58] prepared AgCl cubic cages using cubic NaCl crystals as a water-soluble sacrificial

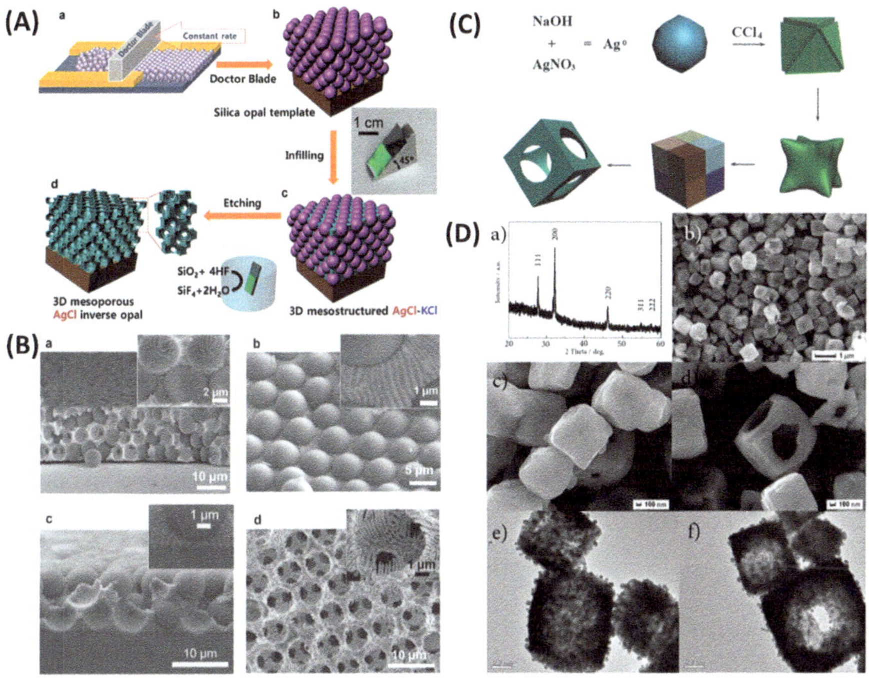

Fig. 13.10 (**a**) Key steps for fabricating 3D mesostructured AgCl–KCl and 3D mesoporous AgCl inverse opal structure; (**b**) SEM images of the air-cooled AgCl–KCl eutectic templated by colloid template. (*a*) cross-sectional, (*b*) plan view, (*c*) a partially infilled template and (*d*) mesostructured AgCl inverse opal [63]. Reproduced from Ref. [63] by permission of John Wiley & Sons Ltd. (**c**) The formation process of AgCl hollow cubes; (**d**) (*a*) XRD pattern, (*b–d*) FESEM, and (*e*, *f*) TEM images of AgCl hollow cubes [64]. (Reproduced from Ref. [64] by permission of the Royal Society of Chemistry)

agent (Fig. 13.9a, b). Detailedly, $AgNO_3$ was added into the prepared cubic NaCl suspension, along with the assistance of surfactant PVP to prevent the aggregation of AgCl particle. Ion exchange diffusion reaction between NaCl and Ag^+ in the solution led to the heterogeneous nucleation and continued growth of AgCl on the surface of the NaCl template. Finally, NaCl template would be removed by water washing and the AgCl cubic cages would be obtained. Similarly, AgCl nanoframe, AgBr cubic cage, and porous AgBr microsphere can also be synthesized via similar experimental steps [59–62].

3. *Other Methods*

Besides ion exchange method, there still exist other ways to prepare 3D AgX materials. For example, Braun et al. utilized AgCl–KCl eutectic system to prepare 3D mesoporous AgCl inverse opal [63]. As illustrated in Fig. 13.10a, AgCl–KCl eutectic powder was placed on the top of silica opal template. With increasing the temperature, the AgCl–KCl eutectic melted (eutectic temperature = 318 °C) and flowed into the porous opal through a combination of capillary force and gravity, wetting the opal up to its top. The mesoporous AgCl inverse opal structure was obtained by dissolution of the silica colloidal template and KCl with 5% HF. As shown in Fig. 13.10b-*a* and b-*c*, the AgCl–KCl eutectic infilled and solidified in colloidal crystal template. After removing template and KCl, a rather complex porous AgCl mesostructure containing features with characteristic dimensions of 100–200 nm is observed (Fig. 13.10b-*d*). Another feasible way to prepare 3D AgX material is one-pot growing method. Shen et al. synthesized a cuboidal AgCl hollow nanostructure by a simple one-pot reaction using $AgNO_3$ and CCl_4 as precursors (Fig. 13.10c) [64]. The key process in this strategy is dissolution–precipitation process. During this process, the Ag atoms in the core of the cubes gradually diffuse outward and react with CCl_4 on the surface, where Ag atoms could be oxidized by CCl_4 to form AgCl and deposited on the surface of each cube as a shell. In this case, Ag cations or atoms have a higher diffuse speed compared with the incoming Cl species, which causes a void space inside the cube and the formation of a hollow structure (Fig. 13.10d).

13.4.3 Facet Exposed

The crystal structure of AgX is a face-centered cubic belonging to the space group Fm3m. Hence, there exist several kinds of facet-exposed AgX crystal, such as hexahedron with {100}, octahedron with {111}, dodecahedron with {110}, tetrakaidecahedron with both {100} and {111}, trisoctahedron with {331}, icositetrahedron with {211} or {311}, tetrahexahedron with {210}, and hexoctahedron with {321} [25]. However, due to the higher surface energy of others facets, only {100} and {111} facets are usually stable on the crystal surface. Consequently, hexahedron, octahedron, and tetrakaidecahedron are the shapes of stable crystals (Fig. 13.11).

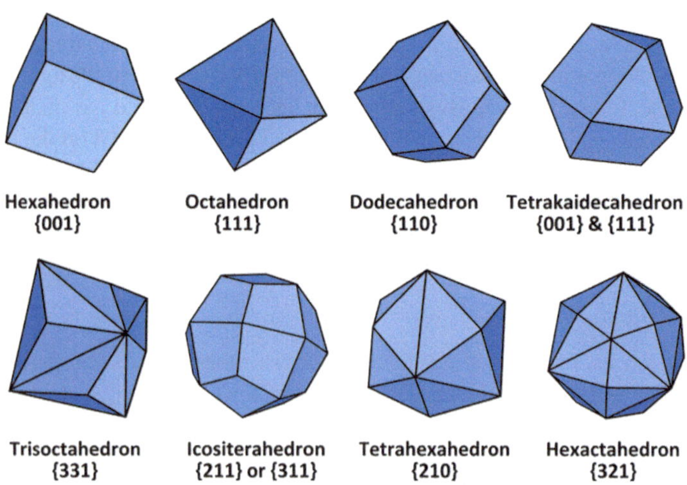

Fig. 13.11 Schematic diagram of different facet-exposed AgX crystal

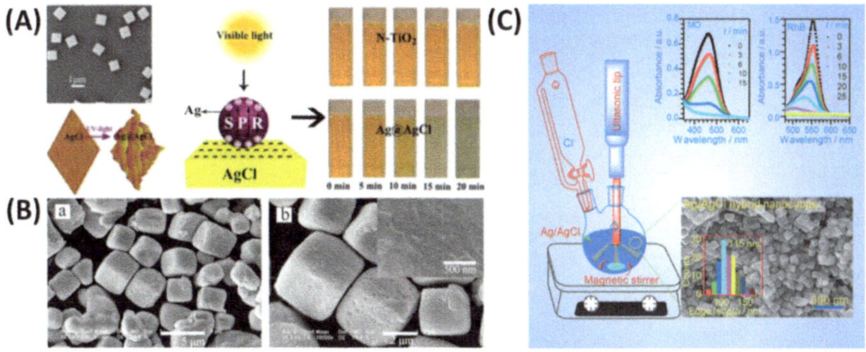

Fig. 13.12 AgCl cube prepared by (**a**) precipitation method [41]. Reprinted with the permission from Ref. [41]. Copyright 2013 American Chemical Society. (**b**) Solvothermal method [70]. Reprinted from Ref. [70], Copyright 2011, with permission from Elsevier. (**c**) Sonochemical method [71]. (Reproduced from Ref. [71] by permission of John Wiley & Sons Ltd)

Normally, it is hard to control the morphology and exposed facets of AgX due to the high reaction rate between Ag^+ ions and X^- ions [65–71]. Therefore, it is necessary to slow down the reaction speed between Ag^+ ions and X^- ions to obtain AgBr crystals with regular morphology and specific exposed facets. By precisely controlling the injection speed of Ag^+ ions and X^- ions using the double-jet equipment, Tian et al. synthesized cubic AgCl and AgBr crystals with {100} exposed facets in the absence of structure-directing agents (Fig. 13.12a) [35, 36, 39–44]. The obtained cubic AgCl and AgBr photocatalytic exhibited excellent photocatalytic activity for organic contaminant degradation. Using methylene dichloride as chlorine source instead of inorganic chloride source, Dong et al. prepared cube Ag/AgCl via a hydrothermal method. In the hydrothermal process, the slow release of Cl^- ions is favorable to the formation of cubic Ag/AgCl morphology (Fig. 13.12b) [70]. Moreover, cubic AgCl can also be obtained with the assistance of structure-directing agents. For instance, Cho et al. [71] synthesized cube-shaped Ag/AgCl photocatalysts by a sonochemical route using PVP as the structure-directing agent (Fig. 13.12c). The obtained Ag/AgCl plasmonic photocatalysts show enhanced photocatalytic activity for the degradation of methyl orange (MO), Rhodamine B (RhB), and methylene blue (MB) under visible light irradiation.

Compared with {100} facet, the surface energy of {111} facet of AgX is a litter higher under common condition. So, the {111} facet easily disappears during the growth of AgX crystals. However, if the surface energy of {111} facet can be decreased to that of {100} facet by adding structure-directing agent, the {111} facet-exposed AgX crystals can be obtained. Enlightened by this principle, we synthesized different facet-exposed AgBr crystals by a double-jet precipitation method with the inherent Br^- as the structure-directing agent [36]. It was found that the morphology and exposed facets of the AgBr crystals were conveniently tailored by adjusting the concentration of Br^- ions, i.e., cubes (C-AgBr) with {100}

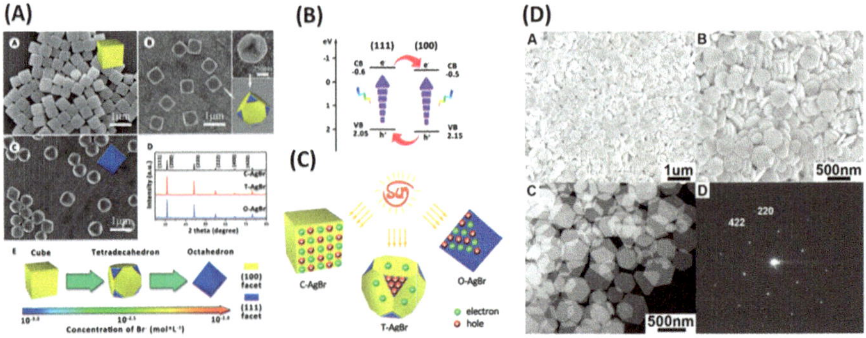

Fig. 13.13 (**a**) FE-SEM images and XRD patterns of different facet-exposed AgBr crystals and the illustration of morphology change as a function of Br^- ion concertation; (**b**) Band energy levels of {111} and {100} facets; (**c**) Illustration of the distribution of photo-generated electrons and holes on the different facet-exposed AgBr [36]. Reproduced from Ref. [36] by permission of the Royal Society of Chemistry. (**d**). SEM images, TEM image, and the corresponding SAED pattern of AgBr nanoplates [72]. (Reproduced from Ref. [72] by permission of the Royal Society of Chemistry)

facets, tetradecahedrons (T-AgBr) with both {100} and {111} facets, and octahedrons (O-AgBr) with {111} facets were synthesized when the concentrations of Br^- ions were $10^{-3.0}$, $10^{-2.5}$, and $10^{-2.0}$ M, respectively (Fig. 13.13a). Br^- ions can clearly decrease the surface energies of the (100) and (111) surfaces, by which the growth rate of AgBr nuclei along the [100, 111] directions can be tuned by adjusting the concentration of Br^- ions, leading to the formation of AgBr crystals with different exposed facets. As shown in Fig. 13.13b, because the conduction band (CB) and valence band (VB) positions of the {111} facets are higher than those of the {100} facets, the {111} and {100} facets can form facet heterojunction structures. Consequently, for the C-AgBr and O-AgBr which only have one kind of facet, the photo-generated electrons and holes will accumulate on the same facets, leading to a high recombination rate of electrons and holes. In the case of T-AgBr, the spatial isolation of photo-generated electrons and holes not only reduces the recombination rate but also effectively prevents the back reaction by isolating the reduction and oxidation reaction sites (Fig. 13.13c). Besides the inherent Br^- ion, organic chemicals, such as PVP and DMSO with –C=O and –S=O functional groups, can also play the role of structure-directing agent to lower the surface energy of {111} facet [72, 73] (Fig. 13.13d).

Although the facets with higher surface energy, such as {110}, {311}, and {15 5 2}, are thermodynamically unstable, they can still be prepared by adding special structure-directing agents. For example, Huang et al. synthesized AgBr microcrystals with different morphologies by ionic liquid (IL)-assisted hydrothermal method (Fig. 13.14a) [74]. In this method, four ionic liquids with different alkyl chains were used as the structure-directing agent. And the existence of ILs restricted step growth of AgBr {001} faces by restraining the diffusion of Ag^+, so the morphology of AgBr microcrystals could be tuned. With the assistant of 3-methylimidazolium bromides (C_4MimBr), AgBr dodecahedron crystals with exposed {110} facet were prepared.

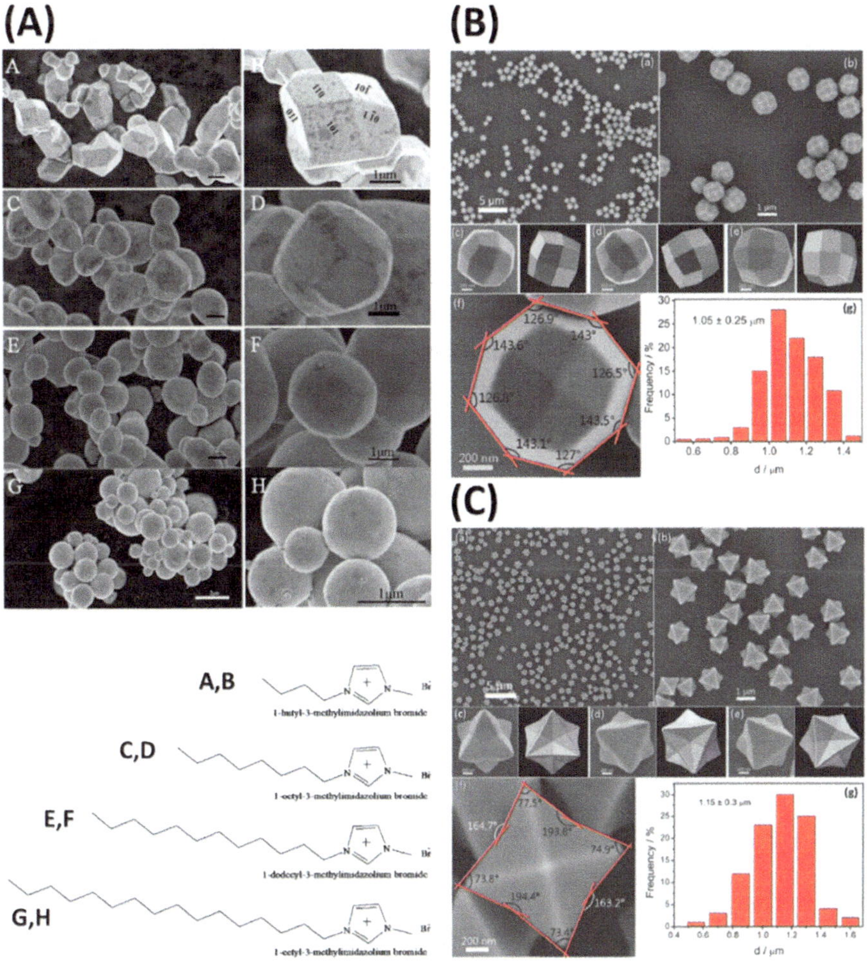

Fig. 13.14 (**a**) SEM images of samples synthesized with C_4MimBr, C_8MimBr, C_{12}MimB, and C_{16}MimBr, respectively [74]. Reproduced from Ref. [74] by permission of the Royal Society of Chemistry. (**b** and **c**). SEM images of AgCl crystals: (*a*) large-area and (*b*) enlarged. (*c–e*) in different orientations, with the corresponding geometrical models shown to the right of each SEM image. (*f*) A single AgCl crystal. (*g*) Size distribution of the as-prepared AgCl crystals [75]. (Reproduced from Ref. [75] by permission of the Royal Society of Chemistry)

Similarly, Zhang et al. found that exposed facets of AgCl crystals could be toiled from {311} to {15 5 2} using poly(diallyldimethylammonium) chloride (PDDA) as the both Cl source and structure-directing agent (Fig. 13.14b, c) [75].

13.5 Synthesis and Application of AgX-Based Heterojunction Structure

The most common heterojunction structure is based on a semiconductor-semiconductor architecture in which a p-type semiconductor usually closely contacts with an n-type semiconductor. This structure will result in a space charge region and an electric field at the interface, causing the directed flow of electrons to the CB of n-type semiconductor and holes to the VB of p-type semiconductor. This charge transfer can enhance the separating efficiency, charge carrier lifetime, and reaction rates [76–83]. Since AgX materials are prone to be reduced by the photo-generated electrons, combining with other semiconductor not only improves the separation rate of charge carriers but also promotes the photostability of AgX materials. Based on the structure, the AgX-based heterojunction structures can be classified into two types: AgX-Y and Ag-AgX-Y (Y is another semiconductor).

13.5.1 AgX–Y

AgX-Y is composed of AgX and another semiconductor Y. The closely contacting interface between AgX and Y can be produced via ion exchange method. For example, Huang et al. fabricated AgI–BiOI hierarchical hybrids by ion exchange between BiOI hierarchical microspheres and $AgNO_3$ (Fig. 13.15a) [77]. It was found that AgI nanoparticles were uniformly anchored on the surface of BiOI nanosheets and the particle size of AgI can be toiled from 55–16 nm by the addition of poly (vinylpyrrolidone) surfactant molecules. Besides ion exchange method, adsorption of organics with halogen or $Ag(NH_3)_2^+$ beforehand can also create the close contact between AgX and semiconductor Y. For instance, $[C_{16}min]Br$ ionic liquid was used to adsorb on the surface of BiPO4 to form $AgBr/BiPO_4$ heterojunction structure [83]. Beneficial from the close contact, the photo-generated charges can be efficiently separated, as shown in Fig. 13.15b, c. Some of the synthesized AgX–Y photocatalysts are summarized in Table 13.2.

13.5.2 Ag–AgX–Y

Compared to AgX–Y, Ag–AgX–Y can more effectively absorb visible light by the surface plasmon resonance effect (SPR) of Ag nanoparticles formed on the surface of AgX. Therefore, Ag–AgX–Y not only can respond to visible light but also has higher electron–hole separation rate. According to the substrates (Y), Ag–AgX–Y can be classified into several classes as follows.

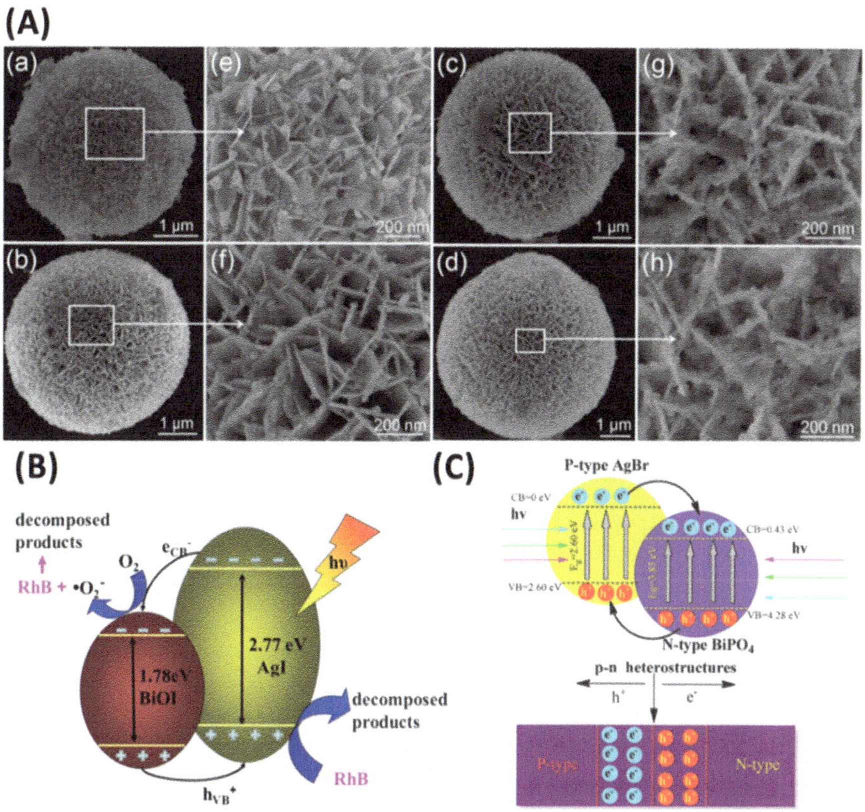

Fig. 13.15 (**a**) SEM images of AgI–BiOI with different sized AgI (55 nm, 36 nm, 24 nm, and 16 nm) [77]. Reproduced from Ref. [77] by permission of the Royal Society of Chemistry. (**b and c**) Photocatalytic mechanism of (**b**) AgI–BiOI [77] and (**c**) AgBr-$BiPO_4$ [83]. (Reproduced from Ref. [83] by permission of the Royal Society of Chemistry)

Table 13.2 Different heterojunction structure of AgX–Y

Name	Application	References
AgI–BiOI	Photodegrading 2,4-dichlorophenol	[77]
AgI–BiOI	Photodegrading RhB	[78]
AgI–TiO_2	Photodegrading RhB	[79]
AgI–N–TiO_2	Photodegrading MO	[80]
AgBr–ZnO	Photodegrading RhB	[81]
AgBr–$BiWO_6$	Photodegrading MB	[82]
AgBr–$BiPO_4$	Photodegrading MB	[83]

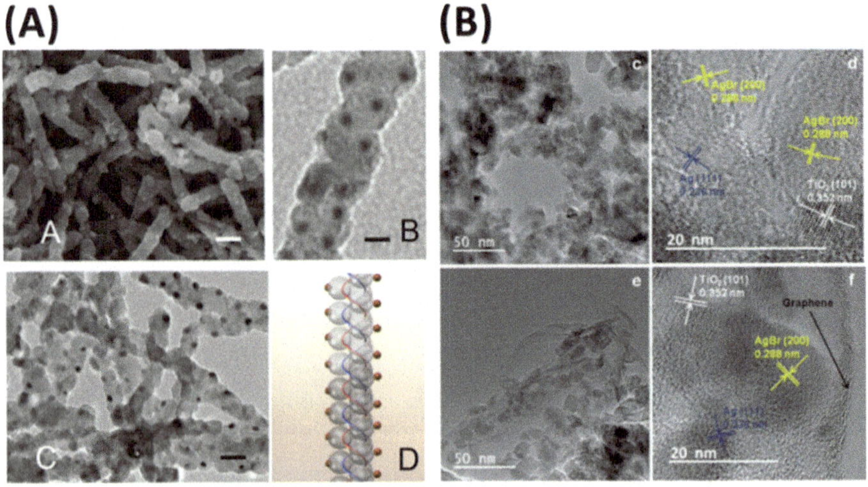

Fig. 13.16 (**a**) SEM and TEM images of Ag–AgCl–TiO_2 [84]. Reprinted from Ref. [84], Copyright 2015, with permission from Elsevier. (**b**) TEM images of Ag–AgBr–TiO_2–GO [87]. (Reproduced from Ref. [87] by permission of the Royal Society of Chemistry)

13.5.2.1 Ag–AgX–TiO_2

Titanium dioxide (TiO_2) has been widely researched as a kind of traditional photocatalytic material with the advantages of low cost and high stability. Generally, Ag–AgX–TiO_2 is synthesized by cation surfactant adsorption and photoreduction technique [84–88]. In detail, a layer of cetyltrimethylammonium chloride (CTAC) or cetyltrimethylammonium bromide (CTAB) is absorbed on the surface of TiO_2. Subsequently, Ag^+ ions are added and react with halogen ions on the surface of TiO_2 to form AgX nanoparticles. Following this synthesis strategy, Ag-AgCl-TiO_2 and Ag-AgBr-TiO_2 were prepared in Fig. 13.16a, b. And thanks to the heterojunction structure, the photocatalytic activity and charge separation ability had been improved greatly [84, 87].

For the establishment of core–shell structure, its main aim is to improve the stability of AgX. Although the Ag nanoparticle can capture photo-generated electrons to prevent AgX been photo-corroded, the chemicals in the reaction system can also destroy AgX and decrease its photocatalytic activity. Therefore, it is a feasible and effective way to prevent the deterioration of stability by covering with a layer of stable semiconductor. Besides with good carriers transfer ability, this layer of semiconductor must be thin enough to transmit visible light. Moreover, it also needs to be porous for the transfer of reactants and products. Taking the above requirements into consideration, TiO_2 is a good choice.

As shown in Fig. 13.17a, we successfully coated TiO_2 shell layer on the cubic AgCl crystals by a gradual temperature rise process [35]. During this process, the pH value of suspension and temperature rise rate are the key steps, which can effectively control the hydrolysis rate of $Ti(SO_4)_2$, or else a mass of TiO_2 would aggregate

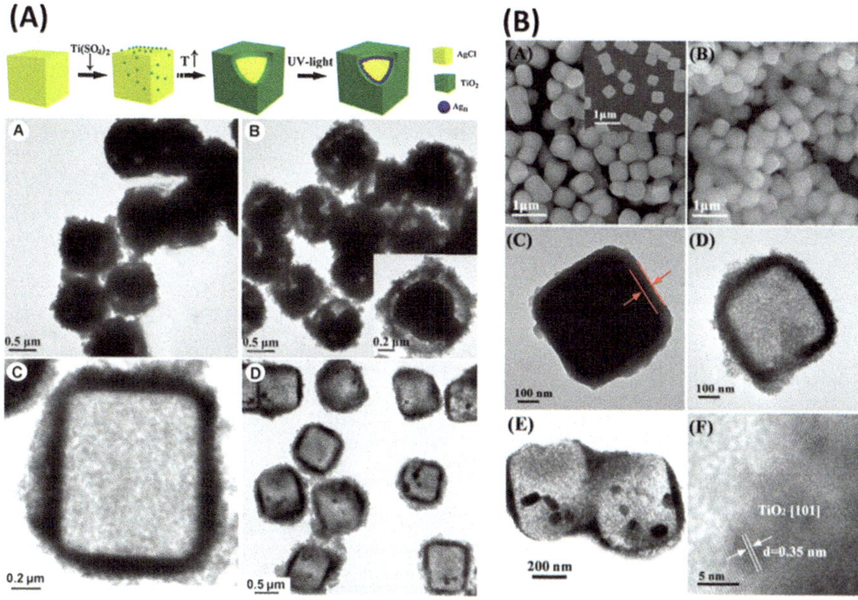

Fig. 13.17 (**a**) Synthetic route of the sandwich-structured AgCl@Ag@TiO_2 plasmonic photocatalyst and TEM images of AgCl@TiO_2 bombed by electron beams for (*A*) 2 min and (*B*) 10 min. TEM images of (*C*) AgCl@TiO_2 and (*D*) AgCl@Ag@TiO_2 treated with $Na_2S_2O_3$ solution [35]. Reprinted from Ref. [35], Copyright 2014, with permission from Elsevier. (**b**) SEM images of AgBr@TiO_2 before (*A*) and after (*B*) hydrothermal treatment. TEM image of AgBr@TiO_2 (*C*), TiO_2 shell (*D*) and Ag@TiO_2 shell. High-resolution TEM image of AgBr@Ag@TiO_2 [44]. (Reprinted from Ref. [44], Copyright 2016, with permission from Elsevier)

together. In Fig. 13.17**a***A–D*, it could be found that the thickness of TiO_2 shell layer is about 100 nm. Moreover, after being etched by $Na_2S_2O_3$ solution, there exist several Ag nanoparticles inside the TiO_2 shell, which is the evidence for the fully coating of TiO_2 shell. In addition, our group also reported the synthesis of AgBr@–Ag–TiO_2 with core–shell structure, in Fig. 13.17**b** [44]. In this work, a series of AgBr@Ag–TiO_2 with different shell thickness (0.01, 0.02, 0.03 and 0.05 μm) has been prepared by similar method, and the addition of NH_3H_2O is found as a key factor to control the thickness of TiO_2 shell layer.

13.5.2.2 Ag–AgX–C_3N_4

Recently, carbon nitride (C_3N_4) has received much attention as a stable, metal-free, and visible light-driven photocatalyst [89–94]. Commonly, the C_3N_4 nanosheets are delaminated from the bulk C_3N_4 by HCl solution. This acid treatment process can increase the abundance of amino functional groups on the surface of the C_3N_4 nanosheets. Furthermore, the amino groups with the lone pairs of electrons on the N atom in the tri-s-triazine ring structure can bind strongly to Ag^+. Therefore, as

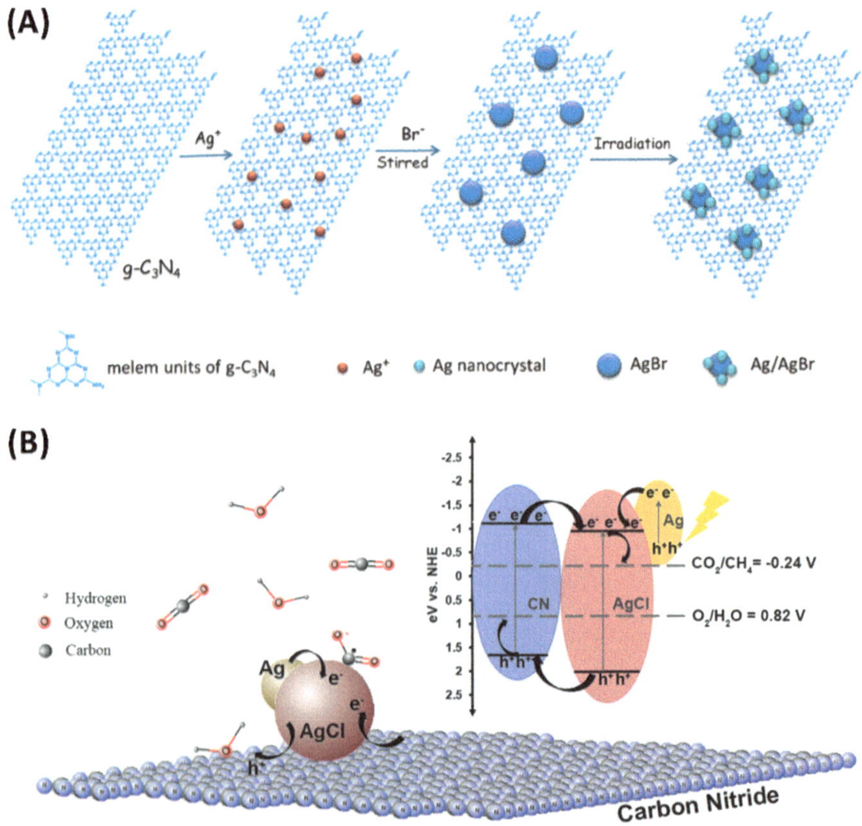

Fig. 13.18 (**a**) Schematic representation of Ag–AgBr–C_3N_4 nanocomposites [89]. Reproduced from Ref. [89] by permission of John Wiley & Sons Ltd. (**b**) Photocatalytic mechanism of Ag-AgCl-C_3N_4 [90]. (Reproduced from Ref. [90] by permission of the Royal Society of Chemistry)

shown in Fig. 13.18a, during the synthesis of Ag–AgX–C_3N_4, Ag^+ is added into suspension and adsorbed on the surface of C_3N_4 nanosheets beforehand. Subsequently, the added X^- will react with the Ag^+ in situ to form AgX nanoparticles. For example, Zhang et al. prepared Ag-AgBr-C_3N_4 by this method and this photocatalytic material exhibited 28-fold and sixfold enhancements in photodegrading RhB than bare C_3N_4 and Ag–AgBr nanoparticles [89]. Chai et al. reported that the amount of AgCl in Ag–AgCl–C_3N_4 can be adjusted with the different addition of $AgNO_3$ [90]. Generally, the CB of C_3N_4 is more negative than CB of AgX, so that the photo-generated electrons will transfer from C_3N_4 to AgX and finally be trapped by the Ag nanoparticles on the surface of AgX due to the lower Fermi level. And the photo-generated holes will transfer to C_3N_4 and take participate in the degrading reactions (Fig. 13.18b).

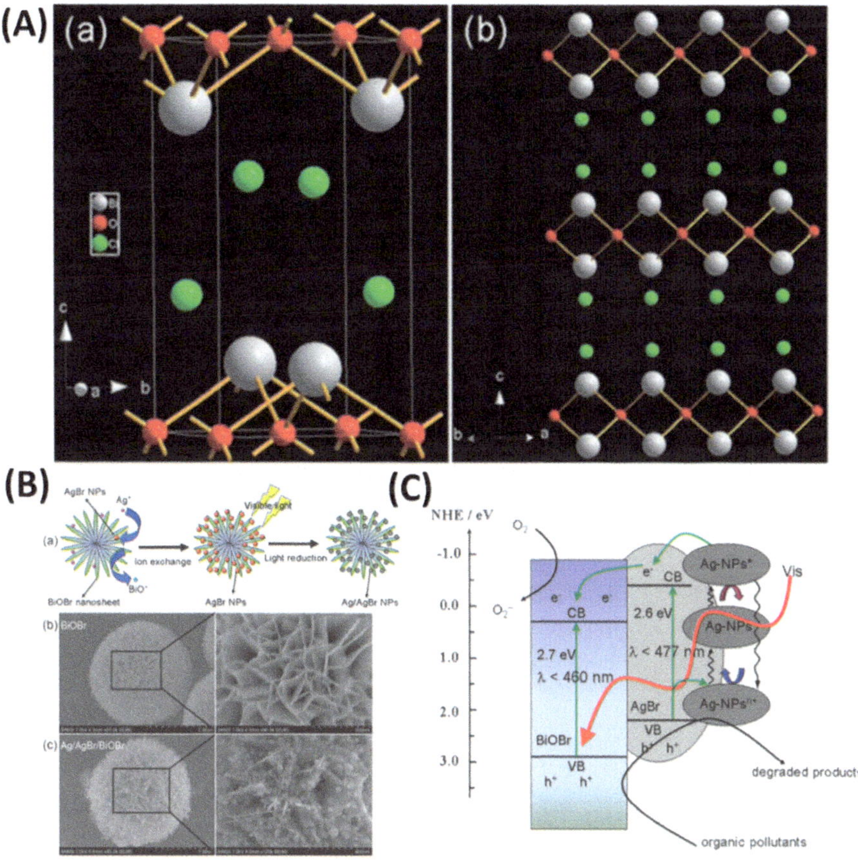

Fig. 13.19 (**a**) Schematic diagram of (*a*) unit cell and (*b*) crystal structure of BiOCl. (**b**) Synthetic route of Ag/AgBr/BiOBr hybrid and SEM images of BiOBr and Ag/AgBr/BiOBr hybrid; (**c**) photocatalytic reaction mechanism of the Ag/AgBr/BiOBr hybrid [98]. (Reproduced from Ref. [98] by permission of the Royal Society of Chemistry)

13.5.2.3 Ag–AgX–BiOX

Bismuth oxyhalides BiOX (X = Cl, Br, I) have been paid close attention for their outstanding optical and photocatalytic properties [95–97]. From the literatures, we know that all BiOX crystal is ascribed to tetragonal matlockite structure. Taking BiOCl as an example, BiOX are characterized by the layered structure that are composed of [Bi_2O_2] slabs interleaved with double halogen atom slabs along the [001] direction (Fig. 13.19a) [98]. Utilizing this structured characteristic, Huang et al. synthesized a series of AgX–BiOX (AgBr–BiOBr and AgI–BiOI) via ion exchange method [99, 100]. In this process, flowerlike BiOX particles were prepared firstly. Then, through the ion exchange reaction between BiOX and Ag^+ in ethylene glycol, AgX nanoparticles deposited on the surface of BiOX nanosheets with high dispersity, while the produced BiO^+ ions would dissolve in the solvent ($BiOBr + Ag^+ = AgBr + BiO^+$) (Fig. 13.19b). By this process, the AgX would be in situ formed on the surface of BiOX. This intimate connection between BiOX and AgX is beneficial to the transfer of photo-generated carriers (Fig. 13.19c). Besides this method, AgX–BiOX can also be obtained via the direct reaction of Bi^{3+}, Ag^+ and X^- in the hydrothermal process [101–103].

13.5.2.4 Ag–AgX–AgY

Similar to AgX, the other Ag salts (Ag–AgX–AgY), such as Ag_3PO_4, Ag_2O, Ag_2CO_3, $AgVO_3$, etc., are also visible light-driven photocatalysts with high photocatalytic activity. Combining with these Ag salts to construct heterojunction can greatly improve the photocatalytic performance of AgX [104–107]. As we all know, the solubility product (Ksp) of AgX is much lower than the most of other Ag salts. Consequently, it is feasible to construct the heterojunction of AgX–AgY by the means of ion exchange method. For example, Wu et al. utilized NaBr solution to exchange the PO_4^{3+} ions in Ag_3PO_4 to form AgBr–$AgPO_4$ (Fig. 13.20a). In addition, Ag–AgCl–Ag_2O [105], Ag–AgBr–$AgVO_3$ [106], and Ag–AgBr–Ag_2CO_3 [107] can also be prepared by this ion exchange method (Fig. 13.20b, c). Being benefited from this heterojunction, photo-generated carriers can be effectively separated (Fig. 13.20d).

13.6 Z-Scheme Structure

Constructing Z-scheme structure is another effective way to promote the separation rate of photo-generated carriers. Moreover, beneficial from the special transfer process of electrons and holes, the higher redox ability of photocatalytic composites can be reserved via Z-scheme structure [108–113]. In Z-scheme structured AgX-Ag-Y photocatalytic material, the semiconductor Y has the same metallic element with AgX, which can facilely be in situ generated via photo-reduction process. The in situ formed Ag nanoparticles can be stably anchored on the surface of AgX inhibiting the exfoliation of Ag nanoparticles from AgX surface [111]. Moreover, the tight solid-solid contact interface between Ag nanoparticles and AgX can reduce the electric resistance, beneficial to forming Ohmic contact. Interestingly, different from the conventional Z-scheme structure, the role of Ag nanoparticles, electron mediator or photosensitizer, is determined by the energy band structure of AgX and semiconductor Y. According to the energy levels of the two semiconductors, the visible light-driven Z-scheme AgX–Ag–Y can be classified into three types, as shown in Fig. 13.21.

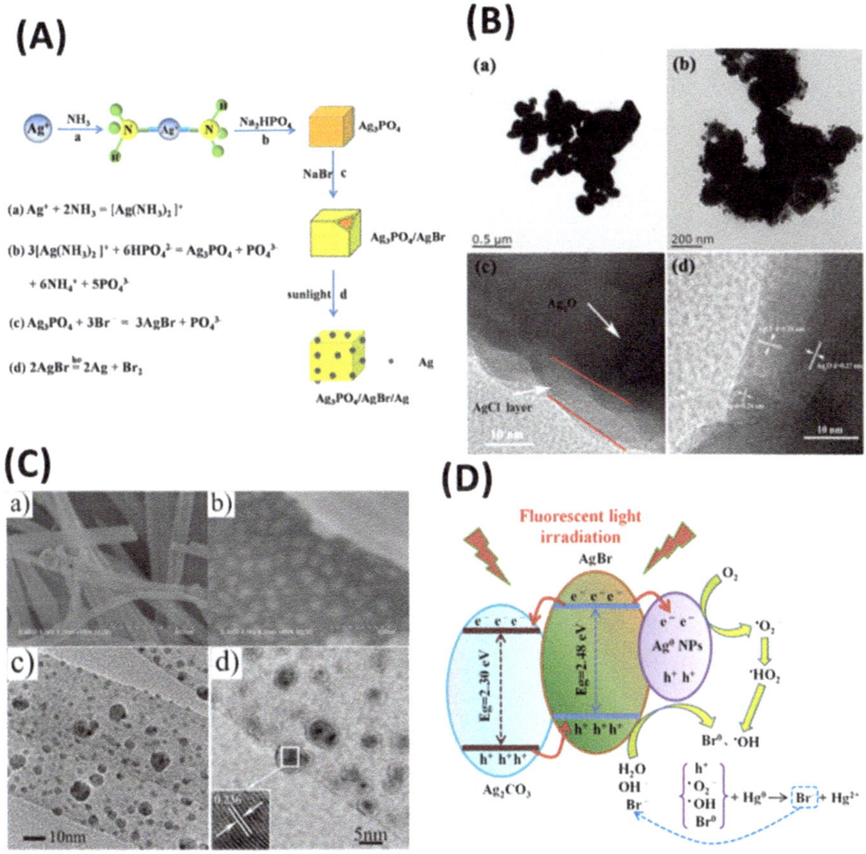

Fig. 13.20 (**a**) Schematic illustration of the growth process of Ag–AgBr–$AgPO_4$ [104]. Reproduced from Ref. [104] by permission of the Royal Society of Chemistry. (**b**) TEM images of Ag–AgCl–Ag_2O [105]. Reprinted from Ref. [105], Copyright 2017, with permission from Elsevier. (**c**) SEM and TEM images of Ag–AgBr–$AgVO_3$ [106]. Reprinted from Ref. [106], Copyright 2014, with permission from Elsevier. (**d**) photocatalytic mechanism and charge transfer in Ag–AgBr–Ag_2CO_3 [107]. (Reprinted with the permission from Ref. [107]. Copyright 2016 American Chemical Society)

1. *Type A*

In type A, the Z-scheme AgX-Ag-Y photocatalytic material is composed of AgCl ($E_g = 3.2$ eV) and another semiconductor Y with narrower band gap ($E_g < 3.0$ eV). As a result, under visible light irradiation, semiconductor Y can adsorb photons and generate electron–hole pairs, while AgCl cannot. Furthermore, to meet the requirement of Z-scheme structure, the CB of semiconductor Y should be lower than that of AgCl ($E_{CB,\ AgCl} = -0.05$ eV). In this case, the Ag nanoparticles simultaneously serve as the electron mediator and photosensitizer, differing from the role of noble metal nanoparticles in common Z-scheme structured photocatalytic materials.

As illustrated in Fig. 13.21, Ag nanoparticles can produce photo-generated electrons and holes via SPR effect under visible light irradiation. Due to the store

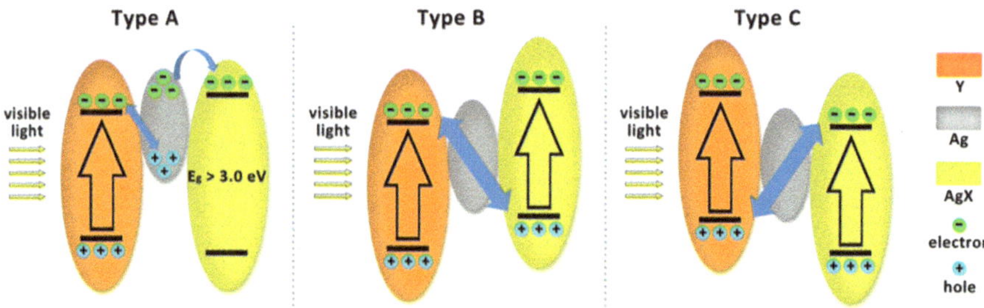

Fig. 13.21 Schematic of the three types of Z-scheme structured AgX-Ag-Y photocatalytic materials

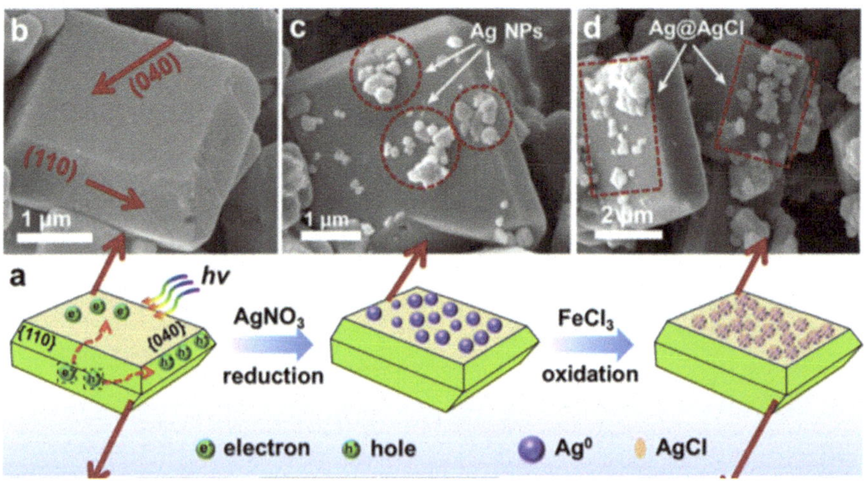

Fig. 13.22 (**a**) Schematic illustration of the formation process of the Ag@AgCl/$BiVO_4$ heterojunction photocatalyst. SEM images of the as-prepared $BiVO_4$ (**b**), Ag/$BiVO_4$ (**c**), and Ag@AgCl/$BiVO_4$ (**d** and **e**) [114]. (Reprinted from Ref. [114], Copyright 2015, with permission from Elsevier)

of electrons in the lowest unoccupied orbital of Ag, the Fermi energy level of Ag nanoparticle will be lifted. Afterward, the superfluous electrons will migrate into the CB of the neighboring AgCl. Simultaneously, the photo-generated electrons in CB of semiconductor Y will transfer to Ag nanoparticle and combine with the holes in the highest occupied orbital of Ag, while the photo-generated holes in VB of semiconductor Y can participate in the photocatalytic reaction. Beneficial from this special mechanism for separating photo-generated carriers, Z-scheme structured AgX-Ag-Y photocatalytic material exhibit superior photocatalytic activity for organic compound degradation [114–122]. For example, Wu et al. designed a hierarchical Z-scheme photocatalyst Ag@AgCl/$BiVO_4$ (Fig. 13.22) [114]. In this system, the metallic Ag species not only act as the solid-state electron mediator but also absorb the photons by SPR effect. Compared with pristine $BiVO_4$, the photocatalytic performance for degrading RhB was enhanced about 300 times. Moreover, Yu et al. prepared $H_2WO_4{\cdot}H_2O/Ag/AgCl$ composite nanoplates by a one-step ionic reaction between $Ag_8W_4O_{16}/Ag$ nanorods and HCl aqueous solution [119]. The photocatalytic experiments indicated that the $H_2WO_4{\cdot}H_2O/Ag/AgCl$ composite nanoplates exhibited a much higher photocatalytic activity than the one-component ($H_2WO_4{\cdot}H_2O$) or two-component (such as Ag/AgCl and $H_2WO_4{\cdot}H_2O/Ag$) photocatalysts.

2. *Type B*

For type B photocatalytic material, the Z-scheme AgX-Ag-Y is composed of AgX (X = Br or I) and another semiconductor Y with narrower band gap ($E_g < 3.0$ eV). Moreover, both the CB and VB energy levels of semiconductor Y are lower than those of AgBr ($E_{CB\text{-}AgBr} = -0.3$ eV, $E_{VB\text{-}AgBr} = 2.3$ eV) or AgI ($E_{CB\text{-}AgI} = -0.4$ eV, $E_{VB\text{-}AgBr} = 2.36$ eV). Therefore, the electrons in the VB of AgX and semiconductor Y can be excited to the corresponding CB under visible light illumination. The photo-generated electrons in the CB of semiconductor Y can migrate into Ag nanoparticles. Simultaneously, the photo-generated holes in VB of AgX will flow into Ag nanoparticles easily and recombine with the photo-generated electrons from semiconductor Y. Beneficial from the above carriers transfer process, the photo-generated electrons with stronger reduction power in CB of AgX and photo-generated holes with higher oxidation ability in VB of semiconductor Y can be reserved. This advantage will lead to higher photocatalytic activity of Z-scheme structured AgX based photocatalysts [37, 101, 123–129]. For instance, Fan et al. synthesized $Ag/AgBr/AgIn(MoO_4)_2$ nanosheets by in situ photoreduction of $AgBr/AgIn(MoO_4)_2$ composites (Fig. 13.23a) [123]. It was found that $Ag/AgBr/AgIn(MoO_4)_2$ composites exhibited dramatic enhanced photocatalytic activity for tetracycline degradation when compared with $AgIn(MoO_4)_2$ nanosheets, Ag/AgBr nanoparticles and $Ag/AgIn(MoO_4)_2$ composites. Yang et al. reported a facile in

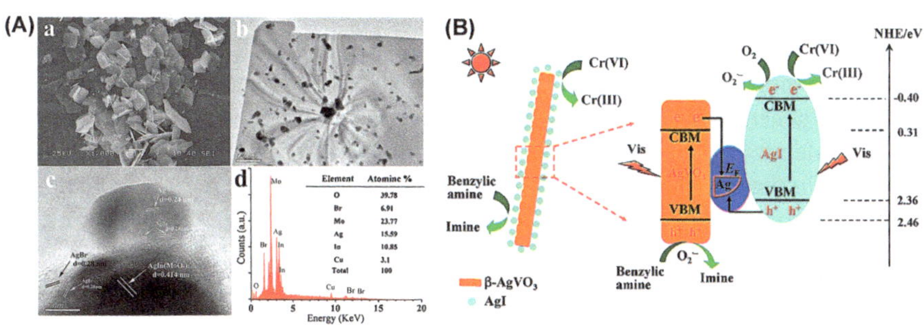

Fig. 13.23 (**a**) (*a*) FESEM image of $AgIn(MoO_4)_2$ nanosheets. (*b–d*) TEM, HRTEM, EDX of $Ag/AgBr/AgIn(MoO_4)_2$ [123]. Reprinted from Ref. [123], Copyright 2015, with permission from Elsevier. (**b**) Photocatalytic mechanism of $AgI/AgVO_3$ toward the selective oxidation of benzylic amine and reduction of Cr(VI) under visible light irradiation [124]. (Reproduced from Ref. [124] by permission of the Royal Society of Chemistry)

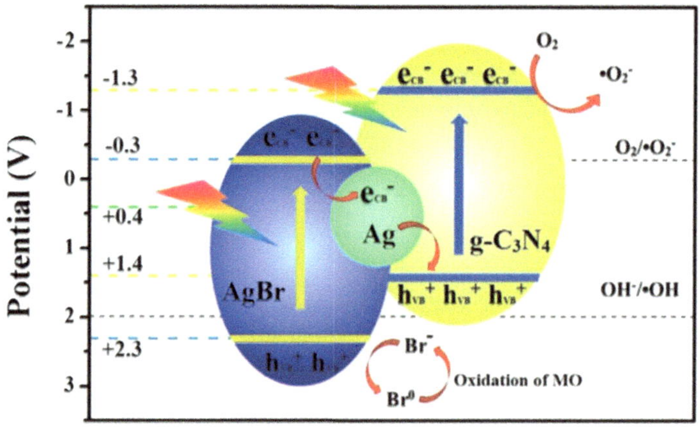

Fig. 13.24 Generation, separation and transportation of h_{VB}^+ and e_{CB}^- at the interface of the visible light-driven Ag@AgBr/g-C_3N_4 plasmonic photocatalyst [130]. (Reprinted from Ref. [130], Copyright 2014, with permission from Elsevier)

situ ion exchange approach to prepared AgI/$AgVO_3$ nanocomposites with β-$AgVO_3$ nanoribbons as the Ag source and support to immobilize AgI (Fig. 13.23b) [124]. The as-prepared composites can serve as highly efficient visible light-driven photocatalysts toward selective oxidation of benzylic amine to imine and reduction of toxic Cr (VI) ions. It also found that the considerable improvement in the photocatalytic redox properties of AgI/$AgVO_3$ is mainly ascribed to the efficient separation of photoinduced electrons/holes via a Z-scheme bridge mechanism of formed Ag/AgI/$AgVO_3$, in which Ag nanoparticles serve as the charge migration bridge.

3. *Type C*

Similar to Type B, the Z-scheme structured AgX-Ag-Y photocatalytic material in Type C is also composed of AgX and semiconductor Y, in which both of them can absorb visible light. However, as shown in Fig. 13.21c, the CB and VB of semiconductor Y is higher than that of AgX. Consequently, after the recombination process, the photo-generated electrons with higher reduction ability in CB of semiconductor Y and holes with better oxidation property in VB of AgX will be reserved. For example, in the system of AgBr–Ag–C_3N_4, photo-generated carriers are efficiently separated via Z-scheme structure Fig. 13.24) [93, 130]. Afterward, the electrons in CB of C_3N_4 will further react with O_2 to form. O_2^- radicals and holes in VB of AgBr will generate Br^0 to oxide organic contaminants.

13.7 Recoverable AgX Photocatalytic Materials

The key problem that restrains the application of nano- or micro-sized AgX-based photocatalytic materials in waste water treatment is how to effectively separate and recycle the used photocatalysts. Traditional centrifugation and filtration strategy will not merely cause serious loss of photocatalysts but high energy consumption [131]. In the past 10 years, two strategies were introduced to improve the recoverability of AgX-based photocatalytic materials. One is loading the photocatalysts on the substrates, and another is combining with magnetic components.

13.7.1 Loaded on the Substrates

Although conventional 2D substrates for loading photocatalysts, such as ITO glass and metal foil, can be conveniently recycled, they suffer from the disadvantage of low surface area. Thus, recently, a series of substrates with high surface area (Al_2O_3 mesoporous microsphere, 3D graphene aerogels, and nylon mesh) were utilized to load AgX. For instance, He et al. immobilized the plasmonic Ag–AgI with photo-induced self-stability on mesoporous Al_2O_3 by a wet impregnation–precipitation and in situ photoreduction method [132]. Niu et al. synthesized a novel 3D structure AgX/GA (X = Br, Cl) composites with a macroscopic block appearance (Fig. 13.25a). Owing to the unique structure, this bulk composite material could just be recycled by directly clipping out using tweezers and washed with deionized water several times [133]. Tian et al. successfully fabricated Ag@AgBr gelatin film by embedding Ag@AgBr particles into a gelatin matrix and constructing 3D network structures via the cross-linking reaction between gelatin and cross-linking agent 1,3-bis(vinylsulfonyl) propanol, in Fig. 13.25b–e [40]. Beneficial from the SPR effect of Ag and excellent penetrability of 3D network structure for reactants

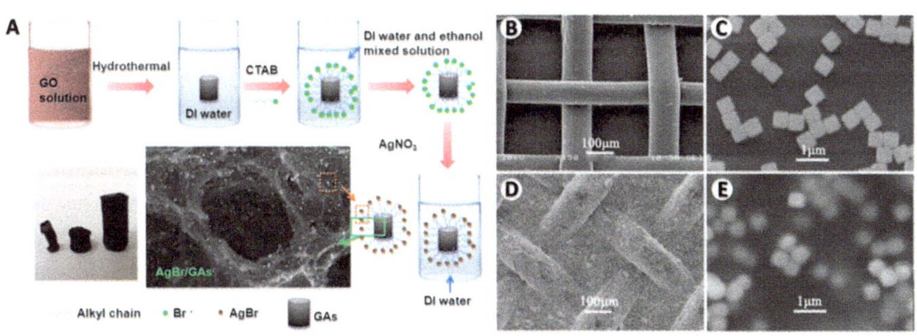

Fig. 13.25 (**a**) Fabrication of the AgBr/gas 3D photocatalytic composite materials [133]. Reproduced from Ref. [133] by permission of John Wiley & Sons Ltd. SEM images of (**b**) nylon mesh; (**c**) Ag@AgBr grains, (**d**) and (**e**) Ag@AgBr-gelatin film loaded on nylon mesh [40]. (Reprinted from Ref. [40], with kind permission from Springer Science+Business Media)

and products, Ag@AgBr-gelatin film showed excellent visible light activity for MO degradation. Based on recycle experiments, the film exhibited excellent stability and recyclability in the application of organic contaminant degradation. Moreover, it should be noticed that this synthetic route can be used to fabricate photocatalytic films in a large scale, which is meaningful for practical application.

13.7.2 Combined with Magnetic Components

Besides loading AgX materials on subtracts, immobilizing photocatalysts on the surface of magnetic nano- or microparticles is another effective strategy to enhance the recoverability, by which the photocatalysts can be easily separated by an external magnet under the premise of keeping photocatalytic activity. Recently, $CoFe_2O_4$ [134–136], Fe_3O_4 [137–139], Fe_2O_3 [140–142], and γ-Fe_2O_3 [131] were used as photocatalyst carriers to obtain efficient recyclable AgX-based materials for water treatment. For example, Xu et al. attempted to construct ferromagnetic plasmonic nanophotocatalysts by coupling Ag/AgCl with magnetic material $CoFe_2O_4$, which enhanced the photo-activity of Ag/AgCl/$CoFe_2O_4$ [134]. They also prepared Ag/AgBr@Fe_2O_3 magnetic photocatalyst by solvothermal process. The Ag/AgBr was covered by Fe_2O_3 and formed a uniquely core–shell nanostructure, which would provide a high surface area and numerous active sites for the photocatalytic reaction (Fig. 13.26a) [140]. Zhang et al. reported a magnetic adsorptive photocatalyst composite, Ag/AgCl-magnetic activated carbon (MAC) synthesized via a facile deposition–precipitation–photoreduction method (Fig. 13.26b) [143]. The resulting composites possessed quasi-superparamagnetic behavior and exhibited good visible light-induced photocatalytic activity toward the inactivation of *E. coli* K-12 and degradation of methyl orange and phenol.

Although these composite photocatalysts can be successfully separated by applying an external magnetic field, there still exist some deficiencies, such as wide size distribution and irregular morphological structures. These deficiencies will cause a longer time for completely harvesting photocatalysts. To solve this problem, Tian et al. fabricated core–shell structured γ-Fe_2O_3@SiO_2@AgBr:Ag composite microspheres with narrow size distribution by a versatile multistep route, including solvothermal method to fabricate magnetic core, modified Stöber method to coat SiO_2 interlayer, electrostatic assembly to deposit AgBr shell, and light reduction to form Ag nanoparticles (Fig. 13.26c, d) [131]. Beneficial from the narrow size distribution, high saturation magnetization, and superparamagnetic property, the core–shell structured γ-Fe_2O_3@SiO_2@AgBr:Ag shows excellent magnetic separation and recovery performances (completely harvesting in 30 s), as shown in Fig. 13.26e.

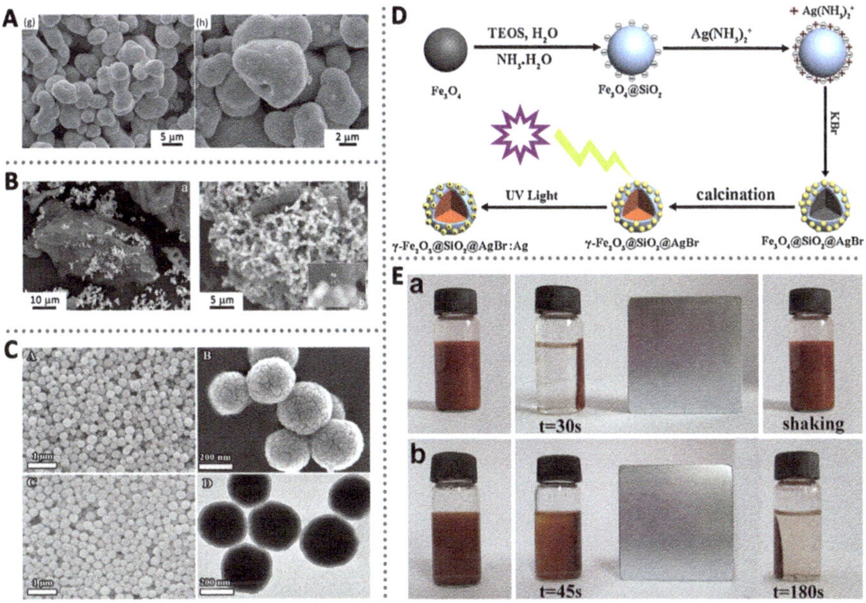

Fig. 13.26 SEM images of (**a**) 5% Ag/AgBr@Fe_2O_3 [140]. Reproduced from Ref. [140] by permission of the Royal Society of Chemistry. (**b**) Ag/AgCl-AC and Ag/AgCl-MAC [143]. Reprinted from Ref. [143], Copyright 2014, with permission from Elsevier. (**c**) Fe_3O_4 and γ-Fe_2O_3@SiO_2@AgBr:Ag [131]. (**d**) Schematic diagram illustrating the synthetic route of core–shell structured γ-Fe_2O_3@SiO_2@AgBr:Ag composite; (**e**) magnetic separation tests for γ-Fe_2O_3@SiO_2@AgBr:Ag and Fe_3O_4@SiO_2@AgBr:Ag using a cubic Nd-Fe-B magnet [131]. (Reprinted from Ref. [131], Copyright 2014, with permission from Elsevier)

References

1. Qiu B-C, Zhu Q-H, Xing M-Y, Zhang J-L (2017) A robust and efficient catalyst of $Cd_xZn_{1-x}Se$ motivated by CoP for photocatalytic hydrogen evolution under sunlight irradiation. Chem Commun 53:897–900
2. Qiu B-C, Zhu Q-H, Du M-M, Fan L-G, Xing M-Y, Zhang J-L (2017) Efficient solar light harvesting CdS/Co_9S_8 hollow cubes for Z-scheme photocatalytic water splitting. Angew Chem 129:2728–2732
3. Wu Q-F, Bao S-Y, Tian B-Z, Xiao Y-F, Zhang J-L (2016) Double-diffusion-based synthesis of $BiVO_4$ mesoporous single crystals with enhanced photocatalytic activity for oxygen evolution. Chem Commun 52:7478–7481
4. Xing M-Y, Zhang J-L, Chen F, Tian B-Z (2011) An economic method to prepare vacuum activated photocatalysts with high photo-activities and photosensitivities. Chem Commun 47:4947–4949
5. Du J, Du Z-L, Hu J-S, Pan Z-X, Shen Q, Sun J-K, Long D-H, Dong H, Sun L-T, Zhong X-H, Wan L-J (2016) Zn-Cu-In-Se quantum dot solar cells with a certified power conversion efficiency of 11.6%. J Am Chem Soc 138:4201–4209
6. Zhao K, Pan Z-X, Mora-Seró I, Wang H, Song Y, Gong X-Q, Wang J, Bonn M, Bisquert J, Zhong X-H (2015) Boosting power conversion efficiencies of quantum-dot-sensitized solar cells beyond 8% by recombination control. J Am Chem Soc 137:5602–5609

7. Ali M, Zhou F-L, Chen K, Kotzur C, Xiao C-L, Bourgeois L, Zhang X-Y, MacFarlane D-R (2016) Nanostructured photoelectrochemical solar cell for nitrogen reduction using plasmon-enhanced black silicon. Nat Commun 11335:1–5
8. Li Q-Y, Guan Z-P, Wu D, Zhao X-G, Bao S-Y, Tian B-Z, Zhang J-L (2017) Z-Scheme BiOCl-Au-CdS heterostructure with enhanced sunlight-driven photocatalytic activity in degrading water dyes and antibiotics. ACS Sustain Chem Eng 5:6958–6968
9. Li Q-Y, Li T-Y, Chang S-Z, Tao Q-S, Tian B-Z, Zhang J-L (2016) Enlarging {110} exposed facets of anatase TiO_2 by the synergistic action of capping agents. CrystEngComm 18:5074–5078
10. Li T-Y, Tian B-Z, Zhang J-L, Dong R-F, Wang T-T, Yang F (2013) Facile tailoring of anatase TiO_2 morphology by use of H_2O_2: from microflowers with dominant {101} facets to microspheres with exposed {001} facets. Ind Eng Chem Res 52(20):6704–6712
11. Weon S-H, Choi J-M, Park T-H, Choi W-Y (2017) Freestanding doubly open-ended TiO_2 nanotubes for efficient photocatalytic degradation of volatile organic compounds. Appl Catal B 205:386–392
12. Ren L, Li Y-Z, Hou J-T, Bai J-L, Mao M-Y, Zeng M, Zhao X-J, Li N (2016) The pivotal effect of the interaction between reactant and anatase TiO_2 nanosheets with exposed {001} facets on photocatalysis for the photocatalytic purification of VOCs. Appl Catal B 181:625–634
13. Wu D, Yue S-T, Wang W, An T-C, Li G-Y, Yip H-Y, Zhao H-J, Wong P-K (2016) Boron doped BiOBr nanosheets with enhanced photocatalytic inactivation of Escherichia coli. Appl Catal B 192:35–45
14. Wang W-J, An T-C, Li G-Y, Xia D-H, Zhao H-J, Yu J-C, Wong P-K (2017) Earth-abundant $Ni_2P/g\text{-}C_3N_4$ lamellar nanohydrids for enhanced photocatalytic hydrogen evolution and bacterial inactivation under visible light irradiation. Appl Catal B 217:570–580
15. Rtimi S, Giannakis S, Sanjines R, Pulgarin C, Bensimon M, Kiwi J (2016) Insight on the photocatalytic bacterial inactivation by co-sputtered TiO_2-Cu in aerobic and anaerobic conditions. Appl Catal B 182:277–285
16. Kuehnel M-F, Orchard K-L, Dalle K-E, Reisner E (2017) Selective photocatalytic CO_2 reduction in water through anchoring of a molecular Ni catalyst on CdS nanocrystals. J Am Chem Soc 139:7217–7223
17. Takeda H, Ohashi K, Sekine A, Ishitani O (2016) Photocatalytic CO_2 reduction using Cu (I) photosensitizers with a Fe(II) catalyst. J Am Chem Soc 138:4354–4357
18. Dong C-Y, Xing M-Y, Zhang J-L (2016) Double-cocatalysts promote charge separation efficiency in CO_2 photoreduction: spatial location matters. Mater Horiz 3:608–612
19. Dong C-Y, Xing M-Y, Zhang J-L (2016) Economic hydrophobicity triggering of CO_2 photoreduction for selective CH_4 generation on noble-metal-free TiO_2–SiO_2. J Phys Chem Lett 7:2962–2966
20. Fang W-Z, Dappozze F, Guillard C, Zhou Y, Xing M-Y, Mishra S, Daniele S, Zhang J-L (2017) Zn-assisted TiO_{2-x} photocatalyst with efficient charge separation for enhanced photocatalytic activities. J Phys Chem C 121:17068–17076
21. Schneider J, Matsuoka M, Takeuchi M, Zhang J-L, Horiuchi Y, Anpo M, Bahnemann D-W (2014) Understanding TiO_2 photocatalysis: mechanisms and materials. Chem Rev 114:9919–9986
22. Qiu B-C, Xing M-Y, Zhang J-L (2014) Mesoporous TiO_2 nanocrystals grown in situ on graphene aerogels for high photocatalysis and lithium-ion batteries. J Am Chem Soc 136:5852–5855
23. Qi D-Y, Lu L-J, Wang L-Z, Zhang J-L (2014) Improved SERS sensitivity on plasmon-free TiO_2 photonic microarray by enhancing light-matter coupling. J Am Chem Soc 136:9886–9889
24. Li X, Liu P-W, Mao Y, Xing M-Y, Zhang J-L (2015) Preparation of homogeneous nitrogen-doped mesoporous TiO_2 spheres with enhanced visible-light photocatalysis. Appl Catal B 164:352–359

25. Tani T (1995) Photographic sensitivity theory and mechanisms. Oxford University Press, New York
26. Tian B-Z, Zhang J-L (2012) Morphology-controlled synthesis and applications of silver halide photocatalytic materials. Catal Surv Jpn 16:210–230
27. Tani T (2011) Photographic sensitivity advances in nanoparticles, J-aggregates, dye sensitization, and organic devices. Oxford University Press, Oxford/New York
28. Cox R-J (1973) Photographic sensitivity. Academic Press Inc (London) LT, London
29. Kakuta N, Goto N, Ohkita H, Mizushima T (1999) Silver bromide as a photocatalyst for hydrogen generation from CH_3OH/H_2O solution. J Phys Chem B 103:5917–5919
30. Schürch D, Currao A, Sarkar S, Hodes G, Calzaferri G (2002) The silver chloride photoanode in photoelectrochemical water splitting. J Phys Chem B 106:12764–12775
31. Pfanner K, Gfeller N, Calzaferri G (1996) Photochemical oxidation of water with thin AgCl layers. J Photochem Photobiol A 95:175–180
32. Hu C, Lan Y-Q, Qu J-H, Hu X-X, Wang A-M (2006) Ag/AgBr/TiO_2 visible light photocatalyst for destruction of azodyes and bacteria. J Phys Chem B 110:4066–4072
33. Wang P, Huang B-B, Qin X-Y, Zhang X-Y, Dai Y, Wei J-Y, Whangbo M-H (2008) Ag@AgCl: a highly efficient and stable photocatalyst active under visible light. Angew Chem Int Ed 47:7931–7933
34. Wikipedia. www.en.wikipedia.org
35. Tian B-Z, Dong R-F, Zhang J-M, Bao S-Y, Yang F, Zhang J-L (2014) Sandwich-structured AgCl@Ag@TiO_2 with excellent visible-light photocatalytic activity for organic pollution degradation and E. coli K12 inactivation. Appl Catal B 158–159:76–84
36. Bao S-Y, Wang Z-Q, Gong X-Q, Zeng C-Y, Wu Q-F, Tian B-Z, Zhang J-L (2016) AgBr tetradecahedrons with co-exposed {100} and {111} facets: simple fabrication and enhancing spatial charge separation using facet heterojunction. J Mater Chem A 4:18570–18577
37. Lin H-L, Cao J, Luo B-D, Xu B-Y, Chen S-F (2012) Synthesis of novel Z-scheme AgI/Ag/AgBr composite with enhanced visible light photocatalytic activity. Catal Commun 21:91–95
38. Tao Q-S, Yang F, Teng F, Wu P-Y, Tian B-Z, Zhang J-L (2015) Study of the factors influencing the photo-stability of Ag@AgBr plasmonic photocatalyst. Res Chem Intermed 41:7285–7297
39. Zeng C-Y, Guo M, Tian B-Z, Zhang J-L (2013) Reduced graphene oxide modified Ag/AgBr with enhanced visible light photocatalytic activity for methyl orange degradation. Chem Phys Lett 575:81–85
40. Zhu J, Li C-J, Teng F, Tian B-Z, Zhang J-L (2015) Recyclable Ag@AgBr-gelatin film with superior visible-light photocatalytic activity for organic degradation. Res Chem Intermed 41:9715–9730
41. Dong R-F, Tian B-Z, Zeng C-Y, Li T-Y, Wang T-T, Zhang J-L (2013) Ecofriendly synthesis and photocatalytic activity of uniform cubic Ag@AgCl plasmonic photocatayst. J Phys Chem C 117:213–220
42. Zeng C-Y, Tian B-Z, Zhang J-L (2013) Silver halide/silver iodide@silver composite with excellent visible light photocatalytic activity for methyl orange degradation. J Colloid Interface Sci 405:17–21
43. Dong R-F, Tian B-Z, Zhang J-L, Wang T-T, Tao Q-S, Bao S-Y, Yang F, Zeng C-Y (2013) AgBr@Ag/TiO_2 core-shell composite with excellent visible light photocatalytic activity and hydrothermal stability. Catal Commun 38:16–20
44. Zhang P, Wu P-Y, Bao S-Y, Wang Z, Tian B-Z, Zhang J-L (2016) Synthesis of sandwich-structured AgBr@Ag@TiO_2 composite photocatalyst and study of its photocatalytic performance for the oxidation of benzyl alcohols to benzaldehydes. Chem Eng J 306:1151–1161
45. Yang F, Tian B-Z, Zhang J-L, Xiong T-Q, Wang T-T (2014) Preparation, characterization, and photocatalytic activity of porous AgBr@Ag and AgBrI@Ag plasmonic photocatalysts. Appl Surf Sci 292:256–261
46. Bi Y-P, Ye J-H (2010) Direct conversion of commercial silver foils into high aspect ratio AgBr nanowires with enhanced photocatalytic properties. Chem Eur J 16:10327–10331

47. Cao Q, Liu X, Yuan K-P, Yu J, Liu Q-H, Delaunay J-J, Che R (2017) Gold nanoparticled decorated Ag(Cl, Br) micro-necklaces for efficient and stable SERS detection and visible-light photocatalytic degradation of Sudan I. Appl Catal B 201:607–616
48. Bi Y-P, Ye J-H (2009) In situ oxidation synthesis of Ag/AgCl core-shell nanowires and their photocatalytic properties. Chem Commun 43:6551–6553
49. Bi Y-P, Ye J-H (2010) Heteroepitaxial growth of platinum nanocrystals on AgCl nanotubes via galvanic replacement reaction. Chem Commun 46:1532–1534
50. Jia C-C, Yang P, Huang B-B (2014) Uniform Ag/AgCl necklace-like nano-heterostructures: fabrication and highly efficient plasmonic photocatalysis. ChemCatChem 6:611–617
51. Ge J, Wang X, Yao H-B, Zhu H-W, Peng Y-C, Yu S-H (2015) Durable Ag/AgCl nanowires assembled in a sponge for continuous water purification under sunlight. Mater Horiz 2:509–513
52. Sun Y-G (2010) Conversion of Ag nanowires to AgCl nanowires decorated with Au nanoparticles and their photocatalytic activity. J Phys Chem C 114:2127–2133
53. Zhu M-S, Chen P-L, Liu M-H (2012) Highly efficient visible-light-driven plasmonic photocatalysts based on graphene oxide-hybridized one-dimensional Ag/AgCl heteroarchitectures. J Mater Chem 22:21487–21494
54. Sun L, Zhang R-Z, Wang Y, Chen W (2014) Plasmonic Ag@AgCl nanotubes fabricated from copper nanowires as high-performance visible light photocatalyst. ACS Appl Mater Interfaces 6:14819–14826
55. Li B, Wang H, Zhang B-W, Hu P-F, Chen C, Guo L (2013) Facile synthesis of one dimensional AgBr@Ag nanostructures and their visible light photocatalytic properties. ACS Appl Mater Interfaces 5:12283–12287
56. Chen D-L, Liu M-N, Chen Q-Q, Ge L-F, Fan B-B, Wang H-L, Lu H-X, Yang D-Y, Zhang R, Yan Q-S, Shao G-S, Sun J, Gao L (2014) Large-scale synthesis and enhanced visible-light-driven photocatalytic performance of hierarchical Ag/AgCl nanocrystals derived from freeze-dried PVP-Ag^+ hybrid precursors with porosity. Appl Catal B 144:394–407
57. Ai L-H, Zhang C-H, Jiang J (2013) Hierarchical porous AgCl@Ag hollow architectures: self-templating synthesis and highly enhanced visible light photocatalytic activity. Appl Catal B 142–143:744–751
58. Tang Y-X, Jiang Z-L, Xing G-C, Li A-R, Kanhere P-D, Zhang Y-Y, Sum T-C, Li S-Z, Chen X-D, Dong Z-L, Chen Z (2013) Efficient Ag@AgCl cubic cage photocatalysts profitfrom ultrafast plasmon-induced electron transfer processes. Adv Funct Mater 23:2932–2940
59. Li H-Y, Wu T-S, Cai B, Ma W-G, Sun Y-J, Gan S-Y, Han D-X, Niu L (2015) Efficiently photocatalytic reduction of carcinogenic contaminant Cr(VI) upon robust AgCl:Ag hollow nanocrystals. Appl Catal B 164:344–351
60. Han C-C, Ge L, Chen C-F, Li Y-J, Zhao Z, Xiao X-L, Li Z-L, Zhang J-L (2014) Site-selected synthesis of novel Ag@AgCl nanoframes with efficient visible light induced photocatalytic activity. J Mater Chem A 2:12594–12600
61. Xiao X-L, Ge L, Han C-C, Li Y-J, Zhao Z, Xin Y-J, Fang S-M, Wu L-N, Qiu P (2015) A facile way to synthesize Ag@AgBr cubic cages with efficient visible-light-induced photocatalytic activity. Appl Catal B 163:64–572
62. Lou S-Y, Jia X-B, Wang Y-Q, Zhou S-M (2015) Template-assisted in-situ synthesis of porous AgBr/Ag composite microspheres as highly efficient visible-light photocatalyst. Appl Catal B 176–177:586–593
63. Kim J-W, Agesen L-K, Choi J-H, Choi J, Kim H-S, Liu J-Y, Cho C-R, Kang J-G, Ramazane A, Thornton K, Braun P-V (2015) Template-directed directionally solidified 3D mesostructured AgCl–KCl eutectic photonic crystals. Adv Mater 27:4551–4559
64. Wu S-K, Shen X-P, Ji Z-Y, Zhu G-X, Chen C-J, Chen K-M, Bu R, Yang L-M (2015) Synthesis of AgCl hollow cubes and their application in photocatalytic degradation of organic pollutants. CrystEngComm 17:2517–2522

65. Zhu M-S, Chen P-L, Liu M-H (2013) High-performance visible-light-driven plasmonic photocatalysts Ag/AgCl with controlled size and shape using graphene oxide as capping agent and catalyst promoter. Langmuir 29:9259–9268
66. Shahzad A, Kim W-S, Yu T (2016) A facile synthesis of Ag/AgCl hybrid nanostructures with tunable morphologies and compositions as advanced visible light plasmonic photocatalysts. Dalton Trans 45:9158–9165
67. Wang H, Li Y, Li C, He L, Guo L (2012) Facile synthesis of AgBr nanocubes for highly efficient visible light photocatalysts. CrystEngComm 14:7563–7566
68. An C-H, Peng S, Sun Y-G (2010) Facile synthesis of sunlight-driven AgCl:Ag plasmonic nanophotocatalyst. Adv Mater 22:2570–2574
69. Zhu M-S, Chen P-L, Ma W-H, Lei B, Liu M-H (2012) Template-free synthesis of cube-like Ag/AgCl nanostructures via a direct-precipitation protocol: highly efficient sunlight-driven plasmonic photocatalysts. ACS Appl Mater Interfaces 4:6386–6392
70. Han L, Wang P, Zhu C-Z, Zhai Y-M, Dong S-J (2011) Facile solvothermal synthesis of cube-like Ag@AgCl: a highly efficient visible light photocatalyst. Nanoscale 3:2931–2935
71. Chen D-L, Yoo S-H, Huang Q-S, Ali G, Cho S-O (2012) Sonochemical synthesis of Ag/AgCl nanocubes and their efficient visible-light-driven photocatalytic performance. Chem Eur J 18:5192–5200
72. Wang H, Gao J, Guo T-Q, Wang R-M, Guo L, Liu Y, Li J-L (2012) Facile synthesis of AgBr nanoplates with exposed {111} facets and enhanced photocatalytic properties. Chem Commun 48:275–277
73. Shen Y-F, Chen P-L, Xiao D, Chen C-C, Zhu M-S, Li T-S, Ma W-G, Liu M-H (2015) Spherical and sheetlike Ag/AgCl nanostructures: interesting photocatalysts with unusual facet-dependent yet substrate-sensitive reactivity. Langmuir 31:602–610
74. Lou Z-Z, Huang B-B, Qin X-Y, Zhang X-Y, Wang Z-Y, Zheng Z-K, Cheng H-F, Wang P, Dai Y (2011) One-step synthesis of AgBr microcrystals with different morphologies by ILs-assisted hydrothermal method. CrystEngComm 13:1789–1793
75. Zhang H-B, Lu Y-G, Liu H, Fang J-Z (2015) One-pot synthesis of high-index faceted AgCl nanocrystals with trapezohedral, concave hexoctahedral structures and their photocatalytic activity. Nanoscale 7:11591–11601
76. Moniz S-J-A, Shevlin S-A, Martin D-J, Guo Z-X, Tang J-W (2015) Visible-light driven heterojunction photocatalysts for water splitting-a critical review. Energy Environ Sci 8:731–759
77. Cheng H-F, Wang W-J, Huang B-B, Wang Z-Y, Zhan J, Qin X-Y, Zhang X-Y, Dai Y (2013) Tailoring AgI nanoparticles for the assembly of AgI/BiOI hierarchical hybrids with size-dependent photocatalytic activities. J Mater Chem A 1:7131–7136
78. Chen L-L, Jiang D-L, He T, Wu Z-D, Chen M (2013) In-situ ion exchange synthesis of hierarchical AgI/BiOI microsphere photocatalyst with enhanced photocatalytic properties. CrystEngComm 15:7556–7563
79. An C-H, Jiang W, Wang J-Z, Wang S-T, Ma Z-Z, Li Y-P (2013) Synthesis of three-dimensional AgI@TiO_2 nanoparticles with improved photocatalytic performance. Dalton Trans 42:8796–8801
80. Wu D-Y, Long M-C (2011) Realizing visible-light-induced self-cleaning property of cotton through coating N-TiO_2 film and loading AgI particles. ACS Appl Mater Interfaces 3:4770–4774
81. Song J-M, Zhang J, Ni J-J, Niu H-L, Mao C-J, Zhang S-Y, Shen Y-H (2014) One-pot synthesis of ZnO decorated with AgBr nanoparticles and its enhanced photocatalytic properties. CrystEngComm 16:2652–2659
82. Wang D-J, Guo L, Zhen Y-Z, Yue L-L, Xue G-L, Fu F (2014) AgBr quantum dots decorated mesoporous Bi_2WO_6 architectures with enhanced photocatalytic activities for methylene blue. J Mater Chem A 2:11716–11727

83. Xu H, Xu Y-G, Li H-M, Xia J-X, Xiong J, Yin S, Huang C-J, Wan H-L (2012) Synthesis, characterization and photocatalytic property of AgBr/$BiPO_4$ heterojunction photocatalyst. Dalton Trans 41:3387–3394
84. Wang D-W, Li Y, Puma G-L, Wang C, Wang P-F, Zhang W-L, Wang Q (2015) Dye-sensitized photoelectrochemical cell on plasmonic Ag/AgCl@chiral TiO_2 nanofibers for treatment of urban wastewater effluents, with simultaneous production of hydrogen and electricity. Appl Catal B 168–169:25–32
85. Andersson M, Birkedal H, Franklin N-R, Ostomel T, Boettcher S, Palmqyist A-E-C, Stucky G-D (2005) Ag/AgCl-loaded ordered mesoporous anatase for photocatalysis. Chem Mater 17:1409–1415
86. Hayashido Y, Naya S, Tada H (2016) Local electric field-enhanced plasmonic photocatalyst: formation of Ag cluster-incorporated AgBr nanoparticles on TiO_2. J Phys Chem C 120:19663–19669
87. Wang P-H, Tang Y-X, Dong Z-L, Chen Z, Lim T-T (2013) Ag–AgBr/TiO_2/RGO nanocomposite for visible-light photocatalytic degradation of penicillin G. J Mater Chem A 1:4718–4727
88. Elahifard M-R, Rahimnejad S, Haghighi S, Gholami M-R (2007) Apatite-coated Ag/AgBr/TiO_2 visible-light photocatalyst for destruction of bacteria. J Am Chem Soc 129:9552–9553
89. Xu Y-S, Zhang W-D (2013) Ag/AgBr-grafted graphite-like carbon nitride with enhanced plasmonic photocatalytic activity under visible light. ChemCatChem 5:2343–2351
90. Putri L-K, Ong W-J, Chang W-S, Chai S-P (2016) Enhancement in the photocatalytic activity of carbon nitride through hybridization with light sensitive AgCl for carbon dioxide reduction to methane. Cat Sci Technol 6:744–754
91. Zhang S-W, Li J-X, Wang X-K, Huang Y-S, Zeng M-Y, Xu J-Z (2014) In situ ion exchange synthesis of strongly coupled Ag@AgCl/g-C_3N_4 porous nanosheets as plasmonic photocatalyst for highly efficient visible-light photocatalysis. ACS Appl Mater Interfaces 6:22116–22125
92. Yao X-X, Liu X-H, Hu X-L (2014) Synthesis of the Ag/AgCl/g-C_3N_4 composite with high photocatalytic activity under visible light irradiation. ChemCatChem 6:3409–3418
93. Li H-Y, Gan S-Y, Wang H-Y, Han D-X, Niu L (2015) Intercorrelated Superhybrid of AgBr supported on graphitic-C_3N_4-decorated nitrogen-doped graphene: high engineering photocatalytic activities for water purification and CO_2 reduction. Adv Mater 27:6906–6913
94. Chen D-M, Wang Z-H, Du Y, Yang G-L, Ren T-Z, Ding H (2015) In situ ionic-liquid-assisted synthesis of plasmonic photocatalyst Ag/AgBr/g-C_3N_4 with enhanced visible-light photocatalytic activity. Catal Today 258:41–48
95. Jiang J, Zhao K, Xiao X-Y, Zhang L-Z (2012) Synthesis and facet-dependent photoreactivity of BiOCl single-crystalline nanosheets. J Am Chem Soc 134:4473–4476
96. Ye L-Q, Zhan L, Tian L-H, Peng T-Y, Zhang J-J (2011) The {001} facets-dependent high photoactivity of BiOCl nanosheets. Chem Commun 47:6951–6953
97. Li H, Shang J, Ai Z-H, Zhang L-Z (2015) Efficient visible light nitrogen fixation with BiOBr nanosheets of oxygen vacancies on the exposed {001} facets. J Am Chem Soc 137:6393–6399
98. Cheng H-F, Huang B-B, Dai Y (2014) Engineering BiOX (X=Cl, Br, I) nanostructures for highly efficient photocatalytic applications. Nanoscale 6:2009–2026
99. Cheng H-F, Huang B-B, Dai Y, Qin X-Y, Zhang X-Y (2010) One-step synthesis of the nanostructured AgI/BiOI composites with highly enhanced visible-light photocatalytic performances. Langmuir 26(9):6618–6624
100. Cheng H-F, Huang B-B, Wang P, Wang Z-Y, Lou Z-Z, Wang J-P, Qin X-Y, Zhang X-Y, Dai Y (2011) In situ ion exchange synthesis of the novel Ag/AgBr/BiOBr hybrid with highly efficient decontamination of pollutants. Chem Commun 47:7054–7056
101. Ye L-Q, Liu J-Y, Gong C-Q, Tian L-H, Peng T-Y, Zan L (2012) Two different roles of metallic Ag on Ag/AgX/BiOX (X=Cl, Br) visible light photocatalysts: surface plasmon resonance and Z-scheme bridge. ACS Catal 2:1677–1683

102. Xiong W, Zhao Q-D, Li X-Y, Zhang D-K (2011) One-step synthesis of flower-like Ag/AgCl/BiOCl composite with enhanced visible-light photocatalytic activity. Catal Commun 16:229–233
103. Kong L, Jiang Z, Lai H-H, Nicholls R-J, Xiao T-C, Jones M-O, Edwards P-P (2012) Unusual reactivity of visible-light-responsive AgBr-BiOBr heterojunction photocatalysts. J Catal 293:116–125
104. Wang W-S, Du H, Wang R-X, Wen T, Xu A-W (2013) Heterostructured Ag_3PO_4/AgBr/Ag plasmonic photocatalyst with enhanced photocatalytic activity and stability under visible light. Nanoscale 5:3315–3321
105. Yang S-B, Xu D-B, Chen B-Y, Luo B-F, Shi W-D (2017) In-situ synthesis of a plasmonic Ag/AgCl/Ag_2O heterostructures for degradation of ciprofloxacin. Appl Catal B 204:602–610
106. Sang Y, Kuai L, Chen C-Y, Fang Z, Geng B-Y (2014) Fabrication of a visible-light-driven plasmonic photocatalyst of $AgVO_3$@AgBr@Ag nanobelt heterostructures. ACS Appl Mater Interfaces 6:5061–5068
107. Zhang A-C, Zhang L-X, Lu H, Chen G-Y, Liu Z-C, Xiang J, Sun L-S (2016) Facile synthesis of ternary Ag/AgBr-Ag_2CO_3 hybrids with enhanced photocatalytic removal of elemental mercury driven by visible light. J Hazar Mater 314:78–87
108. Bao S-Y, Wu Q-F, Chang S-Z, Tian B-Z, Zhang J-L (2017) Z-scheme CdS-Au-$BiVO_4$ with enhanced photocatalytic activity for organic contaminant decomposition. Cat Sci Technol 7:124–132
109. Tada H, Mitsui T, Kiyonaga T, Akita T, Tanaka K (2006) All-solid-state Z-scheme in CdS–Au–TiO_2 three-component nanojunction system. Nat Mater 5:782–786
110. Wang X-W, Liu G, Wang L-Z, Chen Z-G, Lu G-Q, Cheng H-M (2012) ZnO-CdS@Cd heterostructure for effective photocatalytic hydrogen generation. Adv Energy Mater 2:42–46
111. Zhou P, Yu J-G, Jaroniec M (2014) All-solid-state Z-scheme photocatalytic systems. Adv Mater 26:4920–4935
112. Iwase A, Ng Y-H, Ishiguro Y, Kudo A, Amal R (2011) Reduced graphene oxide as a solid-state electron mediator in Z-scheme photocatalytic water splitting under visible light. J Am Chem Soc 133:11054–11057
113. Maeda K (2013) Z-scheme water splitting using two different semiconductor photocatalysts. ACS Catal 3:1486–1503
114. Li H-Y, Sun Y-J, Cai B, Gan S-Y, Han D-X, Niu L, Wu T-S (2015) Hierarchically Z-scheme photocatalyst of Ag@AgCl decorated on $BiVO_4$ (040) with enhancing photoelectrochemical and photocatalytic performance. Appl Catal B 170–171:206–214
115. Qiao R, Mao M-M, Hu E-L, Zhong Y-J, Ning J-Q, Hu Y (2015) Facile formation of mesoporous $BiVO_4$/Ag/AgCl heterostructured microspheres with enhanced visible-light photoactivity. Inorg Chem 54:9033–9039
116. Zhang J, Niu C-G, Ke J, Zhou L-F, Zeng G-M (2015) Ag/AgCl/Bi_2MoO_6 composite nanosheets: a plasmonic Z-scheme visible light photocatalyst. Catal Commun 59:30–34
117. Hou J-G, Wang Z, Yang C, Zhou W-L, Jiao S-Q, Zhu H-M (2013) Hierarchically plasmonic Z-scheme photocatalyst of Ag/AgCl nanocrystals decorated mesoporous single-crystalline metastable $Bi_{20}TiO_{32}$ nanosheets. J Phys Chem C 117:5132–5141
118. Hou J-G, Yang C, Wang Z, Ji Q-H, Li Y-T, Huang G-C, Jiao S-Q, Zhu H-M (2013) Three-dimensional Z-scheme AgCl/Ag/γ-TaON heterostructural hollow spheres for enhanced visible-light photocatalytic performance. Appl Catal B 142–143:579–589
119. Wang X-F, Li S-F, Ma Y-Q, Yu H-G, Yu J-G (2011) $H_2WO_4 \cdot H_2O$/Ag/AgCl composite nanoplates: a plasmonic Z-scheme visible-light photocatalyst. J Phys Chem C 115:14648–14655
120. Yao X-X, Liu X-H (2014) One-pot synthesis of ternary Ag_2CO_3/Ag/AgCl photocatalyst in natural geothermal water with enhanced photocatalytic activity under visible light irradiation. J Hazard Mater 280:260–268

121. Cheng H-J, Hou J-G, Zhu H-M, Guo X-M (2014) Plasmonic Z-scheme α/β-Bi_2O_3-Ag-AgCl photocatalyst with enhanced visible-light photocatalytic performance. RSC Adv 4:41622–41630
122. Zhou T, Xu Y-G, Xu H, Wang H-F, Da Z-L, Huang S-Q, Ji H-Y, Li H-M (2014) In situ oxidation synthesis of visible-light-driven plasmonic photocatalyst Ag/AgCl/g-C_3N_4 and its activity. Ceram Int 40:9293–9301
123. Yan X, Wang X-Y, Gu W, Wu M-M, Yan Y, Hu B, Che G-B, Han D-L, Yang J-H, Fan W-Q, Shi W-D (2015) Single-crystalline $AgIn(MoO_4)_2$ nanosheets grafted Ag/AgBr composites with enhanced plasmonic phtotcatalytic activity for degradation of tetracycline under visible light. Appl Catal B 164:297–304
124. Wang X-H, Yang J, Ma S-Q, Zhao D, Dai J, Zhang D-F (2016) In situ fabrication of AgI/$AgVO_3$ nanoribbon composites with enhanced visible photocatalytic activity for redox reactions. Cat Sci Technol 6:243–253
125. Li J-J, Xie Y-L, Zhong Y-J, Hu Y (2015) Facile synthesis of Z-scheme Ag_2CO_3/Ag/AgBr ternary heterostructured nanorods with improved photostability and photoactivity. J Mater Chem A 3:5474–5481
126. Zhang L-S, Wong K-H, Chen Z-G, Yu J-C, Zhao J-C, Hu C, Chan C-Y, Wong P-K (2009) AgBr-Ag-Bi_2WO_6 nanojunction system: a novel and efficient photocatalyst with double visible-light active components. Appl Catal A General 363:221–229
127. Katsumata H, Hayashi T, Taniguchi M, Suzuki T, Kaneco S (2014) Highly efficient visible-light driven AgBr/Ag_3PO_4 hybrid photocatalysts with enhanced photocatalytic activity. Mater Sci Semicond Process 25:68–75
128. Cao J, Zhao Y-J, Lin H-L, Xu B-Y, Chen S-F (2013) Facile synthesis of novel Ag/AgI/BiOI composites with highly enhanced visible light photocatalytic performances. J Solid State Chem 206:38–44
129. Xie R-Y, Zhang L-P, Xu H, Zhong Y, Sui X-F, Mao Z-P (2015) Fabrication of Z-scheme photocatalyst Ag-AgBr@$Bi_{20}TiO_{32}$ and its visible-light photocatalytic activity for the degradation of isoproturon herbicide. J Mol Catal A Chem 406:194–203
130. Yang Y-X, Wan G, Guo Y-N, Zhao Y-H, Yuan X, Guo Y-H (2014) Fabrication of Z-scheme plasmonic photocatalyst Ag@AgBr/g-C_3N_4 with enhanced visible-light photocatalytic activity. J Hazard Mater 271:150–159
131. Tian B-Z, Wang T-T, Dong R-F, Bao S-Y, Yang F, Zhang J-L (2014) Core-shell structured γ-Fe_2O_3@SiO_2@AgBr:Ag composite with high magnetic separation efficiency and excellent visible light activity for acid orange 7 degradation. Appl Catal B 147:22–28
132. Xia D-H, Hu L-L, Tan X-Q, He C, Pan W-Q, Yang T-S, Huang Y-L, Shu D (2016) Immobilization of self-stabilized plasmonic Ag-AgI on mesoporous Al_2O_3 for efficient purification of industrial waste gas with indoor LED illumination. Appl Catal B 185:295–306
133. Fan Y-Y, Ma W-G, Han D-X, Gan S-Y, Dong X-D, Niu L (2015) Convenient recycling of 3D AgX/Graphene aerogels (X= Br, Cl) for efficient photocatalytic degradation of water pollutants. Adv Mater 27:3767–3773
134. Xu Y-G, Zhou T, Huang S-Q, Xie M, Li H-P, Xu H, Xia J-X, Li H-M (2015) Preparation of magnetic Ag/AgCl/$CoFe_2O_4$ composites with high photocatalytic and antibacterial ability. RSC Adv 5:41475–41483
135. Jing L-Q, Xu Y-G, Huang S-Q, Xie M, He M-Q, Xu H, Li H-M, Zhang Q (2016) Novel magnetic $CoFe_2O_4$/Ag/Ag_3VO_4 composites: highly efficient visible light photocatalytic and antibacterial activity. Appl Catal B 199:11–22
136. Li Z-L, Ai J-Z, Ge M (2017) A facile approach assembled magnetic $CoFe_2O_4$/AgBr composite for dye degradation under visible light. J Environ Chem Eng 5:1394–1403
137. An C-H, Ming X-J, Wang J-Z, Wang S-T (2012) Construction of magnetic visible-light-driven plasmonic Fe_3O_4@SiO_2@AgCl:Ag nanophotocatalyst. J Mater Chem 22:5171–5176
138. Guo J-F, Ma B-W, Yin A-Y, Fan K-N, Dai W-L (2011) Photodegradation of rhodamine B and 4-chlorophenol using plasmonic photocatalyst of Ag-AgI/Fe_3O_4@SiO_2 magnetic nanoparticle under visible light irradiation. Appl Catal B 101:580–586

139. Li G-T, Wong K-H, Zhang X-W, Hu C, Yu J-C, Chan R-C-Y, Wong P-K (2009) Degradation of Acid Orange 7 using magnetic AgBr under visible light: the roles of oxidizing species. Chemosphere 76:1185–1191
140. Huang S-Q, Xu Y-G, Chen Z-G, Xie M, Xu H, He M-Q, Li H-M, Zhang Q (2015) A core-shell structured magnetic Ag/AgBr@Fe_2O_3 composite with enhanced photocatalytic activity for organic pollutant degradation and antibacterium. RSC Adv 5:71035–71045
141. Zhao H-H, Zhang L-S, Gu X-D, Li S-J, Li B, Wang H-L, Yang J-M, Liu J-S (2015) Fe_2O_3-AgBr nonwoven cloth with hierarchical nanostructures as efficient and easily recyclable macroscale photocatalysts. RSC Adv 5:10951–10959
142. Xu Y-G, Huang S-Q, Xie M, Li Y-P, Jing L-Q, Xu H, Zhang Q, Li H-M (2016) Core-shell magnetic Ag/AgCl@Fe_2O_3 photocatalysts with enhanced photoactivity for eliminating bisphenol A and microbial contamination. New J Chem 40:3413–3422
143. McEvoy J-G, Zhang Z-S (2014) Synthesis and characterization of magnetically separable Ag/AgCl-magnetic activated carbon composites for visible light induced photocatalytic detoxification and disinfection. Appl Catal B 160–161:267–278

Chapter 14
Synthesis and Modifications of Mesoporous $g\text{-}C_3N_4$ Photocatalyst

14.1 Introduction

In recent years, graphitic carbon nitride ($g\text{-}C_3N_4$), as a metal-free polymeric semiconductor with the merits of high chemical stability, narrow bandgap (~2.7 eV), excellent abrasion resistance, as well as good biological compatibility, has become a hot research topic in the field of nanomaterials [1]. Similar to graphite, $g\text{-}C_3N_4$ also possesses sheet structure, which contains C_3N_3 rings and C_6N_7 rings connected with each other by the N to form an unlimited plane. With the unique bandgap structure and the highly conjugated electron pair of N [2], $g\text{-}C_3N_4$ has shown potential application prospects in many aspects, such as photocatalytic hydrogen production with water [3], reduction of carbon dioxide [4], degradation of organic pollutants [5, 6], oxidation of alcohols [7], and so on. There are many reported raw materials, which can be used to synthesize $g\text{-}C_3N_4$, such as dicyandiamide, cyanamide, urea, melamine, hexamethylenetetramine, and ammonium thiocyanate. The preparation methods of bulk $g\text{-}C_3N_4$ are also numerous, including high-temperature sintering [8, 9], shock wave compression [10, 11], high-pressure pyrolyzation [12], ion implantation [13], low-energy ion radiation [14], ion beam deposition [15, 16], sputtering [17], chemical vapor deposition (CVD) [18, 19], laser pulse [20, 21], and so on. With the development of researches, the synthesis methods and the characterization means of bulk $g\text{-}C_3N_4$ have been quite mature. However, general drawbacks of bulk materials such as small specific surface area as well as less active sites hinder the further development of bulk $g\text{-}C_3N_4$.

Mesoporous material is a research hot spot in recent decades. Mesoporous materials are famous for the pore channels in diameter sizes of 2–50 nm inside them, which contribute them to possessing more excellent properties and more potential applications than bulk materials [22–24]. Therefore, mesoporous materials with many super properties of high specific surface area, more active sites, ordered pore structure, lower density, and strong adsorption ability [25] have attracted widespread interest in multidisciplinary researches. Fabricating $g\text{-}C_3N_4$ into

J. Zhang et al., *Photocatalysis*, Lecture Notes in Chemistry 100,
https://doi.org/10.1007/978-981-13-2113-9_14

mesoporous structure favors the increase of the active sites and the utilization of photon, thus improving its photocatalytic activity and further promoting the research and the extensive applications of g-C_3N_4 materials. Therefore, great efforts have been devoted to synthesizing and studying mesoporous g-C_3N_4 (MCN). In this chapter, the development of MCN in recent years has been reviewed, including the preparation and modifications of it. Firstly, we introduced four main different preparation methods of MCN, which are soft-template method [25], hard-template method [26], template-free method [27], and sol–gel method [28]. These synthesis methods using different precursors and strategies make the as-prepared products possess different pore regularity and other properties. After that, the modification methods of MCN, which can lead to further improvement of photocatalytic activity and the broaden of applications, were presented. In the end, a short summary of MCN as well as its application prospects was provided.

14.2 The Preparation of MCN

The synthesis of bulk g-C_3N_4 is quite simple by calcinating precursors at high temperature to create a thermal polymerization reaction. The kinds of the precursors are various and abundant, including urea, cyanamide, dicyandiamide, melamine, carbon tetrachloride, ethylenediamine, ammonium thiocyanate, hexamethylenetetramine, and so on. In the process of high-temperature polymerization, the main reactions for the precursor represented by cyanamide are shown in Fig. 14.1 [27].

Figure 14.1 shows that the reactions contain the two processes of polyaddition and polycondensation. Firstly, the precursor polymerizes to melamine. Secondly, a condensation occurs through removing ammonia from melamine to form the g-C_3N_4 polymer. The results of X-ray analysis and differential scanning calorimetry (DSC) demonstrated that in the process of the thermal polymerization, melamine can be found in the products at 350 °C, while the melamine rearranges to form tris-s-triazine when the temperature arises to around 390 °C. The condensation process of the tris-s-triazine unit to polymers, networks, and potentially the final g-C_3N_4 occurs at the temperature of ~520 °C, with the material becoming unstable slightly above 600 °C. Heating to 700 °C causes the residue-free disappearance of the material via generating nitrogen and cyano fragments. This reaction process may relate to the cohesive energy, which was proved by the theoretical calculations results. As shown in Fig. 14.2, the cohesive energy of the molecules increases along with the polyaddition path [27].

The preparation of MCN includes not only the abovementioned reactions for the preparation of bulk g-C_3N_4 but also an additional step to generate the mesoporous structure. The reported preparation methods of MCN are not as abundant as that of bulk g-C_3N_4, which are summarized as follows.

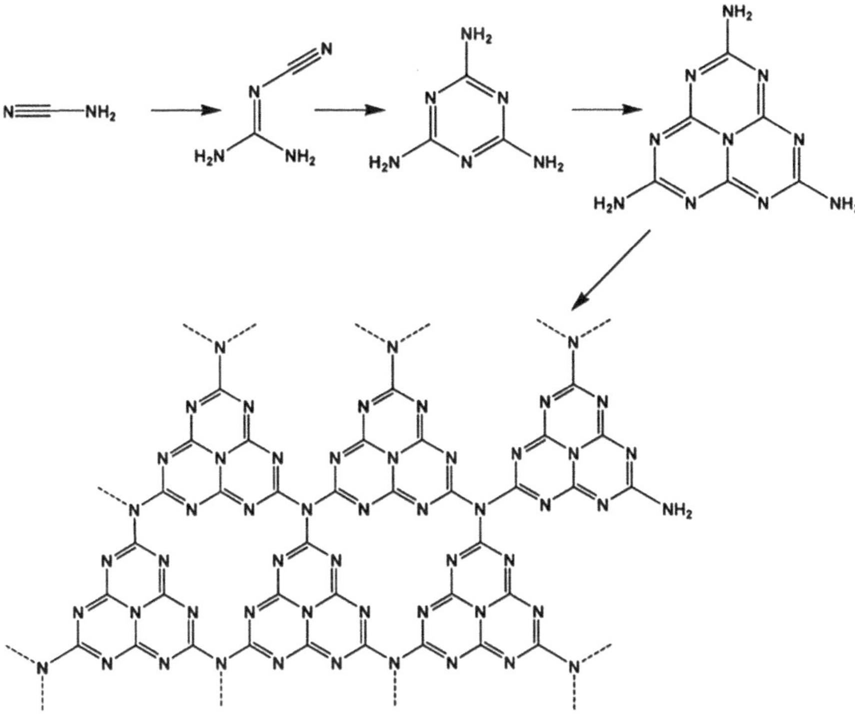

Fig. 14.1 Represented reactions for the formation of bulk g-C_3N_4 using cyanamide as the precursor [27]. (Reprinted with permission from Ref. [27]. Copyright 2008, Royal Society of Chemistry)

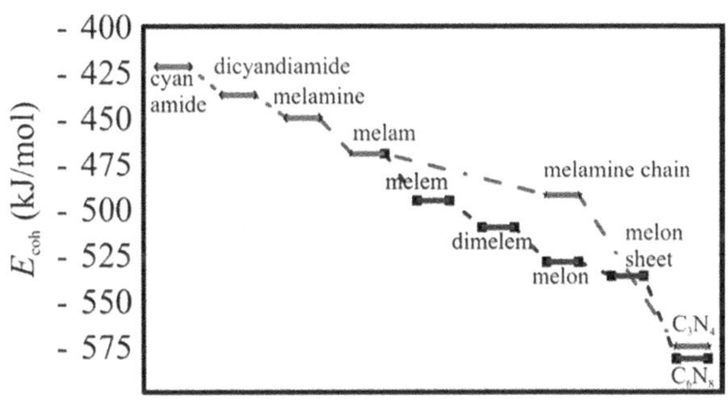

Fig. 14.2 Calculated energy diagram for the synthesis of g-C_3N_4. Cyanamide as the starting precursor is condensed into melamine at first. Further condensation can then proceed through the triazine route (dash-dot line) to C_3N_4, or melamine forms melem and then follows the tri-s-triazine route (dashed line) to form C_6N_8 [27]. (Reprinted with permission from Ref. [27]. Copyright 2008, Royal Society of Chemistry)

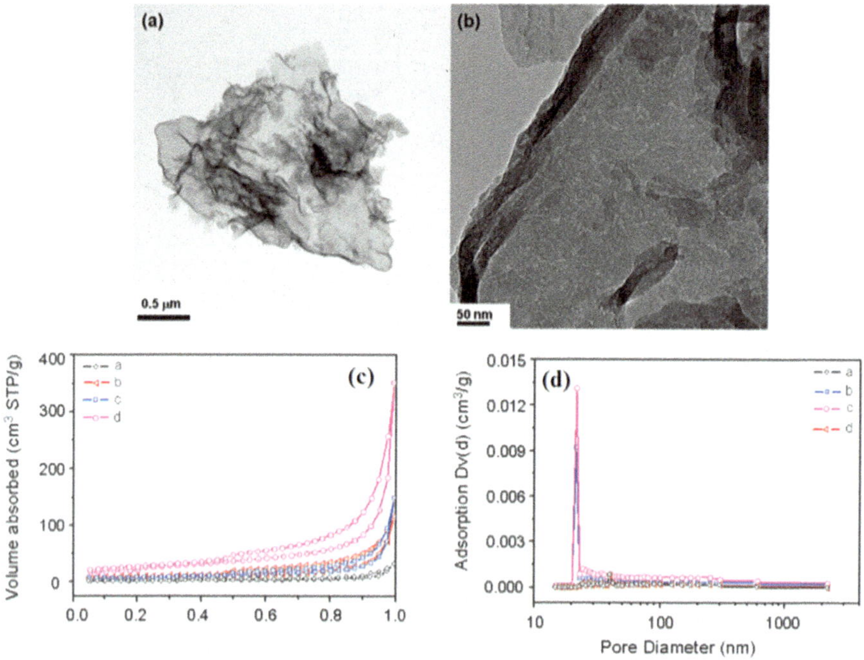

Fig. 14.3 (**a**) Low-magnification and (**b**) high-magnification TEM images of the MCN synthesized by using Pluronic P123 as the soft template, (**c**) N_2 adsorption/desorption isotherm, and (**d**) corresponding pore size distributions of MCN products prepared without (line a, b) and with (line c, d) P123 [29]. (Reprinted with permission from Ref. [29]. Copyright 2010, Wiley Online Library)

14.2.1 Soft-Template Method

Soft-template method generally uses surfactant, amphiphilic polymer, or other organic molecules as the templates. The mesophase was usually formed by the self-assembly of the template and the used precursor in the reaction solution system.

Weibin Fan et al. synthesized MCN material with bimodal pore distribution using a surfactant Triton X-100 as the soft template through the polymerization and carbonization of the melamine and glutaraldehyde precursors. The sizes of the mesopores in the as-prepared product were centered at 3.8 nm and 10–40 nm [25]. Hongjian Yan et al. developed a novel method to synthesize MCN by using the block polymer Pluronic P123 as the soft template and melamine as the precursor. Firstly, the melamine and Pluronic P123 were simultaneously dispersed in the distilled water. After those processes, including the perfusion, addition of sulfuric acid solution, drying, and the high-temperature calcination, the MCN materials with wormlike mesopores were obtained, as observed by the HRTEM image shown in Fig. 14.3b. The pore size distributions, which were calculated by the Barret–Joyner–Halender (BJH) method, displayed different characters for the products synthesized without and with P123 template, as shown in Fig. 14.3d. The pore size distributions of the former were very broad, while those of the latter were highly uniform and narrow [29].

Haoran Li et al. reported a facile soft-template method for the preparation of boron- and fluorine-doped MCN materials. It was found that 1-butyl-3-methylimidazoliumtetrafluoroborate ($BmimBF_4$), as a commercially available room-temperature ionic liquid, was a unique soft template for the preparation of boron- and fluorine-enriched MCN, in which an organic precursor, such as dicyandiamide (DCDA), self-condensed to form carbon nitride in the presence of $BmimBF_4$. Their experimental results revealed that ionic liquids were good soft templates, which were generally defined as organic salts with a <100 °C melting point. They inherited many features of inorganic molten salts, including the excellent chemical and thermal stability as well as negligibly small vapor pressure and the convenience of being liquid in ambient conditions [30].

Soft-template method possesses the advantages such as easy operation, simple equipment, and low cost. The exploration of new soft templates is essential for the further applications of the soft-template method for the synthesis of MCN materials. Meanwhile, the soft-template method has drawbacks. For example, it cannot always strictly control the size and morphology of the as-prepared products, limiting the catalytic performance of the final MCN materials.

14.2.2 *Hard-Template Method*

In the hard-template method, the templates with mesopores are firstly synthesized, which are usually mesoporous silica materials like SBA-15. Then, the precursor solution of $g\text{-}C_3N_4$ such as urea solution is perfused into the mesopores of the as-prepared template, and the precursor will transfer into $g\text{-}C_3N_4$ in the pore channels by the high-temperature calcination. After removing the template, MCN is obtained.

Ajayan Vinu et al. prepared MCN by a typical hard-template method. In their synthesis process, mesoporous silica SBA-15 [31] was added to a mixture of ethylenediamine and carbon tetrachloride. After the reflux of the silica framework, MCN with uniform mesopores and high photocatalytic activity was obtained [32]. Zhongkui Zhao et al. also synthesized MCN by using SBA-15 as a hard template. Their experimental results showed the specific surface area and pore volume and N content of the obtained MCN had a tight relationship with the chosen precursor. By using hexamethylenetetramine as the precursor, the obtained MCN was highly ordered, with ultrahigh specific surface area of 1116 $m^2\ g^{-1}$ and pore volume of 1.54 $cm^3\ g^{-1}$. The procedure for the synthesis is presented in Fig. 14.4 [33]. In addition to changing the precursor, it is also demonstrated that the specific pore volume as well as the pore diameter can also be controlled by simply adjusting the structure of the SBA-15 template. Ajayan Vinu et al. successfully prepared MCN with tunable pore diameters by using SBA-15 materials with different pore diameters as templates via a simple polymerization reaction between carbon tetrachloride

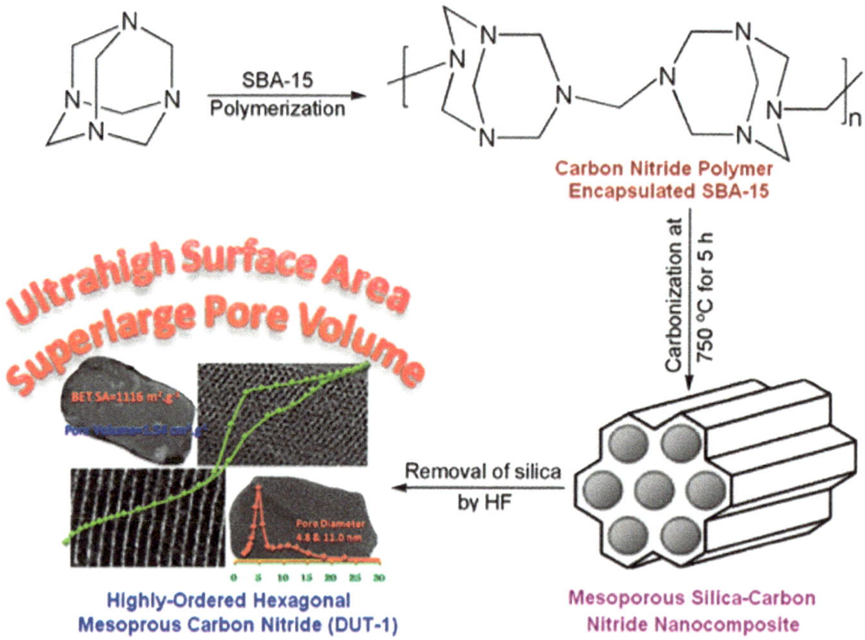

Fig. 14.4 Scheme for the synthesis of highly ordered MCN material by using SBA-15 as a template and hexamethylenetetramine as the precursor [33]. (Reprinted with permission from Ref. [33]. Copyright 2014, American Chemical Society)

and ethylenediamine. Their results showed that the pore diameter of the MCN materials can be easily controlled from 4.2 to 6.4 nm without affecting their structural order. Moreover, they also tuned the nitrogen content of the MCN materials by simply adjusting the weight ratio of ethylenediamine (EDA) to carbon tetrachloride (CTC) [34]. With the increase of EDA to CTC weight ratio from 0.3 to 0.9, the carbon to nitrogen ratio of the MCN decreased from 4.3 to 3.3. The results showed that the pore diameter of the MCN materials could be easily tuned from 4.2 to 6.4 nm without impairing their structural order.

In addition to SBA-15, other kinds of mesoporous silica materials have also been used as template for the synthesis of MCN. Ajayan Vimu et al. synthesized MCN with cage-type mesopores through a straight forward polymerization of carbon tetrachloride and ethylenediamine inside the pore channels of the mesoporous silica material FDU-12. The obtained MCN materials, which were used as adsorbents, were applied to capture CO_2 molecules under different high pressures and temperatures. The MCN materials exhibited excellent affinity toward CO_2 molecules due to the strong acid–base interactions [35]. Sang-Eon Park used disk-type mesoporous silica (INC-2) as hard template and melamine as precursor to synthesize MCN material with hexagonal platelet morphology. This material possessed high nitrogen content in the framework as well as the surface, which provided potential Lewis base sites for Knoevenagel condensation [36]. By using a mesoporous silica KIT-6 as the

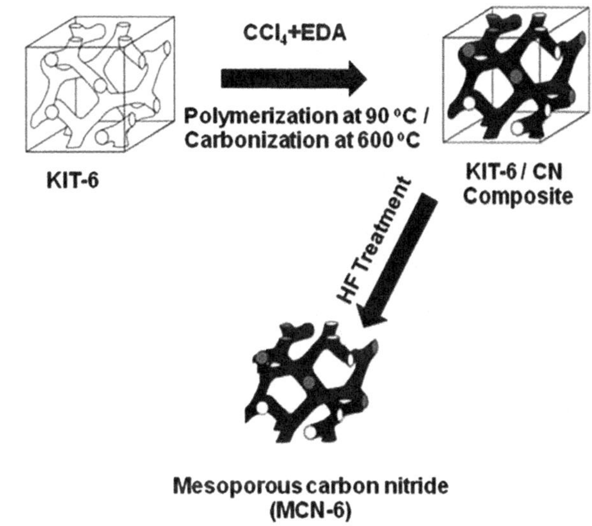

Fig. 14.5 Scheme for the synthesis of MCN with 3D structure using mesoporous silica KIT-6 as templates [38]. (Reprinted with permission from Ref. [38]. Copyright 2012, Royal Society of Chemistry)

hard template and a cyclic aromatic compound 3-amino-1,2,4-triazine as the precursor, MCN material with 3D porous structure and high nitrogen content was prepared by Toshiyuki Mori et al. Their work showed that the cyclic aromatic precursor benefited to preserve the nitrogen in the carbon matrix of the final product even after the process of the carbonization [37]. Ajayan Vinu et al. also synthesized 3D MCN by using mesoporous silica KIT-6 as templates and ethylenediamine and carbon tetrachloride as the N and C sources, respectively. The obtained materials possessed bimodal pores that can be controlled by simply adjusting the pore diameter of the KIT-6 templates [38]. The experimental steps are shown in Fig. 14.5.

Since most researchers used amorphous mesoporous silica materials as hard template, Murugulla A. Chari et al. reported the utilization of mesoporous silica nanoparticles as the template for the first time. The obtained sample was well-ordered MCN nanoparticles with a size of ~150 nm and high nitrogen content (C_4N_2) which is twice that of the MCN synthesized by using mesoporous silica SBA-15 as hard templates [7].

The hard-template method for synthesizing MCN can keep the original structure properties of the template materials. It is easy to design the morphology and the pore structure of the target MCN through tuning the hard template. The MCN materials synthesized by this method generally possess highly ordered 2D or 3D nanopores and excellent thermal stability. Therefore, the hard-template method is the most widely used method at present. But there are still some inherent disadvantages of this method which are difficult to overcome. For instance, the precursors of the products must be liquid or soluble in some polar solvent. Otherwise, the perfusion process will be difficult. The products are easy to form on the surface of the template rather than in the pore channels. Moreover, the removal of the hard templates usually needs to use strong base or hydrofluoric acid which is highly corrosive and requires for high experimental safety.

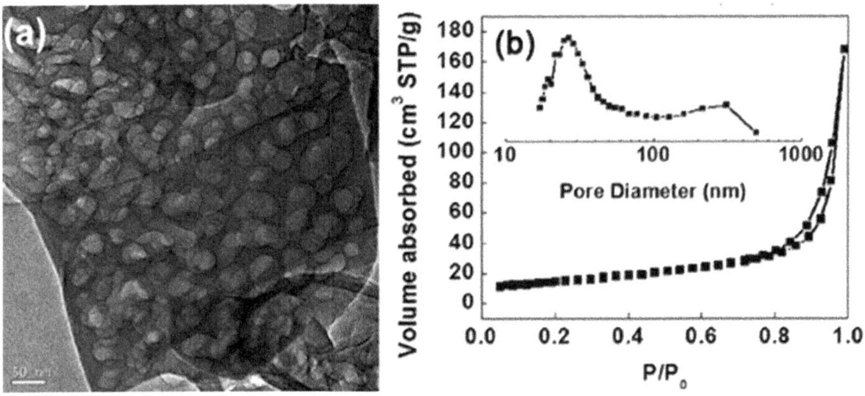

Fig. 14.6 (**a**) TEM image of MCN synthesized by pyrolyzing urea without templates. (**b**) N_2 adsorption/desorption isotherm of the obtained MCN; inset: pore size distribution [39]. (Reprinted with permission from Ref. [39]. Copyright 2012, American Chemical Society)

14.2.3 Template-Free Methods

This method doesn't require other substances to serve as templates. By simply controlling the experimental parameters or selecting an appropriate precursor, the mesopores can be obtained in the products.

It was reported that MCN materials could be synthesized without using a template through directly pyrolyzing urea at ambient conditions by Gongxuan Lu et al. In their experimental procedures, urea powder was put into an alumina crucible with a cover and calcinated at the temperature of 600 °C. The resultant powder was washed with deionized water and anhydrous ethanol thoroughly, followed by the filtration and the drying process. As shown in Fig. 14.6a, the obtained MCN materials possessed layered structure and abundant mesopores. The pore size was in the range of 20–40 nm [39] (Fig. 14.6b).

Different from producing mesopores in the preparation process of g-C_3N_4 as mentioned above, Vishnu Shanker et al. reported a template-free strategy to synthesize MCN materials through a post-ultrasonic treatment of bulk g-C_3N_4. In their method, the bulk g-C_3N_4 was dispersed in an ethanol aqueous solution and then sonicated at room temperature for 5 h. Porous g-C_3N_4 sheet with high surface area and large pore volume was obtained. More attractively, the photocatalytic performance of the MCN was much better than that of bulk g-C_3N_4 under visible light irradiation. Besides, the high reusability of the prepared MCN makes the materials meet the requirement of suitable candidates for practical applications in the field of photocatalysis [40].

Template-free method is the most ideal synthesis method of MCN. First of all, it is a quite simple method, because the operation procedure does not need the lengthy preparation process, saving time and raw materials. Secondly, because there is no template in this method, no toxic template etchants need to be used, making it also a kind of environment-friendly method. Therefore, it is a kind of simple and green synthesis method. However, there are still some shortcomings in template-free method. For examples, similar to the soft-template method, the mesopore structure in the products prepared by this method cannot achieve a high degree of order, and it is also difficult to design and tune the properties of the pores. In addition, there are more stringent requirements in the calcination process and the selection of raw materials.

14.2.4 Sol–Gel Method

Sol–gel method is a commonly used method for the synthesis of nanomaterials. The precursors of g-C_3N_4 are usually mixed with the precursor of silica or a sol of silica in the sol–gel method for the synthesis of MCN. After those processes including consecutive sol mixing, gel formation, calcinated polymerization, and silica removal, the target MCN materials can be obtained. Similar to the hard-template method, this method requires a procedure for removing the template. But the silica template is formed in the preparation process of g-C_3N_4 rather than pre-synthesized, which is different from the hard-template method.

Jianmin Sun et al. prepared MCN materials through adding tetraethyl orthosilicate (TEOS) which served as the silica precursor into the cyanamide hydrochloride solution and forming g-C_3N_4/SiO_2 composite in situ by thermal polymerization. As shown in Fig. 14.7, after the removal of the silica in the composite, MCN with a great number of mesopores (labeled by red circles) in pore diameter sizes of ~5 nm was obtained. The as-prepared MCN with a large surface area exhibited enhanced photocatalytic degradation performance for rhodamine B (RhB) pollutant under visible light irradiation [41].

Arne Thomas et al. also synthesized MCN by a sol–gel method, using cyanamide and TEOS as the precursor for g-C_3N_4 and silica, respectively. In their experimental procedures, cyanamide and TEOS were firstly mixed in an acidic ethanol solution.

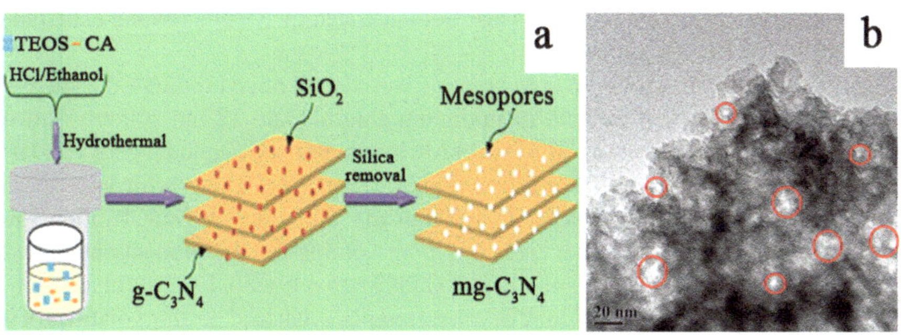

Fig. 14.7 (**a**) Scheme for the synthesis of MCN products by a sol–gel method using cyanamide hydrochloride and TEOS; (**b**) TEM image of the obtained MCN [41]. (Reprinted with permission from Ref. [41]. Copyright 2015, Elsevier)

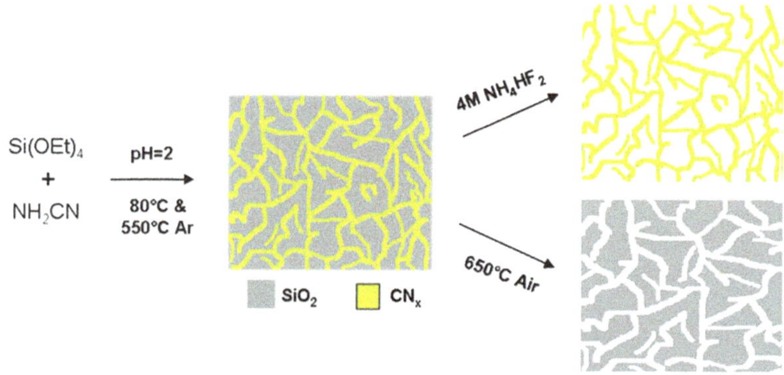

Fig. 14.8 Synthesis of the mesoporous carbon nitride and silica by the sol–gel route [42]. (Reprinted with permission from Ref. [42]. Copyright 2011, Royal Society of Chemistry)

After the processes of condensation and heating, it could be found that the carbon nitride and silica formed highly interpenetrating mesophases, leading to the formation of MCN or mesoporous silica after selectively removing either of the phases. The fabrication process of MCN by sol–gel method is shown in Fig. 14.8 in detail. Importantly, the carbon nitride retained its graphitic stacking even in the spatial constraint introduced by the surrounding silica phase. Because TEOS is liquid and cyanamide can be dissolved to be liquids, this approach allowed convenient adjusting into thin and thick films or monoliths of mesoporous carbon nitrides [42].

In addition to the alkoxy silane precursor, silica sol was also directly used as raw material in the sol–gel method. Markus Antonietti et al. prepared MCN by using Ludox-HS 40 silica dispersion and cyanamide as the precursor. The cyanamide was completely dissolved into Ludox-HS 40 silica dispersion to form a homogeneous and viscous solution and dried at 70 °C for 1 h to obtain a transparent gel, which was then transferred into a porcelain boat and calcinated at the temperature of 550 °C for 4 h under N_2 atmosphere. After the removal treatment for the silica template with a 4 M NH_4HF_2 solution, yellow powders of MCN were obtained. The mechanism for the formation of the MCN is shown in Fig. 14.9 [43]. Similar approach using Ludox-HS 40 silica to synthesize MCN was also demonstrated by Xinchen Wang et al. [44] and Yasuhiro Shiraishi et al. [45].

MCN product with bimodal pore systems was also obtained through the sol–gel method [46]. In this experiment, guanidinium chloride and 12 nm colloidal silica were used as the raw materials. A series of MCN samples with bimodal pore systems varying from 1.8 nm to 13.4 nm and high surface areas along with large pore volumes were successfully prepared by adjusting the silica to GndCl ratio. The results revealed that the surface areas and pore volumes of the products increased progressively with the increase of silica to GndCl ratios from 0.4 to 1.0. In addition, the utilization of a higher amount of silica template favored to the volatilization process in calcinations of GndCl. As a result, lower nitrogen content in the MCN material was achieved as the weight ratio of silica to GndCl increased continuously. Therefore, it was demonstrated that, in this sol–gel method, it was easy to tune the

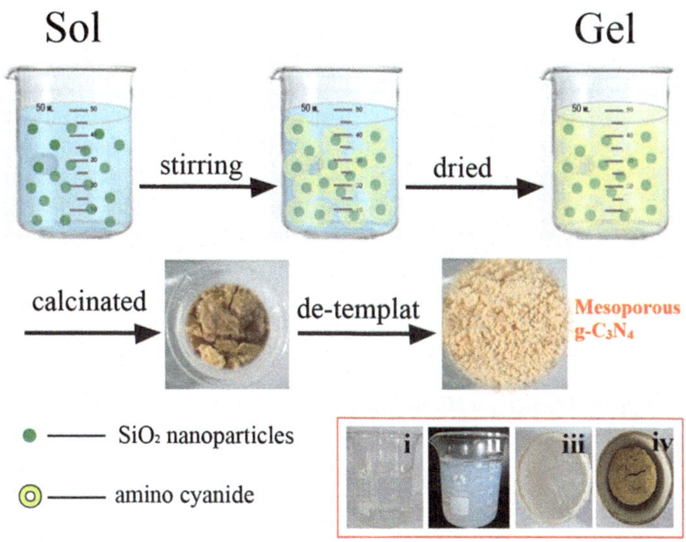

Fig. 14.9 Scheme for the formation mechanism of MCN by the sol–gel method using Ludox-HS 40 silica dispersion and cyanamide as the precursor [43]. (Reprinted with permission from Ref. [43]. Copyright 2013, Royal Society of Chemistry)

surface area, pore volume, and C/N ratio of the MCN product by simply adjusting the adding ratio of silica to GndCl.

The sol–gel method is a simple, efficient, and viable approach to the large-scale production of MCN materials. This proposed method also allows to develop and design the controllable pore structures, which is of crucial importance for the applications of MCN materials. Meanwhile, some shortcomings have to be overcome for this method, for example, it is usually time-consuming in the whole sol–gel process. Moreover, the removal of the silica template makes the overall preparation procedure become relatively tedious and may cause the risk of environmental pollution.

Carbon nitride, as a new kind of environment-friendly catalyst, is in the rapid development period. Many researchers are doing comprehensive study from all aspects. With time goes by and the development of new technologies, it is believed that more excellent methods for the synthesis of MCN will appear in the near future.

14.3 The Modifications of MCN

The unique electronic structure and the advantages from the mesoporous texture make MCN be widely recognized as efficient heterogeneous catalysts for many reaction systems. In situ or post-modifications to the MCN can further enhance the catalytic performance or broaden the application scope of the material. For instance, when MCN is applied to the field of photocatalysis as a photocatalyst, proper modifications can greatly improve its photocatalytic activity. The modifications can not only reduce the recombination rate of photo-generated electron and hole pairs, which can improve the quantum efficiency of the photocatalysis, but also make the absorption wavelength have a redshift and thus extend the absorption spectrometry, leading to better absorption and utilization of solar light. Nowadays, the modification methods for MCN are rich. The representative and typical modification methods include noble metal deposition, metallic oxide loading, nonmetal doping, dye photosensitization, as well as polyoxometalate immobilization.

14.3.1 Noble Metal Deposition

Noble metal deposition can change the distribution and transmission of electrons in the MCN catalyst, which is because it has different Fermi level from MCN. After the two substances noble metal and MCN contact with each other, electrons can transfer from MCN with the higher Fermi level to the noble metal whose Fermi level is relatively lower, and the holes can transfer from noble metal to MCN to balance the Fermi levels. So the photo-generated electrons and holes concentrate on noble metal and MCN surface, respectively. Then the photo-generated electrons and holes can be used for promoting different redox reactions.

Ajayan Vinu et al. deposited Au nanoparticles into MCN for the first time (Fig. 14.10a) [47]. Au nanoparticles with a size of <7 nm were highly dispersed on the inner surface of MCN product. It was observed that there was no agglomeration of the Au nanoparticles in the composite (Fig. 14.10b). The MCN was demonstrated to serve as stabilizing, size-controlling, and reducing agent for the Au nanoparticles. The experimental results demonstrated that the Au@MCN composite was a selective, highly active, and recyclable catalyst for the three-component

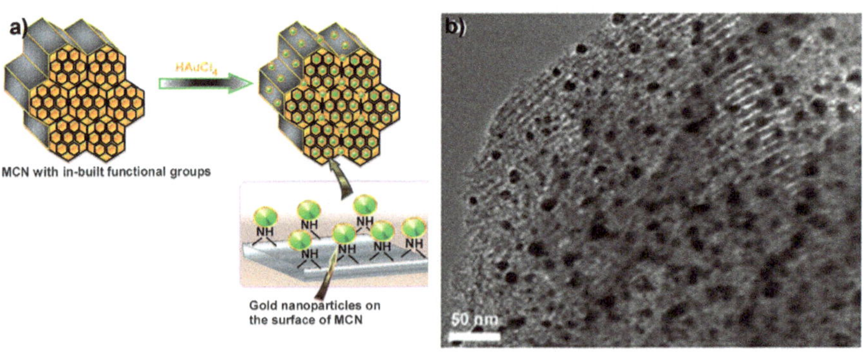

Fig. 14.10 (**a**) Scheme for the growth of Au nanoparticles into MCM; (**b**) HRTEM images of the Au@MCN composite [47]. (Reprinted with permission from Ref. [47]. Copyright 2010, Wiley Online Library)

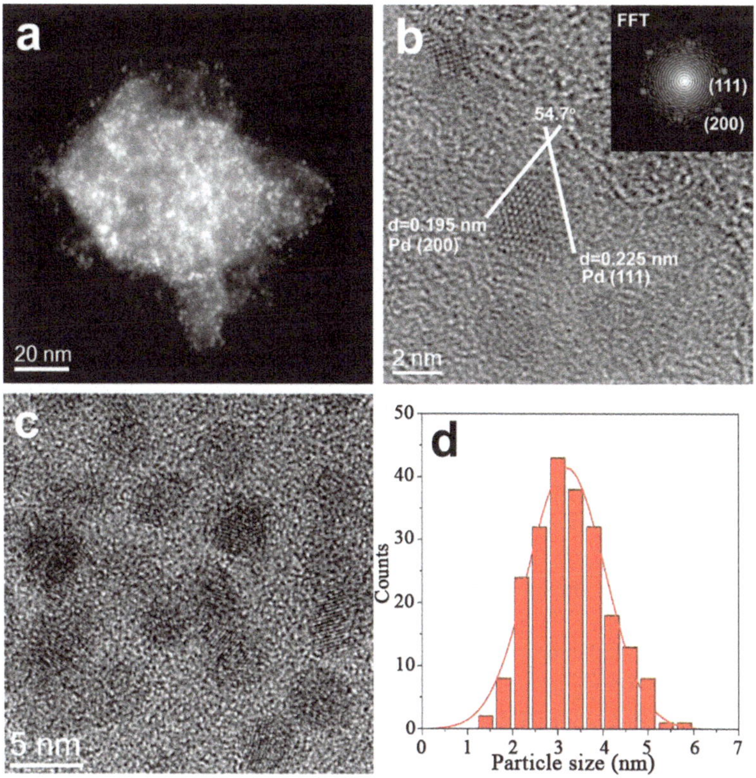

Fig. 14.11 (**a**) STEM, (**b**) HRTEM, and (**c**) TEM images of and (**d**) particle size distribution of Pd@MCN catalyst. The inset in (**b**) is the local fast Fourier transform [48]. (Reprinted with permission from Ref. [48]. Copyright 2011, American Chemical Society)

coupling reaction of piperidine, benzaldehyde, and phenylacetylene for the synthesis of propargylamine.

Through loading Pd nanoparticles on MCN, Yong Wang et al. synthesized a novel catalyst [48], which was highly active and selective for the hydrogenation of phenol to cyclohexanone. It can be observed that in the product, the Pd nanoparticles were highly dispersed on the MCN from the STEM, HRTEM, as well as TEM images presented in Fig. 14.11. As demonstrated by the size distribution in Fig. 14.11d, the average size of the Pd nanoparticles was ~5 nm. The as-prepared catalyst possessed a high catalytic activity for the hydrogenation of phenol to cyclohexanone, which was ascribed to the special structure of the semiconductor–metal heterojunction between MCN and Pd. The special heterojunction structure led not only to a highly uniform dispersion of Pd nanoparticles but also to additional electronic activation of the metal nanoparticles as well as a "nonplanar" to adsorb phenol. Therefore, the fast and highly selective hydrogenation of phenol to cyclohexanone was achieved.

Apart from Au and Pd nanoparticles, Pt nanoparticles were also homogenously loaded in the mesoporous channels of MCN by Wen-sheng Dong et al. [49] The synthesized materials were used for the selective oxidation of glycerol with molecular oxygen in base-free aqueous solution, showing high conversion efficiency and selectivity. Shijiao Sun et al. also synthesized similar composites of MCN and Pt nanoparticles [50]. The Pt nanoparticles with a diameter of 3–4 nm were loaded on the MCN support. The as-prepared composites were applied as bifunctional air electrodes, showing improved electrochemical performances with good round-trip efficiency as high as 87% through the examination with an all-solid-state Li-air battery.

14.3.2 *Metallic Oxide Loading*

In addition to the noble metal deposition, the metallic oxide has also been loaded into the ordered mesoporous channels of MCN material. The metallic oxide improves the photocatalytic activity of MCN via the formation of the heterojunction structure, which can efficiently increase the separation rate of photo-generated electrons and holes.

Jie Xu et al. prepared vanadia-loaded MCN catalysts by a wet impregnation method using NH_4VO_3 as a precursor and mesoporous carbon nitride as a support. The modified MCN materials exhibited remarkable catalytic performance for the hydroxylation of benzene to phenol in the presence of H_2O_2, which was attributed to the high dispersion of vanadia species and the benzene-activation capability of MCN. Due to the inbuilt and unique tri-s-triazine moieties, benzene was chemically adsorbed and then activated on the catalytic surface of MCN through an electron transfer from HOMO of C_3N_4 to LUMO of benzene. In this path, the high surface areas and the rich mesopores of MCN materials upgraded the amount of adsorbed benzene molecules. On the other hand, V^{4+} species that dispersed on the surface of MCN were oxidized to radical-containing V^{5+} species by H_2O_2, simultaneously generating H_2O. And then the as-produced V^{5+} species reacted with the activated benzene, thus yielding the target phenol. The possible path of the reaction is shown in Fig. 14.12 [51].

Kamalakannan Kailasam et al. reported mesoporous carbon nitride–tungsten oxide composites (CN/WO_3) and applied it to photocatalytic hydrogen evolution. MCN/WO_3 composite showed very high photocatalytic activity for the evolution of hydrogen from water with sacrificial electron donors under the irradiation of visible light. The higher activity could be ascribed to the high surface area as well as synergetic effect between the WO_3 and the carbon nitrides which led to improved charge separation rate through a photocatalytic solid-state Z-scheme mechanism, as shown in Fig. 14.13 [52]. After modification with WO_3, the photo-generated holes in the valance band (VB) of g-C_3N_4 and the photo-generated electrons in the conduction band (CB) of WO_3 would recombine at the interface, leaving the electrons in the CB of g-C_3N_4 and holes in the VB of WO_3. Therefore, the overall separation

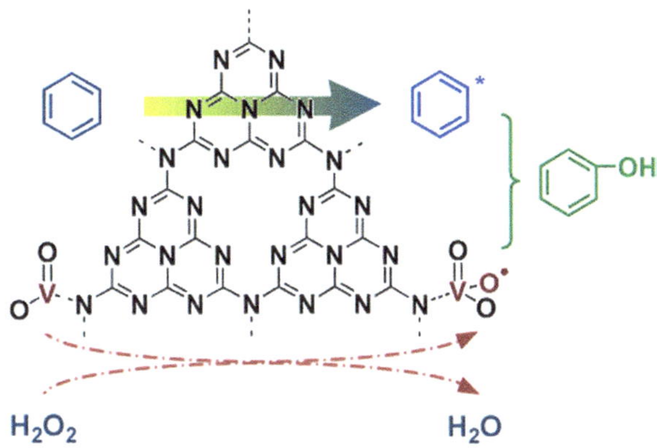

Fig. 14.12 A proposed reaction mechanism for the hydroxylation of benzene catalyzed by vanadia-modified MCN [51]. (Reprinted with permission from Ref. [51]. Copyright 2015, Royal Society of Chemistry)

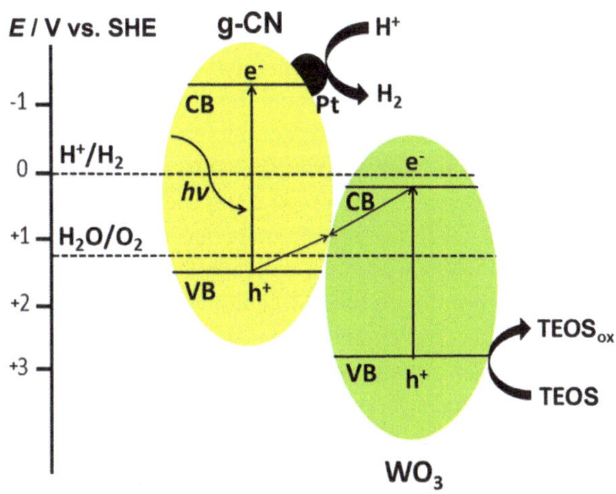

Fig. 14.13 Schematic illustration of the photocatalytic mechanism for MCN/WO_3 composites as the solid-state Z-scheme [52]. (Reprinted with permission from Ref. [52]. Copyright 2015, Wiley Online Library)

efficiency of photo-generated electrons and holes was improved, causing the improved photocatalytic activity of MCN material [52].

14.3.3 Nonmetal Doping

Nonmetal doping method refers to doping the nonmetal elements into the C–N matrix of the MCN material to modify the properties of the product. After the modification process, the material keeps the advantage of metal-free, and its catalytic performance is improved.

Haoran Li et al. reported boron- and fluorine-rich MCN material by using 1-butyl-3-methylimidazoliumtetrafluoroborate ($BmimBF_4$), which is a kind of room-temperature ionic liquid, as a soft template for the mesopores and a source for the boron and fluorine dopant [53]. As shown in Fig. 14.14a, the obtained materials

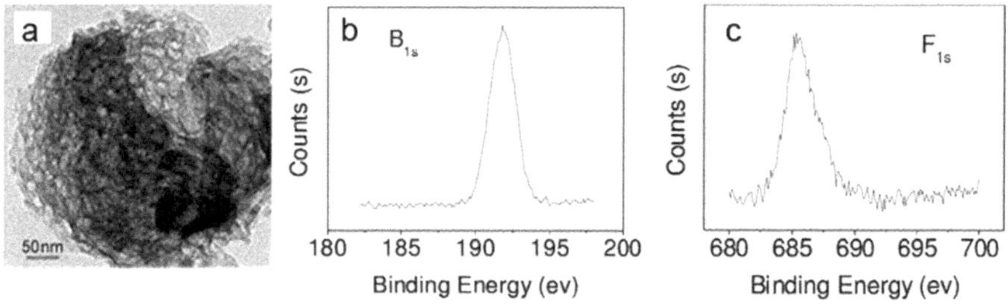

Fig. 14.14 (**a**) Typical TEM image of boron- and fluorine-co-doped MCN sample; (**b**, **c**) XPS patterns of B1s and F1s in the doped MCN product [53]. (Reprinted with permission from Ref. [53]. Copyright 2012, Wiley Online Library)

possessed a well-ordered, sponge-like mesoporous structure. The results of X-ray photoelectron spectroscopy (XPS) analysis (Fig. 14.14c) revealed that both boron and fluorine atoms were incorporated into the C–N matrix. The boron content in the product was about 20 mol%, and the fluorine content was only about 4 mol%. It was proposed that the boron atoms entered carbon sites in polymeric C–N structures (thus balancing stoichiometry), with fluorine saturating residual bonds. The as-prepared material possessed an excellent photoconductivity under visible light and efficient catalytic oxidation for the degradation of cyclohexane. Furthermore, these doped MCN materials, which had large surface area and suitable pore volume with a large number of boron functional groups on the surface acting as strong Lewis acid sites and complementing the basic nitrogen sites, are expected to show excellent results in other organic reactions. In addition, the ionic liquid-based strategy could be applied to incorporate other heteroatoms in MCN materials by altering the anion or cation in the ionic liquids.

Although there are a large number of studies on the nonmetal doping of bulk g-C_3N_4 materials [54, 55], the researches on this modification method for MCN are still in the infancy, which is probably because it is difficult to realize the generation of mesopores and the doping of nonmetal elements into the C–N matrix at the same time.

14.3.4 Dye Photosensitization

Dye photosensitization refers to the dye molecules being adsorbed on the surface of the photocatalyst through chemical or physical interactions, making the light absorption of the photocatalyst extend to a longer wavelength range, thereby expanding the response range of the excitation wavelength [56]. Generally, dye sensitization involves three basic processes:

1. The adsorption process of the dye molecules on the surface of the photocatalyst.
2. The adsorbed dye molecules excite a photon.

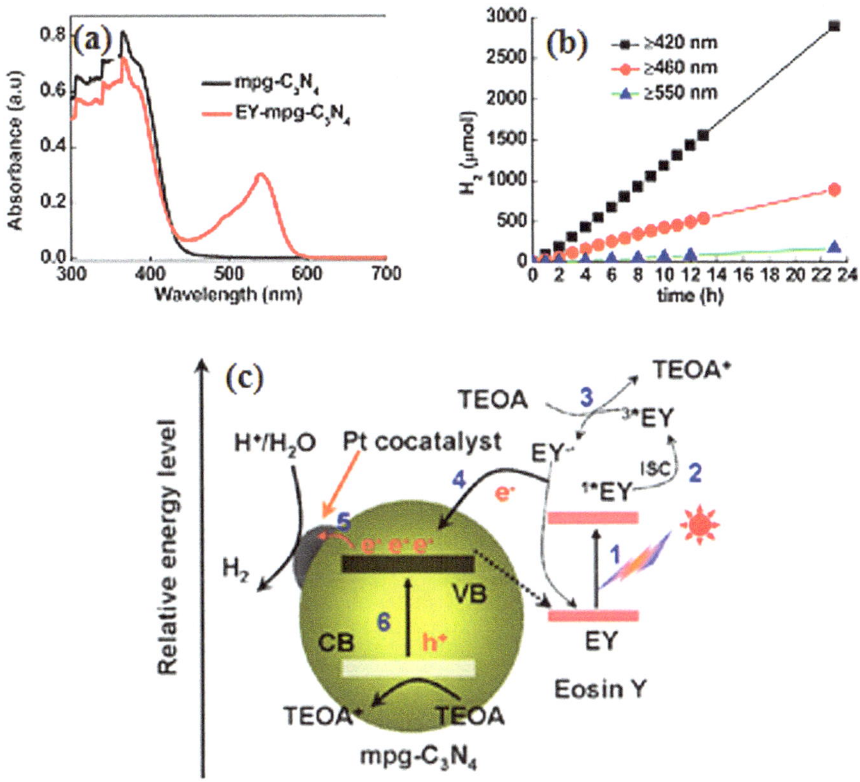

Fig. 14.15 (**a**) UV–vis diffused reflectance spectra of MCN and EY-MCN samples. (**b**) Time courses of photocatalytic H_2 evolution with water over EY-MCN/Pt photocatalyst under different wavelengths of light irradiation. (**c**) Proposed photocatalytic mechanism for H_2 evolution under visible light, using EY-MCN/Pt as photocatalyst [39]. (Reprinted with permission from Ref. [39]. Copyright 2012, American Chemical Society)

3. The excited dye molecules transfer electrons into the CB of the photocatalyst.

If the CB energy level of semiconductor is more positive than the oxidation level of dye molecules, dye molecules in excited states can transfer electrons to the CB of the semiconductor. Then semiconductor can correspondingly accept an electron from the excited dye molecule, followed by transferring it to the O_2 adsorbed on the surface to produce active species. Through the above processes, the semiconductor obtains the enhanced photocatalytic activity.

Shixiong Min et al. reported that the photo response of the MCN material can be greatly extended to nearly 600 nm after sensitization with eosin Y (EY), as shown in Fig. 14.15a. This sensitized MCN was proved to exhibit high photocatalytic activity for H_2 evolution under the irradiation of visible light, especially in the longer wavelength regions (450–600 nm), as shown in Fig. 14.15b. In addition, the light absorption of the photosensitizer determined the reaction rate. The experimental

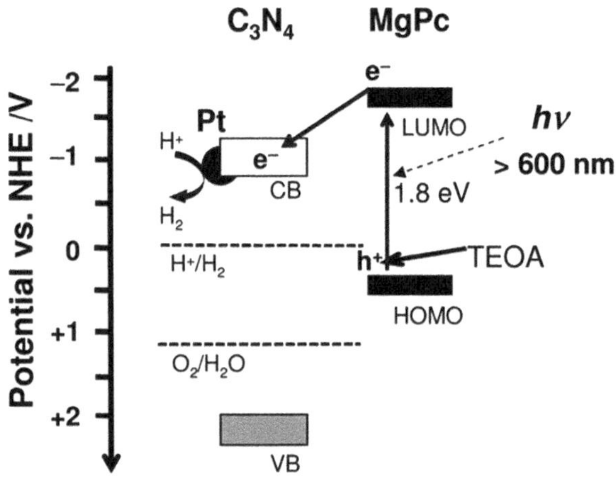

Fig. 14.16 Transportation of electrons and holes in MgPc/Pt/MCN photocatalysts [57]. (Reprinted with permission from Ref. [57]. Copyright 2010, Royal Society of Chemistry)

results indicated that the efficient electron transfer between excited EY molecules and MCN was realized. The mechanism for photocatalytic H_2 evolution on the EY-MCN photocatalyst is shown in Fig. 14.15c. Moreover, the mesoporous structure and high surface area of the MCN material facilitated the adsorption of EY molecules on the surface, thus promoting the photocatalytic activity via improved light harvesting [39].

Markus Antonietti et al. carried out a research on the spectral sensitization of MCN catalyst by depositing magnesium phthalocyanine (MgPc) to extend the absorption wavelength to that longer than those of the original MCN. The obtained catalyst, MgPc/Pt/MCN (Pt as a cocatalyst), exhibited stable photocatalytic activity for the evolution of H2 from water in the presence of sacrificial reagents (triethanolamine), even under the light irradiation of >600 nm [57]. As shown in Fig. 14.16, the photo-excited electrons on MgPc were most likely transferred through the CB of MCN to Pt. It was therefore reasonable to regard that the MgPc molecules which were not in contact with MCN would be incapable of the efficient charge transfer from MgPc to MCN. Besides, dye (MgPc) to dye (MgPc) charge transfer would not take place smoothly. Therefore, a monolayer of dye molecules on the photocatalyst enhanced the overall photocatalytic efficiency.

14.3.5 *Polyoxometalate Immobilization*

Polyoxometalates belonging to inorganic compounds have attracted wide attention owing to their special functions [58, 59]. One of their most notable features is that they simultaneously possess acidic and oxidizing capabilities. In addition, by altering the elements of the polyoxometalate cluster, the catalytic properties of the product can be managed and tuned. Therefore, polyoxometalates, as the acidic or oxidized or dual function catalysts, are widely used in many reaction systems. The polyoxometalate immobilized to the surface of MCN catalyst can form a synergistic effect from the combination of catalytic merits of polyoxometalates and MCN, greatly enhancing the photocatalytic performance.

Through a vacuum-assisted impregnation method, Yi Xie et al. [59] successfully immobilized polyoxometalate complex $[Co_4(H_2O)_2(PW_9O_{34})2]_{10}$ (CoPOM) to the ordered MCN support as Fig. 14.17a shows. The high surface area of MCN material advanced a perfect and stable dispersion of CoPOM. The synthesized composite was demonstrated to act as highly efficient water-oxidation catalyst. This might be because anchoring CoPOM to the MCN material could improve the electrical contact of the redox-active centers and the surface of the electrode. Besides, it was likely that the carbon nitride environment played an important role for sustaining activity through protecting the active cobalt centers of the composite from deactivation by surface restructuring.

Yunfeng Zhu et al. synthesized phosphotungstic acid/MCN photocatalyst by immobilizing phosphotungstic acid ($H_3PW_{12}O_{40}$, HPW) on MCN. The HPW/MCN photocatalyst exhibited an excellent photocatalytic performance in the oxidative desulfurization process. Under optimal reaction conditions, dibenzothiophene could be removed completely, and there was no significant loss of the photocatalytic activity after 15 recycles [60].

14.4 Summery and Outlook

In this chapter, the synthesis and the modifications of MCN in recent years have been reviewed briefly, which is expected to provide some guidance for the future research on MCN. MCN with the superior advantages such as large surface area, narrow pore

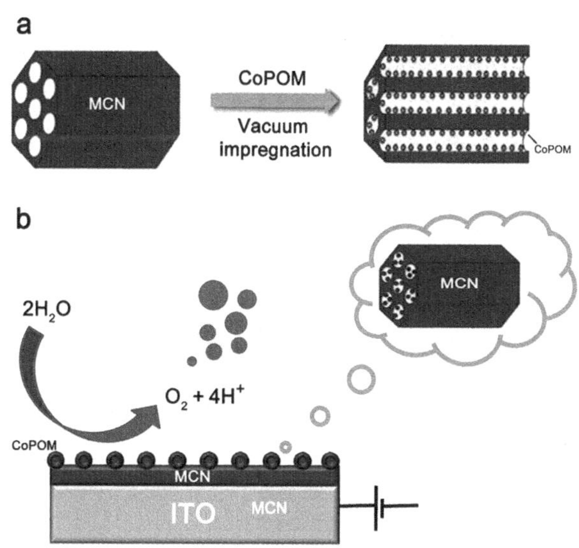

Fig. 14.17 (**a**) Synthesis of a CoPOM/MCN composite through vacuum-assisted impregnation. (**b**) A scheme for using CoPOM/MCN composite as water-oxidation photocatalyst [59]. (Reprinted with permission from Ref. [59]. Copyright 2012, Wiley Online Library)

size distribution, convenient preparation, easy modification, nontoxicity, and environmental friendliness has attracted increasingly attentions of researchers. It is believed that with the development of the research, MCN will be used in more aspects, helping to solve the problems of environment and energy.

References

1. Zhao Z, Sun Y, Dong F (2015) Graphitic carbon nitride based nanocomposites: a review. Nanoscale 7(1):15–37
2. Gu Y, Zhang Y, Chang X et al (2000) Synthesis and characterization of C_3N_4 hard films. Sci China Ser A Math 43(2):185
3. Laribi M, Chenouf R, Bachari K et al (2013) Properties and reactivity of zinc-folded sheet mesoporous materials. Res Chem Intermed 39(4):1771–1780
4. Yu W, Xu D, Peng T (2015) Enhanced photocatalytic activity of g-C_3N_4 for selective CO_2 reduction to CH_3OH via facile coupling of ZnO: a direct Z-scheme mechanism. J Mater Chem A 3(39):19936–19947
5. Lei J, Chen Y, Wang L et al (2015) Highly condensed g-C_3N_4-modified TiO_2 catalysts with enhanced photodegradation performance toward acid orange 7. J Mater Sci 50(9):3467
6. Lei J, Chen Y, Shen F et al (2015) Surface modification of TiO_2 with g-C_3N_4 for enhanced UV and visible photocatalytic activity. J Alloys Compd 631:328–334
7. Jin X, Balasubramanian VV, Selvan ST et al (2009) Highly ordered mesoporous carbon nitride nanoparticles with high nitrogen content: a metal-free basic catalyst. Angew Chem 121 (42):8024–8027
8. Li H, Zhou L, Wang L et al (2015) In situ growth of TiO_2 nanocrystals on g-C_3N_4 for enhanced photocatalytic performance. Phys Chem Chem Phys 17(26):17406–17412
9. Cui Y, Zhang J, Zhang G et al (2011) Synthesis of bulk and nanoporous carbon nitride polymers from ammonium thiocyanate for photocatalytic hydrogen evolution. J Mater Chem 21 (34):13032–13039
10. Wixom MR (1990) Chemical preparation and shock wave compression of carbon nitride precursors. J Am Ceram Soc 73(7):1973–1978
11. Liu J, Sekine T, Kobayashi T (2006) A new carbon nitride synthesized by shock compression of organic precursors. Solid State Commun 137(1):21–25
12. Ma HA, Jia XP, Chen LX et al (2002) High-pressure pyrolysis study of $C_3N_6H_6$: a route to preparing bulk C_3N_4. J Phys Condens Matter 14(44):11269
13. Galan L, Montero I, Rueda F (1996) An X-ray photoelectron spectroscopy study of carbon nitride films grown by low energy ion implantation. Surf Coat Technol 83(1–3):103–108
14. Hoffman A, Gouzman I, Brener R (1994) Possibility of carbon nitride formation by low-energy nitrogen implantation into graphite: in situ electron spectroscopy studies. Appl Phys Lett 64 (7):845–847
15. Hofsäss H, Ronning C, Griesmeier U et al (1995) Cubic boron nitride films grown by low energy B^+ and N^+ ion beam deposition. Appl Phys Lett 67(1):46–48
16. Song HW, Cui FZ, He XM et al (1994) Carbon nitride films synthesized by NH_3-ion-beam-assisted deposition. J Phys Condens Matter 6(31):6125
17. Zhou ZF, Bello I, Lei MK et al (2000) Synthesis and characterization of boron carbon nitride films by radio frequency magnetron sputtering. Surf Coat Technol 128:334–340
18. Yen TY, Chou CP (1995) Growth and characterization of carbon nitride thin films prepared by arc-plasma jet chemical vapor deposition. Appl Phys Lett 67(19):2801–2803
19. Bousetta A, Lu M, Bensaoula A et al (1994) Formation of carbon nitride films on Si (100) substrates by electron cyclotron resonance plasma assisted vapor deposition. Appl Phys Lett 65 (6):696–698

20. Geretovszky Z, Kántor Z, Bertóti I (2000) Pulsed laser deposition of carbon nitride films by a sub-ps laser. Appl Phys A 70(1):9
21. Mihailescu IN, Gyorgy E, Alexandrescu R et al (1998) Optical studies of carbon nitride thin films deposited by reactive pulsed laser ablation of a graphite target in low pressure ammonia. Thin Solid Films 323(1):72–78
22. Jabbari A, Mahdavi H, Nikoorazm M et al (2015) Salen copper (II) complex heterogenized on mesoporous MCM-41 as nano-reactor catalyst for the selective oxidation of sulfides using urea hydrogen peroxide (UHP). Res Chem Intermed 41(8):5649–5663
23. Lei J, Yang L, Lu D et al (2015) Carbon Dot-incorporated PMO nanoparticles as versatile platforms for the design of ratiometric sensors, multichannel traceable drug delivery vehicles, and efficient photocatalysts. Adv Opt Mater 3(1):57
24. Hu J, Zou Y, Liu J et al (2015) Immobilization of Cu-chelate onto SBA-15 for partial oxidation of benzyl alcohol using water as the solvent. Res Chem Intermed 41(8):5703–5712
25. Shen W, Ren L, Zhou H et al (2011) Facile one-pot synthesis of bimodal mesoporous carbon nitride and its function as a lipase immobilization support. J Mater Chem 21(11):3890–3894
26. Fellinger T-P, Hasché F, Strasser P et al (2012) Mesoporous nitrogen-doped carbon for the electrocatalytic synthesis of hydrogen peroxide. J Am Chem Soc 134(9):4072
27. Thomas A, Fischer A, Goettmann F et al (2008) Graphitic carbon nitride materials: variation of structure and morphology and their use as metal-free catalysts. J Mater Chem 18 (41):4893–4908
28. Kuhn P, Antonietti M, Thomas A (2008) Porous, covalent triazine-based frameworks prepared by ionothermal synthesis. Angew Chem Int Ed 47(18):3450–3453
29. Yan H (2012) Soft-templating synthesis of mesoporous graphitic carbon nitride with enhanced photocatalytic H_2 evolution under visible light. Chem Commun 48(28):3430–3432
30. Wang Y, Zhang J, Wang X et al (2010) Boron-and fluorine-containing mesoporous carbon nitride polymers: metal-free catalysts for cyclohexane oxidation. Angew Chem Int Ed 49 (19):3356–3359
31. Hartmann M, Vinu A (2002) Mechanical stability and porosity analysis of large-pore SBA-15 mesoporous molecular sieves by mercury porosimetry and organics adsorption. Langmuir 18 (21):8010–8016
32. Vinu A, Ariga K, Mori T et al (2005) Preparation and characterization of well-ordered hexagonal mesoporous carbon nitride. Adv Mater 17(13):1648–1652
33. Zhao Z, Dai Y, Lin J et al (2014) Highly-ordered mesoporous carbon nitride with ultrahigh surface area and pore volume as a superior dehydrogenation catalyst. Chem Mater 26 (10):3151–3161
34. Vinu A (2008) Two-dimensional hexagonally-ordered mesoporous carbon nitrides with tunable pore diameter, surface area and nitrogen content. Adv Funct Mater 18(5):816–827
35. Lakhi KS, Cha WS, Joseph S et al (2015) Cage type mesoporous carbon nitride with large mesopores for CO_2 capture. Catal Today 243:209–217
36. Ansari MB, Jin H, Parvin MN et al (2012) Mesoporous carbon nitride as a metal-free base catalyst in the microwave assisted Knoevenagel condensation of ethylcyanoacetate with aromatic aldehydes. Catal Today 185(1):211–216
37. Mane GP, Dhawale DS, Anand C et al (2013) Selective sensing performance of mesoporous carbon nitride with a highly ordered porous structure prepared from 3-amino-1, 2, 4-triazine. J Mater Chem A 1(8):2913–2920
38. Talapaneni SN, Anandan S, Mane GP et al (2012) Facile synthesis and basic catalytic application of 3D mesoporous carbon nitride with a controllable bimodal distribution. J Mater Chem 22 (19):9831–9840
39. Min S, Lu G (2012) Enhanced electron transfer from the excited eosin Y to mpg-C_3N_4 for highly efficient hydrogen evolution under 550 nm irradiation. J Phys Chem C 116 (37):19644–19652
40. Kumar S, Baruah A, Tonda S et al (2014) Cost-effective and eco-friendly synthesis of novel and stable N-doped ZnO/g-C_3N_4 core–shell nanoplates with excellent visible-light responsive photocatalysis. Nanoscale 6(9):4830–4842

41. Shi L, Liang L, Wang F et al (2015) Tetraethylorthosilicate induced preparation of mesoporous graphitic carbon nitride with improved visible light photocatalytic activity. Catal Commun 59:131–135
42. Kailasam K, Epping JD, Thomas A et al (2011) Mesoporous carbon nitride–silica composites by a combined sol–gel/thermal condensation approach and their application as photocatalysts. Energy Environ Sci 4(11):4668–4674
43. Yang S, Zhou W, Ge C et al (2013) Mesoporous polymeric semiconductor materials of graphitic-C_3N_4: general and efficient synthesis and their integration with synergistic AgBr NPs for enhanced photocatalytic performances. RSC Adv 3(16):5631–5638
44. Cui Y, Huang J, Fu X et al (2012) Metal-free photocatalytic degradation of 4-chlorophenol in water by mesoporous carbon nitride semiconductors. Cat Sci Technol 2(7):1396–1402
45. Shiraishi Y, Kofuji Y, Sakamoto H et al (2015) Effects of surface defects on photocatalytic H_2O_2 production by mesoporous graphitic carbon nitride under visible light irradiation. ACS Catal 5(5):3058–3066
46. Yang Q, Wang W, Zhao Y et al (2015) Metal-free mesoporous carbon nitride catalyze the Friedel–Crafts reaction by activation of benzene. RSC Adv 5(68):54978
47. Datta KKR, Reddy BV, Ariga K et al (2010) Gold nanoparticles embedded in a mesoporous carbon nitride stabilizer for highly efficient three-component coupling reaction. Angew Chem Int Ed 49(34):5961–5965
48. Wang Y, Yao J, Li H et al (2011) Highly selective hydrogenation of phenol and derivatives over a Pd@ carbon nitride catalyst in aqueous media. J Am Chem Soc 133(8):2362–2365
49. Wang FF, Shao S, Liu CL et al (2015) Selective oxidation of glycerol over Pt supported on mesoporous carbon nitride in base-free aqueous solution. Chem Eng J 264:336–343
50. Lu Y, Wen Z, Jin J et al (2012) Mesoporous carbon nitride loaded with Pt nanoparticles as a bifunctional air electrode for rechargeable lithium-air battery. J Solid State Electrochem 16 (5):1863
51. Xu J, Jiang Q, Chen T et al (2015) Vanadia supported on mesoporous carbon nitride as a highly efficient catalyst for hydroxylation of benzene to phenol. Cat Sci Technol 5(3):1504–1513
52. Kailasam K, Fischer A, Zhang G et al (2015) Mesoporous carbon nitride-tungsten oxide composites for enhanced photocatalytic hydrogen evolution. ChemSusChem 8(8):1404–1410
53. Wang Y, Wang X, Antonietti M (2012) Polymeric graphitic carbon nitride as a heterogeneous organocatalyst: from photochemistry to multipurpose catalysis to sustainable chemistry. Angew Chem Int Edit 51(1):68
54. Shalom M, Inal S, Fettkenhauer C et al (2013) Improving carbon nitride photocatalysis by supramolecular preorganization of monomers. J Am Chem Soc 135(19):7118–7121
55. Wei J (2002) Formation of hydrogenated carbon nitride films by reactive sputtering. J Appl Phys 92(11):6525
56. Najafian A, Rahimi R, Zargari S et al (2016) Synthesis and photocatalytic activity of V-doped mesoporous TiO_2 photosensitized with porphyrin supported by SBA-15. Res Chem Intermed 42(4):3441–3458
57. Takanabe K, Kamata K, Wang X et al (2010) Photocatalytic hydrogen evolution on dye-sensitized mesoporous carbon nitride photocatalyst with magnesium phthalocyanine. Phys Chem Chem Phys 12(40):13020–13025
58. Maleki B, Eshghi H, Barghamadi M et al (2016) Silica-coated magnetic $NiFe_2O_4$ nanoparticles-supported H3PW12O40; synthesis, preparation, and application as an efficient, magnetic, green catalyst for one-pot synthesis of tetrahydrobenzo[*b*]pyran and pyrano[2,3-*c*]pyrazole derivatives. Res Chem Intermed 42(4):3071
59. Wu J, Liao L, Yan W et al (2012) Polyoxometalates immobilized in ordered mesoporous carbon nitride as highly efficient water oxidation catalysts. ChemSusChem 5(7):1207–1212
60. Zhu Y, Zhu M, Kang L et al (2015) Phosphotungstic acid supported on mesoporous graphitic carbon nitride as catalyst for oxidative desulfurization of fuel. Ind Eng Chem Res 54 (7):2040–2047

Chapter 15
MoS_2 Applications in Photo-Fenton Technology

15.1 The Brief Introduction of Fenton Technology

Wastewater arising from different chemical industries like resin manufacturing, petrochemical, oil refineries, paper making, and textile industry has high concentration of organics and their derivatives which are extremely toxic and refractory to the environment. The presence of organic compounds in wastewater, surface water, and groundwater poses serious threat to humans' health. Thus, its removal from contaminated water is of high priority. The conventional treatment methods like biotechnology are not effective in some fields. Consequently, the need of more efficient treatment methods is imperative. In recent years, oxidation processes are preferred to degrade pollutants.

Advanced oxidation processes (AOPs) have been described as a promising choice to remove persistent pollutants from contaminated water. Additionally, AOPs require less energy than direct oxidation [1–3]. Typically, AOPs are operated under ambient temperature and pressure, which involve the generation of hydroxyl radicals in sufficient quantity to achieve water purification [4]. The hydroxyl radicals as highly reactive species are able to break the structure of most organic molecules with rate constants in the order of 10^6–10^9 mol/L• [5]. The reduction potential of various oxidants is shown in Table 15.1. Hydroxyl radical is the second strongest oxidant followed fluorine, and it reacts about 1000 times faster than ozone depending on the substrate to be degraded [6]. Classified by the reactive phase, AOPs contain homogeneous and heterogeneous phases. One of the most commonly used AOPs for dealing contaminants is Fenton process. The generated hydroxyl radicals in Fenton process possess inherent properties that enable it to achieve a complete mineralization of organic molecules into CO_2, water, and mineral acids such as sulfuric, hydrochloric, and nitric acids [7–9].

Fenton's reagent was discovered about 100 years ago, but its application as an oxidation process for decomposing toxic organics was not applied until the late

J. Zhang et al., *Photocatalysis*, Lecture Notes in Chemistry 100,
https://doi.org/10.1007/978-981-13-2113-9_15

Table 15.1 Standard reduction potential of common oxidants [10]

Oxidant	Oxidation potential (eV)
Fluorine (F_2)	3.03
Hydroxyl radical (OH)	2.80
Atomic oxygen (O)	2.42
Ozone (O_3)	2.07
Hydrogen peroxide (H_2O_2)	1.77
Potassium permanganate ($KMnO_4$)	1.67
Chlorine dioxide (ClO_2)	1.5
Hypochlorous acid (HClO)	1.49
Chlorine (Cl_2)	1.36
Oxygen (O_2)	1.23
Bromine (Br_2)	1.09

1960s [11]. Commonly, the accepted Fenton mechanism is presented by Eqs. (15.1), (15.2), and (15.3), and its reaction rates were well reported in the literature [12].

$$Fe^{2+} + H_2O_2 \rightarrow Fe^{3+} + OH^- + \bullet OH k_1 = \left(40\tilde{}80 L \cdot mol^{-1} \cdot s^{-1}\right) \quad (15.1)$$

$$Fe^{3+} + H_2O_2 \rightarrow Fe^{2+} + \bullet O_2H + H^+ k_2 = \left(9.1 \times 10^{-7} L \cdot mol^{-1} \cdot s^{-1}\right) \quad (15.2)$$

$$Fe^{3+} + \bullet O_2H \rightarrow Fe^{2+} + O_2 + H^+ k_3 = \left(0.33\tilde{}2.1 \times 10^6 L \cdot mol^{-1} \cdot s^{-1}\right) \quad (15.3)$$

Typically, Eq. (15.1) is considered as the core step in Fenton chemistry and implies the oxidation of ferrous to ferric ions to decompose H_2O_2 into hydroxyl radicals. Eq. (15.2) is usually called Fenton-like reaction. This step allows Fe^{2+} regeneration in an effective cyclic mechanism. Apart from ferrous ion regeneration, hydroperoxyl radicals ($^{\cdot}O_2H$) are also produced.

15.2 Photo-Fenton Technology

15.2.1 UV Light-Assisted Fenton Process

Photo-Fenton process has attracted much attention among researchers. Several studies have reported that Photo-Fenton process has good influence on degradation of organic pollutants under UV irradiation, for instance, anisole [13], phenols [14–16], Orange II [17], and xylidine [18]. Because the majority of sunlight is visible light, therefore, it must be of great importance in the environmental field to utilize visible light for the degradation of pollutants.

In 2000, Chen et al. reported the photodegradation of methyl orange in methylated solution under both UV and visible irradiations in the presence of ferric ions and H_2O_2 [19]. It was found that different radicals were generated under UV and visible light irradiations. Besides the organic pollutants, in our daily life, coffee is necessary for many people. At the same time, coffee effluent contains a large amount of dark brown pigments. When coffee effluent is released in aquatic system, it cuts off light and affects photosynthesis due to its intense dark brown color. Furthermore, its high organic load causes eutrophication. Interestingly, Masahiro Tokumura et al. [20] investigated the influence of various operating parameters on the color removal of coffee effluents and found that the UV/Fe^{2+}/H_2O_2 system is very efficient.

15.2.2 Visible Light-Assisted Fenton Process

In previous published work, it is well known that the oxidizing power of the Fenton-type systems can be enhanced by visible light. Zhao et al. [21] examined the Photo-Fenton process of malachite green (MG) under irradiation by visible light. It was found that the Fenton degradation of dyes could be accelerated greatly by visible irradiation because of strong absorbance of visible light. This possibly provides another good approach to the treatment of dyes pollutants. Furthermore, Zhao et al. [22] found the mechanism of visible light-induced Fenton reaction. The specific mechanism is shown in Eqs. (15.4), (15.5), (15.6), and (15.7). It is found that the effective electron transfer from the visible light-excited dyes to Fe^{3+} leads to regeneration of Fe^{2+} and easy cycles of Fe^{3+}/Fe^{2+}. This result represents that much faster degradation and mineralization of various dyes have been achieved in the Photo-Fenton reaction under visible light irradiation, comparing with that in the dark.

$$dye + \text{visible light} \rightarrow \text{dye}^* \tag{15.4}$$

$$\text{dye}^* + Fe^{3+} \rightarrow Fe^{2+} + \text{dye}^{+\bullet} \tag{15.5}$$

$$Fe^{2+} + H_2O_2 \rightarrow Fe^{3+} + HO\bullet + OH^- \tag{15.6}$$

$$HO\bullet + dye(or\,dye^{+\bullet}) \rightarrow \text{degraded}\,products \tag{15.7}$$

15.2.3 Fenton-Like Process

Apart from conventional Fenton degradation, Bokare et al. reviewed the recent experimental attempts to establish the iron-free AOPs based on Fenton reaction chemistry [23]. For example, the use of Ce^{4+}/Ce^{3+} redox cycle has also been extended to the oxidation of organic pollutants using ceria as a heterogeneous Fenton-like catalyst. In the presence of H_2O_2, Heckert et al. [24] first confirmed the generation of HO•. In terms of its reactivity toward H_2O_2, copper shows similar redox properties like iron [25]. In this work, the performance of the solar

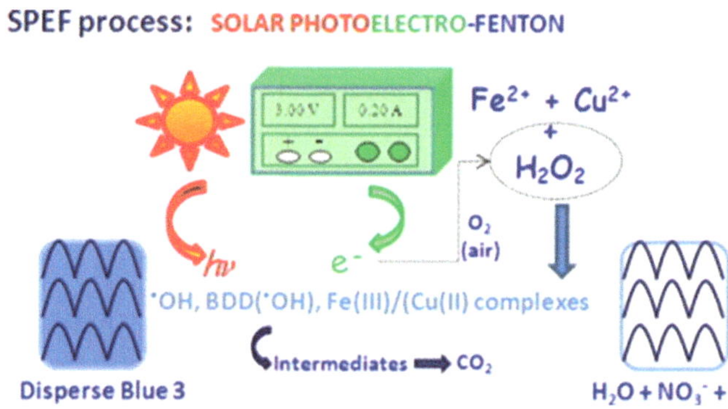

Fig. 15.1 The solar photoelectro-Fenton (SPEF) process. Reprinted with the permission from ref. [25]. Copyright 2012 American Chemical Society

photoelectro-Fenton (SPEF) process with Fe^{2+} and Cu^{2+} as metal cocatalysts and their application in the treatment of Disperse Blue 3 (DB3) dye solution simulating effluents are reported. The whole process is illustrated in Fig. 15.1.

15.3 Transition Metal Catalysts as Fenton Reagents

Among the transition metal catalysts, ruthenium (Ru) is the only member of the platinum group metals that exhibits Fenton-like activity in the presence of H_2O_2. Ruthenium complexes have been widely investigated for various organic transformation reactions like olefin hydroxylation, alcohol dehydrogenation, water oxidation, and alkene epoxidation [26]. Moreover, transition metal dichalcogenides have aroused great interest among the scientists. As to MoS_2 (molybdenum disulfide), a few studies have elucidated its properties in Fenton reaction. In 2010, single-layer MoS_2 was pointed to be as effective as graphene when used as transistor [27]. This finding provoked a stir among scientists. Dai et al. prepared MoS_2/RGO hybrid materials [28], which exhibited superior electrocatalytic activity in the hydrogen evolution reaction (HER), owing to the strong chemical and electronic coupling and the exposed edges of molybdenum disulfide. In 2014, Li et al. reported that electrochemically induced Fenton (electro-Fenton) reaction was used for efficient and controllable preparation of hydroxyl radicals, leading to the generation of luminescent quantum dots through etching of as-exfoliated MoS_2 nanosheets [29]. Apart from these applications, MoS_2 can act as a cocatalyst in the conventional heterogeneous Fenton process. In 2016, Yang et al. [30] synthesized a novel three-dimensional (3D) α-Fe_2O_3/MoS_2 hierarchical nanoheterostructure via a facile hydrothermal method. The zero-dimensional (0D) Fe_2O_3 nanoparticles guide the growth of two-dimensional (2D) MoS_2 nanosheets and formed 3D flowerlike structures, while MoS_2 facilitates the good dispersion of porous Fe_2O_3 with abundant oxygen vacancies. Figure 15.2(a) illustrated the formation process of the 3D α-Fe_2O_3/MoS_2

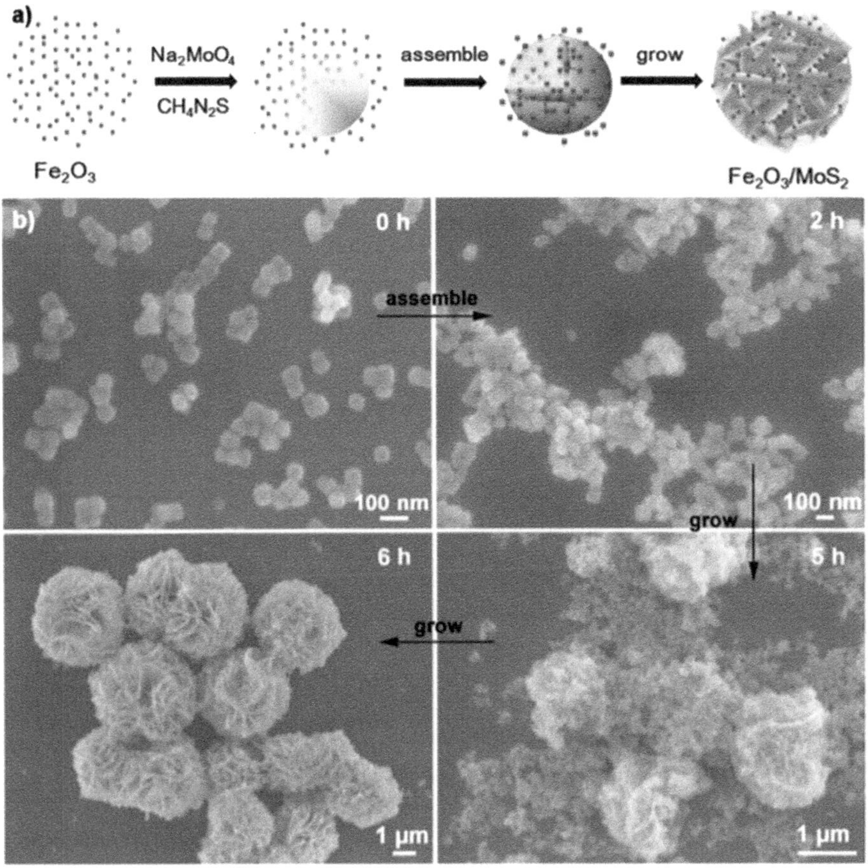

Fig. 15.2 (**a**) Schematic illustration of formation of the 3D α-Fe_2O_3/MoS_2 heterostructures; (**b**) morphological characterizations of the formation process of the 3D Fe_2O_3/MoS_2 heterostructures (3.0MF) with different reaction time [30]

heterostructures. And Fig. 15.2(b) shows the 3D flowerlike structure and the time-dependent growth morphology.

Figure 15.3(a) shows the change of the MO concentration (C/C_0) during the photodegradation process; C_0 and C are the initial concentration and the measured concentration after photodegradation for a certain time, respectively. To quantitatively estimate the reaction kinetics of the MO degradation, the degradation rate is calculated based on Fig. 15.3(a), and the constants k of Fe_2O_3, MoS_2, 1.4MF, 2.0MF, 3.0MF, and 4.0MF are 0.01485, 0.00262, 0.175, 0.18021, 0.2301, and 0.16411 per min, respectively [31]. It can be clearly seen that all the Fe_2O_3/MoS_2 heterostructures exhibit much higher photocatalytic activities than the pure Fe_2O_3 and MoS_2. Additionally, the 3.0MF was found to exhibit the highest rate, about 87 times higher than that of MoS_2 and 15 times higher than that of Fe_2O_3. Figure 15.3 (b) shows the normalized chemical oxygen demand (COD) removal during the

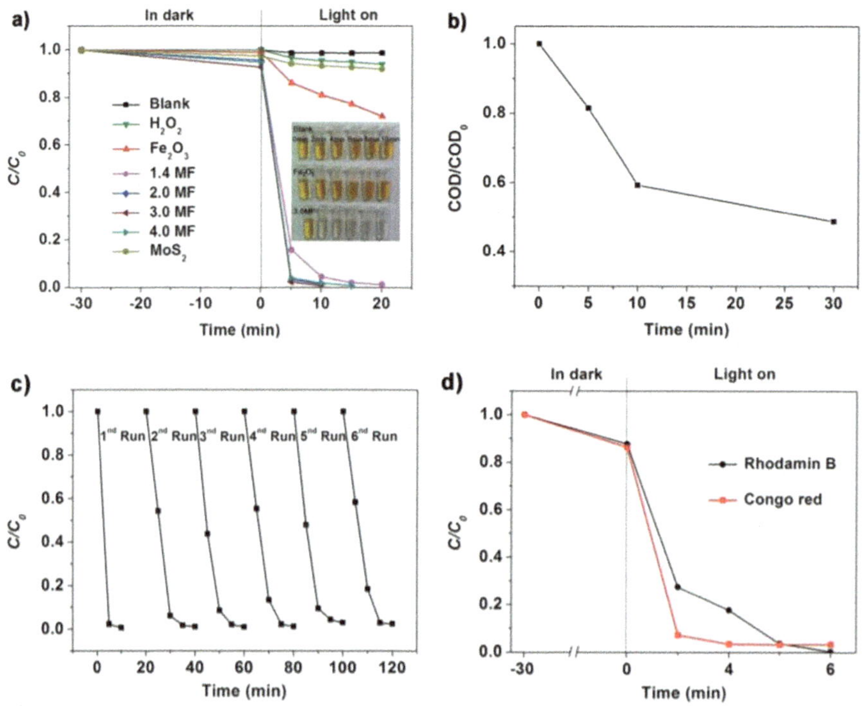

Fig. 15.3 Photocatalytic degradation of MO with different catalysts; (**b**) variations in COD during the photocatalytic (3.0MF) degradation of MO in 30 min. The inset in (**a**) is the photo of the fade of the MO; (**c**) six cycles of the photocatalytic reduction of MO using sample 3.0MF as the photocatalyst under simulated solar light irradiation for 20 min; (**d**) photocatalytic degradation of CR (50 mg L^{-1}) and RhB (20 mg L^{-1}) under simulated solar light with the presence of 3.0MF [30]

photocatalytic treatment with 10 mg of the catalysts. It is observed that more than 56% of COD in the wastewater can be reduced only after 30 min of exposure to the simulated solar light in the presence of 3.0MF. And Fig. 15.3(c) shows that after six cyclic tests, the degradation rate can still keep above 95%. Besides, the generalizability of the Fe_2O_3/MoS_2 heterostructures was also evaluated by the degradation of Congo red (CR, 50 mg L^{-1}) and Rhodamine B (RhB, 20 mg L^{-1}) (Fig. 15.3(d)). This charming 3D structure with perfect match of non-equal dimension exhibits excellent recyclable Photo-Fenton catalytic activity for methyl orange pollutant and other organic pollutants.

In this work, H_2O_2 is added to further enhance the contamination degradation. When irradiated with sunlight, Fe^{3+} would be reduced to Fe^{2+} and generate the •OH. And then, Fe^{2+} would react with the adsorbed oxygen molecules on the surface of the heterostructures to form oxidizing species (O_2•−). The superoxide anion radicals (O_2•−) generate hydroperoxy (HO_2•) radicals and subsequently produce hydroxyl radicals •OH. Meanwhile, in the presence of H_2O_2, Fe^{2+} is easily oxidized to Fe^{3+}

and •OH is generated at the same time. So the regeneration of Fe^{3+}/Fe^{2+} cycles would make more strong oxidant •OH for the photocatalytic degradation.

15.4 Conclusions

In summary, we have briefly introduced the Fenton technology and specifically clarified the mechanism of Fenton process. Besides that, the application of transition metal dichalcogenides (such as MoS_2) in transistor, hydrogen evolution reaction (HER), and heterogeneous Photo-Fenton process is also introduced. As a unique representative of transition metal dichalcogenides, MoS_2 has been widely explored as a cocatalyst for photodegradation of contaminants. MoS_2 has a broad promising future in Fenton technology.

References

1. Malato S, Blanco J, Vidal A et al (2002) Photocatalysis with solar energy at a pilot-plant scale: an overview. Appl Catal B Environ 37(1):1–15
2. Levec J, Pintar A (2007) Catalytic wet-air oxidation processes: a review. Catal Today 124 (3):172–184
3. Cybulski A (2007) Catalytic wet air oxidation: are monolithic catalysts and reactors feasible? Ind Eng Chem Res 46(12):4007–4033
4. Babuponnusami A, Muthukumar K (2014) A review on Fenton and improvements to the Fenton process for wastewater treatment. J Environ Chem Eng 2(1):557–572
5. Hoigné J (1997) Inter-calibration of OH radical sources and water quality parameters. Water Sci Technol 35(4):1–8
6. Munter R (2001) Advanced oxidation processes-current status and prospects. Proc Estonian Acad Sci Chem 50(2):59–80
7. Rodrıguez M, Abderrazik NB, Contreras S et al (2002) Iron(III) photoxidation of organic compounds in aqueous solutions. Appl Catal B Environ 37(2):131–137
8. Joseph JM, Varghese R, Aravindakumar CT (2001) Photoproduction of hydroxyl radicals from Fe (III)-hydroxy complex: a quantitative assessment. J Photochem Photobio A Chem 146 (1):67–73
9. Contreras S, Rodrıguez M, Chamarro E et al (2001) UV- and UV/Fe(III)-enhanced ozonation of nitrobenzene in aqueous solution. J Photochem Photobio A Chem 142(1):79–83
10. Weast RC (1986) Handbook of physics and chemistry. CRC Press, Boca Raton, pp 1983–1984
11. Huang C, Dong C, Tang Z (1993) Advanced chemical oxidation: its present role and potential future in hazardous waste treatment. Waste Manag 13(5–7):361–377
12. Sychev AY, Isak VG (1995) Iron compounds and the mechanisms of the homogeneous catalysis of the activation of O_2 and H_2O_2 and of the oxidation of organic substrates. Russ Chem Rev 64(12):1105–1129
13. Zepp RG, Faust BC, Hoigne J (1992) Hydroxyl radical formation in aqueous reactions (pH 3-8) of iron (II) with hydrogen peroxide: the photo-Fenton reaction. Environ Sci Technol 26 (2):313–319
14. Kiwi J, Pulgarin C, Peringer P (1994) Effect of Fenton and photo-Fenton reactions on the degradation and biodegradability of 2 and 4-nitrophenols in water treatment. Appl Catal B Environ 3(4):335–350

15. Chen R, Pignatello JJ (1997) Role of quinone intermediates as electron shuttles in Fenton and photoassisted Fenton oxidations of aromatic compounds. Environ Sci Technol 31 (8):2399–2406
16. Maletzky P, Bauer R (1998) The Photo-Fenton method-degradation of nitrogen containing organic compounds. Chemosphere 37(5):899–909
17. Nadtochenko V, Kiwi J (1996) Dynamics of light-induced excited state quenching of ferrioxalate complexes by peroxides. Fast kinetics events and interaction with toxic pollutants. J Photochem Photobiol A Chem 99(2):145–153
18. Nadtochenko V, Kiwi J (1998) Primary photochemical reactions in the photo-Fenton system with ferric chloride. 1. A case study of xylidine oxidation as a model compound. Environ Sci Technol 32(21):3273–3281
19. Chen F, Xie Y, He J et al (2001) Photo-Fenton degradation of dye in methanolic solution under both UV and visible irradiation. J Photochem Photobiol A Chem 138(2):139–146
20. Tokumura M, Ohta A, Znad HT et al (2006) UV light assisted decolorization of dark brown colored coffee effluent by photo-Fenton reaction. Water Res 40(20):3775–3784
21. Wu K, Xie Y, Zhao J et al (1999) Photo-Fenton degradation of a dye under visible light irradiation. J Mol Catal A Chem 144(1):77–84
22. Ma J, Song W, Chen C et al (2005) Fenton degradation of organic compounds promoted by dyes under visible irradiation. Environ Sci Technol 39(15):5810–5815
23. Bokare AD, Choi W (2014) Review of iron-free Fenton-like systems for activating H_2O_2 in advanced oxidation processes. J Hazard Mater 275:121–135
24. Heckert EG, Seal S, Self WT (2008) Fenton-like reaction catalyzed by the rare earth inner transition metal cerium. Environ Sci Technol 42(13):5014–5019
25. Salazar R, Brillas E, Sirés I (2012) Finding the best Fe^{2+}/Cu^{2+} combination for the solar photoelectro-Fenton treatment of simulated wastewater containing the industrial textile dye Disperse Blue 3. Appl Catal B Environ 115-116:107–116
26. Pagliaro M, Campestrini S, Ciriminna R (2005) Ru-based oxidation catalysis. Chem Soc Rev 34 (10):837–845
27. Radisavljevic B, Radenovic A, Brivio J et al (2011) Single-layer MoS_2 transistors. Nat Nano 6 (3):147–150
28. Li Y, Wang H, Xie L et al (2011) MoS_2 nanoparticles grown on graphene: an advanced catalyst for the hydrogen evolution reaction. J Am Chem Soc 133(19):7296–7299
29. Li BL, Chen LX, Zou HL et al (2014) Electrochemically induced Fenton reaction of few-layer MoS_2 nanosheets: preparation of luminescent quantum dots via a transition of nanoporous morphology. Nanoscale 6(16):9831–9838
30. Yang X, Sun H, Zhang L et al (2016) High efficient photo-Fenton catalyst of α-Fe_2O_3/MoS_2 hierarchical nanoheterostructures: reutilization for supercapacitors. Sci Rep 6:31591
31. Herrmann JM, Tahiri H, Ait-Ichou Y et al (1997) Characterization and photocatalytic activity in aqueous medium of TiO_2 and Ag-TiO_2 coatings on quartz. Appl Catal B Environ 13 (3):219–228

Chapter 16
Transition Metal Phosphide As Cocatalysts for Semiconductor-Based Photocatalytic Hydrogen Evolution Reaction

16.1 Introduction

The increasing global energy supply and extreme climate change and related environmental issues are complex technological challenges that are being faced by scientists in the twenty-first century [1]. The rate of global energy consumption averaged 15–17 TW in 2010, and this number will likely increase to 25–27 TW due to the increasing global population and production [2, 3]. It is especially noteworthy that human productions and livings heavily rely on nonrenewable fossil fuels, such as coal, oil, and natural gas [4]. Furthermore, the combustion of these fossil fuels has generated a series of environmental issues, including air and water contamination and global warming [5]. Thus, it is extremely urgent to search for sustainable and environmentally friendly alternative energy to replace exhaustible fossil resource. Solar energy is a clean, stable, and renewable energy source [6–9]. Today resource-poor around the world, the highly efficient conversion of solar energy into usable energy is increasingly showing a bright future. Photocatalytic water reduction represents a promising strategy for a clean, non-contaminative, and low-cost production of H_2 [10–12]. As an energy medium, hydrogen still has too much potential (highest gravimetric energy density compared with other fuels) to shuffle quietly off the dependence on coal [13, 14].

In general, the photocatalytic hydrogen reaction is composed of three major steps: (i) absorption of light by semiconductor to generate electron–hole pairs, (ii) charge separation and migration to semiconductor's surface, and (iii) surface reduction reaction for H_2 evolution (Fig. 16.1). The efficiency of photocatalytic hydrogen evolution reaction is determined by the synergetic effect of thermodynamics and kinetics of the above three steps [4]. To date, much significant progress has been achieved to develop photocatalysts with the broad absorption of solar light (step i) and high charge separation efficiency (step ii). For instance, a plenty of strategies have been employed to extend the light absorption of photocatalyst, such as bandgap reduction (metal or nonmetal element doping) [15–17], coupling with narrow

J. Zhang et al., *Photocatalysis*, Lecture Notes in Chemistry 100,
https://doi.org/10.1007/978-981-13-2113-9_16

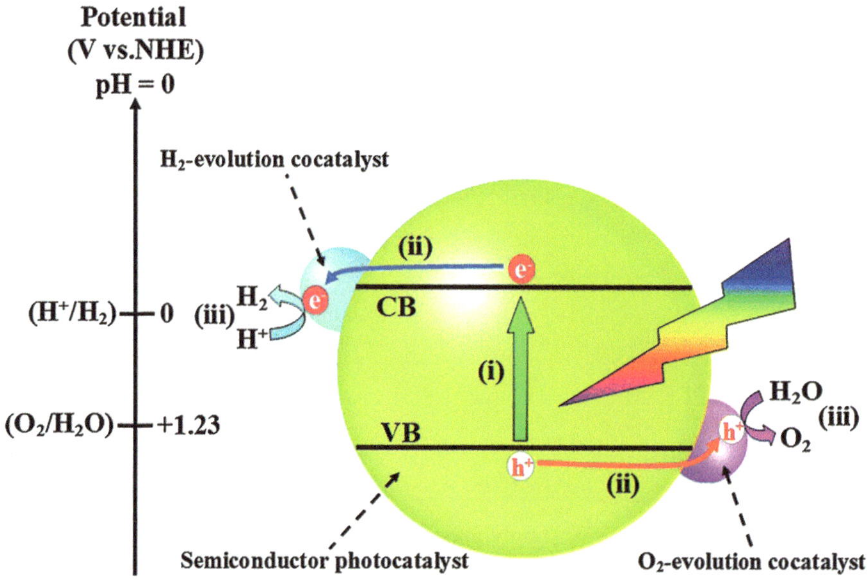

Fig. 16.1 Schematic illustration of photocatalytic hydrogen evolution reaction over a semiconductor photocatalyst loaded with H_2 evolution cocatalysts. (Reproduced from Ref. [4] by permission of The Royal Society of Chemistry (RSC) on behalf of the Centre National de la Recherche Scientifique (CNRS) and the RSC)

bandgap semiconductors [18], dye sensitization [19], vacancy introduction [20], and surface plasmon enhancement [21]. On the other hand, in order to obtain high efficiency of photo-generated charge separation, various semiconductor heterojunctions have been designed [22]. Up to now, much research work has been investigated to optimize the first two steps in the process of photocatalytic hydrogen evolution. However, much less achievements have been done to improve the efficiency of the third step which demands the utilization of cocatalyst. As shown in Fig. 16.1, the third step is realized in the presence of H_2 evolution cocatalyst, which can release the trapped photo-generated electrons for catalytic H_2 evolution and improve the stability of photocatalyst by suppressing photocorrosion. Thus, cocatalysts play an important role in improving the activity and stability of catalyst.

Now, the development of photocatalytic systems for achieving high photocatalytic activity and stability relies on the utilization of noble metal-based cocatalysts. For instance, Ru [23], Au [24], Pt [25], Pd [26], Rh [27], and Ag [28] have been intensively used as cocatalysts in photocatalytic hydrogen evolution reactions. Among them, Pt is one of the most efficient cocatalysts, owing to its largest work function and lowest overpotential [29]. After loading moderate amount of Pt on the surface of semiconductor, the obvious activity enhancement in photocatalytic hydrogen evolution can be observed [30]. However, these abovementioned noble metal-based cocatalysts are very difficult to realize large-scale practical applications due to their limited reserves and high cost. Therefore, the development of noble metal-free cocatalysts with high efficiency, abundant storage, and low cost are highly required.

Recently, transition metal phosphides (TMPs) have been widely studied as typical representatives of burgeoning noble metal-free cocatalysts due to their intriguing properties, such as approximate zero-valent metallic feature, high stability, low-cost, and highly capturing electron ability [31, 32]. In this chapter, we will give a short overview of the recent research progress of transition metal phosphides acted as cocatalysts in photocatalytic hydrogen evolution reactions. The preparation methods, effects of P element, and photocatalytic applications of TMPs are comprehensively discussed.

16.2 Preparation Methods

As the first TMP, Ni_2P was prepared for the application of vapor phase catalysis in the 1950s [33]; it has been gradually depleted for half a century. With the development of synthetic and characteristic techniques, these TMPs materials have been flourished as efficient earth-abundant cocatalysts in photocatalysis. In this section, the synthesis strategies of TMPs will be categorized.

16.2.1 Organophosphorus Sources

Currently, there are two main preparation methods to synthesize TMPs. One is realized in the presence of organophosphorus precursors. The other is to use organophosphorus sources to obtain TMPs. In the latter method, tri-n-octylphosphine (TOP), trioctylphosphine oxide (TOPO), and their analogue tris (trimethylsilyl)phosphine (TMSP) and triphenylphosphine (TPP) are common and attractive phosphorus sources in the synthesis of TMPs [34, 35].

Richard A. Jones and his coworkers [36] used ruthenium phosphite hydride complexes $H_2Ru(P(OR)_3)_4$ (R = Me (1), Et (2), ^{i}Pr (3)) as the chemical vapor deposition (CVD) precursors to deposit the films of amorphous ruthenium–phosphorus alloys. The crystal structure of as-prepared sample cis-$H_2Ru(P(OMe)_3)_4$, cis-$H_2Ru(P(OEt)_3)_4$, and $Cl_2Ru(P(OMe)_3)_4$ is demonstrated in Fig. 16.2a–c. In the structure of these complexes, the two P atoms of the mutually link phosphite ligands at Ru, while the other two $P(OMe)_3$ ligands leave room for the coordination of the relatively small hydride ligands. The synthetic strategies of ruthenium-P(OEt) ligand complex are similar to that of these three complexes by using triethylphosphite as the phosphorus containing ligand. For the single metal/phosphorus organic precursors, the films of RuP are achieved via the chemical process by which both the Ru and P attain a zero oxidation state upon deposition. It is noted that the thermal CVD decomposition of organic phosphorus source to form TMPs shows more potential in controlling the P concentration in TMPs than the phosphidation pathway.

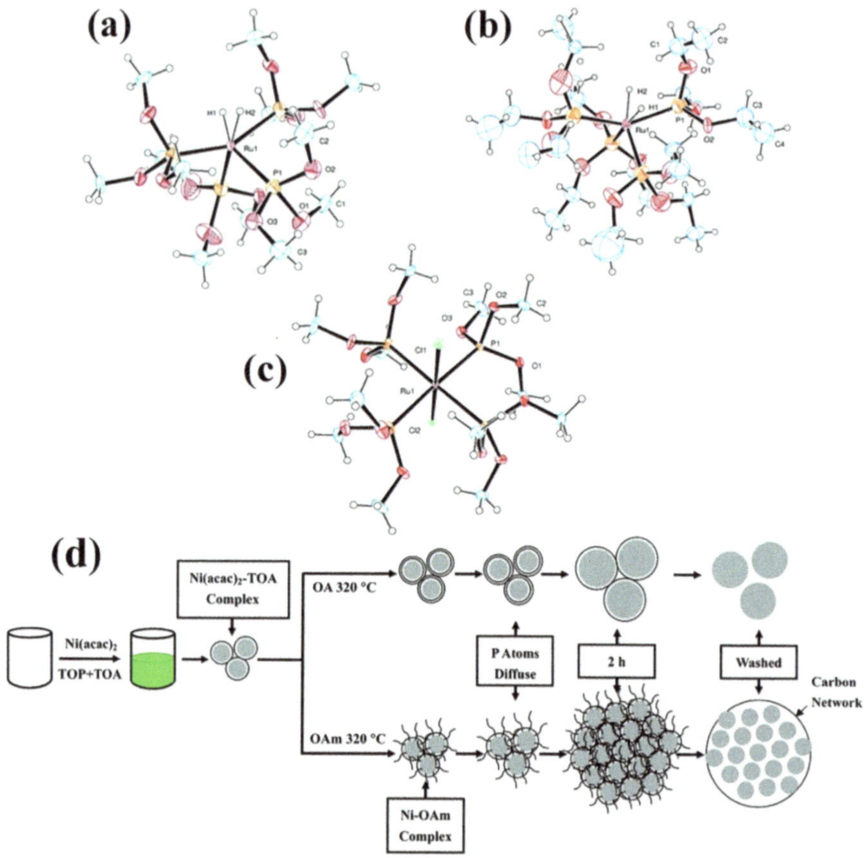

Fig. 16.2 Typical examples of organophosphorus precursors. Partial atom numbering scheme of (**a**) cis-$H_2Ru(P(OMe)_3)_4$, (**b**) cis-$H_2Ru(P(OEt)_3)_4$, (**c**) $Cl_2Ru(P(OMe)_3)_4$. (**d**) Schematic illustration of the formation of hierarchical nickel phosphide (h-Ni_2P) nanospheres. (Reprinted with the permission from Ref. [38]. Copyright 2011 American Chemical Society)

In the organophosphorus route, TMPs with different morphologies can be prepared through a thermal CVD decomposition of metal phosphide or via a heating reflux conversion of metal particles to TMPs in TOP solvent at high temperature (>220 °C) [37]. Hierarchical nickel phosphide (h-Ni_2P) nanospheres with an average size of 5–10 nm were prepared via a one-step strategy from the original reactant nickel acetylacetonate, trioctylphosphine, tri-n-octylamine, and oleylamine (OAm) (Fig. 16.2) [38]. The OAm molecules are introduced as the surfactant to control the morphology of h-Ni_2P. Compared with the oleic acid (OA) process, the thermal treatment with $Ni(acac)_2$-TOA complex at 320 °C can lead to cleaving the P–C bonds located at the surface of Ni–TOP, thus forcing P atoms chemically linking with Ni to generate Ni_2P nanospheres.

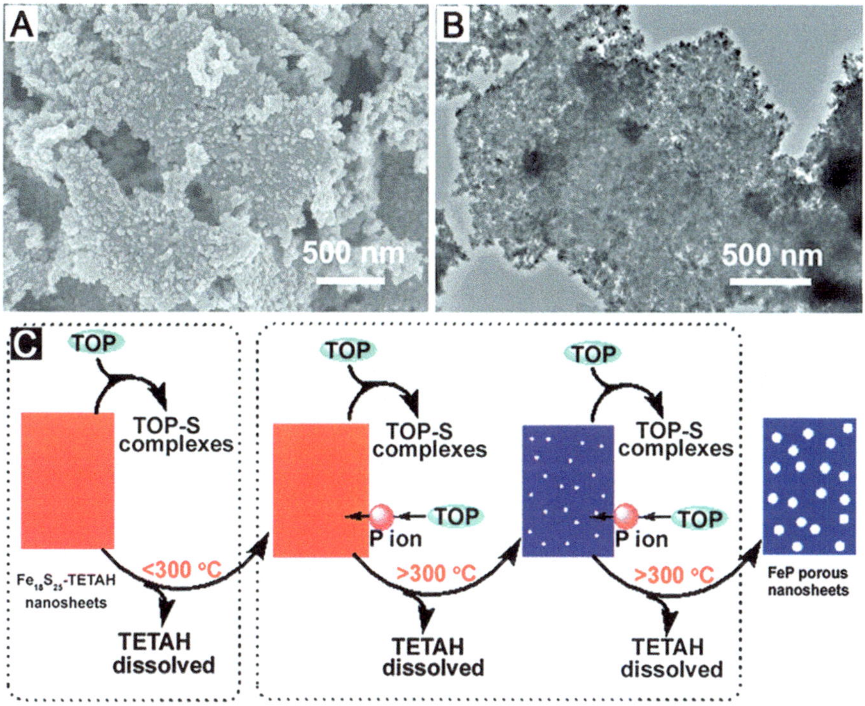

Fig. 16.3 (**a**) SEM image and (**b**) TEM image of porous FeP nanosheets. (**c**) Scheme illustrating the synthesis of the porous FeP nanosheets through the anion-exchange reaction of the inorganic–organic hybrid $Fe_{18}S_{25}$–TETAH nanosheets with P ions. (Reproduced from Ref. [39] by permission of The Royal Society of Chemistry (RSC) on behalf of the Centre National de la Recherche Scientifique (CNRS) and the RSC

TOP is of low toxicity phosphorus source. Since TOP has a very strong coordination effect, the use of TOP can efficiently promote the reaction and create some uncommon nanostructures. For example, Zhang et al. [39] successfully prepared nanoporous FeP nanosheets via anion-exchange reaction of inorganic–organic hybrid $Fe_{18}S_{25}$–TETAH (TETAH = protonated triethylenetetramine) nanosheets (Fig. 16.3). It is noted that low solubility of TOP in water and its high decomposition temperature retard this system only to be finished in some high boiling point solvents (octyl ether and 1-octadecylene), which leads to this reaction to be highly flammable and corrosive. Besides, the TOP oxide (tri-*n*-octylphosphine oxide, TOPO) and other organic phosphines (tri-phenylposphine, TPP) can play a similar role in preparation of TMPs [40, 41].

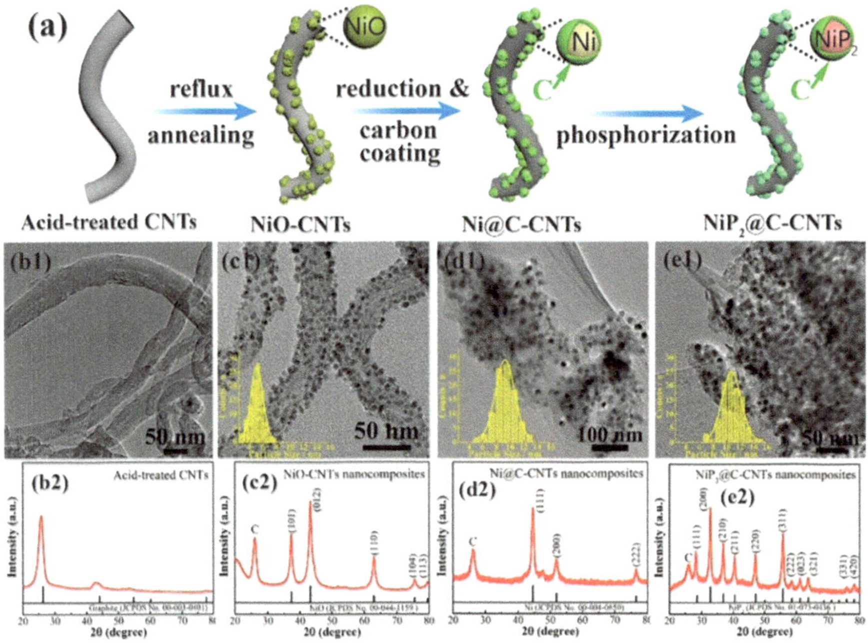

Fig. 16.4 (**a**) Schematic diagram for the synthesis of the NiP_2@C-CNTs nanocomposites. The representative TEM images of (**b1**) acid-treated CNTs, (**c1**) NiO-CNTs, (**d1**) Ni@C-CNTs, and (**e1**) NiP_2@C-CNTs and the XRD patterns of (**b2**) acid-treated CNTs, (**c2**) NiO-CNTs, (**d2**) Ni@C-CNTs, and (**e2**) NiP_2@C-CNTs. The insets of **c1**, **d1**, and **e1** show the particle size distribution of the NiO, Ni, and NiP_2, respectively. (Reprinted with the permission from Ref. [46]. Copyright 2017 American Chemical Society)

16.2.2 Inorganic Phosphorus Sources

Except for these organophosphorus sources, some inorganic phosphorus precursors, such as white phosphorus (WP) [42] and red phosphorus (RP) [43, 44], have been widely studied in liquid phase synthesis of TMPs. Du et al. [45] prepared CdS/Ni_2P composites through two-step hydrothermal method, in which $Cd(CH_3COO)_2$ and thioacetamide are used as cadmium and sulfur source and WP and $Ni(NO_3)_2$ as phosphorus and nickel source, respectively. After hydrothermal treatment, Ni_2P nanoparticles can be formed and loaded on the surface of CdS. Due to its low water solubility, some solvents, such as ethylenediamine and carbon disulfide, will be selected to dissolve WP. Furthermore, WP can react with metal salt under hydrothermal condition at relatively low temperature (ca. 140–180 °C) to form TMPs with various morphologies. What we should note in the reaction in the presence of WP is that it needs to be treated in a glovebox due to its low flash point and high toxicity.

Compared with WP, RP is a relatively safe and inexpensive phosphorus source. Guo et al. [46] designed a NiP_2@C-CNTs nanocomposite via three steps reaction

(Fig. 16.4a). Firstly, acid-treated carbon nanotube (CNT) was used as substrate (Fig. 16.4b1–b2). Afterward, NiO nanoparticles anchored on the surface CNT composites (NiO-CNTs) were prepared simply via the reflux method followed by annealing under an argon atmosphere (Fig. 16.4c1–c2). Then, in order to avoid aggregation during phosphorization and get highly dispersed NiP_2 nanoparticles, NiO@CNTs were pre-coated with a thin layer of carbon by reacting with polyvinylidene fluoride (PVDF) (Fig. 16.4d1–d2). It is noted that NiO nanoparticles can be reduced to Ni after the treatment of PVDF. Besides as a reductive reagent, PVDF is commonly acted as an excellent fluorinating agent for the synthesis of inorganic fluorides. Finally, NiP_2@C-CNTs can be obtained by directly phosphorizing Ni@C-CNTs by using RP as the phosphorus source at 700 °C (Fig. 16.4e1–e2). The phosphorization process involved in RP needs to be heated beyond 700 °C, which is similar with the previous references [47, 48].

Besides, the strategy of hydrothermal treatment of RP is intensively used to form TMPs. Yin et al. [44] prepared Sn_4P_3/reduced graphene oxide (RGO) hybrids through an in situ low-temperature solution-based phosphorization chemical transformation route from Sn/RGO (Fig. 16.5a). First, Sn/RGO composites are synthesized by reducing $SnCl_2$ to Sn with the presence of $NaBH_4$. Then, Sn_4P_3/RGO can be prepared via in situ phosphorization chemical reaction method, in which RP and ethanediamine were employed as phosphorus source and solvent, respectively. TEM and EDX analyses further confirm the successful conversion of Sn/RGO to Sn_4P_3/GRO (Fig. 16.5b–f). The widespread adoption method involved in other inorganic phosphorus sources for the preparation of TMPs is a gas–solid reaction way. The PH_3 gas plays a significant role in formation of TMPs due to its highly active property. However, it is highly flammable and extremely toxic (the gas content at lever of ppm can be fatal). To this end, some substituents which can in situ release PH_3, such as $NH_4H_2PO_2$, $(NH_4)_2HPO_2$, and NaH_2PO_2, are widely investigated. To generate PH_3, these hypophosphites need to be heated beyond 200 °C.

The releasing PH_3 gas can easily react with metal oxide, metal hydroxide, and metal–organic frameworks to form TMPs. Taking NaH_2PO_2, for instance, according to the below reaction Eq. (16.1), NaH_2PO_2 can decompose into two substances.

$$2NaH_2PO_2 \rightarrow Na_2HPO_4 + PH_3 \tag{16.1}$$

The gas–solid reaction strategy is a surfactant-free method and can remain the morphology of metal precursor to the utmost extent. Furthermore, PH_3 can not only achieve the transformation from metal precursor to metal phosphide but also reduce the metal ions. Lou et al. [49] prepared Ni–Co mixed metal phosphides and amorphous composite nanoboxes (NiCoP/C) through a metal–organic framework (MOF)-based strategy (Fig. 16.6a). Phase-pure ZIF-67 nanocubes are selected as the initial reactant. After reacting with a mount of $Ni(NO_3)_2$ at room temperature, the solid ZIF-67 nanocubes are converted into ZIF-67@LDH nanoboxes (Fig. 16.6b–d). Afterward, the NiCoP/C nanoboxes are obtained through a simple one-step phosphidation process (Fig. 16.6e–g). This process is achieved by thermal treating

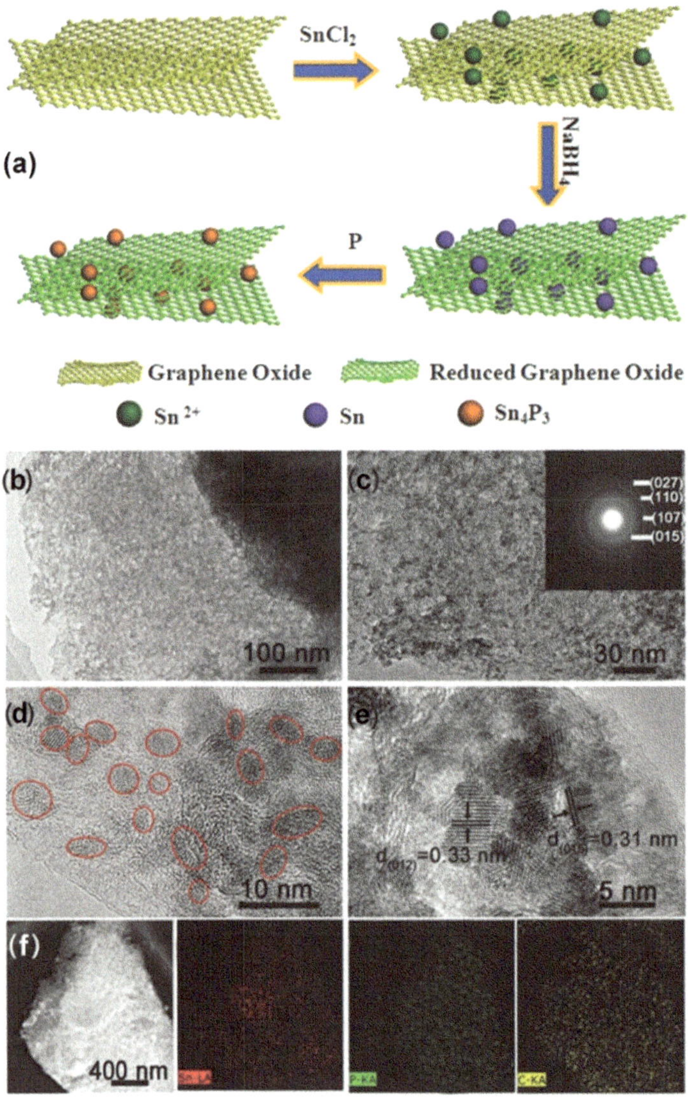

Fig. 16.5 (**a**) Schematic illustration of the synthesis process of the Sn_4P_3/RGO hybrid sample. (**b**, **c**) TEM image. The diffraction rings in the upper inset of (**c**) are an electron diffraction pattern. (**d** and **e**) HRTEM image of Sn_4P_3/RGO. (**f**) A STEM image and its corresponding EDX elemental mapping images of Sn, P, and C of the Sn_4P_3/RGO sample. (Reproduced from Ref. [44] by permission of John Wiley & Sons Ltd)

with ZIF-67@LDH nanoboxes and NaH_2PO_2 at 300 °C. The element mapping images further confirm the successful phosphidation (Fig. 16.6h–k).

Fu et al. [50] successfully prepared Ni_2P nanoparticles via this phosphating reaction with $Ni(OH)_2$ nanoparticles. In this process, Ni^{2+} can easily be reduced to

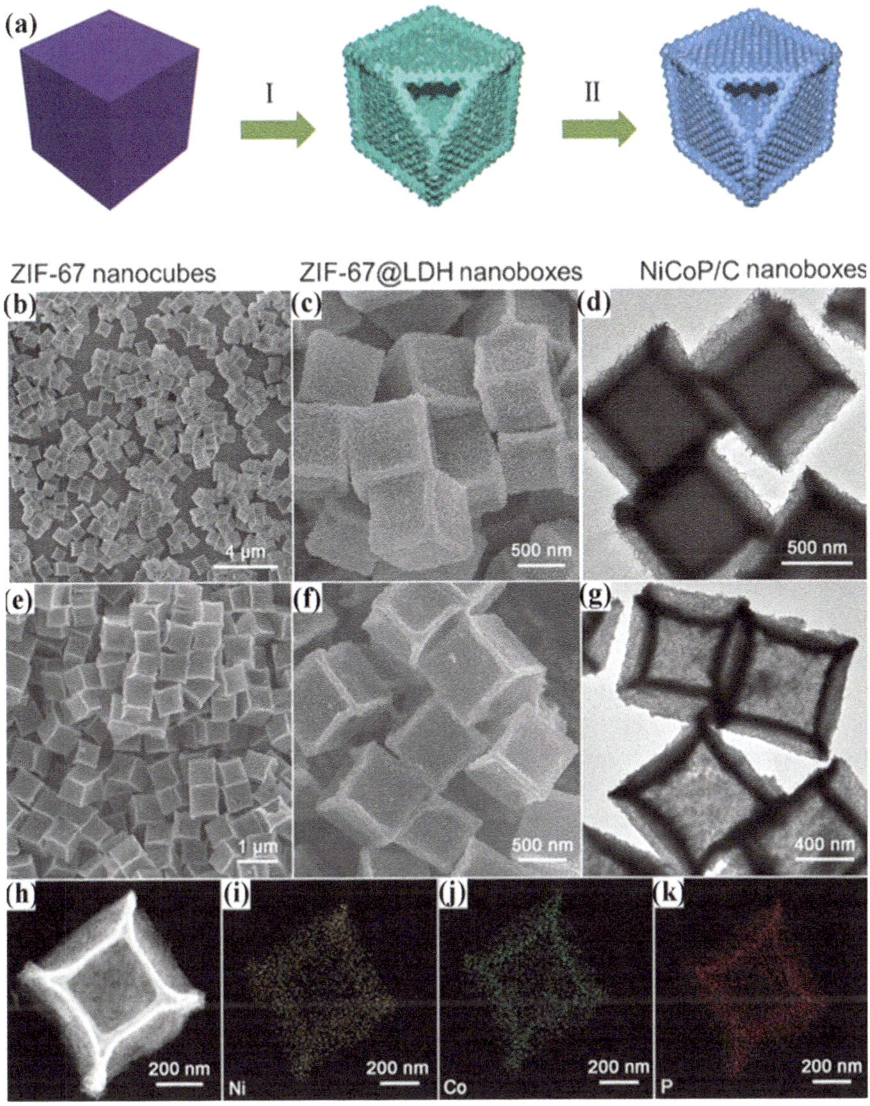

Fig. 16.6 (**a**) Schematic illustration of the formation process of NiCoP/C nanoboxes: (I) reaction of ZIF-67 nanocubes with $Ni(NO_3)_2$ to form ZIF-67@LDH nanoboxes and (II) phosphidation of ZIF-67@LDH nanoboxes to obtain NiCoP/C nanoboxes. (**b**, **c**) FESEM and (**d**) TEM images of ZIF-67@LDH nanoboxes. (**e** and **f**) FESEM images, (**g**) TEM image, and (**h**) HAADF-STEM image of the NiCoP/C nanoboxes. Elemental mapping images of Ni (**i**), Co (**j**), and P (**k**) of an individual nanobox. (Reproduced from Ref. [49] by permission of John Wiley & Sons Ltd)

Ni^+ by PH_3 treatment. The synthetic process of TMPs from $NH_4H_2PO_2$ or $(NH_4)_2HPO_2$ precursors is similar to that of the phosphate route.

In addition, there is another method to form TMPs from inorganic phosphate. In general, the forming of TMPs from a specific metal phosphate precursor needs to go through a deoxidization of metal phosphate from P–O bond to metal–phosphorus bond [51]. Generally speaking, this deoxidization process is achieved under the H_2 or H_2/Ar atmosphere, in which H_2 atmosphere can facilitate the reduction process from metal phosphate to metal phosphide. Taking NaH_2PO_2 source as an example, TMPs can be synthesized via a first hydrothermal method to prepare metal phosphate by mixing up NaH_2PO_2 with metal chloride or nitrate. In these cases, some surfactants can be selected to form specific morphology in this step. In the following step, the metal phosphate was heated to 150–500 °C to form TMPs. In this way, CoP, MoP, and RuP with different nanostructures can be obtained for the potential application in photocatalysis [52].

16.3 Effect of P on Photocatalytic Hydrogen Evolution Reactions

TMPs can be viewed as P element doped into the crystal lattice of transition metal. Until now, six transition metal (Ni, Co, Fe, Mo, Mn, Cu) can be formed into TMPs, and these TMPs can be applied into photocatalytic hydrogen evolution reactions. Other transition metals (Zn, Ti) also can be formed into TMPs, but these TMPs are not suitable for photocatalytic hydrogen evolution reactions owing to their property of easy hydrolysis in an aqueous solution.

16.3.1 The Role of P

It has been certified that P element plays a positive role in photocatalytic hydrogen evolution reactions. Fu et al. [53] prepared phosphorus-doped hexagonal tubular carbon nitride (P-TCN) with a layered microstructure through the hydrothermal method (Fig. 16.7a). In this case, P from phosphorus acid squeezed into graphitic carbon nitride (GCN) skeleton to obtain the P-TCN. Compared with pure TCN, P-TCN has a more narrow bandgap and negative conductive band position (Fig. 16.7b–c). Thus, an enhanced photocatalytic hydrogen activity over P-TCN can be observed (Fig. 16.7d). Photochemical measurement reveals that P element introduces a low hydrogen evolution over potential (Fig. 16.7d).

In addition, density functional theory (DFT) calculations have been proved that P atoms play vital roles in hydrogen evolution reactions [54, 55]. P atoms with more electronegativity can withdraw the electrons from metal atoms [56]. Moreover, it is

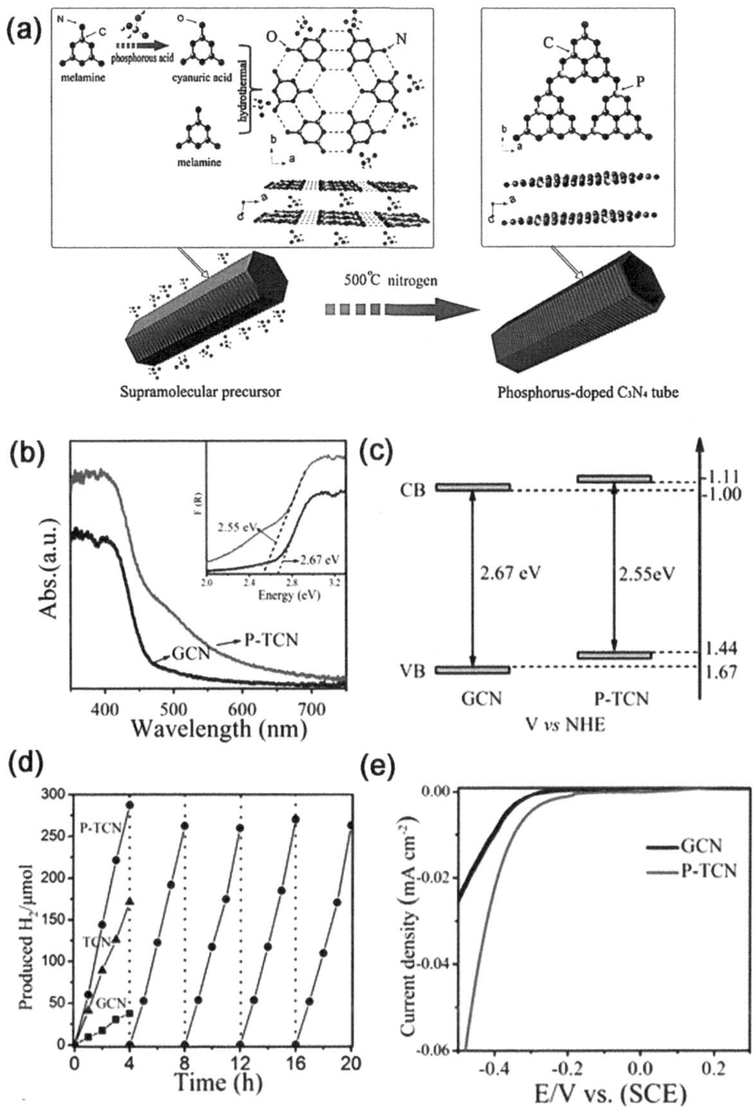

Fig. 16.7 (**a**) The formation process of phosphorus-doped tubular carbon nitride. (**b**) UV–vis light absorption spectra and bandgap energies (inset). (**c**) Band structure alignments of GCN and P-TCN (pH 1), respectively. (**d**) Time course of H_2 evolution for GCN, TCN, and P-TCN (pH 1) under visible light irradiation ($\lambda > 420$ nm). (**e**) Polarization curves of GCN and P-TCN. (Reproduced from Ref. [53] by permission of John Wiley & Sons Ltd)

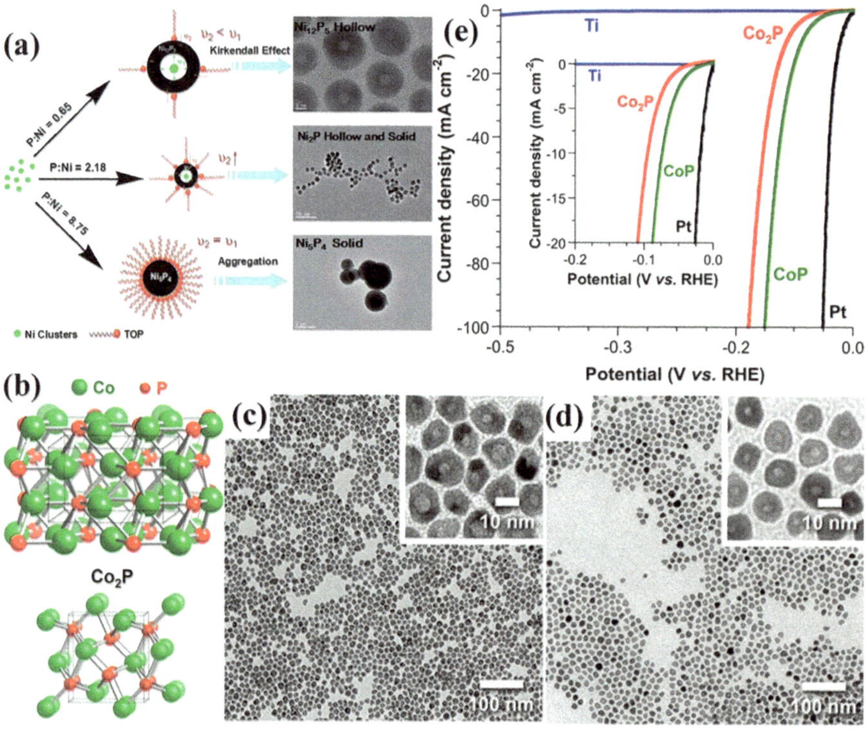

Fig. 16.8 (**a**) Possible formation mechanism of the as-synthesized nickel phosphide NCs with different phases and morphologies. (**b**) Crystal structures of (up) Co_2Si-type Co_2P and (down) MnP-type CoP. Unit cells are shown as dashed black lines. TEM images of Co_2P (**c**) and CoP (**d**), with enlarged regions in the insets. (**e**) Polarization data (plots of current density vs potential) in 0.5 M H_2SO_4 for Co_2P/Ti and CoP/Ti electrodes, along with Pt mesh and bare Ti foil for comparison. (Reproduced from Ref. [57] by permission of The Royal Society of Chemistry (RSC) on behalf of the Centre National de la Recherche Scientifique (CNRS) and the RSC)

noted that negatively charged P atoms can trap the positively charged hydrogen proton, thus achieving highly efficient water reduction.

16.3.2 *The Effect of P Content*

The atomic percentage of P in TMPs is a crucial factor in hydrogen evolution reactions. It is reported that with the increment of atomic percentage of P in TMPs, the hydrogen evolution activity rises accordingly. Liu et al. [57] prepared phase-controlled monodispersed nickel phosphide via the thermal decomposition method using nickel acetylacetonate as nickel source, trioctylphosphine as phosphorus source, and oleylamine as reductive reagent. The different morphologies are realized by changing the molar ratio of P/Ni precursor. The result demonstrated that among Ni_2P, Ni_5P_4, and $Ni_{12}P_5$, the Ni_5P_4 displays the best catalytic activity due to its high P content (44%) (Fig. 16.8a). Similarly, Schaak et al. [58] synthesized morphologically equivalent Co_2P and CoP hollow nanoparticles by adjusting reaction time (Fig. 16.8b–d). The result showed that CoP nanoparticles have greatly decreased the overpotential than Co_2P to produce the same current density (Fig. 16.8e). Furthermore, the same result could be obtained from the hydrogen evolution between Mo_3P and MoP [55]. All abovementioned experiments indicate a same result that TMPs with higher P content would demonstrate better performance in hydrogen evolution reactions.

16.3.3 Conductivity

With P atoms coupling with metal, electrically conductive metals change to semiconductive. The P atoms with more electronegativity can strongly restrict the electron delocalization in metal, thus lowering the conductivity [59]. Through adjusting the difference in electronegativity and atomic ratio of metal/P, TMPs can exhibit metallic character and even superconductivity, especially for metal-rich phosphides. Mar et al. [60] pointed out that the di- and tri-metal phosphides show the same electronic structure with the corresponding pure metal, confirming the excellent conductivity of these TMPs. Based on this consideration, TMPs with higher metal content have the better conductivity. However, as mentioned in Sect. 16.3.2, TMPs with higher P content may perform more excellent activity in hydrogen evolution reactions. Obviously, these two just contradict to each other, that is, activity and conductivity, and we cannot have both of them. To this end, some substances with outstanding conductivity, such as graphene [61], carbon nanotubes [46], and carbon clothes [62], are introduced to improve the conductivity of TMPs.

16.4 Applications of TMPs in Photocatalytic Hydrogen Evolution Reactions

16.4.1 The Origin of TMPs Acted As Cocatalysts in Photocatalytic Hydrogen Evolution Reactions

The discovery of TMPs dates back to the last century. However, due to the limitation of theory and technology, no obvious applications of TMPs can be observed for the following days. The recent research studies of TMPs mainly focus on electrocatalytic hydrogen evolution. In 2005, Liu and his coworkers [54] employed density functional theory (DFT) to predict that Ni_2P is a promising candidate for the electrocatalytic hydrogen evolution. This prediction is based on the theoretical calculation that strong H–Ni interaction on Ni_2P (001) can lead to poisoning of the highly active sites of the surface, which enhances the rate of the HER and makes it comparable to that of the [NiFe] hydrogenase. The first experimental work for TMPs is reported by Zhang's group [39]. They prepared nanoporous FeP nanosheets through anion-exchange method, and the resulting sample shows high electrocatalytic activity toward hydrogen evolution reaction with low overpotential and a small Tafel slope. After that, various TMPs (Cu_3P, CoP, Ni_2P, Mo_3P) as highly efficient electrocatalysts have been investigated in hydrogen evolution reactions [57, 63].

In photocatalytic hydrogen evolution process, the semiconductors have large surface overpotential, which need to be compensated by extra photovoltage. According to the excellent performance of TMPs in electrocatalysis, it can be deduced that TMPs can lower down the overpotential of semiconductor and have great potential as cocatalyst in photocatalytic hydrogen evolution reactions.

16.4.2 Explanation of TMPs As Cocatalysts

A highly efficient cocatalyst can not only rapidly extract the photo-generated electrons from photocatalyst to its surface but also catalyze H_2 evolution on its surface by using these trapped electrons. In general, the overall hydrogen evolution reactions pathway can be summarized as three-state diagram, including an initial state $H^+ + e^-$, an intermediate state adsorbed H*, and a final product $1/2H_2$ [13, 64]. The Gibbs free energy of the intermediate state, $|\Delta G_{H*}|$, is regarded as a key indicator of the hydrogen evolution reaction activity for various cocatalysts. The most desirable value of $|\Delta G_{H*}|$ should be zero. For instance, platinum (Pt), as a well-known and excellent cocatalyst in photocatalytic hydrogen evolution reaction, shows a near zero of $\Delta G_{H*} \approx -0.09$ eV [65]. According to TMPs, the Gibbs free energy value of $|\Delta G_{H*}|$ can be calculated by DFT calculations.

As early as 2005, Liu et al. [54] used DFT calculations to indicate that the P atoms with a small negative charge from the surface of TMPs can not only trap hydrogen protons but also provide high activity for the dissociation of H_2. Wang' s group [55] presented comparative analysis of Mo, Mo_3P, and MoP as cocatalyst for hydrogen evolution reaction, which indicates that phosphorization can potentially modify the properties of the metal and different degrees of phosphorization lead to distinct activities and stabilities (Fig. 16.9). Gibbs free energy of ΔG_{H*} of the P-terminated surface on (001)-MoP is rounded to the nearest zero. When H coverage increases from 1/4 ML to full coverage, ΔG_{H*} gradually turns from −0.36 to 0.54 eV, which suggests that P can bond hydrogen at low coverage when H desorbed at high coverage. This behavior is similar to hydrogen deliverer, which also existed in the S-edges of MoS_2. This good hydrogen evolution activity is determined by the low value of $|\Delta G_{H*}|$.

Wu et al. [66] unraveled that the electrocatalytic process for TMPs plays a vital role in the heterostructure photocatalytic H_2 evolution system for the first time. Electrocatalysts were used as the excellent cocatalysts to facilitate the charge

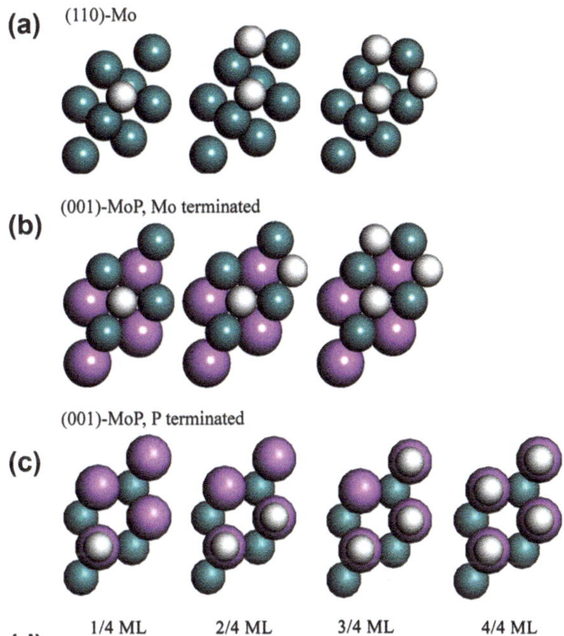

(d)

H Coverage	(110)-Mo		(001) MoP, Mo terminated		(001) MoP, P terminated	
	ΔE_H (eV)	ΔG°_H (eV)	ΔE_H (eV)	ΔG°_H (eV)	ΔE_H (eV)	ΔG°_H (eV)
1/4 ML	-0.77	-0.53	-0.87	-0.63	-0.64	-0.36
2/4 ML	-0.71	-0.47	-0.84	-0.60	-0.29	-0.03
3/4 ML	-0.70	-0.46	-0.82	-0.59	0.08	0.34
4/4 ML	-	-	-	-	0.28	0.54

Fig. 16.9 Top view of H-adsorbed (110)-Mo (**a**), (001)-MoP, Mo (**b**), and P (**c**) terminated surfaces. Only first two atomic layers are shown. (d) Calculated binding energy and Gibbs free energy of H adsorption on (110) Mo and (001) MoP, Mo, and P-terminated surfaces. (Reproduced from Ref. [55] by permission of The Royal Society of Chemistry (RSC) on behalf of the Centre National de la Recherche Scientifique (CNRS) and the RSC)

separation and provide the dominated active sites for hydrogen proton reduction, thus addressing the long-standing controversy on the role of electrocatalysts in photocatalysis (Fig. 16.10a). More specifically, as shown in Fig. 16.10b, an additional bias voltage was required to raise the energy of Fermi lever (E_F) of Co_2P, thus providing the external driving force to splitting water in electrocatalysis. However, in photocatalysis, since the energy value of E_F is lower than the conductive band value of CdS, photo-generated electrons from CdS can be transferred to Co_2P. More importantly, the injection of photo-generated electrons into TMPs can upshift the position of E_F, which is similar to noble metals (Fig. 16.10c).

16.4.3 Semiconductors Motivated by TMPs for Photocatalytic Hydrogen Evolution

16.4.3.1 Nickel Phosphides As Cocatalyst

In electrocatalysis, a series of nickel phosphides (Ni_xP) with different morphologies and dimensionality, such as nanorod, nanowires, nanoplates, nanospheres, and

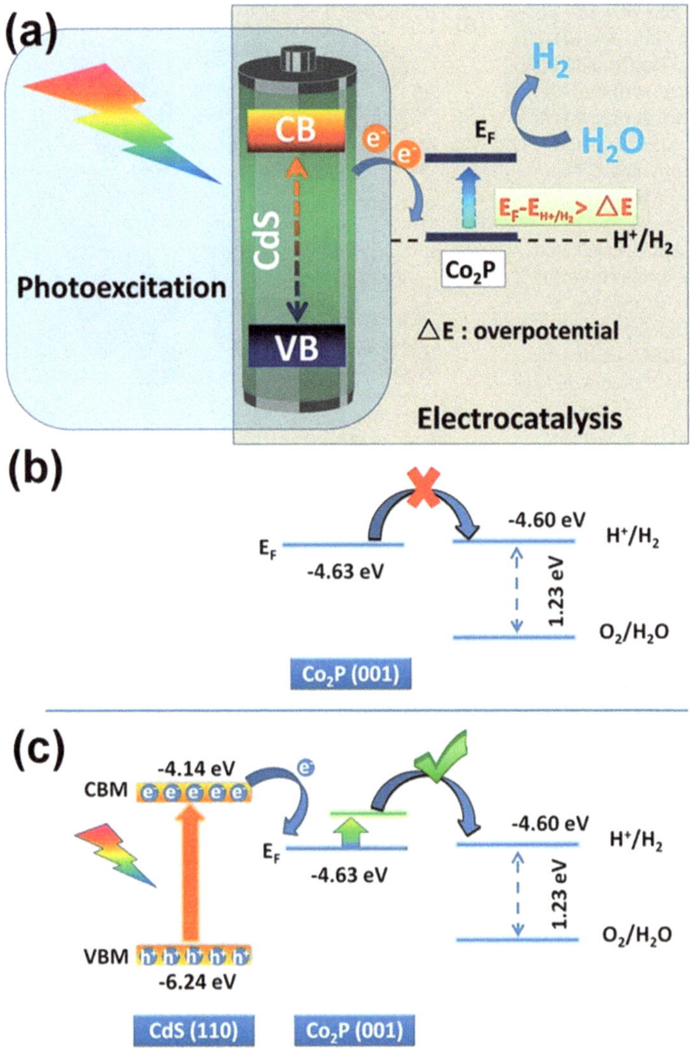

Fig. 16.10 (**a**) Schematic illustration of metallic phosphides electrocatalysts as cocatalysts for photocatalytic H_2 production. (**b**) Comparison of the Fermi level (EF) of Co_2P and EH^+/H_2, depicting the need of an external bias to drive the electrocatalysis H_2 production, and (**c**) representation of the proposed mechanism for photocatalytic H_2 production with the CdS/Co_2P hybrid system described in this work. (Reprinted with the permission from Ref. [66]. Copyright 2016 American Chemical Society)

nanodots, have demonstrated to be satisfied performance. In photocatalysis, the morphology of Ni_xP mainly focused on nanodots due to its high surface area and easy loading properties [67]. Du et al. [45] reported a one-pot hydrothermal method to in situ deposit Ni_2P nanoparticles on CdS nanorods (Fig. 16.11a–d). They found that Ni_2P nanoparticles can act as cocatalyst to improve the photocatalytic hydrogen evolution activity under visible light irradiation. Furthermore, after loading Ni_2P, the catalyst showed highly stable photocatalytic activity due to the photo-generated

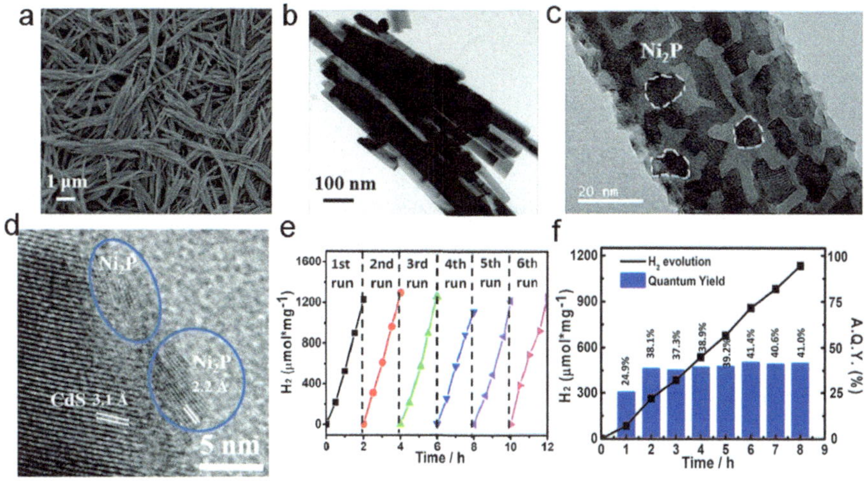

Fig. 16.11 (**a**) SEM and (**b**) TEM image of 0.5 wt% Ni_2P/CdS NRs. (**c**) HRTEM image and (**d**) high-magnification HRTEM image of 0.5 wt% Ni_2P/CdS NRs. (**e**) Cycling runs for photocatalytic hydrogen evolution in the presence of 1.0 mg 0.5 wt% Ni_2P/CdS NRs photocatalyst in a 50 mL aqueous solution containing 0.83 M Na_2S and 1.16 M Na_2SO_3 at room temperature. (**f**) The time courses of H_2 evolution and apparent quantum yield on 0.5 wt% Ni_2P/CdS NRs photocatalyst under monochromatic 450 nm light irradiation using 1.0 mg photocatalyst in a 20 mL aqueous solution containing 0.75 M Na_2S and 1.05 M Na_2SO_3. The bars denote the apparent quantum yield. (Reproduced from Ref. [45] by permission of The Royal Society of Chemistry (RSC) on behalf of the Centre National de la Recherche Scientifique (CNRS) and the RSC)

electrons trapped by Ni_2P and holes reacted with sacrificial agent (Fig. 16.11e). The highest quantum efficiency reached 41% after 6 h, which is higher than recently reported Ni/CdS and Ni/CdSe systems (Fig. 16.11f).

Driess et al. [68] reported an integrated Ni_2P-C_3N_4 catalyst system by mixing $NiCl_2$, NaH_2PO_2, and sol–gel prepared graphitic carbon nitride (sg-C_3N_4), which was heated in Ar atmosphere (Fig. 16.12a–c). In this case, the optimum loading amount of Ni_2P can be confirmed by measuring the photocatalytic hydrogen activity (Fig. 16.12d). The maximum value of charge recombination was detected in the charge separation efficiency (black boxes) as well as the amount of free charges after recombination (red dots) (Fig. 16.12e). The scheme of Ni_2P as cocatalysts can be summarized in Fig. 16.12f. The photo-generated electrons from sg-C_3N_4 can be trapped by Ni_2P due to the fact that unoccupied molecular orbital (LUMO) of Ni_2P has lower energy than the conductive band of sg-C_3N_4 but higher energy than the reduction potential of H^+/H_2.

Besides, Ni_2P as a cocatalyst has also been widely investigated by various research groups [69–72]. Considerable research reveals that the efficiency of charge separation, migration, and transfer can be highly enhanced, after introduction of Ni_2P on the surface of semiconductor. Moreover, the long-term stability during continuous H_2 evolution in the presence of light irradiation promotes the system for a practical implementation.

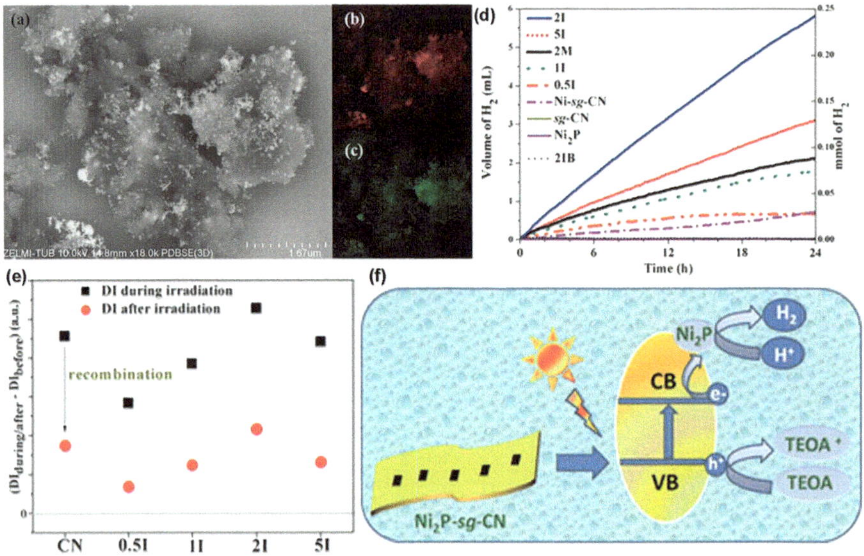

Fig. 16.12 (**a**) SEM image showing the homogeneous distribution of Ni_2P on carbon nitride surface; (**b** and **c**) represent the homogeneous distribution of Ni (red) and P (green). (**d**) Photocatalytic H_2 evolution with different catalysts in the presence of visible light irradiation (>420 nm) from 300 W Xe lamp using TEOA as the sacrificial agent. (**e**) Double integration of the conduction band electron signal under visible light (>420 nm) irradiation from 300 W Xe lamp (black boxes) and after irradiation (red dots). (**f**) Schematic representation of the charge separation and transfer in the valence band (VB) and conduction band (CB) in integrated Ni_2P–sg-CN during visible light photocatalytic H_2 evolution. (Reproduced from Ref. [68] by permission of John Wiley & Sons Ltd)

16.4.3.2 Cobalt Phosphides As Cocatalyst

Traditional Cd-based materials (CdS, CdSe) are attractive visible light-active photocatalysts due to their wide light absorption range and high photocatalytic activity. However, they offer suffer photocorrosion, which leads to inactivation of catalysts.

Zhang's group developed a novel and robust $Cd_xZn_{1-x}Se$/CoP photocatalyst [73]. In this case, CoP nanoparticles are anchored on the surface of $Cd_xZn_{1-x}Se$ hollow nanobelts through an easy physical mixed method (Fig. 16.13a–d). Through adjusting the molar ratio of Cd/Zn, we can obtain $Cd_{0.25}Zn_{0.75}Se$ nanobelts with the most negative conductive band position. That is, the photo-generated electrons with the strongest reduction potential can be trapped by CoP, which achieves the highest photocatalytic hydrogen evolution activity (Fig. 16.13e). It is noted that the $Cd_{0.25}Zn_{0.75}Se$/CoP catalyst shows a high hydrogen evolution rate at 36.6 $mmol{\cdot}g^{-1}{\cdot}h^{-1}$ in photocatalytic splitting artificial seawater (Fig. 16.13f). Moreover, when it was exposed to solar light, the bubbles can be observed obviously, which

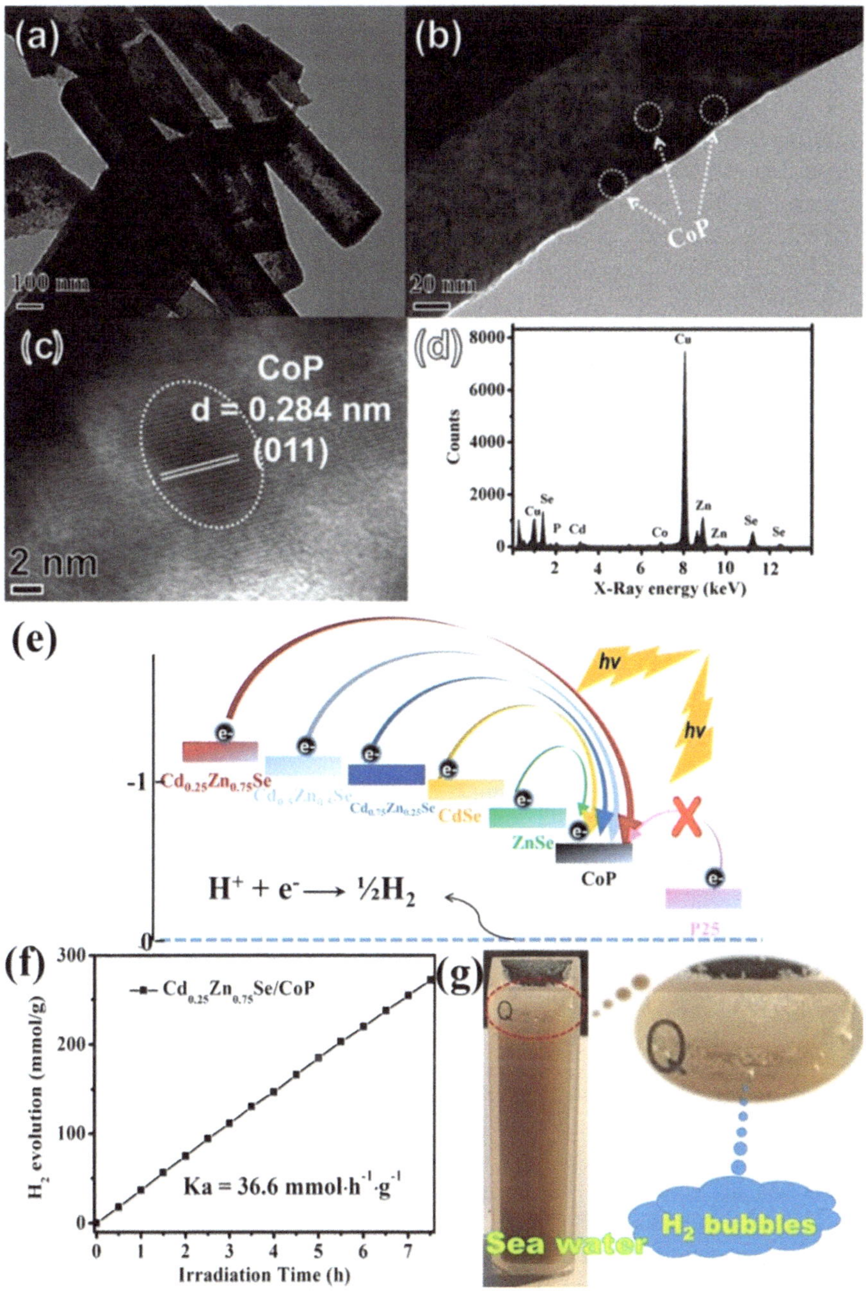

Fig. 16.13 (**a** and **b**) TEM and (**c**) HRTEM images for $Cd_{0.25}Zn_{0.75}Se$/CoP (amplification of white cycle in (**b**)). (**d**) EDX spectrum of $Cd_{0.25}Zn_{0.75}Se$/CoP (analysis of region in (**a**)). (**e**) Structural model of energy states for different catalysts. (**f**) Photocatalytic H_2 evolution over $Cd_{0.25}Zn_{0.75}Se$/CoP in artificial seawater system under simulated solar light irradiation. (**g**) Picture of H_2 evolution in artificial seawater system over $Cd_{0.25}Zn_{0.75}Se$/CoP under outdoor sunlight irradiation. (Reproduced from Ref. [73] by permission of The Royal Society of Chemistry (RSC) on behalf of the Centre National de la Recherche Scientifique (CNRS) and the RSC)

demonstrates the catalyst has hopeful prospects in the field of splitting seawater (Fig. 16.13g).

Zhang et al. [74] reported for the first time on preparation of CoP/TiO_2 hybrid for application in photocatalytic hydrogen evolution reaction. The apparent quantum efficiency (AQE) for 0.5 wt% CoP/TiO_2 at 350 nm is calculated to be 3.8% in 10 vol % triethanolamine solution under simulated solar light irradiation (Fig. 16.14a). It is noted that the CoP/TiO_2 composite catalyst showed the highly stable activity in long-standing photocatalytic experiment (Fig. 16.14b). From the Mott–Schottky analysis (Fig. 16.14c–d), it is noted that the flat band potentials for TiO_2 and CoP are −0.51 and −0.12 V versus Ag/AgCl, respectively, which reveals the injection of photo-generated electrons from TiO_2 into conductive band of CoP that is allowed in the energetic level. Different from the other TMPs photocatalytic systems, both TiO_2 and CoP in this system can be easily excited to form charges under solar light irradiation, and the electrons from the CB of TiO_2 can easily transfer to the CB of CoP. Meanwhile, holes from the VB of TiO_2 can inject into the VB of CoP. Therefore, the two kinds of photo-generated charge carriers are accumulated on CoP, and then they, respectively, go through two processes. Taking the CB electrons on CoP for instance, on the one hand, these electrons are used for participating H_2 evolution reaction; on the other hand, they combine with holes to accelerate charge-carrier recombination (Fig. 16.14e).

Except as the cocatalyst, cobalt phosphide also can be used as photocatalytic semiconductor to be application in photocatalytic water splitting. Sun et al. [75] used CoP nanowires to achieve highly photocatalytic hydrogen evolution with the presence of the probe of human immunodeficiency virus (P_{HIV}) through dye-sensitized process. An obvious enhancement on activity with the addition of P_{HIV} can be observed. An energy level diagram illustrates that under light irradiation, the electron from P_{HIV} can inject into the CB of CoP, and TEOA can react with the holes, thus improving the amount of electrons and charge separation efficiency.

16.4.3.3 Other TMPs As Cocatalyst

Except for these two commonly used cocatalysts (Ni_2P and CoP), other TMPs such as Cu_3P [76], FeP [77], and MoP [32] have been intensively used to drive the photocatalytic water reduction.

Du and his coworkers [78] prepared MoP-modified CdS nanorods through a mechanical commixture method. In order to understand the formation of MoP/CdS, the DFT calculations are used to calculate the surface energy of MoP (001), (100), and (101) surfaces (Fig. 16.15a). The Mo-terminated surfaces have higher surface energy than P-terminated surfaces. Furthermore, the MoP (100) surface has the largest surface energy of 0.509 and 0.242 eV Å^{-2} for Mo- and P-terminated surfaces, which suggests the high energy surface is unstable and easily disappears. However, after introducing CdS, the surface with high surface energy can be remained by the intimate attachment of MoP and CdS. In the first cycle, the apparent quantum yields (AQYs) gradually increase with the increment of irradiation time (Fig. 16.15b). It is

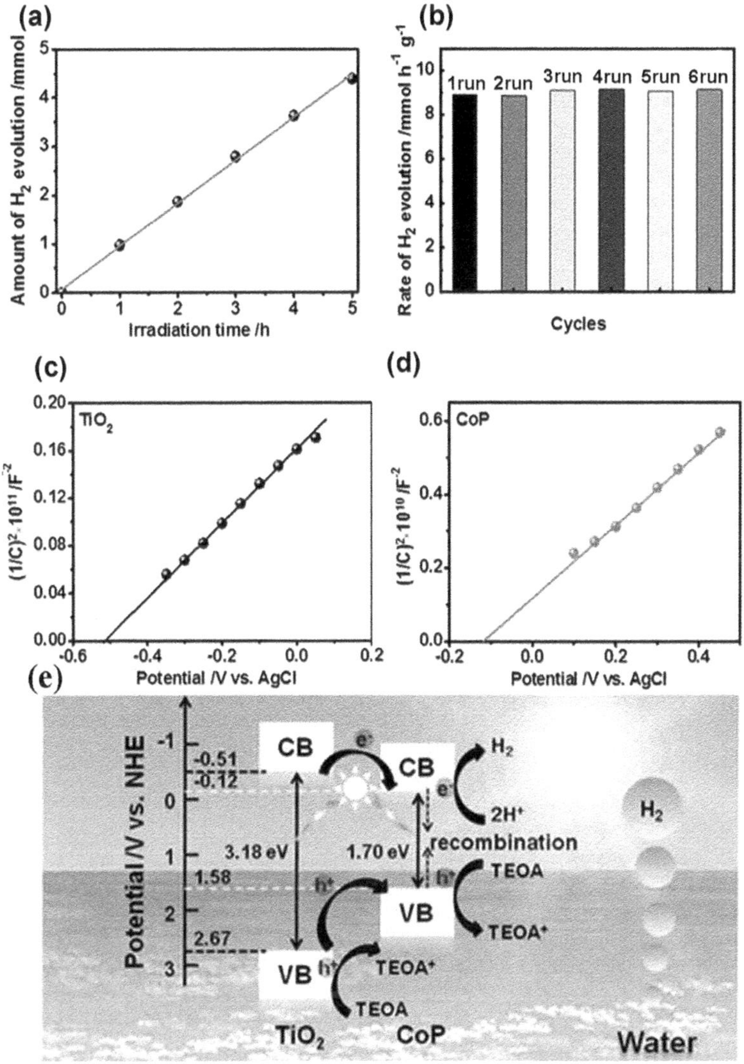

Fig. 16.14 (**a**) Photocatalytic H_2 evolution measured for 0.5 wt% CoP/TiO_2 at 350 nm in 10 vol% of triethanolamine aqueous solution. (**b**) Recycle test of photocatalytic H_2 evolution for 0.5 wt% CoP/TiO_2 with every cycling time for 4 h. Mott–Schottky plots of (**c**) TiO_2 and (**d**) CoP samples. (**e**) Schematics of the band structure of TiO_2 and CoP at the interface and the proposed photocatalytic mechanism over CoP/TiO_2 hybrid structure under simulated solar light irradiation. (**f**) Mott–Schottky plot of CoP NWs. (Reproduced from Ref. [74] by permission of John Wiley & Sons Ltd)

caused by the unattached MoP nanoparticles that are collision with CdS nanorods during the stirring process. The photocatalytic reaction in cycle 2 is very stable and can keep an average AQY of 5.6%. However, the decrement of activity in cycle 3 and 4 is caused by the reduction of the content of sacrificial agent. Due to the good

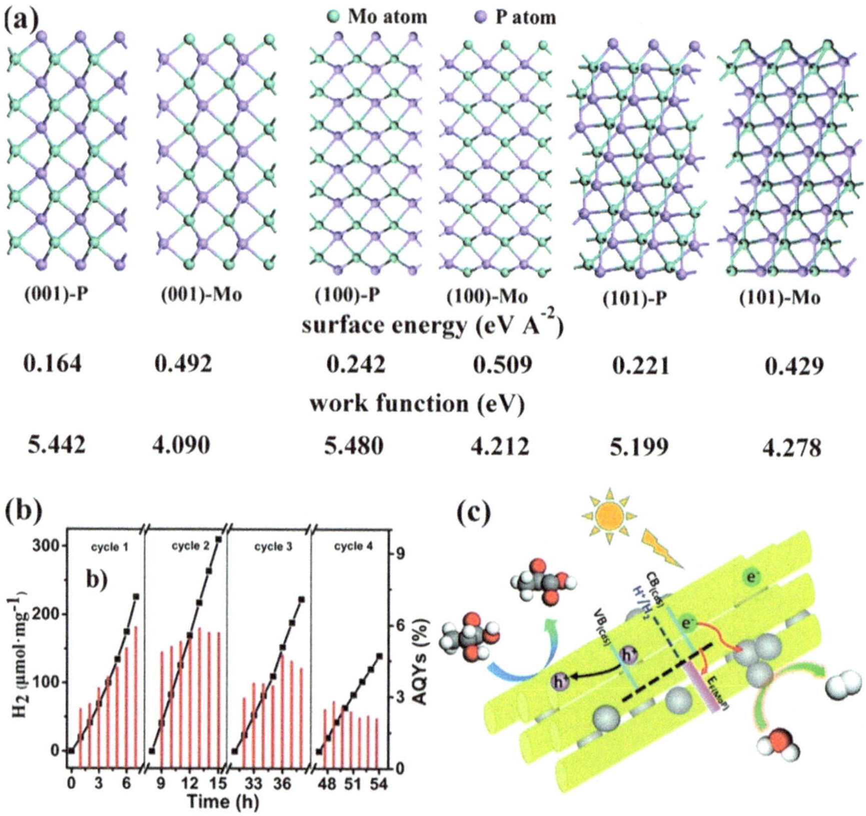

Fig. 16.15 (**a**) The side view of (001), (100), and (101) surfaces with different atoms terminated on MoP and the calculated surface energies and work functions of (001), (100), and (101) surfaces with different atoms terminated on MoP. (**b**) Time courses of H_2 production and apparent quantum yields (AQYs) of the 16.7 wt% MoP/CdS NR photocatalyst under monochromatic 450 nm light irradiation using 1.0 mg photocatalyst in a 20 mL 10% (v/v) lactic acid aqueous solution at room temperature. Fresh N_2 was bubbled into the vessel in the interval of each cycle to exhaust H_2 gas generated with no additional lactic acid. The reaction suspension was stored in darkness for 1 h, 16 h, and 9 h after cycle 1, cycle 2, and cycle 3, respectively. (**c**) The reaction mechanism for photocatalytic H_2 evolution using the MoP/CdS NR hybrid. (Reproduced from Ref. [78] by permission of The Royal Society of Chemistry (RSC) on behalf of the Centre National de la Recherche Scientifique (CNRS) and the RSC)

Fermi level alignment between MoP and CdS, the photo-generated electrons located at the CB of CdS will facilely transfer to the Fermi level of MoP, which results in highly efficient separation of photo-generated electron–hole pairs and decrement of charge recombination rate (Fig. 16.15c).

Zhang et al. [79] used Cu_3P as cocatalyst to couple with TiO_2 nanoparticles to form an efficient photocatalyst Cu_3P/TiO_2. The photocatalytic hydrogen evolution rate over the optimum Cu_3P/TiO_2 is 11 times higher than that of bare TiO_2 owing to the strong interaction existing between Cu_3P and TiO_2, providing a charge separation pathway across the catalyst's interface.

16.5 Conclusion and Prospects

In conclusion, robust and active cocatalyst of TMPs composed with earth-rich and inexpensive elements are desperately needed for achieving highly efficient, stable, and low-priced photocatalytic water reduction. Indeed, TMPs as cocatalysts play multiple roles in photocatalytic process. Firstly, the introduction of TMPs can improve light absorption of semiconductor; secondly, TMPs can trap the photogenerated electrons and improve the separation efficiency of charges; and thirdly, TMPs as active site can lower down activation energy of H_2 evolution. Up to now, a plenty of TMPs, such as Ni_2P, CoP, Cu_3P, FeP, and MoP, have been widely investigated. Among them, the highest hydrogen evolution rate and quantum efficiency reported are 553 $\mu mol \cdot h^{-1} \cdot mg^{-1}$ at visible light irradiation and 41% at 450 nm for the optimized Ni_2P/CdS system, respectively. With the compressive progress in earth-abundant transition metal phosphides nanoparticles available for photocatalytic hydrogen evolution reaction, an overall review of these TMPs as cocatalyst in photocatalysis for deep understanding and optimal utilization is urgently required. In this review, the recent development of TMPs as cocatalysts in photocatalytic hydrogen evolution reaction, including preparation, the role of P atom, and the photocatalytic application, has been extensively reviewed.

It should be noted that although continuous progress has been made in constructing TMPs as cocatalyst in photocatalysis, a wide range of scientific and technical challenges still exists to achieve the commercialization of TMP cocatalysts in photocatalysis. (1) It is very vital to refine the cocatalysts loading condition and develop more effective method to uniform and controllable loading cocatalysts on the surface of photocatalysts for achieving atomically well-bond nanojunction. (2) Development of highly efficient, selective, and stable cocatalysts for semiconductor-based photocatalytic water splitting is very desirable. (3) The fundamental research about physicochemical properties, electrochemical performance, electron transfer, and multiple-electron coupled reaction mechanism of loaded cocatalysts are very rare, and it will be very necessary to enrich this field.

References

1. Habisreutinger SN, Schmidt-Mende L, Stolarczyk JK (2013) Photocatalytic reduction of CO_2 on TiO_2 and other semiconductors. Angew Chem Int Ed 52(29):7372–7408
2. Fontecave M (2011) Energy for a sustainable world. From the oil age to a sun-powered future. By Nicola Armaroli and Vincenzo Balzani. Angew Chem Int Ed 50(30):6704–6705
3. Pasten C, Santamarina JC (2012) Energy and quality of life. Energy Policy 49:468–476

4. Ran J, Zhang J, Yu J, Jaroniec M, Qiao SZ (2014) Earth-abundant cocatalysts for semiconductor-based photocatalytic water splitting. Chem Soc Rev 43(22):7787–7812
5. Du P, Eisenberg R (2012) Catalysts made of earth-abundant elements (Co, Ni, Fe) for water splitting: recent progress and future challenges. Energy Environ Sci 5(3):6012–6021
6. Qiu B, Xing M, Zhang J (2014) Mesoporous TiO_2 nanocrystals grown in situ on graphene aerogels for high photocatalysis and lithium-ion batteries. J Am Chem Soc 136(16):5852–5855
7. Qiu B, Zhou Y, Ma Y, Yang X, Sheng W, Xing M, Zhang J (2015) Facile synthesis of the Ti^{3+} self-doped TiO_2-graphene nanosheet composites with enhanced photocatalysis. Sci Rep 5:8591
8. Qiu B, Deng Y, Du M, Xing M, Zhang J (2016) Ultradispersed cobalt ferrite nanoparticles assembled in graphene aerogel for continuous Photo-Fenton reaction and enhanced lithium storage performance. Sci Rep 6:29099–29108
9. Qiu B, Li Q, Shen B, Xing M, Zhang J (2016) Stöber-like method to synthesize ultradispersed Fe_3O_4 nanoparticles on graphene with excellent Photo-Fenton reaction and high-performance lithium storage. Appl Catal B Environ 183:216–223
10. Qiu B, Zhu Q, Du M, Fan L, Xing M, Zhang J (2017) Efficient solar light harvesting CdS/Co_9S_8 hollow cubes for Z-scheme photocatalytic water splitting. Angew Chem 129(10):2728–2732
11. Xing M, Zhang J, Qiu B, Tian B, Anpo M, Che M (2015) A brown mesoporous TiO_{2-x}/MCF composite with an extremely high quantum yield of solar energy photocatalysis for H_2 evolution. Small 11(16):1920–1929
12. Zhou Y, Yi Q, Xing M, Shang L, Zhang T, Zhang J (2016) Graphene modified mesoporous titania single crystals with controlled and selective photoredox surfaces. Chem Commun 52 (8):1689–1692
13. Jiao Y, Zheng Y, Jaroniec M, Qiao S (2015) Design of electrocatalysts for oxygen-and hydrogen-involving energy conversion reactions. Chem Soc Rev 44(8):2060–2086
14. Shi Y, Zhang B (2016) Recent advances in transition metal phosphide nanomaterials: synthesis and applications in hydrogen evolution reaction. Chem Soc Rev 45(6):1529–1541
15. Iqbal W, Qiu B, Lei J, Wang L, Zhang J, Anpo M (2017) One-step large-scale highly active g-C_3N_4 nanosheets for efficient sunlight-driven photocatalytic hydrogen production. Dalton T 46(32):10678–10684
16. Qiu B, Zhong C, Xing M, Zhang J (2015) Facile preparation of C-modified TiO_2 supported on MCF for high visible-light-driven photocatalysis. RSC Adv 5(23):17802–17808
17. Wang Y, Lkhamjav S, Qiu B, Dong C, Dong C, Zhou Y, Shen B, Xing M, Zhang J (2017) A facile strategy to prepare Fe^{3+} modified brookite TiO_2 with high photocatalytic activity under ultraviolet light and visible light. Res Chem Intermediat 43(4):2055–2066
18. Ai G, Li H, Liu S, Mo R, Zhong J (2015) Solar water splitting by TiO_2/CdS/Co-Pi nanowire array photoanode enhanced with Co-Pi as hole transfer relay and CdS as light absorber. Adv Funct Mater 25(35):5706–5713
19. Yuan Y, Yin L, Cao S, Xu G, Li C, Xue C (2015) Improving photocatalytic hydrogen production of metal-organic framework UiO-66 octahedrons by dye-sensitization. Appl Catal B Environ 168:572–576
20. Li H, Shang J, Ai Z, Zhang L (2015) Efficient visible light nitrogen fixation with BiOBr nanosheets of oxygen vacancies on the exposed {001} facets. J Am Chem Soc 137 (19):6393–6399
21. Hou W, Cronin SB (2013) A review of surface plasmon resonance-enhanced photocatalysis. Adv Funct Mater 23(13):1612–1619
22. Qiu B, Deng Y, Li Q, Shen B, Xing M, Zhang J (2016) Rational design of a unique ternarystructure for highly photocatalytic nitrobenzene reduction. J Phys Chem C 120 (22):12125–12131
23. Tsuji I, Kato H, Kudo A (2005) Visible-light-induced H_2 evolution from an aqueous solution containing sulfide and sulfite over a ZnS-$CuInS_2$-$AgInS_2$ solid-solution photocatalyst. Angew Chem 117(23):3631–3634
24. Zhang J, Wang Y, Zhang J, Lin Z, Huang F, Yu J (2013) Enhanced photocatalytic hydrogen production activities of Au-loaded ZnS flowers. ACS Appl Mater Inter 5(3):1031–1037

25. Yoshida M, Yamakata A, Takanabe K, Kubota J, Osawa M, Domen K (2009) ATR-SEIRAS investigation of the fermi level of Pt cocatalyst on a GaN photocatalyst for hydrogen evolution under irradiation. J Am Chem Soc 131(37):13218–13219
26. Sreethawong T, Yoshikawa S (2005) Comparative investigation on photocatalytic hydrogen evolution over Cu-, Pd-, and Au-loaded mesoporous TiO_2 photocatalysts. Catal Commun 6 (10):661–668
27. Maeda K, Xiong A, Yoshinaga T, Ikeda T, Sakamoto N, Hisatomi T, Takashima M, Lu D, Kanehara M, Setoyama T (2010) Photocatalytic overall water splitting promoted by two different cocatalysts for hydrogen and oxygen evolution under visible light. Angew Chem 122(24):4190–4193
28. Iizuka K, Wato T, Miseki Y, Saito K, Kudo A (2011) Photocatalytic reduction of carbon dioxide over Ag cocatalyst-loaded $ALa_4Ti_4O_{15}$ (A= Ca, Sr, and Ba) using water as a reducingreagent. J Am Chem Soc 133(51):20863–20868
29. Yang J, Wang D, Han H, Li C (2013) Roles of cocatalysts in photocatalysis and photoelectrocatalysis. Accounts Chem Res 46(8):1900–1909
30. Li X, Bi W, Zhang L, Tao S, Chu W, Zhang Q, Luo Y, Wu C, Xie Y (2016) Single-atom Pt as co-catalyst for enhanced photocatalytic H_2 evolution. Adv Mater 28(12):2427–2431
31. Callejas JF, McEnaney JM, Read CG, Crompton JC, Biacchi AJ, Popczun EJ, Gordon TR, Lewis NS, Schaak RE (2014) Electrocatalytic and photocatalytic hydrogen production from acidic and neutral-pH aqueous solutions using iron phosphide nanoparticles. ACS Nano 8 (11):11101–11107
32. Reddy DA, Kim HK, Kim Y, Lee S, Choi J, Islam MJ, Kumar DP, Kim TK (2016) Multicomponent transition metal phosphides derived from layered double hydroxide double-shelled nanocages as an efficient non-precious co-catalyst for hydrogen production. J Mater Chem A 4(36):13890–13898
33. Sweeny NP, Rohrer CS, Brown O (1958) Dinickel phosphide as a heterogeneous catalyst for the vapor phase reduction of nitrobenzene with hydrogen to aniline and water. J Am Chem Soc 80 (4):799–800
34. Sun J, Liu C, Yang P (2011) Surfactant-free, large-scale, solution-liquid-solid growth of gallium phosphide nanowires and their use for visible-light-driven hydrogen production from water reduction. J Am Chem Soc 133(48):19306–19309
35. Lo CT, Kuo PY (2010) Synthesis and magnetic properties of iron phosphide nanorods. J Phys Chem C 114(11):4808–4815
36. McCarty WJ, Yang X, DePue Anderson LJ, Jones RA (2012) Chemical vapour deposition of amorphous Ru(P) thin films from Ru trialkylphosphite hydride complexes. Dalton T 41 (43):13496–13503
37. Shi Y, Xu Y, Zhuo S, Zhang J, Zhang B (2015) Ni_2P nanosheets/Ni foam composite electrode for long-lived and pH-tolerable electrochemical hydrogen generation. ACS Appl Mater Inter 7 (4):2376–2234
38. Lu Y, Tu J, Xiang J, Wang X, Zhang J, Mai Y, Mao S (2011) Improved electrochemical performance of self-assembled hierarchical nanostructured nickel phosphide as a negative electrode for lithium ion batteries. J Phys Chem C 115(48):23760–23767
39. Xu Y, Wu R, Zhang J, Shi Y, Zhang B (2013) Anion-exchange synthesis of nanoporous FeP nanosheets as electrocatalysts for hydrogen evolution reaction. Chem Commun 49 (59):6656–6658
40. Popczun EJ, Roske CW, Read CG, Crompton JC, McEnaney JM, Callejas JF, Lewis NS, Schaak RE (2015) Highly branched cobalt phosphide nanostructures for hydrogen-evolution electrocatalysis. J Mater Chem A 3(10):5420–5425
41. Zhang H, Ha DH, Hovden R, Kourkoutis LF, Robinson RD (2010) Controlled synthesis of uniform cobalt phosphide hyperbranched nanocrystals using tri-n-octylphosphine oxide as a phosphorus source. Nano Lett 11(1):188–197

42. Sun Z, Lv B, Li J, Xiao M, Wang X, Du P (2016) Core-shell amorphous cobalt phosphide/cadmium sulfide semiconductor nanorods for exceptional photocatalytic hydrogen production under visible light. J Mater Chem A 4(5):1598–1602
43. Fullenwarth J, Darwiche A, Soares A, Donnadieu B, Monconduit L (2014) NiP_3: a promising negative electrode for Li-and Na-ion batteries. J Mater Chem A 2(7):2050–2059
44. Li Q, Li Z, Zhang Z, Li C, Ma J, Wang C, Ge X, Dong S, Yin L (2016) Low-temperature solution-based phosphorization reaction route to Sn_4P_3/reduced graphene oxide nanohybrids as anodes for sodium ion batteries. Advanced Energy Mater 6(15):1600376
45. Sun Z, Zheng H, Li J, Du P (2015) Extra-ordinarily efficient photocatalytic hydrogen evolution in water using semiconductor nanorods integrated with crystalline Ni_2P cocatalysts. Energy Environ Sci 8(9):2668–2676
46. Lou P, Cui Z, Jia Z, Sun J, Tan Y, Guo X (2017) Monodispersed carbon-coated cubic NiP_2 canoparticles anchored on carbon canotubes as ultra-long-life anodes for reversible lithium storage. ACS Nano 11(4):3705–3715
47. Wu T, Chen S, Zhang D, Hou J (2015) Facile preparation of semimetallic MoP_2 as a novel visible light driven photocatalyst with high photocatalytic activity. J Mater Chem A 3 (19):10360–10367
48. Pu Z, Liu Q, Asiri AM, Sun X (2014) Tungsten phosphide nanorod arrays directly grown on carbon cloth: a highly efficient and stable hydrogen evolution cathode at all pH values. ACS Appl Mater Inter 6(24):21874–21879
49. He P, Yu X, Lou X (2017) Carbon-incorporated nickel-cobalt mixed metal phosphide nanoboxes with enhanced electrocatalytic activity for oxygen evolution. Angew Chem 129 (14):3955–3958
50. Cao S, Chen Y, Wang C, He P, Fu W (2014) Highly efficient photocatalytic hydrogen evolution by nickel phosphide nanoparticles from aqueous solution. Chem Commun 50 (72):10427–10429
51. Feng L, Li K, Chang J, Liu C, Xing W (2015) Nanostructured PtRu/C catalyst promoted by CoP as an efficient and robust anode catalyst in direct methanol fuel cells. Nano Energy 15:462–469
52. Tong Y, Gu C, Zhang J, Huang M, Tang H, Wang X, Tu J (2015) Three-dimensional astrocyte-network Ni-P-O compound with superior electrocatalytic activity and stability for methanol oxidation in alkaline environments. J Mater Chem A 3(8):4669–4678
53. Guo S, Deng Z, Li M, Jiang B, Tian C, Pan Q, Fu H (2016) Phosphorus-doped carbon nitride tubes with a layered micro-nanostructure for enhanced visible-light photocatalytic hydrogen evolution. Angew Chem Int Ed 55(5):1830–1834
54. Liu P, Rodriguez JA (2005) Catalysts for hydrogen evolution from the [NiFe] hydrogenase to the Ni_2P (001) surface: the importance of ensemble effect. J Am Chem Soc 127 (42):14871–14878
55. Xiao P, Sk MA, Thia L, Ge X, Lim RJ, Wang J, Lim KH, Wang X (2014) Molybdenum phosphide as an efficient electrocatalyst for the hydrogen evolution reaction. Energy Environ Sci 7(8):2624–2629
56. Yan H, Tian C, Wang L, Wu A, Meng M, Zhao L, Fu H (2015) Phosphorus-modified tungsten nitride/reduced graphene oxide as a high-performance, non-noble-metal electrocatalyst for the hydrogen evolution reaction. Angew Chem Int Ed 54(21):6325–6329
57. Pan Y, Liu Y, Zhao J, Yang K, Liang J, Liu D, Hu W, Liu D, Liu Y, Liu C (2015) Monodispersed nickel phosphide nanocrystals with different phases: synthesis, characterization and electrocatalytic properties for hydrogen evolution. J Mater Chem A 3(4):1656–1665
58. Callejas JF, Read CG, Popczun EJ, McEnaney JM, Schaak RE (2015) Nanostructured Co_2P electrocatalyst for the hydrogen evolution reaction and direct comparison with morphologically equivalent CoP. Chem Mater 27(10):3769–3774
59. Carenco S, Portehault D, Boissiere C, Mezailles N, Sanchez C (2013) Nanoscaled metal borides and phosphides: recent developments and perspectives. Chem Rev 113(10):7981–8065

60. Blanchard PE, Grosvenor AP, Cavell RG, Mar A (2008) X-ray photoelectron and absorption spectroscopy of metal-rich phosphides M_2P and M_3P (M= Cr-Ni). Chem Mater 20 (22):7081–7088
61. Jiao L, Zhou Y, Jiang H (2016) Metal-organic framework-based CoP/reduced graphene oxide: high-performance bifunctional electrocatalyst for overall water splitting. Chem Sci 7 (3):1690–1695
62. Tian J, Liu Q, Asiri AM, Sun X (2014) Self-supported nanoporous cobalt phosphide nanowire arrays: an efficient 3D hydrogen-evolving cathode over the wide range of pH 0-14. J Am Chem Soc 136(21):7587–7590
63. Ma L, Shen X, Zhou H, Zhu G, Ji Z, Chen K (2015) CoP nanoparticles deposited on reduced graphene oxide sheets as an active electrocatalyst for the hydrogen evolution reaction. J Mater Chem A 3(10):5337–5343
64. Nørskov JK, Bligaard T, Rossmeisl J, Christensen CH (2009) Towards the computational design of solid catalysts. Nat Chem 1(1):37–46
65. Hinnemann B, Moses PG, Bonde J, Jørgensen KP, Nielsen JH, Horch S, Chorkendorff I, Nørskov JK (2005) Biomimetic hydrogen evolution: MoS_2 nanoparticles as catalyst for hydrogen evolution. J Am Chem Soc 127(15):5308–5309
66. Bi W, Zhang L, Sun Z, Li X, Jin T, Wu X, Zhang Q, Luo Y, Wu C, Xie Y (2016) Insight into electrocatalysts as co-catalysts in efficient photocatalytic hydrogen evolution. ACS Catal 6 (7):4253–4257
67. Xing M, Qiu B, Du M, Zhu Q, Wang L, Zhang J (2017) Spatially separated CdS shells exposed with reduction surfaces for enhancing photocatalytic hydrogen evolution. Adv Funct Mater 27 (35):1702624
68. Indra A, Acharjya A, Menezes PW, Merschjann C, Hollmann D, Schwarze M, Aktas M, Friedrich A, Lochbrunner S, Thomas A, Driess M (2017) Boosting visible-light-driven photocatalytic hydrogen evolution with an integrated nickel phosphide-carbon nitride system. Angew Chem Int Ed 56(6):1653–1657
69. Choi J, Reddy DA, Han NS, Jeong S, Hong S, Kumar DP, Song JK, Kim TK (2017) Modulation of charge carrier pathways in CdS nanospheres by integrating MoS_2 and Ni_2P for improved migration and separation toward enhanced photocatalytic hydrogen evolution. Cat Sci Technol 7(3):641–649
70. Ye L, Han C, Ma Z, Leng Y, Li J, Ji X, Bi D, Xie H, Huang Z (2017) Ni_2P loading on $Cd_{0.5}Zn_{0.5}S$ solid solution for exceptional photocatalytic nitrogen fixation under visible light. Chem Eng J 307:311–318
71. Kudo A, Kato H, Tsuji I (2004) Strategies for the development of visible-light-driven photocatalysts for water splitting. Chem Lett 33(12):1534–1539
72. Ye P, Liu X, Iocozzia J, Yuan Y, Gu L, Xu G, Lin Z (2017) Highly stable non-noble metal Ni_2P co-catalyst for increased H_2 generation by g-C_3N_4 under visible light irradiation. J Mater Chem A 5:8493–8498
73. Qiu B, Zhu Q, Xing M, Zhang J (2017) A robust and efficient catalyst of $Cd_xZn_{1-x}Se$ motivated by CoP for photocatalytic hydrogen evolution under sunlight irradiation. Chem Commun 53 (5):897–900
74. Yue X, Yi S, Wang R, Zhang Z, Qiu S (2017) Cobalt phosphide modified titanium oxide nanophotocatalysts with significantly enhanced photocatalytic hydrogen evolution from water splitting. Small 13(14):1603301
75. Tian J, Cheng N, Liu Q, Xing W, Sun X (2015) Cobalt phosphide nanowires: efficient nanostructures for fluorescence sensing of biomolecules and photocatalytic evolution of dihydrogen from water under visible light. Angew Chem Int Ed 54(18):5493–5497
76. Sun Z, Yue Q, Li J, Xu J, Zheng H, Du P (2015) Copper phosphide modified cadmium sulfide nanorods as a novel p-n heterojunction for highly efficient visible-light-driven hydrogen production in water. J Mater Chem A 3(19):10243–10247
77. Cheng H, Lv X, Cao S, Zhao Z, Chen Y, Fu W (2016) Robustly photogenerating H2 in water using FeP/CdS catalyst under solar irradiation. Sci Rep 6:19846–19855

78. Yue Q, Wan Y, Sun Z, Wu X, Yuan Y, Du P (2015) MoP is a novel, noble-metal-free cocatalyst for enhanced photocatalytic hydrogen production from water under visible light. J Mater Chem A 3(33):16941–16947
79. Yue X, Yi S, Wang R, Zhang Z, Qiu S (2016) A novel and highly efficient earth-abundant Cu_3P with TiO_2 "P-N" heterojunction nanophotocatalyst for hydrogen evolution from water. Nanoscale 8(40):17516–17523

Chapter 17
Novel Porous Metal–Organic Frameworks (MOFs) for Water Splitting

17.1 Metal–Organic Frameworks (MOFs)

Metal–organic frameworks (MOFs), known as porous coordination polymers (PCPs) as well, are emerging as a class of promising crystalline microporous materials and receive great research interest. MOFs are hybrid materials composed of organic linkers and metal-oxo clusters. The outstanding structural features of MOFs are the ultrahigh porosity (up to 90% free volume) and incredibly high internal surface areas, extending beyond a Langmuir surface area of 10,000 m^2/g [1, 2]. This plays a crucial role in many functional applications such as storage, separation [3, 4], sensing [3], proton conduction [5–7], and drug delivery [8]. Among various highly porous materials, MOFs are unique in their degree of tunability and structural diversity as well as their range of chemical and physical properties. Owing to their unique attributes, open structures with periodic dual composition, MOFs are amenable to change secondary building blocks into a desired framework expanding or decorating a specific blueprint network topology [9, 10]. So they have been considered as a promising type of materials.

17.2 Hydrogen Production

In recent years, visible light-responsive porous metal–organic framework photocatalysts have been investigated deeply. As a kind of porous material, MOFs have shown semiconductor-like characters in photocatalysis [11–15]. Commonly, the photochemical reduction of water into hydrogen molecules using a photocatalyst does not rely on fossil fuels, but it is an ideal method for producing clean energy. MOFs provide versatile means to integrate functional components for solar energy utilization.

J. Zhang et al., *Photocatalysis*, Lecture Notes in Chemistry 100,
https://doi.org/10.1007/978-981-13-2113-9_17

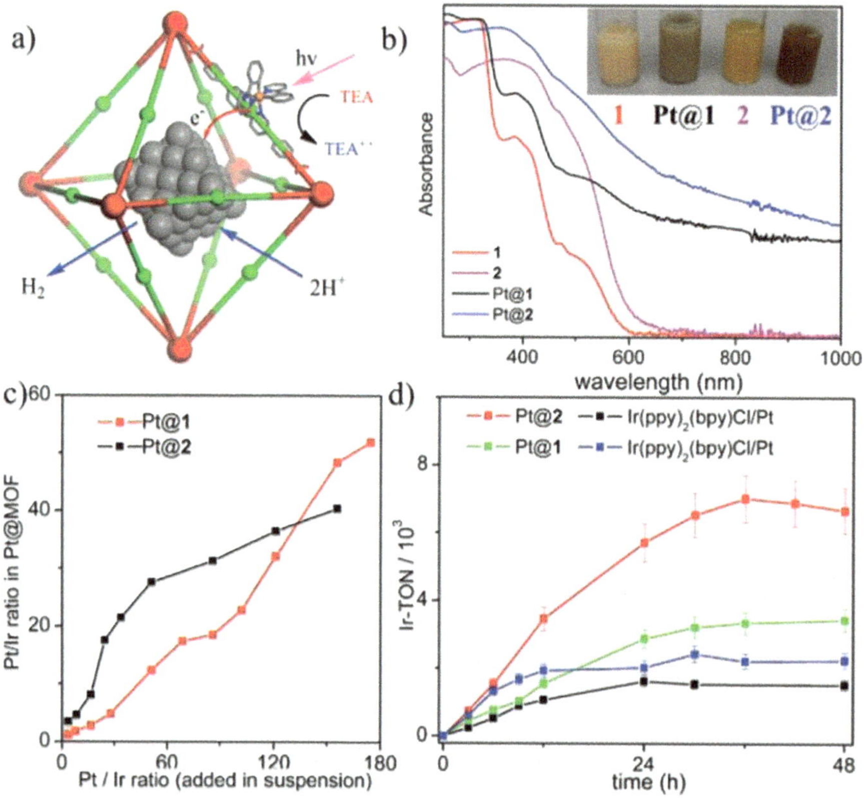

Fig. 17.1 (**a**) Scheme showing the synergistic photocatalytic hydrogen evolution process via photo-injection of electrons from the light-harvesting MOFs (3 and 4) into the Pt NPs. (**b**) Diffuse reflectance spectra and a photograph of suspensions of these samples. (**c**) Relationship between the amount of K_2PtCl_4 added in the reaction solution and the amount of Pt deposited inside the MOF. (**d**) Time-dependent hydrogen evolution curves of samples. (Reprinted with permission from Ref. [13] Copyright 2012, American Chemical Society)

Interestingly, the effective hydrogen evolution from water containing a sacrificial electron donor under visible light irradiation can be achieved by the Pt-loaded MOFs constructed from light-harvesting organic linkers. In 2012, Lin et al. [13] prepared Pt@MOFs by loading Pt NPs in photoactive MOFs, $Zr_6(\mu_3\text{-}O)_4(\mu_3\text{-}OH)_4(bpdc)_{5.94}(L1)_{0.06}$(3) and $Zr_6(\mu_3\text{-}O)_4(\mu_3\text{-}OH)_4(L_2)_6{\cdot}64DMF$ (4), which were found to be effective photocatalysts for hydrogen evolution by synergistic photoexcitation of the frameworks and electron injection into the Pt NPs. These two MOFs were built from two $[Ir(ppy)_2(bpy)]^+$-derived dicarboxylate ligands. Meanwhile, the radicals, $[Ir(ppy)_2(bpy)]$, generated in the MOFs by TEA-mediated photoreduction can transfer electrons to the entrapped Pt NPs to reduce protons of water (Fig. 17.1a). Pt@3 and Pt@4 photocatalysts gave the hydrogen evolution turnover numbers (TONs) of 3400 and 7000 based on Ir phosphors (Fig. 17.1d), 1.5 and 4.7 times

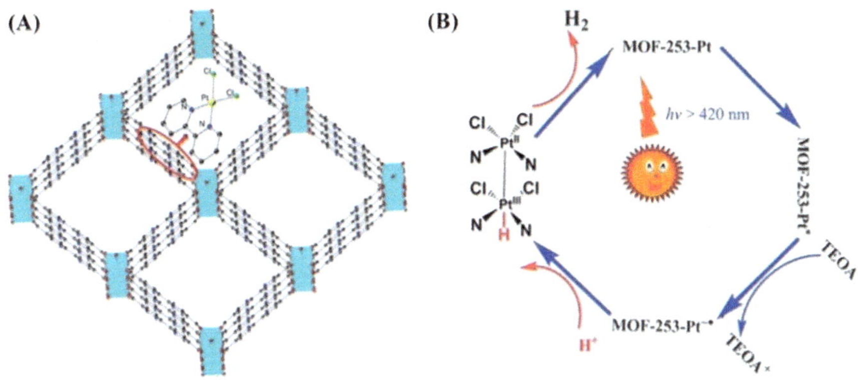

Fig. 17.2 (**a**) Model structure of MOF-253-Pt, through post-synthetic modification of MOF-253 with $PtCl_2$. Key: Cyan octahedron represents Al atoms; while yellow, green, red, blue, and blank circles represent Pt, Cl, O, N, and C atoms, respectively; H atoms are omitted for clarity. (**b**) Proposed reaction mechanism for the photocatalytic H_2 evolution over MOF-253-Pt under visible light irradiation. (Reprinted with the permission from Ref. [11], Copyright 2013, Royal Society of Chemistry)

the values afforded by the homogeneous control $[Ir(ppy)_2(bpy)]Cl/K_2PtCl_4$ under their respective conditions under the illumination of a 450 W Xe-lamp with a 420 nm cutoff filter for 48 h.

In 2009, Kataoka et al. [15] detected the first example of open porous MOFs that functioned as an activity site for the reduction of water into hydrogen molecules in the presence of $Ru(bpy)_3^{2+}$, MV^{2+}, and EDTA–2Na under visible light irradiation. Zhou et al. [11] synthesized and characterized a new metal–organic framework (MOF-253-Pt) material through immobilizing a platinum complex in 2,20-bipyridine-based microporous MOF (MOF-253) using a post-synthesis modification strategy. The functionalized MOF-253-Pt serves both as a photosensitizer and a photocatalyst for hydrogen evolution under visible light irradiation. The structure and proposed mechanism are presented as shown in Fig. 17.2a, b, respectively. Based on light irradiation, the MOF with TEOA's presence firstly generates a one-electron-reduced species MOF*(3MLCT) and holes. The MOF*(^{3}MLCT) species is reductively quenched to form MOF^- with the electrons stored on the bpy^- ligands. After that, the reduced MOF^- further forms a Pt(III)-hydride intermediate via a proton-coupled electron transfer (PCET), which contributes to the formation of the hydride-diplatinum (II, III) intermediate by the synergistic effect of the nearing anchored $Pt(bpy)Cl_2$ complex on framework, leading to H_2 production by a heterolytic coupling pathway.

The efficiency of electron–hole separation and charge-carrier utilization play a central role in photocatalysis. Jiang et al. [16] prepared a representative MOF UiO-66-NH_2 where Pt nanoparticles of ca. 3 nm are incorporated inside or supported on. The resulting products are denoted as Pt@UiO-66-NH_2 and Pt/UiO-66-NH_2, respectively. Finally, these materials are especially applied in photocatalytic

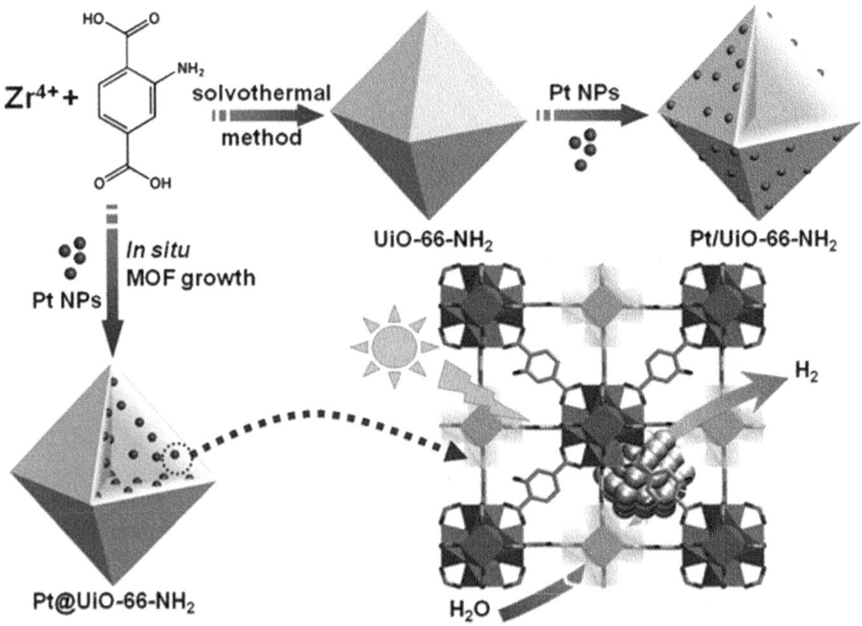

Scheme 17.1 Schematic illustration for the synthesis of Pt@UiO-66-NH_2 and Pt/UiO-66-NH_2, with the photocatalytic hydrogen production process over Pt@UiO-66-NH_2 being highlighted. (Reprinted with permission from Ref. [16], Copyright 2016, John Wiley and Sons)

hydrogen production via water splitting. Scheme 17.1 simply clarified the pathway of synthesizing these two different materials. Pt@UiO-66-NH_2 greatly shortens the electron-transport distance and hence suppresses the electron–hole recombination, which is expected to have an enhanced catalytic activity compared to Pt/UiO-66-NH_2. In addition, the Pt NPs embedded in the MOF do not undergo aggregation or leaching during the reaction, which leads to better catalytic recyclability of Pt@UiO-66-NH_2 than that of Pt/UiO-66-NH_2.

In 2010, Garcia and his coworkers synthesized Zr-containing MOFs that exhibit photocatalytic activity for hydrogen generation upon irradiation at wavelength longer than 300 nm [17]. In 2012, Anpo et al. [12] employed 2-amino-benzenedicarboxylic acid as an organic linker to synthesize amino-functionalized Ti (IV) metal–organic framework (Ti–MOF–NH_2) by a facile solvothermal method, and it described the hydrogen produced from an aqueous medium under visible light. The structure of the Ti–MOF–NH_2, its mechanism, and the yield of hydrogen are shown in Fig. 17.3.

With the exception of the application in water splitting, the Ti–MOFs can also be applied to CO_2 reduction. At the same year, Li et al. [18] successfully prepared a targeted photoactive catalyst $Ti_8O_8(OH)_4(bdc\text{-}NH_2)_6$ (NH_2-MIL-125 (Ti)) for the first time, which reduced CO_2 even under visible light irradiation. In Fig. 17.4a, MIL-125 (Ti) shows an absorption edge to 350 nm, whereas NH_2-MIL-125

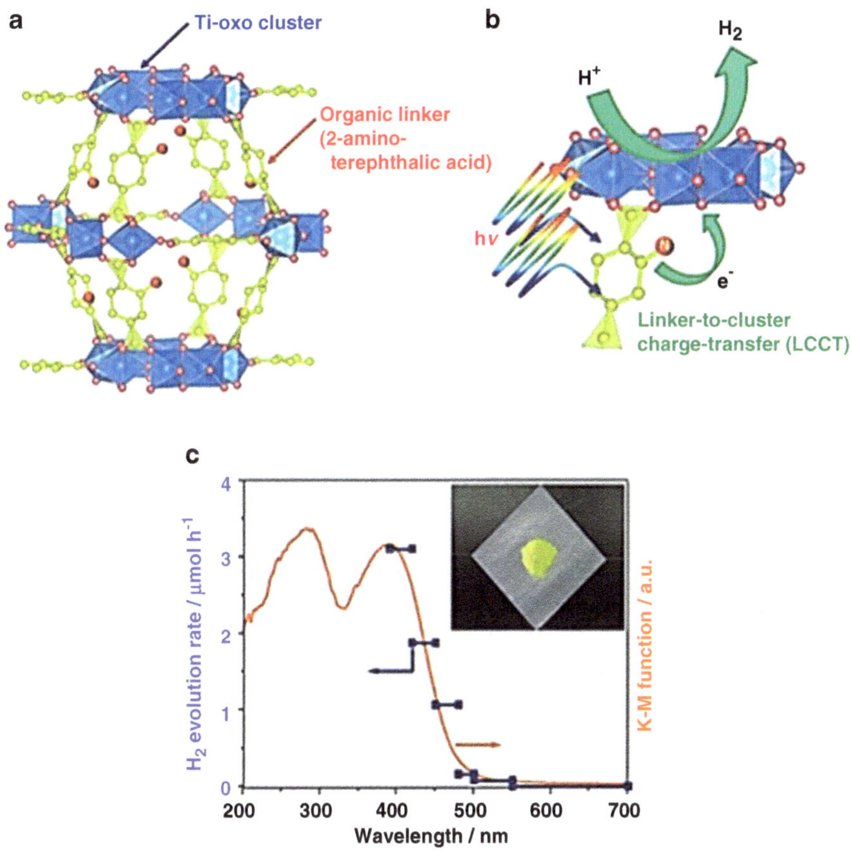

Fig. 17.3 Schematic illustrations of (**a**) the structure of Ti–MOF–NH_2 and (**b**) the reaction mechanism for hydrogen evolution over Ti–MOF–NH_2 induced by visible light irradiation. (**c**) Action spectrum for hydrogen evolution from water containing TEOA as a sacrificial electron donor over Ti–MOF–NH_2. Inset shows the photograph of Ti–MOF–NH_2. (Reprinted with the permission from Ref. [12], Copyright 2012, American Chemical Society)

(Ti) shows an extra absorption band in the visible light region with the absorption edge extending to around 550 nm, in agreement with the bright yellow color. An interesting photochromic phenomenon was observed over NH_2-MIL-125 (Ti) during the photocatalytic reaction. When the solution of NH_2-MIL-125 (Ti) and TEOA in MeCN was irradiated with visible light in the presence of N_2, the color of the solution changed from the original bright yellow to green. If assuming that CO_2 or O_2 was introduced into the reaction system, the green color of the solution changed gradually back to the original bright yellow (Fig. 17.4b). The photochromic phenomenon is ascribed to the presence of the inter-valence electron transfer from the optically induced hopping of electrons from Ti (III) to Ti (IV) sites in the titanium-oxo clusters.

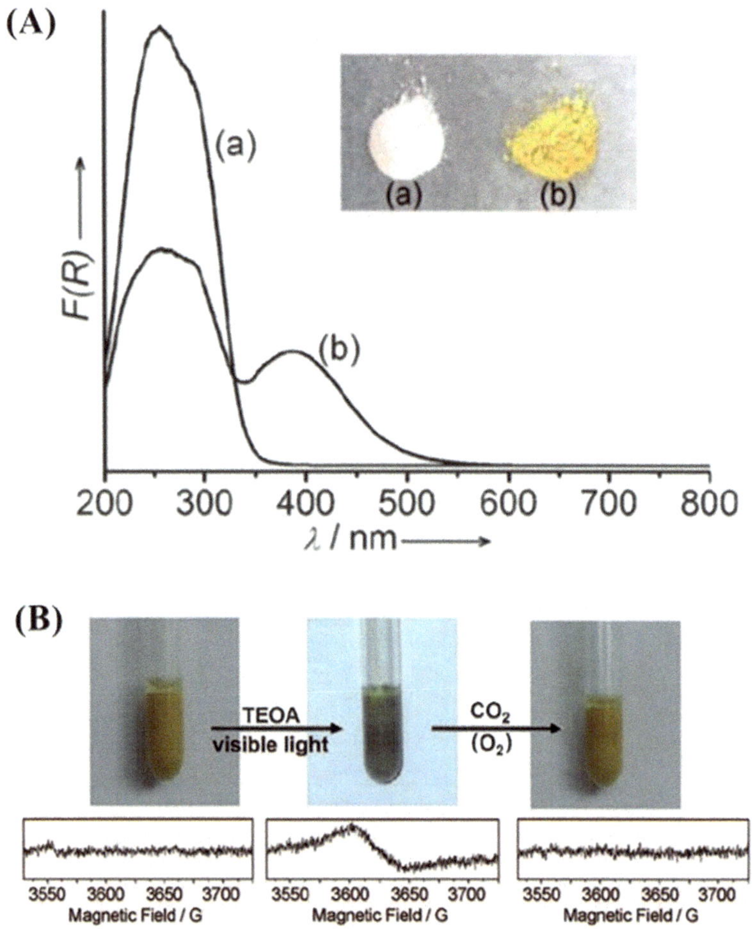

Fig. 17.4 (**a**) UV–vis spectra of (a) MIL-125(Ti) and (b) NH_2-MIL-125(Ti). The inset shows the samples. (**b**) Photos and corresponding ESR spectra of NH_2-MIL-125(Ti) under different conditions: (a) fresh NH_2-MIL-125(Ti); (b) TEOA, visible light, and N_2; and (c) after the introduction of CO_2 (or O_2). (Reprinted with permission from Ref. [18], Copyright 2012, John Wiley and Sons)

17.3 Conclusions

In summary, as abovementioned, the Ti-based MOFs or other metal-based MOFs are able to photo-generate hydrogen from water. Meanwhile, if precious metals are loaded on MOFs, higher efficiency will be exhibited under visible light, which possess enormous potential in developing visible light-responsive MOFs in water splitting. All the reported literatures provide us with new ideas in the further development of hydrogen production.

References

1. Wang TC, Bury W, Gómez-Gualdrón DA et al (2015) Ultrahigh surface area zirconium MOFs and insights into the applicability of the BET theory. J Am Chem Soc 137(10):3585–3591
2. Furukawa H, Cordova KE, O'Keeffe M, Yaghi OM (2013) The chemistry and applications of metal-organic frameworks. Science 341(6149):1230444
3. Zhao X, Xiao B, Fletcher AJ, Thomas KM, Bradshaw D, Rosseinsky MJ (2004) Hysteretic adsorption and desorption of hydrogen by nanoporous metal-organic frameworks. Science 306 (5698):1012
4. Rosi NL, Eckert J, Eddaoudi M, Vodak DT, Kim J, Keeffe M, Yaghi OM (2003) Hydrogen storage in microporous metal-organic frameworks. Science 300(5622):1127
5. Taylor JM, Mah RK, Moudrakovski IL, Ratcliffe CI, Vaidhyanathan R, Shimizu GKH (2010) Facile proton conduction via ordered water molecules in a phosphonate metal-organic framework. J Am Chem Soc 132(40):14055–14057
6. Sadakiyo M, Yamada T, Kitagawa H (2009) Rational designs for highly proton-conductive metal-organic frameworks. J Am Chem Soc 131(29):9906–9907
7. Yamada T, Otsubo K, Makiura R, Kitagawa H (2013) Designer coordination polymers: dimensional crossover architectures and proton conduction. Chem Soc Rev 42(16):6655–6669
8. Della Rocca J, Liu D, Lin W (2011) Nanoscale metal-organic frameworks for biomedical imaging and drug delivery. Acc Chem Res 44(10):957–968
9. Seo JS, Whang D, Lee H, Im Jun S, Oh J, Jeon YJ, Kim K (2000) A homochiral metal-organic porous material for enantioselective separation and catalysis. Nature 404(6781):982–986
10. Hill RJ, Long DL, Champness NR, Hubberstey P, Schröder M (2005) New approaches to the analysis of high connectivity materials: design frameworks based upon 44- and 63-subnet tectons. Acc Chem Res 38(4):335–348
11. Zhou T, Du Y, Borgna A, Hong J, Wang Y, Han J, Zhang W, Xu R (2013) Post-synthesis modification of a metal-organic framework to construct a bifunctional photocatalyst for hydrogen production. Energy Environ Sci 6(11):3229–3234
12. Horiuchi Y, Toyao T, Saito M, Mochizuki K, Iwata M, Higashimura H, Anpo M, Matsuoka M (2012) Visible-light-promoted photocatalytic hydrogen production by using an amino-functionalized Ti (IV) metal-organic framework. J Phys Chem C 116(39):20848–20853
13. Wang C, DeKrafft KE, Lin W (2012) Pt nanoparticles@photoactive metal-organic frameworks: efficient hydrogen evolution via synergistic photoexcitation and electron injection. J Am Chem Soc 134(17):7211–7214
14. Fateeva A, Chater PA, Ireland CP, Tahir AA, Khimyak YZ, Wiper PV, Darwent JR, Rosseinsky MJ (2012) A water-stable porphyrin-based metal-organic framework active for visible-light photocatalysis. Angew Chem 124(30):7558–7562
15. Kataoka Y, Sato K, Miyazaki Y, Masuda K, Tanaka H, Naito S, Mori W (2009) Photocatalytic hydrogen production from water using porous material $[Ru_2(p\text{-}BDC)_2]_n$. Energy Environ Sci 2 (4):397–400
16. Xiao J, Shang Q, Xiong Y, Zhang Q, Luo Y, Yu S, Jiang H (2016) Boosting photocatalytic hydrogen production of a metal-organic framework decorated with platinum nanoparticles: the platinum location matters. Angew Chem 128(32):9535–9539
17. Gomes Silva C, Luz I, Llabrés i Xamena FX, Corma A, García H (2010) Water stable Zr-benzenedicarboxylate metal-organic frameworks as photocatalysts for hydrogen generation. Chem A Eur J 16(36):11133–11138
18. Fu Y, Sun D, Chen Y, Huang R, Ding Z, Fu X, Li Z (2012) An amine-functionalized titanium metal-organic framework photocatalyst with visible-light-induced activity for CO_2 reduction. Angew Chem 124(14):3420–3423

Printed by Printforce, the Netherlands